ROUTLEDGE HANDBOOK OF URBAN FORESTRY

More than half of the world's population now lives in cities. Creating sustainable, healthy and aesthetic urban environments is therefore a major policy goal and research agenda. This comprehensive handbook provides a global overview of the state of the art and science of urban forestry.

It describes the multiple roles and benefits of urban green areas in general and the specific role of trees, including issues such as air quality, human well-being and stormwater management. It reviews the various stresses experienced by trees in cities and tolerance mechanisms, as well as cultural techniques for either pre-conditioning or alleviating stress after planting. It sets out sound planning, design, species selection, establishment and management of urban trees. It shows that close interactions with the local urban communities who benefit from trees are key to success.

By drawing upon international state-of-art knowledge on arboriculture and urban forestry, the book provides a definitive overview of the field and is an essential reference text for students, researchers and practitioners.

Francesco Ferrini is Dean of the School of Agriculture and Professor of Arboriculture at the University of Florence, Italy.

Cecil C. Konijnendijk van den Bosch is Professor of Urban Forestry at the University of British Columbia, Canada.

Alessio Fini is a researcher in the Department of Production and Agri-Environment at the University of Florence, Italy.

ROUTLEDGE HANDBOOK OF URBAN FORESTRY

Edited by Francesco Ferrini,
Cecil C. Konijnendijk van den Bosch
and Alessio Fini

Routledge
Taylor & Francis Group
LONDON AND NEW YORK

earthscan
from Routledge

First published 2017 by Routledge

2 Park Square, Milton Park, Abingdon, Oxfordshire OX14 4RN
52 Vanderbilt Avenue, New York, NY 10017

Routledge is an imprint of the Taylor & Francis Group, an informa business

First issued in paperback 2019

British Library Cataloguing-in-Publication Data
A catalogue record for this book is available from the British Library

Library of Congress Cataloging in Publication Data
Names: Ferrini, Francesco, editor. | Bosch, Cecil C. Konijnendijk
van den, editor. | Fini, Alessio, editor.
Title: Routledge handbook of urban forestry / edited by Francesco
Ferrini, Cecil C. Konijnendijk van den Bosch, and Alessio Fini.
Other titles: Handbook of urban forestry
Description: London; New York : Routledge, 2017. | Includes
bibliographical references and index.
Identifiers: LCCN 2016043232| ISBN 9781138647282 (hbk) |
ISBN 9781315627106 (ebk)
Subjects: LCSH: Urban forestry--Handbooks, manuals, etc. | Trees
in cities. | Urban ecology (Biology)
Classification: LCC SB436 .R68 2017 | DDC 635.9/77--dc23
LC record available at https://lccn.loc.gov/2016043232

ISBN: 978-1-138-64728-2 (hbk)
ISBN: 978-0-367-35238-7 (pbk)

Typeset in Bembo
by HWA Text and Data Management, London

CONTENTS

LIST OF FIGURES

LIST OF TABLES

CONTRIBUTORS

Noor Azlin Yahya holds a PhD in forest recreation management. At FRIM she is the head of the ecotourism and urban forestry programme. Her research activities focus on ecotourism, urban forestry, environmental education, natural resource management, park management and recreation ecology. She is actively participating in FAO networking in the Asia-Pacific region.

Sara Barron received her BA in environmental geography from Simon Fraser University and her masters of landscape architecture from the University of British Columbia (UBC). She has worked with many communities to advance sustainability and climate change planning. Funded by the Future Forests Fellowship, Sara is currently pursuing a PhD in forestry at UBC.

Nina Bassuk is a professor and programme leader of the Urban Horticulture Institute at Cornell University. She has authored over 100 papers on the physiological problems of plants growing in urban environments, including improved plant selections for difficult sites, soil modification including the development of 'CU-Structural Soil' and improved transplanting technology.

Simone Borelli has a first degree in forest science, while he also holds an MSc in watershed management and postgraduate diploma in public management. He has worked for FAO for 18 years in different positions and is currently the officer responsible for urban forestry and agroforestry. In addition to FAO he has also worked for WWF, IPGRI (now Bioversity) and as a private consultant.

Cecilia Brunetti has a field of study comprising physiological and biochemical responses of plants to environmental constraints with an emphasis on secondary metabolites production (phenylpropanoids, volatile and non-volatile isoprenoids, hormones).

Jill Butler has been the ancient tree, wood pasture and parkland specialist for the Woodland Trust, the UK's largest woodland conservation charity, for 15 years. The Ancient Tree Inventory, an online, citizen science database of ancient and other veteran trees in the UK, is one aspect of her work.

Carlo Calfapietra (www.carlocalfapietra.com) is a senior researcher at CNR-IBAF, Italy, and lecturer in urban forestry at Tuscia University. His main interests are the biosphere-atmosphere interactions both in urban and rural environments. He has published several research papers on these topics (H-index: 36). He is chair of the COST Action 'GreenInUrbs'.

Claudia Canedoli is currently involved in a research PhD project studying multiple ecosystem services in urban areas. Her scientific interests and experience focus on ecosystem services (particularly related to biodiversity and soils) and conservation ecology of the endemic fauna (mammals and amphibians).

Yujuan Chen obtained a PhD in urban forestry from Virginia Tech. She worked for New Jersey State Forestry Services in the United States as a forestry specialist. From 2014 to 2016 she was a junior professional officer at FAO of Urban and Peri-urban Forestry and Trees Outside of Forests.

Galina Churkina is a senior fellow at the Institute for Advanced Sustainability Studies in Potsdam, Germany. She uses state-of-the-art computer models along with ground and satellite measurements to analyse interactions between urbanization and global environmental change as well as the implications of these interactions for society.

Michela Conigliaro holds a first degree in Biological Science and an MSc in biodiversity conservation and ecosystem management. She has worked for FAO as a Consultant for four years supporting the implementation of the urban forestry and agroforestry programme. She has been involved in the IUCN-Med programmes and worked for the Italian Federation of Parks and Nature Reserves.

Owen Croy is manager of parks for the City of Surrey, Canada. He has worked in the forest industry, in federal plant protection programmes and in the management of municipal parks and urban forests. Owen is an instructor with the Municipal Forestry Institute and the University of British Columbia, and is active on committees of the Society of Municipal Arborists and International Society of Arboriculture.

Clive Davies is an international consultant on urban forestry and green infrastructure and visiting research and teaching fellow at Newcastle University, UK. His principle interests in urban forestry and green infrastructure are related to planning, strategy, delivery and governance issues. He is involved with international projects and COST actions and is heavily involved with the European Forum on Urban Forestry (EFUF).

Johanna Deak Sjöman recently completed her PhD at the Swedish University of Agricultural Sciences, Alnarp. Her work concentrates on landscape planning with regards to green infrastructure and regulating ecosystem services.

Cynnamon Dobbs is an urban ecologist specializing in forestry, using the socio-ecological framework approach to understand the dynamics of vegetation and ecosystem services spatially and temporally, how these are affected by policy making and urbanization.

Susan Downing Day studies tree–soil interactions and their role in urban soil quality and stormwater mitigation. Her contributions helped shape the standards for sustainable soil management in the Sustainable Sites Initiative (SITES). She holds a BA from Yale University, an MSc from Cornell University, and a PhD from Virginia Tech.

Peter N. Duinker has degrees in agriculture, environmental studies, and forest management. He specializes in scholarship on forest management and policy as well as environmental assessment. His forest-related work is now dominated by urban forests. He and his students examine trees in the city from both biophysical and sociopolitical perspectives.

Neville Fay is director and principal consultant at Treework Environmental Practice Consultancy. He is a chartered arboriculturist, consultant and expert witness. Neville is past chairman of the Ancient Tree Forum and founded the charity Tree Aid. He co-authored the Specialist Survey Method and directs the international Innovations in Arboriculture seminar series. He was honoured with the AA Award for Services to Arboriculture.

Francesco Ferrini is a professor of arboriculture, while also serving as the dean of the School of Agriculture of the University of Florence. His research concerns physiological and growth aspects of different tree species as affected by the urban environment. He is also involved in several national and international projects regarding arboriculture and urban forestry.

Alessio Fini has been working on projects regarding sustainable cultivation techniques for nursery production and urban forestry, and on stress tolerance of urban trees since 2005. He is member of the International Society of Arboriculture (ISA), and vice president of the Arboricultural Research and Education Academy (2016–2017). In 2014, he was awarded with the Early Career Scientist Award by ISA. He is associate editor of the scientific journals *Urban Forestry and Urban Greening* and *Frontiers in Plant Science*. He has authored more than 70 publications in international peer-reviewed journals, books and conference proceedings.

Stanislaw W. Gawronski's research has focused on phytoremediation of heavy metals, organic contaminants and particulate matter in the urban environment. During the last 10 years, his research has been directed towards the number one world air pollutant: particulate matter.

Paolo Gonthier is associate professor at the University of Turin, where he teaches forest pathology and protection of ornamental plants. His scientific activity is documented by over 230 papers, focused on diseases of forest trees, with emphasis on wood decay. He is president of the Italian Society of Arboriculture (SIA).

Jason Grabosky is the John and Eleanor Kuser Faculty Scholar in urban and community forestry, and the director of the general honours programme for the School of Environmental and Biological Sciences. He is also the co-director of the Center For Resilient Landscapes (http://crl.rutgers.edu).

Rüdiger Grote is a senior researcher at KIT who has developed several process-based vegetation models and published about 60 peer-reviewed journal articles. He is particularly interested in the mechanisms of biogenic volatile organic compounds (BVOCs) emission. Other fields of his research are air qualtiy, bioenergy plantations and forest development under stress.

Gabriele Guidolotti is researcher at CNR-IBAF, Italy. His main fields of interests are forest ecology and ecophysiology applied to the study of greenhouse gases, biogenic volatile organic compounds, carbon pools and fluxes in relation to global changes.

Rieke Hansen is a landscape architect with focus on strategic landscape planning and green infrastructure planning. She is a research and teaching associate at the chair for strategic landscape planning and management, Technical University of Munich, and participated in European and German urban planning and greening research projects.

Hans Martin Hanslin has main research focuses on ecological processes, plant ecophysiology and biodiversity in both constructed and natural systems to improve sustainability and resilience of green infrastructure.

J. Roger Harris is professor and head of the Department of Horticulture at Virginia Tech. He specializes in tree establishment and nursery production. He holds a BSc from Georgia Tech, a BSc from Michigan State University, an MSc from the University of Florida, and a PhD from Cornell University.

Richard J. Hauer is a professor of urban forestry at the University of Wisconsin–Stevens Point teaching courses in urban forestry, nursery management, woody plants, and dendrology. He received his BSc from the University of Wisconsin–Stevens Point, MSc from the University of Illinois, and PhD from the University of Minnesota.

Andrew Hirons is a senior lecturer in arboriculture and urban forestry at Myerscough College, UK. He teaches a wide range of arboricultural modules and is actively involved with research focusing on using plant traits to aid tree selection for urban environments.

Sara Janhäll has focused her research on particle-related air quality related to the transport sector for 20 years. Special emphasis of her research is on abatement strategies related to airborne particles, emissions and effects of vegetation on air quality near roads.

C. Y. Jim is chair professor at the University of Hong Kong. His research concentrates on the nature-in-city theme, encompassing urban ecology, urban forestry, urban green infrastructure, urban soil science, green roof, green wall, urban nature conservation, and climate change adaptation. He adopts an interdisciplinary approach and focuses on compact and south cities.

Brian Kane is the Massachusetts Arborists Association Professor of Commercial Arboriculture at the University of Massachusetts–Amherst. Prior to entering academia, he worked as an arborist in New York. He is an International Society of Arboriculture (ISA)-certified arborist and placed fourth overall in the 2006 New England ISA Tree Climbing Competition.

Dave Kendal is an interdisciplinary scientist drawing on social psychology, ecology and horticulture to better understand the vegetation we live with, why we manage ecosystems the way we do, and what the general public thinks about green spaces and the way we manage them.

Andrew K. Koeser is an Assistant Professor of Landscape Management at the University of Florida Gulf Coast Research and Education Center near Tampa, Florida (United States). Andrew is an ISA-certified master arborist and holds the ISA Tree Risk Assessment Qualification.

Cecil Konijnendijk van den Bosch has been professor of urban forestry at UBC since July 2016, after past employments including head of the landscape department at the Swedish University of Agricultural Sciences. Cecil's research and teaching focus on urban forestry, green space governance, and the linkages between nature and people in urban settings. He is founding editor-in-chief of the journal *Urban Forestry and Urban Greening* and series editor of Springer's Future City book series.

Raffaele Lafortezza is senior lecturer at the University of Bari, Italy, and adjunct professor at the Center for Global Change and Earth Observations (CGCEO), Michigan State University, USA. He has accumulated considerable experience in green infrastructure and ecosystem services issues by participating in numerous research projects and scientific collaborations worldwide.

Susanna Lehvävirta has run several research projects on urban ecology. Her studies are interdisciplinary and problem-driven and she has also worked in cooperation with artistic projects. She combines different disciplines to understand the nature–society interplay, to explore values, and to help develop decision-making and regulation.

Sharon J. Lilly spent 18 years as director of educational goods and services at the International Society of Arboriculture, and 25 years in commercial arboriculture as a climber and tree service owner. She is a past president of ISA and former chair of the ISA Certification Board. She is the author of several books, including the *ISA Arborist Certification Study Guide* and *The Tree Climbers' Guide*. Sharon was instrumental in the development of the Tree Risk Assessment Manual and Qualification.

Jari Lyytimäki works as a senior researcher at the Finnish Environment Institute, Environmental Policy Centre. His research interests span from environmental management and sustainability indicators to environmental communication. He is a founding member of the FIDEA research group (www.fidea.fi) and holds an adjunct professorship at the University of Helsinki.

Ana Macias is a PhD graduate in forestry. In her doctoral work, Ana studied the role of visualizations in urban forestry management and planning from the managers' perspectives in London, UK. Her current research explores the use of landscape visualizations, also focusing on citizen science and public engagement in urban forestry.

Maria Jose Martinez-Harms is a conservation scientist interested in the development of methods to synthesize ecosystem services research to provide evidence that can be used in the science-policy interface to inform ecosystem services and biodiversity management decisions.

Luciano Massetti graduated in Engineering, with a background on data analysis and modelling. His research interest is in meteorology and climate and their influence in the rural and urban environment.

Nelda P. Matheny is founder and president of HortScience, an arboriculture, urban forestry, and horticulture consulting firm in California, USA. She completed her bachelor's and master's degrees in horticulture at University of California Davis. She is co-author of several books including *Arboriculture: Integrated Care of Trees, Shrubs and Vines*, and the *Photographic Guide to the Evaluation of Hazard Trees in Urban Areas*.

Andreas Matzarakis is head of the Research Center Human Biometeorology of the German Meteorological Service in Freiburg and extraordinary Professor at the Albert-Ludwigs-University Freiburg. His research is mainly focused on human-biometeorology, urban climatology, tourism climatology and climate impact research. He is the developer of several models and tools in applied climatology and biometeorology (RayMan, SkyHelios).

Robert W. Miller received a BSc and MSc in forestry from West Virginia University and a PhD in forestry from the University of Massachusetts. Experience includes the US Forest Service, Florida Forest Service and the Forestry Faculty of the University of Wisconsin–Stevens Point. Bob retired as an emeritus professor of urban forestry from the faculty in 2002.

Justin Morgenroth is an urban forestry researcher with interest in tree growth and survival in challenging urban environments. Prior to joining the University of Canterbury's School of Forestry, he obtained a master's degree in forest conservation from the University of Toronto and his PhD in forestry from the University of Canterbury.

Anders Busse Nielsen is a professor of landscape with a focus on planting design and management. His work combines urban ecology, forestry, and landscape architecture. In parallel, he works as an independent advisor where he activates and test-drives research in contextual design, planning, and management of urban green infrastructures.

Karl J. Niklas joined the Cornell faculty in 1978. He is the author of over 345 research articles and five books. He has received numerous awards including being elected as a fellow in the Institute of Advanced Study (Wissenschaftskolleg zu Berlin) and in the American Academy of Arts and Sciences.

Robert J. Northrop is an extension forester for the University of Florida IFAS Extension. The focus of his work involves teaching urban and community forestry; providing urban forest conservation planning assistance to local, state and federal governments; and applied research into the changing character and ecological function of the Tampa Bay Watershed's urbanizing forest.

David J. Nowak is a team leader with the US Forest Service. His research investigates urban forest structure, health, and change, and its effect on human health and environmental quality. He has authored over 275 publications and leads teams developing software tools to quantify ecosystem services from trees (e.g., i-Tree).

Simone Orlandini is a full professor of agronomy. He graduated in agricultural sciences and holds a PhD in agrometeorology. He carries out teaching activity in agronomy, agrometeorology, and agroclimatology. He is presently director of the Interdepartmental Centre for Bioclimatology and coordinator of PhD course in agriculture and environmental sciences.

Johan Östberg is a researcher and teacher at the Swedish University of Agricultural Sciences where he also obtained an MSc and a PhD in landscape planning. Apart from his academic career he is also active as a consultant, running a network of consultants from different professions.

Emilio Padoa-Schioppa is currently associate professor of ecology at the Department of Earth and Environmental Science at the University of Milano Bicocca. His main research interests and experience are in landscape ecology and biogeography of reptiles focusing on the ecological challenges induced by the Anthropocene.

Stephan Pauleit leads the chair for strategic landscape planning and management at the Technical University of Munich. His main fields of interest are urban ecology and green infrastructure planning. He has participated in European research projects and he has worked in urban planning and urban greening projects in China and Africa.

Glynn C. Percival is the plant physiologist and technical support specialist for the Bartlett Tree Expert Company, and manages their UK and Ireland research and diagnostic laboratory. He is an honorary visiting research fellow at the University of Reading and Royal Botanic Gardens, Kew.

Martina Petralli is post-doc research fellow in urban and rural biometeorology. She graduated in forestry and environmental studies and holds a PhD in urban climatology. Her present research interests relate to the influence of weather and climate on human health and the effects of vegetation in the urban environment.

Michael J. Raupp is professor, extension specialist, and fellow of the Entomological Society of America. He has more than 250 publications, including *Managing Insect and Mites on Woody Landscape Plants* (Tree Care Industry Association, 2009). He has received more than a dozen awards including the Richard Harris Authors Citation from International Society of Arboriculture.

Steffen Rust is professor of arboriculture. He worked as a scientist in projects on forest stand and tree–water relations before he joined a tree-care company to develop tools for tree assessment, notably sonic tomography. His current research interests are tree care and non-destructive tree assessment.

Arne Sæbø has his main working field in urban greening, and has been pursuing this for the last 15 years. His research focus is on ecosystem services of woody species, aiming at finding the best plants to optimize the benefits for urban dwellers.

Fabio Salbitano holds a PhD in forest ecology and is an associate professor at the University of Florence. He is involved in research and education projects in urban forestry, landscape ecology, and silviculture in Europe and in developing countries. He is actively participating in FAO's urban forestry programmes in Africa, Latin America, Asia and the Mediterranean. Fabio Salbitano is a member of the editorial board of the journal *Urban Forestry & Urban Greening*.

Giovanni Sanesi is professor of urban forestry and forest planning at the University of Bari, Italy. His research has been related to many aspects of green infrastructure and ecosystem services in urban and peri-urban landscapes. He has participated in several European research projects and COST Action network and in FAO Urban Forestry activities in the Mediterranean and Asia Pacific.

Stephen R. J. Sheppard teaches in landscape and climate change planning, community engagement, and visualization. He currently directs UBC's bachelor of the urban forestry programme and CALP, an interdisciplinary research group working with communities on climate change solutions. He has 35 years' experience in environmental planning, aesthetics, and public involvement.

Alan Simson is a chartered landscape architect, urban designer and urban forester. He has gained extensive professional experience in the UK new towns, in private practice (including his own) and in higher education. He is currently professor of landscape architecture and urban forestry at Leeds Beckett University, involved in research, consultancy and teaching.

Henrik Sjöman is a senior lecturer and researcher at the Swedish University of Agricultural Sciences, Alnarp, and a scientific curator at the Gothenburg botanical garden. Henrik's research focus is towards plant selection for urban environments – understanding trees' capacity of growing under different climate conditions and site situations.

E. Thomas Smiley is a senior arboricultural researcher at the Bartlett Tree Research Laboratory in Charlotte, NC, and an adjunct professor of urban forestry at Clemson University. Dr Smiley is active in the arboriculture industry and has co-authored many of the ISA's Best Management Practices series. His research has led to improved methods of increasing sidewalk longevity near trees, protecting trees from lightning damage, improving tree root growth, and reducing tree risk.

Daniel K. Struve taught plant propagation and nursery management classes at the Ohio State University for 32 years before retiring. He conducted applied research on water- and nutrient-efficient container production systems and explored sexual production systems for woody plants.

Frank W. Telewski pursued both a masters and doctoral research focused on the influence of wind on tree biomechanics, growth and development. In January 1993 Dr Telewski moved to Michigan State University, where he is currently professor of plant biology and curator of the W. J. Beal Botanical Garden and Campus Arboretum. He has published over 50 peer-reviewed articles and book chapters on tree physiology, ecology, dendrochronology, and biomechanics.

Sydney A. Toni holds degrees in botany and urban forestry. She is a multi-faceted biologist, having worked on projects ranging from lichen chemical analyses to wildlife surveys on foot. Her recent research focused on urban-forest naturalness assessment and management in Canada.

Germán Tovar Corzo holds a degree in forest engineering and a postgraduate in administration and planning for regional development. He is a professional expert of the Section of Silviculture, Flora and Wildlife of the Municipality of Bogota where he is coordinating the team of Urban Silviculture. He has been involved in several FAO activities on urban forestry in Latin America.

Matilda van den Bosch is an assistant professor at two departments at the University of British Columbia. She is trained is a medical doctor and holds a PhD in landscape planning. Matilda's research and teaching focus on the importance of nature and outdoor areas for public health. She has worked as a consultant for WHO and is co-editing an Oxford Textbook on nature and public health.

Philip van Wassenaer is founder and principal consultant at Urban Forest Innovations, a Canadian-based consulting firm. He received his BSc in environmental sciences and master of forest conservation at the University of Toronto. Philip is a consultant, expert witness and international speaker. He is past president of the Ontario Urban Forest Council and was honoured with the ISA 'True Professional of Arboriculture' Award for his services to Arboriculture.

Jennifer K. Vanos specializes in human biometeorology within urban microclimates with applications to heat mitigation and thermal comfort. She has authored numerous manuscripts addressing health impacts of heat and air pollution. Prior to UC San Diego, Jennifer worked at Texas Tech University, Health Canada, and completed her PhD at the University of Guelph.

Les P. Werner is a professor of urban forestry at the University of Wisconsin–Stevens Point, teaching courses in arboriculture, functional tree biology, and forest soils. He received his BSc and MSc degrees from the University of Wisconsin–Stevens Point and his PhD from the University of Wisconsin–Madison.

Kathleen L. Wolf is a Research Social Scientist at the University of Washington, and the USDA Forest Service, Pacific Northwest Research Station. Her professional mission is to discover, understand and communicate human behaviour and benefits, as people experience nature in cities and towns. An overview of her research is at www.naturewithin.info.

1

INTRODUCTION

Cecil C. Konijnendijk van den Bosch,
Francesco Ferrini and Alessio Fini

Our urban world needs urban forests

We are living in the Anthropocene (Crutzen and Stoermer, 2000), an epoch in which human activities have started to have a significant global impact on the Earth's geology and ecosystems. Humankind has, in this epoch, moved from being primarily rural to urban. Soon the large majority of us will live in cities and towns of different sizes; already by 2030 this will be 60 per cent of all humans (United Nations, 2015). Urban areas make up only 3 per cent of the Earth's land surface, but they account for no less than 60–80 per cent of energy consumption and 75 per cent of carbon emissions. Urbanization is especially dramatic in the developing world, where 95 per cent of future urban expansion will take place (United Nations, 2015). Continuing urbanization and urban sprawl places tremendous pressures on, for example, fresh water supplies, public health, and on biodiversity.

The importance of urban areas as living environments for most humans is reflected in the 2015 United Nations Sustainable Development Goals (United Nations, 2015). One specific goal, Goal no. 11, highlights urban areas, under the heading 'Making cities inclusive, safe, resilient and sustainable'. Several targets were formulated under Goal 11, such as strengthening efforts to protect and safeguard the world's cultural and natural heritage; reducing the adverse per capita environmental impact of cities; supporting positive economic, social and environmental links between urban, peri-urban and rural areas by strengthening regional and development planning; as well as implementation of integrated policies and plans towards resource efficiency, climate change mitigation and adaptation, social inclusion and the like. The importance of urban green spaces for better cities is stressed in the target of providing (by 2030) universal access to safe, inclusive and accessible, green and public spaces, in particular for women and children, older persons and people with disabilities. Green spaces and urban vegetation, including trees, provide a wide range of essential benefits to urban societies through a range of supporting, provisioning, regulating and cultural ecosystem services. Urban trees and urban green can help cool cities, reduce the impacts of air and water pollution, and assist cities in dealing with floods and other extreme weather events. They can provide food and fodder, while also reducing our stress levels and encouraging us to be more physically active. Moreover, urban nature provides meeting places, inspiration, opportunities for learning, while also stimulating creativity (e.g. Konijnendijk et al., 2013; Roy and Byrne, 2014; Miller et al., 2015).

Ironically, urban green spaces have come under pressure in the quest for better cities. One of the urban planning and development approaches that has gained prominence during recent years is that of densification, thought to bring efficiency gains and technological innovation while reducing resource and energy consumption (e.g. Haaland and Konijnendijk van den Bosch, 2015). Densification is seen as an answer to continuing urban sprawl and its many negative effects. However, from a green space perspective, densification is considered challenging, as both green space quantity and quality will often decrease when cities become more compact (Haaland and Konijnendijk van den Bosch, 2015). When green space declines or even disappears, adverse effects to public health and human wellbeing can be expected.

Densification adds to the major challenges urban environments pose to urban vegetation such as urban trees. When looking at the latter, urban areas constrain tree growth and survival. Drought, poor soil quality, soil compaction, light heterogeneity, transplanting shock, pollutants, salinity, pathogens, and conflicts with human activities often cause premature plant death, thus reducing the net benefit from trees (Bussotti et al., 2014; Ferrini et al., 2014). It is, therefore, important to better understand the dynamics leading to tree decline in the urban environment and to develop strategies and techniques aimed at improving the horticultural tolerance (i.e. the capacity to provide benefits, not only to survive, under stressful conditions) of urban trees (Flowers and Yeo, 1995; Fini et al., 2013). These include nursery pre-conditioning techniques and post-planting management techniques (Franco et al., 2006), but a key role is played by species selection (Fini et al., 2009, 2014). Many different species are used in the urban environment, but selection criteria are frequently based on aesthetics and on whether the species are native or not, rather than on the tolerance to the typical stresses imposed by the built environment and on the capacity to provide substantial benefits therein. This has generated limited knowledge only about the ecophysiology of shade trees, if compared to fruit trees and crop species. We also need to know more about the planning, design and management of urban vegetation and urban green spaces.

How can we make sure that urban vegetation and green spaces are part of our efforts to develop inclusive, safe, resilient and sustainable cities; cities that are not threatening public health, but rather promoting it? Urban forestry, an interdisciplinary approach to the planning and management of all woody and associated vegetation in and around dense human settlements (Miller et al., 2015), can provide at least part of the answer. Based on this premise, the current handbook provides an overview of the current art, science and practice of urban forestry, focusing on state-of-art knowledge and good practice.

Why this urban forestry handbook – and who should read it?

This book in the Routledge Handbook series is timely. It is published at a time when the field of urban forestry is thriving and where research, practice and education are under continuous development. This is reflected, for example, in the rapid growth of journals such as *Urban Forestry and Urban Greening*, which now receives over 500 research and review paper submissions per year. The maturing of the field is also demonstrated by the large number of international, national and local conferences, seminars and workshops taking place annually with specific focus on aspects of urban forestry. Globally, the International Society of Arboriculture (ISA) and the International Association of Forest Research Organizations (IUFRO) have provided leadership. In Europe, the annual European Forum on Urban Forestry (EFUF) has a history of close to 20 years, providing a platform for urban forestry practitioners and academics from Europe and elsewhere to meet. At a more regional level, working groups such as Silva Mediterranea (currently under the Food and Agriculture Organization of the United Nations,

FAO) have also had specific groups and networks of urban forestry. The important urban forestry book by Miller (the latest edition in 2015) has served as a textbook for university students in North America and across the globe. In 2005, publication of the book 'Urban Forests and Trees' (Konijnendijk et al., 2005) was another milestone.

However, urban forestry keeps developing rapidly and there is a need for comprehensive, state-of-art publications that provide guidance to students and experts through the growing forest of studies and practices. Taking a global perspective on urban forestry, this handbook aims to fulfil part of this need, with the following objectives:

- To stimulate interest in the multiple roles and benefits of urban green areas.
- To provide insights about the multiple stresses experienced by trees in cities. This also relates to details about tolerance mechanisms of species with validated and potential ornamental use, as well as to cultural techniques for either pre-conditioning plants in the nursery or alleviating stress after planting.
- To promote sound planning, design, establishment and management of urban trees and urban forests for the optimization of their benefits is the domain of the fields of arboriculture and urban forestry. These fields draw upon a wide range of disciplines and have generated a large amount of knowledge and good practice during the past decades. Close interactions with the local urban communities who benefit from trees – and in some cases also are affected by tree disservices – are key to successful arboriculture and urban forestry.

By drawing upon international state-of-art knowledge on urban forestry and urban arboriculture, this book provides a comprehensive overview of the current state of the field. The book will be an essential handbook for researchers, students and practitioners dealing with urban trees in one way or another. It provides a unique perspective by having the tree and its growing site in focus, looking at biotic, abiotic and anthropogenic interactions between the tree and its environment.

What is urban forestry all about?

Chapter 2 provides some insight into the emergence and development of the concept of urban forestry in different parts of the world. Although its roots are in North America, with professor Eric Jorgensen at the University of Toronto famously coining the term during the 1960s (e.g. Konijnendijk et al., 2006), other continents can build on a rich heritage of planning and managing urban trees and urban woodlands. Konijnendijk et al. (2006) offer a comprehensive review of definitions of urban forestry in both North America and Europe, showing important differences. In North America, street trees and their role as 'shade trees' have often been in focus, while the Europeans tended to focus on their heritage of 'town forestry' which focused on woodlands that often had been in city ownership for centuries. However, most scholars would agree today that urban forestry should be defined in a more comprehensive way, as in the spirit of the definition provided by the Society of American Foresters (Helms, 1998) urban forestry is the 'art, science and technology of managing trees and forest resources in and around urban community ecosystems for the physiological, sociological, economic and aesthetic benefits trees provide society'. Thus urban forestry is about *all* trees in urban and peri-urban areas, ranging from individual trees in private gardens, through street trees and trees in parks, to urban woodland. Moreover, trees should be seen in connection with associated vegetation, such as shrubs and grass. However, this also shows the diversity of approaches needed in urban forestry, from sound arboricultural practices for individual trees to specific silvicultural approaches for urban woodland stands, and from single tree assessments to city-wide assessments of urban forest cover and benefits.

Urban forestry is multi- or even interdisciplinary, combining expertise, theories and methods from a range of disciplines. Obviously forestry is an important contributor, with its long-term perspective on natural resource management and its sustainability thinking, but so are landscape planning and landscape architecture, horticulture and arboriculture, ecology, biology, plant diseases, and the like. Moreover, urban forestry links the natural sciences with the social sciences and humanities. Urban forestry is just as much about people as about trees, and thus important contributions are made by sociology, environmental psychology and economics, as well as by the design fields (architecture, landscape architecture, art). Moreover, urban planning is crucial to successful and strategic urban forestry.

Many definitions of, and scholarly works on urban forestry, have stressed its socially-inclusive nature. This normative perspective recognizes that urban forestry can only be truly successful when it is done in close collaboration with local (urban) communities. Thus governance and public engagement are often given considerable attention. Through close collaboration between municipal and other experts with local residents, non-governmental organizations, but also private businesses, more resilient urban forests that provide a range of relevant benefits can be developed and maintained.

As mentioned above, a key aspect of urban forestry is that it is about growing trees and other vegetation in urban and highly artificial environments. In some cases, urban settings are very dense and growing space is limited. This challenge is very different from other types of natural resource management, with forestry for example typically focusing on natural or rural settings.

Urban forestry in relation to emerging approaches

With the growing recognition of the importance of urban green spaces for the quality of urban life and environment, and in the light of the need for more strategic approaches to green spaces planning and management, several approaches have emerged during recent times. Urban forestry should be seen within the context of these. Approaches related to the planning of greenbelts, green wedges and green structure gained popularity first of all in the Western world and then were also implemented in other parts of the globe (e.g. Xiu et al., 2016). During recent years, green infrastructure is a concept often used as referring to integrated networks / structures of green (and blue) space, either in urban or rural areas. Chapter 14 provides an in-depth perspective on green infrastructure and green infrastructure planning, tracing back its roots and defining some of its key characteristics as relating to, for example, connectivity and multi-functionality. Green infrastructure thinking has gained following at, for example, the European Union level, where it is seen as a potential tool for linking up urban with rural landscapes, and for creating synergies between nature conservation and other policy domains (Lafortezza et al., 2013). Urban planners, engineers and others can more easily relate to green infrastructure as one of the types of essential infrastructure for cities, in line with e.g. road networks, sewage systems and utility networks.

Even more recently, Europe has led the world in promoting the concept of nature-bases solutions, meaning solutions aimed to help societies address a variety of environmental, social and economic challenges in sustainable ways. They are

> actions which are inspired by, supported by or copied from nature. Some involve
> using and enhancing existing natural solutions to challenges, while others are

exploring more novel solutions, for example mimicking how non-human organisms and communities cope with environmental extremes. Nature-based solutions use the features and complex system processes of nature, such as its ability to store carbon and regulate water flow, in order to achieve desired outcomes, such as reduced disaster risk, improved human well-being and socially inclusive green growth. Maintaining and enhancing natural capital, therefore, is of crucial importance, as it forms the basis for implementing solutions.

Horizon 2020 Expert Group, 2015, p. 5

In other words, urban green spaces, urban forests and urban trees are typical nature-based solutions (or even just 'nature solutions') as they provide a range of ecosystem services that help cities meet important challenges, such as those related to climate change and public health and wellbeing.

As also argued elsewhere in this handbook, urban forestry still very much has a role to play today, in a time of emerging concepts and growing attention for green space and 'green solutions'. Building on at least 50 years of history and expertise, urban forestry focuses on trees as key provider of natural benefits in urban areas. It represents a comprehensive and integrative approach, supported by a wide range of disciplines, but increasingly also with a set of theories, methods and practices of its own. Where green infrastructure and nature-based solutions can sound abstract and technical, perhaps less related to specific place and people, urban forestry can serve as a delivery mechanism, building on its key component of linking people and trees at the local level. This urban forestry handbook was compiled in this spirit.

The structure of this book

This book was compiled as a comprehensive handbook covering most aspects of urban forestry, but obviously not all. It aims to discuss urban forestry from different angles, providing information about its background, rationale and current practice.

After this introductory chapter which sets the scene, defines urban forestry in the context of the book and provides insight into the structure of the handbook, focus of Chapter 2 by Richard J. Hauer et al. is on the history and development of urban forestry. The chapter describes the early use of trees in cities, before focusing on European tree landscapes. Unlike many earlier texts on urban forest history, however, the chapter also introduces historical development in the Asia-Pacific region, Central and South America, before it goes into greater detail with a history of shade tree programmes and urban forestry in North America. The chapter identifies the distinct, but also interwoven historical development of the fields of urban forestry and arboriculture. Chapter 3 by Justin Morgenroth and Johan Östberg introduces the urban forest as a resource with different components, and focuses on ways of measuring and monitoring this resource. As the chapter shows, in spite of many tools for measuring and monitoring, we often lack information on the urban forest resource in different places. The chapter introduces common variables for measurement, relating to estimating ecosystems services and long-term monitoring, for example. It also highlights the importance of standardization, which currently is often lacking. An overview is provided of common tools for measuring urban trees and urban forests, with special attention for recent technological advances, such as those related to LiDAR technology. Different types of inventories are presented. The final part of the chapter presents who measures the urban forest – from researchers and municipal authorities to local residents.

Chapter 4 on ecosystem services, by Cynnamon Dobbs and colleagues, sets the scene for a series of chapters on the benefits of urban forests. The chapter introduces the ecosystem service concept, which represents a new way of looking at the products and services that nature provides to us humans, and the main categories of services. Some basic information is provided on how to assess these different services, which is often not easy, for example in the case of cultural ecosystem services. The focus then shifts to urban forests and urban trees specifically. An ecosystem service framework is presented for application of a socio-ecological approach to urban areas. Finally the importance of including information about ecosystem service provision into decision making and management is discussed. Chapter 5, by Kathleen Wolf, looks into social aspects and benefits of urban forests. Cultural ecosystem services of urban forests are important, but often less tangible and less easily assessed than other services. Social benefits of urban forests are discussed at different scales, from individual and households, to neighbourhoods and communities. Focus of the chapter is, among other, on social interaction, social cohesion and social capital. The important contributions of 'metro nature' such as urban forests to life quality and urban experience are presented. Urban forests can help foster community and personal relationships, the connection with place, as well as stronger social connections. Some of these aspects can result in economic benefits, reflected in higher property values, local government revenue and enhanced retail business. In the provision of benefits, environmental equity needs to be considered, as urban forest values are typically not distributed evenly within urban populations. In Chapter 6, Matilda van den Bosch discusses the inherent links and interactions between humans, health, and green areas. The chapter takes a public health perspective; with public health having promoting and protecting health in focus rather than treatment of diseases, and here healthy environments are important assets. Substantial research demonstrates that trees and green spaces are corner stones for healthy urban environments, for instance supporting stress relief and recovery from mental fatigue, increasing physical activity and enhancing social cohesion. Surrounding greenness and accessibility to parks reduce the prevalence of many disorders associated with these health factors. Thus, urban trees may prevent, for example, obesity, heart failure and depression. Vulnerable populations, like children and underprivileged groups, seem to benefit the most from urban green. Access to green and natural environments stimulate children's healthy development, and health inequalities are less pronounced in green areas.

Chapter 7 written by Simone Orlandini et al. deals with the urban climate and describes all important aspects involved. We know that this is a very complex field of study because of the many characteristics that affect climatic factors in the urban environment, as well as due to the different kinds of green areas typologies present and materials used in cities all over the world. The chapter analyses climatic knowledge on the world's urban areas and the impacts of urban trees and forests on urban climate, with a special focus on thermal comfort. Examples are provided of urban forestry and climate protection strategies adopted across the world to facilitate the planning of more sustainable cities. Chapter 8 discusses how urban forestry and green infrastructures can contribute to pollution alleviation. In the chapter Arne Sæbø et al. review current trends and expected development, as well as how human health is affected by urban air pollution. They also analyse the modality of dispersion and deposition of air pollution in the urban landscape. The second part of the chapter gives special attention to the role of vegetation in air and soil pollution, while the final part of the chapter is dedicated to design suggestions for improving the effects of urban vegetation and to the art of designing for multiple functions. To acquire knowledge for designing and on where to place filtering screens with the best plants, maximizing deposition without negatively affecting air exchange in urban areas is probably the most demanding challenge for future planning.

Chapter 9 focuses on urban parks as habitat for plants and animals. The chapter explores the relationship between park features and biodiversity, as well as how management (in particular of urban forests) can have important effects not only at a local but also a national scale. Emilio Padoa-Schioppa and Claudia Canedoli highlight the fact that despite some taxa (such as birds) knowledge about their ecological interactions and requirements within the urban environment is sound, for many other taxa we still don't know enough. Both the scarcity of urban-based ecological research and the discrepancies in some research findings limit the possibility for making generalization about habitat requirements for less common taxa. This in turn makes it difficult to provide effective management recommendations that are not supported by highly specific studies only.

Chapter 10, written by Fabio Salbitano and colleagues, shifts the 'benefit focus' to a developing and industrializing country context. The context for urban forestry in these cities is often quite different from the industrialized world, due to, for example, rapid urbanization, high population density, migration from rural areas and the prevalence of informal settlements, urban poverty, and the specific development patterns of cities as reflected in urban form. The chapter identifies main issues that urban foresters and city planners have to address, such as those related to inequity, the need for specific ecosystem services, and land use and resource conflicts. It provides an overview of ecosystem services and benefits particularly relevant in this context, including food and energy provision, climate change adaptation, and public health. Governance approaches for urban forestry in this context are discussed and suggestions are provided for how to move forward and better integrate urban forestry into sustainable urban development and strategies to reduce poverty and enhance livelihoods.

The final two chapters of the 'benefit' series look into benefits and economic value (Chapter 11, by David Nowak) and ecosystem disservices (Chapter 12, by Jari Lyytimäki). Chapter 11 builds on Chapter 3, briefly touching upon measuring of urban forest ecosystem services and benefits. Then the chapter goes into further depth, looking into the benefits and economic value of urban forests and especially ways to assess these. A four-step approach is presented, starting with understanding and quantifying urban forest structure, followed by quantifying how this impacts ecosystem service provision. Next, the impact of the ecosystem service is to be quantified, followed by estimating the value (economic or otherwise) of this. Specific focus of the chapter is on i-Tree as the most comprehensive model to date to quantify urban forest structure, ecosystem services and their economic value. Chapter 12 shows that urban forests can also provide disservices to local communities, and argues that it is important to be aware of these nuisances and harms. The chapter highlights the wide range of disservices, including health effects, physical damage to infrastructure, emissions influencing air quality, unexpected economic costs, and social and psychological factors related to urban forests as places of fear or inconvenience. Different kinds of disservice may be produced by urban trees depending on their location, growth phase and the intensity of maintenance. Moreover, different disservices may be perceived as relevant depending on the knowledge base, attitudes and expectations of the people making the valuation. In some cases, concerns related to disservices may be amplified and exaggerated by the news media and social media debates. In turn, this may increase the risk of misplaced management actions. A balanced and comprehensive assessment of both ecosystem services and disservices provided by urban trees is needed for successful urban green management.

In the next series of chapters the focus is on the more strategic dimensions of urban forestry; that is, on how urban forestry is done in terms of decision-making and planning. Lafortezza et al. introduce strategic green space planning and urban forestry's place within this in Chapter 13. They present the concept of green infrastructure, which has gained prominence especially since the 1990s. Urban green infrastructure planning and management is differentiated from other green space planning approaches by being based on a specific set of principles that relate to the content as well as the process of planning, such as connectivity, multifunctionality, green–grey integration, and being multi-scalar. A brief introduction is given of green infrastructure in Europe. Synergies between green infrastructure and urban forests (as typically the largest component of green infrastructure) are explored. The chapter concludes with two case studies, one from Melbourne and one from the Greater Milan area. Chapter 14 (by Alan Simson) also places urban forestry in its wider context, focusing on the landscape perspective. The chapter argues that contemporary urban design, defined as the process of designing and shaping cities, towns and villages has, in the twenty-first century, not created the 'liveable places' that it claimed it would. Arguing for a transdisciplinary approach in urbanism it places urban forestry within the effort to bring together built form, grey infrastructure, blue infrastructure and green infrastructure. Landscape thinking is important in this respect, as it connects people and the land, urban and rural. The field of landscape architecture can contribute to urban forestry as it, in the broadest sense, deals with integrating people and the outdoor environment in a manner beneficial to both. The chapter states that landscape architecture and urban forestry are really different sides of the same coin and that, combined, they have the potential to be in the vanguard of the suggested new transdisciplinary approach to urbanism. The strategic urban forestry section is concluded with Chapter 15 on urban forest governance and public engagement, by Stephen Sheppard and colleagues. Urban forest governance addresses how rules, decisions, and actions regarding urban green spaces are implemented in multi-actor contexts. The chapter provides a summary overview of the topic area, paying special attention to public involvement and emerging governance and participatory planning processes in urban forestry. It provides a basic framework for considering the roles and practice of governance, planning, and community engagement around urban forestry in a changing environment. The chapter discusses key dimensions of urban forestry governance and engagement, addressing key actors, dominant discourses, and trends in processes and tools for social inclusiveness and citizen mobilization. It explores the application of both traditional and emerging approaches at selected spatial scales, drawing on international case studies. The chapter concludes with a summary of more effective governance and participatory processes in urban forestry, and identification of promising new approaches.

The next set of chapters zooms in on the urban trees as key component of urban forests. Growing conditions in the urban environment differ significantly from those where tree species have evolved and thus are more adapted to. Also, benefits expected from urban vegetation differ substantially from those expected from trees in forest plantations or agricultural crops. Because benefits/costs of urban plantings largely depend on tree health, and conventional forestry and agricultural productivity parameters (i.e. yield, stem growth) do not necessarily correlate to higher benefit, new tools and methods are needed to explore the 'productivity' of urban trees. Chapter 16 by Carlo Calfapietra et al. discusses several ecophysiological techniques for investigating tree responses to environmental parameters and for assessing the occurrence of physiological stress, such as gas exchange techniques, open-top chambers, eddy covariance, and NDVI. Advantages and good practices are provided for each method, and physiological parameters are recommended which should be primary investigated when dealing with specific stress. These methods and tools are gaining popularity for early evaluation of tree health, as well as for comparing species ability to cope with the typical constraints of urban sites. As said before, growing conditions in urban settings are often much less desirable for trees than those found in their native environment.

Abiotic stresses, exotic pests, and pathogen outbreaks threaten urban trees, sometimes suddenly, sometimes chronically, leading them into a mortality spiral that can be avoided by accurate assessment of planting site conditions, planting site modification, and selection of the right species. Chapter 17 by Glynn Percival presents the main abiotic stresses typically occurring in urban sites, while it also reviews the mechanisms adopted by different tree species to cope with these stressors. Effects of drought, salinity, root hypoxia, heat, air pollution and excess light stress on tree growth, physiology and biochemistry are described. Despite significant progress in anatomical, physiological and biochemical studies, very little is known about response mechanisms of shade tree species. The chapter reviews current knowledge, describing key metabolites and strategies involved in plant acclimation and adaptation to urban condition. Also, key criteria that could be used for identifying superior stress tolerance tree genotypes for urban plantings are presented.

Biotic factors, namely pests and pathogens, often contribute to tree decline, acting as both inciting (e.g. *Lymantria dispar*) or contributing (e.g. *Ips typographus*) factors in the decline spiral. Chapter 18 by Michael Raupp and Paolo Gonthier describes the main biotic threats to urban trees with a particular focus on important issues of urban greening in contemporary cities, such as low tree diversity and spread of exotic pests and pathogens around the globe. In the first section of the chapter, Raupp focuses on insects and mites, suggesting that lack of co-evolved defences against non-native pests by trees, the lack of predators and parasitoids in the simplified urban ecosystem, and the widespread use of a limited number of tree species in urban settings around the world may lead to catastrophic tree loss as soon as a new pest is introduced. Exotic, introduced insects with a significant impact on human health, tree survival and tree benefits (namely *Thaumetopoea processionea*, *Agrilus planipennis* and *Cameraria ohridella*) are taken as examples of biologic invasions. Their life cycles are described in detail, and methods for diagnosis and control are provided. In the second section of Chapter 18, Paolo Gonthier describes how global change affects pathogen outbreak in urban areas. The author shows how specialized pathogens may flourish in urban settings, where abiotic stresses, improper cultural practices weaken tree defences and facilitate the spread of diseases. Significance, diagnosis, biology and management are presented for four types of diseases: root rot and wood decay; vascular diseases, such as Dutch elm disease; canker diseases, such as plane tree canker stain and ash dieback; and foliar diseases.

To conclude the book section about typical constraints to tree survival, Chapter 19 by C. Y. Jim presents possible conflicts between trees, grey infrastructure and the built environment in general. While low density urban areas provide ample room for both green and grey infrastructures, the ongoing trend of more compact cities imposes an intense competition for space between the two infrastructures. The chapter approaches the problem from the perspective of an urban-tree constraint typology, highlighting, respectively, subterranean constraints with respect to soil quantity and quality which could confine root spread and affect their functioning; constraints related to the subaerial environment imposing headroom and lateral room obstructions; and management constraints such as improper pruning, and root severance which can alter the ratio between absorbing and transpiring areas. Ameliorative or preventive measures are proposed to avoid or mitigate the described conflicts.

After a detailed description of typical abiotic, biotic and 'anthropogenic' constraints which are likely to be found in the urban environment, Chapter 20 by Nina Bassuk takes the reader down to the planting site. Given the high spatial heterogeneity of pedological and environmental conditions within cities, a detailed assessment of the planting site is required to avoid catastrophic failures. Following a brief general outlook, Bassuk focuses on the micro-scale of the planting site, providing practical suggestions about soil, climate and vegetation traits that should be assessed preliminary to the planning process. With its thorough focus on soil physico-chemical parameters that deserve attention during the site-assessment process, the chapter introduces the following two chapters about soil modifications.

Chapter 21 by Susan Downing Day and J. Roger Harris underlines differences between native soils, which originate from long-term pedological processes, and urban soils, where soil material has a non-agricultural, man-made surface layer more than 50 cm thick that has been produced by mixing, filling, or contamination of land surfaces in urban and suburban areas. The lack of organic matter in particular is proposed by the authors as the main cause of soil degradation in urban areas. Lack of organic matter weakens and impairs soil structure, enhancing compaction, poor aeration, and crust formation. Concrete and limestone used for buildings and pavements raise soil pH, imposing nutritional issues to trees. Criteria and approaches for 'soil conscious' environmental designs are proposed, as well as methods for rehabilitation of urban soils.

In several cases, however, such as soils supporting pavements, the conflicts between engineering and tree requirements become nearly mutually exclusive. Soil compaction is required by civil engineers to avoid pavement settlement, but compaction limits the soil's ability to support root development and biological activity by destroying structure and crushing macroporosity. Chapter 22 by Jason Grabosky and Nina Bassuk illustrates solutions for integrating load-bearing capacity and soil habitability by tree roots. Two main approaches are described, the first of which encompasses soil approaches, meaning the development of an aggregate skeleton matrix capable of providing acceptable bearing capacity for pavement surface support, with a biological component suspended within the interconnected voids of the skeleton matrix. The second group are grid system approaches; that is, designed modular structures that support the pavements, generally filled with (or placed on) uncompacted native soil free of any load-bearing role that can support plant growth. The effects of structural soils and structural cells on soil colonization by roots, and the consequent effects on tree stability during wind-load are discussed in detail.

After the planting site has been characterized in detail by site assessment, and required site modifications have been prescribed, it is essential to select tree species that can live, not only survive, in the planting site, and thus can provide long-term benefits to the community. Chapter 23 by Henrik Sjöman and colleagues focuses on species selection criteria for future urban plantings. Climate is changing fast, and climate change is anticipated and exacerbated in the urban environment, where plants already experience the environmental conditions expected for rural areas by 2050. Therefore, it is essential to look beyond the eternal discussions on planting native or exotic species, and focus on a benefit/cost planning approach that maximizes the success of urban plantings and the benefits provided by trees. The chapter highlights criteria that should guide future plantings in urban areas: hardiness, health, light-requirements, site tolerance, function expected, maintenance requirements, growth pattern, and aesthetic and social qualities. Sjöman et al., while describing criteria and providing lists of species fulfilling desired requirements, use a 'copy from nature' approach, which interestingly provide the answer to several typical constraints of urban sites.

After selecting the right species for successful urban plantings, it is advisable to use high-quality nursery stock. Nursery cultivation techniques, in fact, are a strong determinant of transplanting success, of future maintenance requirements (e.g. for pruning), and of stress tolerance in the landscape. Chapter 24 by Dan Struve focuses on the importance of selecting high-quality nursery plants and provides an overview of examples of standards for nursery stock. Nursery stock standards provide buyers and sellers of nursery stock with a common terminology in order to facilitate nursery stock transactions, including techniques for measuring plants, specifying and stating the size of plants, determining the proper relationship between height and caliper, or height and width, and determining whether a root ball or container is large enough for a particular size and class of plant.

However, even after selecting the right species and high-quality nursery plants, transplant success is still not warranted. As early as in 1995, the International Society of Arboriculture suggested that is better to spend 100 dollars for the tree and 200 for the making of a good planting pit than *vice versa*. Chapter 25, by Andrew K. Koeser and Robert J. Northrop, provides some key considerations to promote successful tree planting. It includes information on minimum planting space requirements, seasonal impacts, proper tree storage and handling, planting hole excavation and backfilling, and early care practices (e.g. staking, mulching, pruning). The chapter also highlights the effects of different production methods (such as container, B&B, bare root) on transplanting success, and discusses installation techniques to maximize success of trees produced according to the different methods.

The book section on 'planning, producing and planting' is closed by Chapter 26 by J. Roger Harris and Susan Downing Day. It provide a solid basis for navigating newly planted trees out of the critical and stressful establishment period, when roots are still mostly confined in the rootball and have little access to the resources of the native soil. After defining the concept of establishment, the authors propose that speed matters and that the faster roots are regenerated out into the soil, the less the tree is likely to suffer and decline. Species differences in their capacity to regenerate roots, as well as seasonal effects, are discussed in the chapter. Fundamental issues such as planting at grade and irrigation after planting are discussed for container, balled and burlapped, and bare root trees.

It has long been said that successful urban arboriculture and forestry is based on the selection of the right species for the right planting site, with the right management. Therefore, the next section of the book is focused on urban forest management and maintenance, including tree care. This section begins with Chapter 27 in which Brian Kane provides an update of best pruning practices and pruning goals. Proper and improper pruning methods are discussed, and the effects of different pruning techniques on tree physiology and structural stability are reviewed. The author observes that, despite much having been said about sound and inappropriate pruning cuts, bad practices such as topping are still in use. Kane provides rationales for pruning trees according to science-based methods, but also highlights that most research has been carried out under controlled conditions (e.g. nursery) on young trees of few genera and species, which may not be fully representative of mature or senescing trees in residential environments.

Irrigation and fertilization are other management techniques widely used in forestry and agriculture, but findings from these disciplines can hardly be transferred to urban forestry. Chapter 28 by Alessio Fini and Cecilia Brunetti reviews water resource availability and water use in modern cities and tries to answer the question whether irrigation is really needed in urban settings. Focusing mainly on established trees (completing the picture of irrigation during establishment touched upon in Chapters 25 and 26), but also on other environmental side-benefits of irrigation (such as direct microclimate amelioration), the authors provide guidelines to optimize irrigation efficiency in urban settings. After preliminary assessment of soil and weather in the planting site, suggestions for increasing irrigation efficiency through precision scheduling, species selection, and reuse of wastewater are provided.

Plants need water, but also nutrients to trigger their growth. Results of fertilization experiments in urban sites are, however, often contrasting. Chapter 29 by Cecilia Brunetti and Alessio Fini reviews the effects of fertilization on urban trees. A particular focus is given to mineral nutrients in urban soils, and to how global change may affect nutrient availability to plants. Methods to evaluate deficiencies are discussed, and effective fertilization management methods to overcome nutritional limitations described.

Chapter 30, written by Frank W. Telewski and Karl J. Niklas, reminds us that trees are particularly susceptible to physical laws and processes as they are structures that begin and end their lives in the same location, change size over many orders of magnitude, and experience dramatic changes in abiotic and biotic conditions throughout their lifetimes (e.g. rainfall, temperature, and epiphyte or snow/ice loadings). The biomechanical behaviour of trees is reviewed on the basis of a few basic equations, with particular emphasis on the effects of wind on branches, trunks and roots. Understanding the limits of these equations is a critical first lesson in dealing with the biomechanical behaviour of trees, regardless of whether they grow in tropical, temperate or desert conditions. As stated by the authors in their final analysis, every tree will ultimately fail. The challenge then is to anticipate when and how.

The following two chapters (31 and 32) deal with the practical part of tree stability and biomechanics. Chapter 31, by E. Thomas Smiley and colleagues, defines terms associated with tree risk in the urban environment, explores options for systematic evaluation of trees, outlines one commonly used system, and briefly discusses risk mitigation. The first part explains what 'risk' is and introduces the types of risk associated with trees. Next, different approaches to risk assessment and risk categorization are considered. Finally, some measures to mitigate risk are presented. The section on tree stability is concluded with Chapter 32, by Steffen Rust and Philip van Wassenaer, examining the procedures and tools most widely used in tree risk assessment. Traditional and advanced techniques for tree risk assessment are discussed and some practical examples are provided by highlighting advances in assessment techniques and how they may be used to retain trees that might otherwise be removed. The authors remind us that advanced assessment does not necessarily involve high-tech instruments. Often, aerial inspection, either from an aerial lift, a ladder, or by climbing the tree, will help to assess features invisible from the ground, such as cracks, included bark, or decay on the upper side of branches. But most advanced methods are used either to measure the extent of internal decay, to assess root loss and its effects on anchorage, or to investigate pathogens.

In the following Chapter 33, Neville Fay and Jill Butler focus on veteran and heritage trees care and management in urban settings, and the challenges associated with this. A description of the fundamentals of 'conservation arboriculture' as a basis for practical management guidance is an important part of this chapter. The authors also explain why ancient trees are irreplaceable and how understanding their qualities and survival strategies can inform management in mature and younger trees.

The final chapter by Peter Duinker and colleagues addresses the management opportunities and challenges associated with urban woodlands. By analysing the literature, the authors extract principles that can guide management of urban woodlands worldwide. They then examine three cases where urban forest management is particularly noteworthy and innovative. According to Duinker et al. we can conclude that managing urban woodlands is challenging both in technical terms and particularly from a socio-political perspectives. However, urban woodland management can be successful when insights from research and adaptive management are combined with meaningful stakeholder involvement.

References

Bussotti, F., Pollastrini, M., Killi, D., Ferrini, F., Fini, A. (2014) 'Ecophysiology of urban trees in a perspective of climate change', *Agrochimica*, July–September 2014, pp. 247–268.

Crutzen, P. J., Stoermer, E. F. (2000) 'The "Anthropocene"', *Global Change Newsletter*, vol. 41, pp. 17–18.

Ferrini, F., Bussotti, F., Tattini, M., Fini, A. (2014) 'Trees in the urban environment: response mechanisms and benefits for the ecosystem should guide plant selection for future plantings', *Agrochimica*, July–September 2014, pp. 234–246.

Fini, A., Bellasio, C., Pollastri, S., Tattini, M., Ferrini, F. (2013) 'Water relations, growth, and leaf gas exchange as affected by water stress in *Jatropha curcas*', *Journal of Arid Environments*, vol. 89, pp. 21–29.

Fini, A., Ferrini, F., Frangi, P., Amoroso, G., Piatti, R. (2009) 'Withholding irrigation during the establishment phase affected growth and physiology of Norway maple (*Acer platanoides* L.) and linden (*Tilia* spp.)', *Arboriculture and Urban Forestry*, vol. 35, issue 5, pp. 241–251.

Fini, A., Ferrini, F., Di Ferdinando, M., Brunetti, C., Giordano, C., Gerini, F., Tattini, M. (2014) 'Acclimation to partial shading or full sunlight determines the performance of container-grown *Fraxinus ornus* to subsequent drought stress', *Urban Forestry and Urban Greening*, vol. 13, pp. 63–70.

Flowers, T. J., Yeo, A. R. (1995) 'Breeding for salinity resistance in crop plants: where next?', *Australian Journal of Plant Physiology*, vol., 22, pp. 875–884.

Franco, J. A., Martinez-Sanchez, J. J., Fernandez, J. A., Banon, S. (2006) 'Selection and nursery production of ornamental plants for landscaping and xerogardening in semi-arid environments', *Journal of Horticultural Science and Biotechnology*, vol. 81, pp. 3–17.

Haaland, C., Konijnendijk van den Bosch, C. (2015) 'Challenges and strategies for urban green-space planning in cities undergoing densification: a review', *Urban Forestry and Urban Greening*, vol. 14, pp. 760–771.

Helms, J. (ed.) (1998) *The dictionary of forestry*, Society of American Foresters, Bethesha, MD.

Horizon 2020 Expert Group (2015) *Towards an EU research and innovation policy agenda for nature-based solutions and re-naturing cities*, final report of the Horizon 2020 expert group on 'Nature-based solutions and re-naturing cities', Horizon 2020 Expert Group on Nature-Based Solutions and Re-Naturing Cities, European Commission, Brussels, retrieved from http://bookshop.europa.eu/en/towards-an-eu-research-and-innovation-policy-agenda-for-nature-based-solutions-re-naturing-cities-pbKI0215162.

Konijnendijk, C. C., Annerstedt, M., Maruthaveeran, S., Nielsen, A. B. (2013) *Benefits of urban parks – systematic review of the evidence*, A report for International Federation of Parks and Recreation Administration (Ifpra), University of Copenhagen and Swedish University of Agricultural Sciences, Copenhagen and Alnarp.

Konijnendijk, C. C, Nilsson, K., Randrup, T. B. and Schipperijn, J. (eds.) (2005) *Urban forests and trees*, Springer, Heidelberg.

Konijnendijk, C. C., Ricard, R. M., Kenney, A., Randrup, T. B. (2006) 'Defining urban forestry – a comparative perspective of North America and Europe', *Urban Forestry and Urban Greening*, vol. 4, no. 3–4, pp. 93–103.

Lafortezza, R., Davies, C., Sanesi, G., Konijnendijk, C. C. (2013) 'Green infrastructure as a tool to support spatial planning in European urban regions', *iForest – Biogeosciences and Forestry*, vol. 6, pp. 102–108.

Miller, R. W., Hauer, R. J., Werner, L. P. (2015) *Urban forestry: Planning and managing urban greenspaces*, 3rd edn, Waveland Press, Long Grove, Il.

Roy, S., Byrne, J. A. (2014) 'A systematic quantitative review of urban tree benefits, costs, and assessment methods across cities in different climatic zones', *Urban Forestry and Urban Greening*, vol. 11, pp. 351–363.

United Nations (2015) 'Sustainable Development Goals', retrieved on 28 April 2016 from www.un.org/sustainabledevelopment/sustainable-development-goals.

Xiu, N., Ignatieva, M., Konijnendijk van den Bosch, C. (2016) 'The challenges of planning and designing urban green networks in Scandinavian and Chinese cities', *Journal of Architecture and Urbanism*, 40(3): 163–176, September 2016, DOI: 10.3846/20297955.2016.1210047.

PART I

Urban forestry

2

THE HISTORY OF
TREES IN THE CITY

Richard J. Hauer, Robert W. Miller, Les P. Werner
and Cecil C. Konijnendijk van den Bosch

This chapter translates the history of trees and vegetation in cities. Urban forestry and urban greening are still evolving. Our ability to understand why people desire greenspaces continues to expand. The historical story begins with our primate ancestors who evolved in the forests of Africa by living in trees, foraging for food in trees and finding protection and escape from predators in trees. As the African climate grew dryer many forests gave way to vast savannas with our primate ancestors evolving new characteristics for this habitat such as an upright stature, larger brains, higher organization skills, and more sophisticated use of tools. In time humans moved across the planet as hunters and gatherers living in a variety of habitats, many of which were savannas or forests. Eventually the development of agriculture allowed us to build permanent settlements, which has led us to the urban world of today. In spite of our city building we do remain attached to the habitats of our ancestors. African forests and savannas shaped us as a species and this includes our attachment to nature in general and trees in particular.

As we have built settlements we have incorporated trees and other elements of nature for reasons ranging from sources of food and fibre to their aesthetic and emotional values. Throughout the world natural elements have become necessary components of our communities. The historical story shows the thoughts and practice of humans to use trees for architectural design, recreation, human health, and spiritual reasons. This chapter traces the history of urbanization and the use of trees through a regional (e.g. Europe, Asia-Pacific, Africa, Central and South America, and North America) perspective. It ends with the profession of urban forestry that has developed into the twenty-first century.

Early use of trees

The introduction of ideas into society is often borne out of necessity and oftentimes nicety. Humans spent much of their existence hunting and gathering food over large areas. While they could not build permanent settlements, they likely planted trees valued for edible fruit and other uses throughout their range. Near 8,000 BC the selection and cultivation of fruit trees began. Annuals were likely cultivated before perennials, but fruit trees followed soon after. As humans moved away from being nomadic hunters and gatherers, fruit trees played an increasingly important role in the history of civilization (Konijnendijk, 2008).

Since agriculture led to the first permanent settlements, it stands to reason that domesticated plants were a part of the community, including trees cultivated for food. The early Egyptians described trees being transplanted with balls of soil around 2000 BC

(Campana, 1999). Trees were valued for shade and aesthetics, and were included in gardens around temples and palaces for priests and rulers. It is likely most trees were selected for their utilitarian value (fruit) as well as their beauty. For example the royal gardens of Egypt had rows of fig trees, palms, and pomegranates. Olive trees were first domesticated between 4000 and 6000 BC in the eastern Mediterranean region of Levant near the border of present Syria and Turkey (Campana, 1999).

Trees were used for amenity reasons since ancient times. The Hanging Gardens of Babylon were built sometime between 600 and 800 BC (Figure 2.1). Ancient Greece had trees in plazas and along roadways. Rome had severe penalties for anyone who injured a tree. Kublai Khan required tree planting in and around Beijing in thirteenth century along all public roads for shade and to mark them when snow covered (Profous, 1992). Dating to 1283, trees in Thai cities were planted for religious and ceremonial, and ship building purposes (Dumrongthai, undated; Intasen, 2014). Despite these documented instances, it is not likely that trees were abundant in ancient cities. The civilizations of the Maya, Inca, and Aztecs built large cities with monumental architecture, and they supported their cities with agriculture and agroforestry systems. Drawings and descriptions of pre-Columbian America suggest many Native American tribes developed extensive agricultural communities, which included gardens with planted trees.

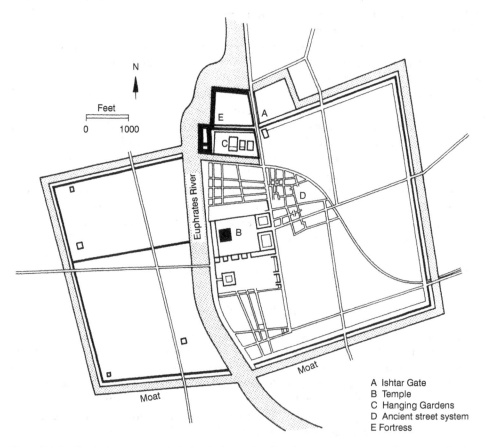

A Ishtar Gate
B Temple
C Hanging Gardens
D Ancient street system
E Fortress

Figure 2.1 By the sixth century BC Babylon had grown to be a large city consisting of a rectangular street system, dwellings for a variety of classes, and monuments, including the Hanging Gardens

Source: Gallion and Eisner (1975)

European urban tree landscapes

Trees played a fundamental part in the historical development of modern Europe (Konijnendijk, 2008). Early Europeans raised crops, hunted, and gathered in the vast forests. Individual trees and forests became a part of sacred pre-Christian religions and rituals. Members of tree cults worshipped in sacred groves with holy trees. Holy groves and sacred trees were regarded as symbols of paganism to the early Christian church, often leading to their destruction in the name of Christianity.

Early Europeans used forests as hunting reserves by the nobility and as commons providing food, fodder, and fuelwood for local communities. Forests also provided refuge for the oppressed and for outlaws. Consider the story of Robin Hood and his merry men lurking in Sherwood Forest waiting to plunder the rich, as well as the tale of the outlaw Dick Turpin roaming the English forests. During the Middle Ages, cities in Europe contained some trees and other plants in the private gardens of the ruling classes, but they were primarily planted for utilitarian purposes such as fruit (Konijnendijk, 2008).

The Renaissance in Italy during the sixteenth century saw the first development of villas on the periphery of cities. These villas had walled gardens, tree-lined paths, called allées, for walking and relaxation, and often had 'wilder' woods as part of the estate. The concept of the villa soon spread to France and Spain. By the seventeenth century the allée moved from the countryside to the city (Figure 2.2). In some towns allées of trees were planted as public promenades along the tops of earthwork fortifications, and the upper classes began to develop allées of trees in and

Figure 2.2 London plane trees (*Platanus* × *acerifolia* (Aiton) Willd.) were used to define an allée in Rosenstein Park, Stuttgart, Germany

Photo: L. Werner

near cities for recreational activities such as archery, early forms of bowling, and the lawn game pall-mall (a precursor to croquet and golf). Allées of trees for both pedestrian and vehicular traffic were planted along and on the fortifications of French cities. The first planting in Paris was on the bulwark which became known as the Grands Boulevards (Lawrence, 2008). Today we often call tree-lined streets with landscaped medians boulevards.

In the Netherlands allées of trees were planted along canals and their adjacent streets. A plan for the expansion of Amsterdam called for one tree for each standard building width. In London trees and lawns were planted in enclosed squares surrounded by new residences for the exclusive use by the occupants. However, by the end of the seventeenth century trees and landscaped spaces were still uncommon in European cities, being primarily available to the upper classes (Lawrence, 2008).

Throughout Europe the lower classes did not have access to parks and gardens, often being excluded by entrance fees or for improper dress (Lawrence, 2008). By the eighteenth century the rising merchant and professional classes were demanding the amenities common to the aristocracy. Entrepreneurs recognized that trees and open space added to the market value of property, and increasingly added these amenities in the construction of residential neighbourhoods. Builders of new upper-class housing in London found they needed to include private squares complete with trees, shrubs, and lawns. New larger-scale developments included carriage ways and avenues lined with trees.

Recreation became a more passive pastime, with parks being used for relaxation and to contemplate nature. Although most parks and gardens were private, some public plazas and large public parks were being constructed. These parks and estates often included trees and other plants brought in from the colonies. Commercial 'pleasure gardens' were also popular as venues of entertainment with a fee for their use (Lawrence, 2008). In the nineteenth century, city walls were removed as they were made obsolete by long-range artillery. Parks and promenades often replaced these walls. Resort towns grew popular, and were extensively landscaped as part of their marketing. Napoleon built promenades and boulevards in the cities he conquered, while the English introduced the naturalistic landscape into parks in contrast to the formal baroque garden. British engineers greatly improved the engineering of city streets and sanitation. Parisians adopted British engineering and added trees to streetscapes. The British House of Commons addressed the need for parks being available to the lower classes, stating they were 'convinced that some Open Places reserved for the humbler classes, would assist to wean them from low and debasing pleasures' (Lawrence, 2008). Members felt that access to public parks would have a civilizing effect on the lower classes, so more parks were constructed and the lower classes were allowed access to them.

Disease at that time was thought to be caused by miasmas, the foul smells coming from polluted water and from sewers. Trees were thought to sanitize the atmosphere and to reduce disease. In a similar fashion more public parks were built in cities throughout much of Europe. Baroque gardens in France were first developed in hunting preserves, with wide pathways radiating from clearings in the forest for shooting game during the hunt (Zube, 1971). These radiating pathways lined with trees came to influence the design of eighteenth-century villages and towns, and ultimately produced the radial street pattern of Washington, DC, USA. Napoleon III and civic planner Georges-Eugène Haussman transformed Paris in the 1850s and 1860s by imposing a radial street pattern of tree-lined boulevards on the existing city, and by building public parks, squares, monuments, and gardens. The new wide boulevards were designed to allow better access to city neighbourhoods for times of both celebration and riot. Cities throughout Europe adopted the renovation designs of Paris, adding more boulevards, tree-lined streets, and parks.

The influence of European urban architecture, complete with trees and public parks, spread throughout much of the world through settlement and colonization by Europeans, and through wide-spread adoption of European-style architecture (Zube, 1971; Lawrence, 2008). Nineteenth-century Great Britain gave rise to the 'romantic landscape' and the detached suburban house for the newly emerging wealthy businessmen of the Industrial Revolution. As cities industrialized and became squalid and polluted, the wealthy classes moved out to the new suburban communities. Large parks, garden squares, and tree-lined residential streets of individual homes became commonplace in the new communities (Johnston, 2015). While the baroque garden was the epitome of the formal landscape (Figure 2.3), the romantic landscape was informal in nature, and consisted of the 'natural' arrangement of plants and structures in the landscape. In its most idealistic sense, the romantic landscape embodied a mix of what was considered the best from the city and nature.

Urban vegetation management and design in Europe has a long history (Konijnendijk et al., 2006). Tree planting led to the early development of arboriculture practices. Arboriculture as we know it today emerged in England between 1400 and 1800. James Lyte's book *Dodens*, published in 1578, uses the term 'arborist'. William Lawson's *A New Orchard and Garden*, published in 1597, explained planting, pruning, fertilizing, wound treatments, and cavity filling. From a more traditional forestry perspective, 'town forestry', as practiced by foresters, managed forests on the periphery of cities and towns for a wood supply, grazing areas and for other forest products. By the late twentieth century town forests became more important for their environmental and social benefits. Urban forestry in Europe emerged in the 1980s as a new discipline, bridging traditional forestry and urban park disciplines such as ornamental horticulture and arboriculture (Forrest and Konijnendijk, 2005; Konijnendijk et al., 2006).

Figure 2.3 Example of a Baroque garden

Photo: R. W. Miller

Asia-Pacific urban tree landscapes

Although the professional growth of urban forestry and urban greening in the Asia Pacific region is recent, historical roots exist for using trees in built environments (Profous, 1992). Tree plantings were recorded as early as 200 BC in China (Jim and Liu, 1997). Trees were used for many functions (e.g. spiritual, food, shade, lining paths, aesthetics). The early introduction and favouring of a variety of trees over time is evident. Even though trees were planted along streets, in parks, and other locations, limited records of tree populations, tree policies, and formal management approaches are seen as barriers towards acceptance that urban forestry and urban greening is formally occurring. The creation of professional societies (e.g. regional chapters of the International Society of Arboriculture) and research journals (e.g. *Journal of Chinese Urban Forestry*) have advanced urban forestry in this region.

Thailand

Urban forest plantings in Thailand occurred as early 1283 for dynastic and religious functions (Dumrongthai, undated). At that time King Ram Khamhaeng recorded that *Borassus flabellifer* L. (Asian palmyra palm) trees were planted in the city for religious and ceremonial purposes. The wood from *Magnolia champaca* (L.) Baill. ex Pierre (champak) was used to construct pavilions and temples. The use of trees and planted locations often followed the desires of the King. Between 1869 and 1890, King Rama V had emphasized the planting of trees along major streets in Bangkok.

Formal goals to incorporate green spaces in urban planning are more recent. A goal to increase green area through park development and tree planting was adopted in 2012 for the Thai capital of Bangkok (Intasen, 2014). The plan should make Bangkok a more liveable city by expanding the green area at an annual approximate rate of 200 ha (Intasen, 2014). A great challenge to achieving this goal is the high cost of land and associate opportunity cost of green space versus commercial or other development.

Hong Kong

Hong Kong has long served as a port of trade and has become a densely populated region since being founded in 1840 (Jim, 1986; Tian et al., 2012). Along with trade includes the development of places for people to hike, bike, exercise, recreate, and to relax. The urban park is one common location that serves citizens in the compact city today (Tian et al., 2012). Historically the development of land was constrained by steep topography that led to a compact city in the more easily buildable areas. Much of the surrounding hillsides of native tropical evergreens were cut over by the 1870s (Tian et al., 2012). These were later transformed by planting into semi-natural forests and through invasion of non-native plants.

The *Ficus microcarpa* L.f. (Chinese banyan) is a large-statured tree planted almost exclusively from the 1840s into the 1930s (Jim, 1986). The use of this species today is more limited due to its size and lack of adequate growing space in the compact locations. Other current challenges to growing trees are many (Tian et al., 2012). Native soils except in alluvial locations tend to be acidic and nutrient poor (Jim, 1986). Development has also led to artificial sites with impervious surfaces to reduce erosion during the monsoon season. A 'concrete jungle' is a common modern challenge for urban forestry in Hong Kong.

The historical stonewall or wall trees of Hong Kong provide a unique and interesting urban greening feature in the city. Since Hong Kong has limited flat land, walls were used to create horizontal locations to build. Several *Ficus* spp. (strangler figs) trees are able to grip and grow upon the vertical walls of the original stone walls and are the dominant tree species (Jim and Chen, 2010). The conservation of the urban ecological heritage is recommended, but faces recent challenges of tree failures that caused human death and property damage.

Australia

Urban forestry is relatively new in Australia (Davison and Kirkpatrick, 2014). Historical roots were influenced by premodern European traditions of Town Forestry and the City Beautiful and Garden City movements. Canberra provides an urban forestry case study of incorporated trees along streets, in parks, and other green spaces since the early 1900s (Banks and Brack, 2003).

Trees transformed a barren Canberra environment into a Garden City and decreased dust and winds (Figure 2.4). The design team of Walter and Marion Griffin developed the original city master plan. They integrated City Beautiful and Garden City ideas and sought to use native species. Charles Weston who implemented tree plans encountered challenges of growing trees in difficult soil and climate (Banks and Brack, 2003). Weston overcame the challenges by:

1 establishing a nursery;
2 raising suitable plants for the area;
3 ameliorating Canberra's harsh environment through vegetation;
4 establishing an arboretum to test new species;
5 beautifying the landscape by using both exotic and native plants; and
6 conservation and afforestation activities on surrounding hills.

Figure 2.4 The planned city of Canberra used trees and other vegetation to line streets and transform greenspaces from a barren environment

Photo: R. J. Hauer

Tree plantings dating to the 1920s have introduced over 400,000 trees from over 200 species (Banks and Brack, 2003). Species composition has changed over time and waves of plantings have occurred based on the desires of decision makers. Initial plantings favoured exotic trees and what followed gave the 'Garden City' appearance. Since the 1970s, native *Eucalyptus* spp. have been favoured. These plantings gave birth to 'The Bush Capital' term for the appearance that replicates the surrounding native landscapes.

African urban tree landscapes

Many of the earliest records of tree cultivation, use, and management can be found within African cultures. Egyptian records from over 4,000 years ago describe the use of trees in temple gardens, private residences, and along travel corridors. Garden designs often included a rectangular pool whose edges were lined with a variety of trees, shrubs, and flowers (Bigelow, 2000). This design concept is likely a precursor and common theme in European gardens. The selection and use of many non-native plants also occurred in landscape designs. Trees were a source of raw materials (fuel, wood, baskets, etc.), food, medicine, shade, and beauty. However, the importance of trees in ancient Egypt went beyond their utilitarian value. There is an intimate association between trees and the Egyptian deities Re, Horus, Isis, and many others. The foundations of modern day agriculture, horticulture, and arboriculture also trace to the Egyptian culture (Campana, 1999).

There is little popular and peer reviewed literature regarding the historical use of trees and/or their management within Sub-Saharan Africa. One example dating to 1924 is the promotion of tree planting and protection in Kenya and creation of Watu wa Miti (People of the Trees). Despite this, the rapidly changing demographics in and around urbanized Sub-Saharan Africa suggest forests and forest management will play an increasingly larger role in the sustainable development of urban areas. Over the past three decades there has been a dramatic increase in the urban growth rate in much of Africa (Dumenu, 2013). Sub-Saharan Africa experienced unparalleled urban growth in the absence of an economic engine to support the influx (Fox, 2012). Widespread poverty is associated with large cities with up to 60% of urbanized Africans estimated living in what can only be called 'slums'. Predictably, natural resources within these areas quickly become degraded as existing trees and forests are the predominant source of fuel for cooking and heating.

Dumenu (2013) suggests the conversion of urban and peri-urban forests to other uses might result from limited information regarding the economic value of urban forests. To reverse this trend and revitalize urban centres, several large cities have created modern landscapes that include well developed and maintained urban forests. Development strategies for the rapidly growing capital city of Kigali, Rwanda, now include green spaces for conserving forest resources and improving quality of life (Seburanga et al., 2014). Fuwape and Onyekwelu (2011) report the success of new sustainable developments within and around African urban centres requires active education programmes and commitments from the private and public sectors to ongoing management.

Despite the fact that urban forestry is a new concept in South Africa (Shackleton, 2006), the benefits of urban trees are well known, tree planting is active, and urban greening is occurring. The city of Johannesburg, South Africa, is locally referred to as the world's largest man-made forest. While this is debatable, Johannesburg officials estimate there are 10 million trees within Johannesburg, 2.5 million are on public property and over 4 million in private gardens (Schäffler and Swilling, 2013). The forest is source of civic pride that is reflected in the annual city planting and maintenance budget. Additionally, a municipal nursery was established to supply trees for planting on public and private property.

Central and South American urban tree landscapes

The civilizations of the Inca, Maya, and Aztec played prominent roles in the pre-Columbian history of Central and South American. These peoples built unprecedented cities, in both scale and design, ranging from Central America to Chile and Argentina. The ancient Incan capital Cusco in the Peruvian Andes is one of the oldest urban areas in the America's, approximately 3,000 years ago, and is estimated to have had a population in excess of 100,000. Urban centres associated with these ancient civilizations were known for their systematic layout, plazas/city centres, flat-topped pyramids, irrigation channels, and well-defined roadways.

Pre-Columbian Amazonian people practised forest management and agriculture. It is likely these civilizations supported 20 million or more people. In Western Amazon, Heckenberger et al. (2008) found evidence of 28, low-density pre-Columbian urban centres. These pre-Columbian Amazonian urban centres were similar in design to Howard's more recent concept of a 'Garden City' in which the networks of urban clusters are connected through well-designed and maintained roads and surrounded by agricultural green belts that gradually diffuse into the rural forest area.

Outside of household gardens for sustenance and medicinal value, public gardens or natural areas were likely not common in the design of the epicentres of these ancient cities. The civilizations had strong ties to the adjacent forests, using them as a source of raw material, fuel, and food. The forests were often intensively managed in combination with agricultural crops, perhaps the precursor to modern day agro-forestry. Neves et al. (2003) indicated there was widespread plant collection, including trees, and subsequent concentration of selected plant species in and around these urban clusters. Soils surrounding the urban clusters were modified, either intentionally or unintentionally, by human activity. These altered soils (i.e. relative to surrounding bulk soils), called Amazonian Dark Earth or *terra petra*, have been theorized to be a sustaining factor in the development and persistence of these urban centres within the Amazon forest.

The Spanish migration into Central and South America resulted in the creation of many new cities, principally economic hubs located along the perimeters of countries or on waterways. The design of these new cities tended to follow a European grid model that could be altered to accommodate the varied geography and were often devoid of vegetation and gardens (Berrizbeitia and Marchant, 2011). The subsequent incursions of France, Germany, and England into this region of the world introduced city design principles that included the creation of parks, gardens, and formal landscapes using trees and other vegetation as accent features. In a number of cities there is historical evidence of a 'city garden' design approach in Sao Paulo, Brazil, and tree-lined streets in Mexico City, Mexico (Ferreira, 2014).

Today there is a continued rapid migration of rural citizens into the cities. Approximately 80 per cent of Brazil's current population resides in urban centres (Choi, 2011). Unplanned expansion is occurring in the vast majority of cities. In many cases, the adjacent forest is cleared to accommodate the influx of people. Ferreira (2014) stated that approximately 5,000 ha of green space in São Paulo was removed between 1997–2011 to accommodate potential development. Accompanying this expansion is widespread environmental degradation, particularly in those areas occupied by the poor.

North American urban tree landscapes

The Native Americans were connected to and dependent upon nature for survival. Native people used tree parts for food, medicines, shelter, recreation, and clothing (Hendee et al.,

2012). They actively used fire to burn underbrush in forests. Native populations developed permanent villages, likely cultivated trees for food (e.g. pecan and plum), and practised intercropping of annual crops underneath cultivated trees.

European colonists were also closely connected with nature and forests in the eastern United States for survival. Overuse of forests for building materials and fuel near new villages led to early principles of governance through tree protection ordinances. As towns developed, the close nature connection continued through the deliberate use of trees. Today the promotion of urban vegetation through planned celebration, design, and professional practice have evolved over several centuries and significantly the term urban forestry was defined over fifty years ago and was first mentioned in an 1894 park report (Konijnendijk et al., 2006).

Colonial towns and cities

Trees and forests in North America were an embedded part of the culture of pioneers and immigrants. Settlers both romanticized and converted forests as they moved inland. They cleared forests for agriculture and used the best wood for building. Settlers also brought the urban values of Europe and applied these values as they built cities. By 1646 the road from Boston to Roxbury in the Massachusetts Bay colony was lined with shade trees (Zube, 1971). Tree removal in and near villages led to enacting tree protection regulations. Cambridge and Boston Massachusetts as early as 1633 and 1635 respectively restricted tree removal and developed tree protection ordinances (Gerhold and Frank, 2002).

Figure 2.5 Village green in the eastern United States

Photo: R. J. Hauer

Early colonial villages in New England were built around a village green. A place to muster militia and keep livestock during times of attack was the function, rather than aesthetics. By the late eighteenth century, trees and lawns were intentionally established in village greens (Figure 2.5). By this time two-thirds of southern New England had been cleared for agriculture and concern was expressed over the treeless landscape and the barren appearance of towns. Over the next century village greens, streets, cemeteries, and public parks were planted with trees and landscaped (Favretti, 1982). The popularity of shade trees forced insurance companies to insure homes with trees near them in 1784, something they were loath to do (Zube, 1971). Philadelphia, on the other hand, was designed in 1682 by William Penn and contained five open spaces of five to ten acres each, filled with trees. These parks still exist today (Zube, 1971).

After the revolution

Following the Revolution, Americans sought to create a new identity for themselves. Thomas Jefferson believed in a country governed by 'sturdy yeoman farmers' and regarded the city dweller with a certain amount of suspicion. This attitude was embraced by the populace, and influenced early attempts to incorporate nature in town design and as a source of moral virtue (Gerhold and Frank, 2002).

Urban vegetation policy was developing in the early 1800s. An 1807 Michigan Territory law specified that trees be planted and established on boulevards and in squares in the city of Detroit. In 1821 the state of Mississippi capital commission recommended that the new capital have every other block filled with native vegetation or be planted with groves of trees. The commission believed vegetation would deliver a healthier environment and provide easier fire control for wood buildings (Zube, 1971).

Arbor Day

Tree planting found national interest by the end of the nineteenth century. Arbor Day was first observed in Nebraska in 1872, following its inception by J. Sterling Morton. The first Arbor Day in Nebraska witnessed the planting of over a million trees (Zube, 1971). The observance of Arbor Day spread to every state and millions of trees are planted annually in cities, suburbs, farms, and forests. Today, over 40 countries celebrate Arbor Day in the spring, typically in March or April.

Three landscape movements

Nineteenth-century America saw three landscape movements that strongly influenced the appearance of our cities, and our concept of urban forestry: the city parks movement, the romantic landscape movement, and the city beautiful movement. A visit to an American city today will provide ample evidence of the three landscape movements of the nineteenth century.

City parks movement

Frederick Law Olmstead is credited with creating the city parks movement, and is best known for Central Park in New York City (Figure 2.6). By the mid-nineteenth century, New York was filling with new European emigrants largely living in squalor and cut off from nature. Olmstead saw the need to bring nature into the city to promote social progress

Figure 2.6 The original 1858 Greensward Plan for Central Park, New York, which was on the urban fringe when originally established

Source: New York Historical Society

(Gerhold and Frank, 2002). The park he envisioned and created was a natural landscape that resembled rural America and Olmstead led his vision, spreading to influence urban park systems throughout the country.

Romantic landscape movement

Industrialization arrived in American cities by the mid-nineteenth century. Cities grew and rapidly industrialized, resulting in the same conditions that spawned the romantic landscape in Europe. The newly emerging wealthy industrialists and industry managers built suburban communities along commuter railroad and trolley lines to escape the squalid cities. Olmstead was instrumental in designing new suburban communities. Suburbs for the emerging middle and upper classes sprang up on the outskirts of cities following the spirit of the romantic landscape. Streets were laid out in curvilinear patterns, and homes built on rolling wooded parcels. Communities such as Llewellyn Park, New Jersey, Roland Park, Maryland, Ridley Park, Pennsylvania, Lake Forest, Illinois, and Riverside, Illinois, were designed in the latter half of the nineteenth century, serving as models for twentieth-century suburbs. As mass housing became common from the 1930s on, land was subdivided into smaller lots and, to accommodate construction, existing trees were removed and not always replaced (Zube, 1971). This trend did not begin to reverse itself until the late 1960s, when home buyers started placing higher premiums on wooded parcels.

City beautiful movement

The Columbian Exposition of 1893 in Chicago influenced city beautification. The fairgrounds featured tree-lined parkways for carriages and walking, landscaped parks and boulevards, and monumental architecture featuring Greek and Roman design. Fair visitors from all of North America experienced an urban life unlike anything at home. Those who could afford to attend the fair were often industrial or civic leaders, and many went home with a mission to transform their communities along the lines of their experience in Chicago to create the City Beautiful.

Urban forestry since the twentieth century

The twentieth century brought the development of urban park systems and tree planting programmes across the United States. Municipal governments hired foresters, landscape architects, horticulturists, arborists, and other professionals to design, plant, and care for vegetation in their communities (Hauer and Peterson, 2016). Most large cities and many medium-sized communities initiated city forestry programmes to plant and care for street and park trees. The city of Milwaukee's forestry programme, started in 1918, continues to plant and maintain trees on city streets. In the 1920s and 1930s, home shading using trees was an important instrument for cooling residences.

The US Department of Agriculture published numerous bulletins dealing with selecting, growing, transplanting, and caring for urban trees. During the Great Depression, government agencies such as the Works Progress Administration and the Civilian Conservation Corps were created to hire the unemployed, who constructed parks and planted trees in communities throughout America. Urban forestry as we know it was flourishing under the shade-tree term (Campana, 1999).

A decline of the urban forest occurred in some large cities following World War II. Suburban sprawl moved the middle class, and a city's tax base, out of the urban core, leading to budget cuts for parks and forestry. Suburban lots were large enough for private trees, so new communities saw little need for public trees planted along their streets. The availability of air conditioning for homes also made shading homes for comfort much less important. The introduction of Dutch elm disease (*Ophiostoma ulmi* (Buisman) Melin & Nannf) added the final blow as communities lost American elms (*Ulmus Americana* L.) along streets and in yards (Campana, 1999). The loss of elms also brought with it the seeds of renewal for urban forestry as the disease denuded community after community. Even in the most economically depressed cities the loss of tree cover brings with it the desire to reforest our streets.

Several events of the 1960s brought urban forestry into its own as a profession and as an academic discipline, initially in North America and later elsewhere (Konijnendijk et al., 2006). The public was becoming increasingly critical of forest management, especially as practised on public lands. Professional foresters responded to that criticism in a variety of ways, one of which was the promotion of urban forestry in cities (Miller, 2004). The USDA Forest Service embraced urban forestry by providing research funding and by providing funds to state forestry agencies to assist communities with urban forestry (Hauer et al., 2008). The first definition for the term 'urban forestry' was by Jorgensen in 1965, followed by a number of scientific conferences in Canada addressing urban forestry in the United States (Campana, 1999). Urban forestry emerged as an academic discipline in the early 1970s in several universities (Miller, 2004). Prior to that, people were trained in tree management through arboriculture, forestry, and horticulture programmes.

The new federalism of Ronald Reagan in the 1980s brought with it a near suspension of urban forestry efforts at the federal level. However, the momentum of the 1970s was carried on through the 1980s both by citizen groups and by American Forests, a national conservation association. The 1988 election of George H. W. Bush as president brought with it renewed effort at the federal level, as President Bush strongly supported tree planting and urban forestry. The USDA Forest Service expanded urban forestry as Congress increased funding ten-fold to support state-level urban and community forestry, and to fund urban forestry research (Hauer et al., 2008).

In North America, urban forestry evolved not only from forestry, but also from the discipline of arboriculture (Campana, 1999). North American practice includes individual tree management on private and public land, management of populations of trees such as street trees, management of park vegetation, and management of peripheral forests. The focus is on all urban forestry on both public and private lands. Unlike Europe, North America has few 'urban forests' in the sense of public forests adjacent to or surrounding cities, thus we include all vegetation in and around our communities (Konijnendijk et al., 2006).

An international urban forestry connection

As also described in Chapter 1, the interconnectedness of urban forestry has truly become an international endeavour over the past several decades. For example, urban forestry came to Europe via Great Britain and Ireland as an interdisciplinary approach during the 1980s. From the mid-1990s, a European urban forestry research community started to emerge, as reflected in a European research network on 'urban forests and trees' with European (COST) funding and the setting up of the annual European Forum on Urban Forestry. Initially European debates included wide-ranging views on how urban forestry should be seen and defined, ranging from a more narrow 'city forestry' focus to an integrative perspective on all tree populations. An important milestone was the launch of a new journal, 'Urban Forestry & Urban Greening', in 2002 (Konijnendijk et al., 2006).

Urban forestry has spread to other parts of the world through academic and professional networking by the International Union of Forest Research Organizations, the International Society of Arboriculture, and the European Forum on Urban Forestry (see Chapter 1). Another example is the Food and Agriculture Organization of the United Nations and Asia-Europe collaboration in urban forestry under the so-called ASEM-programme (Konijnendijk et al., 2006). The first Asia-Pacific Urban Forestry Meeting in 2016 is a recent example that resulted in the Zhuhai Declaration. This declaration summarized the importance of urban forestry by a call to make cities better places through forests and trees. Finally, the interdisciplinary nature of urban forestry is embedded in many schools of higher education through formal programmes and coursework embed in colleges and departments (Vogt et al., 2016).

Conclusions

Urban forestry has a history that has rapidly evolved during the past half-century and further goes back thousands of years. With over half of the global population living in cities today and continuing urbanization, urban forestry can address and be part of the history unfolding to overcome the emerging issues associated with global urbanization. Growing the urban forest today also has challenges from climate change, introduced insects and diseases, wildfires at the urban-rural interface, and in densifying cities. More residents of the world will find themselves disconnected from nature, and urban forestry has and can continue to connect that population back to nature. Urban forestry needs to better understand the complexities of urban ecosystems and the people who live in those systems. We need to better understand what nature means to people and to shape our education of people based on that understanding. We need to understand and educate urban residents, and we need to let them understand and educate us. A sound historical understanding of the world's urban forests can assist in dealing with these challenges, as it provides a base for understanding people–tree and city–forest relations, as well as for current planning, design, and management practices.

Acknowledgements

Substantial parts of this chapter have been adapted from Robert W. Miller, Richard J. Hauer, and Les P. Werner, *Urban Forestry: Planning and Managing Urban Greenspaces*, 3rd edn (Waveland Press, Long Grove, IL, 2015).

References

Banks, J.C.G. and Brack, C.L. (2003) 'Canberra's Urban Forest: Evolution and Planning for Future Landscapes', *Urban Forestry & Urban Greening*, vol. 1, no. 3, 151–160.

Berrizbeitia, A. and Marchant, R.H. (2011) 'Latin American Geographies: A Glance over an Immense Landscape', *Harvard Design Magazine*, vol. 34, 1–7.

Bigelow, J.M.H. (2000) 'Ancient Egyptian Gardens', *The Ostraco*, vol. 11, no. 1, 7–11.

Campana, R.J. (1999) *Arboriculture: History and Development in North America*, Michigan State University Press, East Lansing, MI.

Choi, J.A. (2010) 'Cultivating Urban Forests Policies in Developing Countries', *Sustainable Development Law & Policy*, vol. 11, no.1, 39–40.

Davison, A. and Kirkpatrick, J. (2014) 'Re-inventing the Urban Forest: The Rise of Arboriculture in Australia', *Urban Policy and Research*, vol. 32, no. 2, 145–162.

Dumenu, W.K. (2013) 'What are We Missing? Economic Value of an Urban Forest in Ghana', *Ecosystem Services*, vol. 5, 137–142.

Dumrongthai, P. (undated) *'Urban Forest Management in Thailand'*, Royal Forest Department Bangkok, Thailand.

Favretti, R.J. (1982) 'The Ornamentation of New England Towns 1750–1850', *Journal of Garden History*, vol. 2, no. 4, 323–342.

Ferreira, L.S. (2014) 'Vegetation Management is São Paulo, Brazil: Clearing of Urban Vegetation and Environmental Compensation', in *Trees, People and the Built Environment II*, 2–3 April, 2014, University of Birmingham, Edgbaston, UK, pp. 32–42.

Forrest, M. and Konijnendijk, C.C. (2005) 'A History of Urban Forests and Trees in Europe', in Konijnendijk, C.C., Nilsson, K., Randrup, T.B., and Schipperijn, J. (eds), *Urban Forests and Trees*, Springer, Heidelberg, pp. 23-48.

Fox, S. (2012) 'Urbanization as a Global Historical Process: Theory and Evidence from Sub-Saharan Africa', *Population and Development Review*, vol. 38, no. 2, 285–310.

Fuwape, J.A. and Onyekwelu J.C. (2011) 'Urban Forest Development in West Africa: Benefits and Challenges', *Journal of Biodiversity and Ecological Sciences*, vol. 1, no. 1, 77–94.

Gallion, A., Eisner, S. (1975) *Urban Pattern, Third Edition*, D Van Nostrand, New York.

Gerhold, H.D. and Frank, S.A. (2002) *Our Heritage of Community Trees*, Pennsylvania Urban and Community Forestry Council, Mechanicsburg, PA.

Hauer R.J. and Peterson W.D. (2016) *Municipal Tree Care and Management in the United States: A 2014 Urban & Community Forestry Census of Tree Activities*, Special Publication 16-1, College of Natural Resources, University of Wisconsin–Stevens Point.

Hauer, R.J., Widerstrand C.J., and Miller R.W. (2008) 'Advancement in State Government Involvement in Urban and Community Forestry in the 50 United States: Changes in Program Status from 1986 to 2002', *Arboriculture & Urban Forestry*, vol. 34, no. 1, 5–12.

Heckenberger, M.J., Russell J.C., Fausto C., Toney J.R., Schmidt M.J., Pereira E., Franchetto B., and Kuikuro A. (2008) 'Pre-Columbian Urbanism, Anthropogenic Landscapes, and the Future of the Amazon', *Science*, vol. 321, 1214–1217.

Hendee J.C., Dawson C.P., and Sharpe W.F. (2012) *Introduction to Forests and Renewable Resources*, 8th edn, Waveland Press, Long Grove, IL.

Intasen, M. (2014) Urban Forest Assessment in Bangkok, Thailand, MS thesis, University of Wisconsin–Stevens Point, USA.

Jim, C.Y. (1986) 'Street Trees in High-Density Urban Hong Kong', *Journal of Arboriculture*, vol. 12, no. 10, 257–263.

Jim, C.Y. and Chen W.Y. (2010) 'Habitat Effect on Vegetation Ecology and Occurrence on Urban Masonry Walls', *Urban Forestry and Urban Greening*, vol. 9, no 3. 169–178.

Jim, C.Y. and Liu H.H.T. (1997) 'Storm Damage on Urban Trees in Guangzhou, China', *Landscape and Urban Planning*, vol. 38, no. 1–2, 45–59.

Johnston, M. (2015) *Trees in Towns and Cities: A History of British Urban Arboriculture*, Windgather Press, Oxford.

Konijnendijk, C.C. (2008) *The Forest and the City: The Cultural Landscape of Urban Woodland*, Springer, Berlin.

Konijnendijk, C.C., Richard R.M., Kenney A., and Randrup T.B. (2006) 'Defining Urban Forestry – A Comparative Perspective of North America and Europe', *Urban Forestry and Urban Greening*, vol. 4, no. 3–4, 93–103.

Lawrence, H.W. (2008) *City Trees: A Historical Geography from the Renaissance through the Nineteenth Century*, University of Virginia Press, Charlottesville, VA.

Miller, R.W. (2004) 'Urban Forestry: History and Introduction', in Konijnendijk, C.C., Schipperijn, J., and Hoyer, K.K. (eds), *Forestry Serving Urbanised Societies*, IUFRO World Series vol. 14, pp. 17–23.

Neves, E.G., Petersen J.B., Bartone R.N., and Da Silva C.A. (2003) 'Historical and Socio-Cultural Origins of Amazonian Dark Earths', in Lehman, J., Kern, D.C., Glaser, B., and Woods, W.I. (eds), *Amazonian Dark Earths: Origins, Properties, Management*, Kluwer Academic Publishers, Amsterdam, pp. 29–50.

Profous, G.V. (1992) 'Trees and Urban Forestry in Beijing, China', *Journal of Arboriculture*, vol. 18, no. 3, 145–154.

Schäffler, A. and Swilling M. (2013) 'Valuing Green Infrastructure in an Urban Environment Under Pressure –The Johannesburg Case', *Ecological Economics*, vol. 86, 246–257.

Seburanga, J.L., Kaplin, B.A., Zhang, Q.-X., and Gatesire, T. (2014). 'Amenity Trees and Green Space Structure in Urban Settlements of Kigali', Rwanda. *Urban Forestry and Urban Greening*, vol. 13, no. 1, 84–93.

Shackleton C.M. (2006) 'Urban Forestry – A Cinderella Science in South Africa?', *The Southern African Forestry Journal*, vol. 208, no. 1, 1–4.

Tian, Y., Jim, C., and Tao, Y. (2012) 'Challenges and Strategies for Greening the Compact City of Hong Kong', *Journal of Urban Planning and Development*. vol. 138, no. 2, 101–109.

Vogt, J.M., Fischer, B.C., and Hauer, R.J. (2016) 'Urban Forestry and Arboriculture as Interdisciplinary Environmental Science: Importance and Incorporation of Other Disciplines', *Journal of Environmental Studies and Sciences*, vol. 6, no. 2, 371–386.

Zube, E.H. (1971) *Trees and Woodlands in the Design of the Urban Environment*, Planning and Resource Development Series No. 17, University of Massachusetts, Amherst, MA.

3

MEASURING AND MONITORING URBAN TREES AND URBAN FORESTS

Justin Morgenroth and Johan Östberg

The urban forest and its components

The urban forest includes diverse floristic and physiognomic plant communities, including trees, shrubs, herbaceous plants, and mosses. This chapter focuses solely on measuring and monitoring urban trees, rather than all plant communities within the urban forest. Urban trees can be classified into numerous groups including street trees, residential trees, park trees, and woodland trees. They may have been recently planted or they may be old-growth remnant forests on public or privately owned land.

Measuring and monitoring urban trees has been conducted for over a century (e.g. Solotaroff, 1912), certainly prior to the establishment of urban forestry as a discipline in the late 1960s and early 1970s (Jorgensen, 1986). As knowledge about ecosystem services has developed, and as government responsibility for urban forest management has increased, the need for standardised monitoring has increased. Without detailed data on the location, structure, and condition of city trees, it is not possible to manage them effectively, nor to estimate ecosystem service provision or urban forest value, nor to develop informed policy or strategy.

In this chapter, we will introduce the reader to the topic of urban forest assessment. The chapter begins with a description of some variables measured during urban forest assessment and the tools and approaches used for this task. This is followed by a discussion of the different stakeholders who engage in urban forest assessment, including citizen volunteers, researchers, and governments.

Measuring and monitoring the urban forest

Forest mensuration is among the most fundamental disciplines in forestry and related sciences, including urban forestry. It focuses on the technical aspects of tree and forest measurement and monitoring. The terms 'measuring' and 'monitoring' are often used synonymously, but should not be confused with one another. In an urban forestry context:

- **measuring** refers to determining the number of trees, their structure, their condition, and other quantitative or qualitative characteristics; this yields data for a single point in time.
- **monitoring** involves repeatedly measuring over time to observe and describe changes in the urban forest.

While measuring is useful, monitoring the urban forest allows for better informed policy and decision making with respect to urban forest management. For example, if a city had measured their urban forest species composition ten years ago, they would have known that, at that time, the most abundant species comprised 15 per cent of all planted trees. If the same city had measured their urban forest species composition every ten years between 1950 and 2010, they would know how species diversity changed over that time period. They could correlate these changes with pest and disease outbreaks or climate change to understand how species diversity responded to these, among other factors. The city could develop informed policies and management strategies to achieve desirable future tree species diversity goals and – importantly – the city could monitor how urban forest diversity responds to those policies or management strategies.

Common urban forest variables for measurement

The question of what to measure in an urban forest has many answers depending on the context or purpose of measurement. Miller et al. (2015b) suggest that data collection should aim to provide a minimal level of data to allow intelligent decisions. Measuring the urban forest to inform policy may only require determining tree canopy cover. But this variable fails to differentiate between individual trees and does not describe the health and condition, nor the three-dimensional structure, nor the diversity of the urban forest. A more comprehensive set of variables is required to improve urban forest management. These data may include the location of each tree, its species, and basic structural variables like height and diameter at breast height (DBH). To estimate ecosystem services, more detailed structural variables may be required, including estimation of tree volume and leaf area (Nowak et al., 2008a). To minimise risk in the urban forest, a city may regularly update tree condition and keep records of all maintenance activities conducted on publicly owned trees. An inventory of veteran or notable trees may include planting date as well as details about who planted the tree and why it was planted (e.g. to commemorate an event or person).

The hypothetical examples presented above suggest that the set of potential variables to measure for urban trees and forests will depend on the desired end use of the data. It follows that many different potential end uses will require numerous structural variables to be measured and/or derived. Generally, the measured variables can fall under the categories of location (e.g. latitude, longitude), 1D structure (e.g. DBH, height, crown spread; Figure 3.1), 2D structure (e.g. crown area), 3D structure (e.g. volume), form (e.g. crown shape, lean), condition (e.g. defoliation, leaf chlorosis), and risk (e.g. cracks, weak attachments). While it's impractical to compile all potential urban tree variables, we can examine existing examples of urban forest inventory variables and discuss how they achieve their purpose (Table 3.1 below).

Urban forest variables for estimating ecosystem services

The United States Department of Agriculture (USDA) Forest Service software i-Tree recommends collecting a set of variables that are critical for estimating environmental and

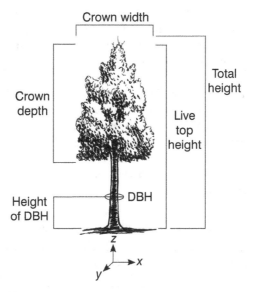

Figure 3.1 Common variables that are measured to describe a tree's structure

Source: Allan McInnes

economic ecosystem services (i-Tree, 2015). These include the tree variables of species, DBH, total tree height, height to live top, height to crown base, crown width, percent missing crown, percent crown dieback, and crown light exposure, as well as the site variable of land use. The tree and crown structural variables are necessary for estimating tree leaf area and biomass and subsequently modelling rainfall interception, carbon sequestration and storage, and energy savings, among other ecosystem services.

Urban forest variables for long-term monitoring

For long-term monitoring of urban trees, the Urban Tree Growth and Longevity working group recommend collecting project variables including measurement date and training level of the field crew, site variables including site type and land use, and tree variables including GPS coordinates, species, mortality status, tree condition, DBH, and the height at which DBH was measured (UTGL, 2014). The GPS coordinates, in conjunction with the species and size data, allow the same tree to be observed over time. The data about field crew training level allows for any spurious data over a time series to be validated or discounted based on the skill level of the field measurement crew.

Urban forest variables for management

The city of Christchurch, New Zealand, collects urban forest inventory for three main reasons:

1 to identify and mitigate risk;
2 to plan future renewal programmes including budgeting; and
3 to inform future strategic directions and planning (e.g. setting species diversity goals).

To meet these goals, the city collects the GPS location of each tree, species, height, crown width, DBH, structural condition, health condition, overall condition, planting date, type of protection, damage, cause of damage, maintenance record, and cost of maintenance activity. The three previous examples highlight that:

1   the possible variables to measure in an urban forest are varied;
2   there are core variables common to many data collection undertakings (e.g. species, DBH), and optional variables used to meet specific desired outcomes (e.g. percentage crown dieback, cost of tree maintenance activity); and
3   the choice of variables to include must be determined by stakeholders to meet a given outcome – and so the purpose of measurement must be clearly defined before the variables can be agreed upon.

Table 3.1 A comparison of the plot and tree variables collected by three different inventory systems, each with a different purpose

	i-Tree	*UTGL*	*Christchurch City Council*
Purpose of data collection	• Ecosystem service modelling	• Long-term monitoring	• Risk mitigation • Budgeting • Strategic planning
Location variables	• N/A	• GPS coordinates	• GPS coordinates
Structural variables	• DBH • Total tree height • Crown width • Height to live top • Height to crown base	• DBH • Height at which DBH was measured	• DBH • Total tree height • Crown width
Condition variables	• Percent missing crown • Percent crown dieback	• Crown dieback • Crown transparency	• Health condition • Overall condition
Risk variables	• N/A	• Wood condition	• Structural condition • Damage • Cause of damage • Maintenance record • Cost of maintenance activity
Other variables	• Species • Crown light exposure • Actual land use	• Species • Tree record • Identification code • Mortality status • Site type • Land use • Date of measurement • Training level of the field crew	• Species • Planting date • Type of protection

The variables listed for each system represent a minimum data collection requirement (see i-Tree, 2015; UTGL, 2014). Other variables can be collected in each system.

Table 3.1 highlights these three points by showing that i-Tree data collection focuses on measuring tree structure, UTGL data collection focuses on ensuring the same tree can be identified and measured consistently over time, while the Christchurch City Council data collection focuses primarily on risk and condition variables.

The importance of standardisation

Management, planning, and strategic directives for urban forests can be informed by comparing variables like canopy cover, structural distribution, or species diversity across different cities or with data previously collected for the same city. This spatial and temporal comparison is most useful if the data collected is available in a common form. While standards for urban forest measurement are widely recommended in the scientific literature (Roman et al., 2013; Nielsen et al., 2014), different approaches to urban forest inventorying are commonly used and different variables are collected. Methods vary widely and efforts to collect data across cities remain uncoordinated (Roman et al., 2013). This is not only true globally, but also locally – cities adjacent to one another are likely to have different standards for data collection and also reporting.

Fortunately, standardised approaches to data collection and reporting are becoming more common in urban forestry. The USDA i-Tree Eco software is arguably the most widely adopted standard globally. Dozens of cities have undertaken such inventories and assessments; while most have been conducted in the USA and Canada, cities in Australia, China, England, Hungary, Portugal, Scotland, Spain, and Switzerland have also used i-Tree Eco to develop knowledge about their urban forests. As of 2015, the US Forest Service began monitoring urban areas as part of their national forest inventory, using i-Tree as the analysis tool. One of the keys to i-Tree's success is that it is in the public domain and is freely available. Not only does i-Tree provide data collection standards, but also reporting is standardised and generally publicly available.

Numerous countries or cities have alternative approaches to data collection and reporting (e.g. the Swedish Tree Portal). But because i-Tree has been used by numerous communities, it is allowing for some of the spatial and temporal comparisons that are best made with standardised data. For example, the cost-benefit ratio of urban trees was calculated for five different cities in the USA using the STRATUM (a precursor to i-Tree Streets) standardised approach to data collection (McPherson et al., 2005). Urban forest functional diversity was compared across seven cities in the Northeastern United States whose inventory data had been collected using the i-Tree Eco plot sampling standard (Nock et al., 2013). Species diversity in 38 cities globally was compared using data collected by the UFORE (a precursor to i-Tree Eco) standardised approach to plot sampling and data collection (Yang et al., 2015). The standardised approach provided by i-Tree has made it possible to find answers to questions that would otherwise be difficult (if not impossible) to solve.

Methods for measuring the urban forest

Once a set of variables has been decided upon, the next step is to undertake the measurements – but how? The detailed methods used to measure trees and forests are described in numerous excellent publications (e.g. Miller et al., 2015b), so this section will not endeavour to provide the reader with another such description. Instead, this section will firstly introduce the reader to common tools and technological advances to measure some common urban forest variables. Secondly a description of general approaches to urban forest inventory will be provided, and thirdly the difference between sample, partial, and complete inventories will be discussed.

Common tools for urban forest measurement

Many of the tools used to measure urban trees were originally used in mensuration of natural and plantation forests. Some of them have been used for decades or even centuries, but new technologies are beginning to replace some of these traditional tools. In particular, remote sensing technologies like satellite imagery and LiDAR are being used more frequently to accurately measure various attributes of urban trees and forests. While remote sensing technologies are being employed more frequently, traditional techniques remain most common for urban forest inventories, due to their simplicity, ease of use, and track record of adequacy. The different tools described in this section have differing accuracy (nearness of the measurement to the actual value), precision (variation when the same measurement is repeated), cost, reliability, and ease of use.

STEM DIAMETER MEASUREMENTS

A nearly ubiquitous tool for forest and urban forest measurement is the diameter tape, used to measure diameter at breast height. The diameter tape generally has a linear scale on one side which can be used to measure the circumference of the tree. On the other side of the tape is a diameter scale, whereby the diameter scale divides the linear scale by pi ($\pi \approx 3.14$). When the tape is placed around the circumference of the tree, the tape measures the circumference and estimates the diameter of the tree for any given circumference value, assuming that the tree stem is circular. Another common tool for measuring tree stem diameter, particularly for small trees, is the calliper. Callipers can be used on any sized tree, but in urban forestry they are typically used on small trees in the nursery before planting out into the landscape.

Both the diameter tape and calliper are considered *contact dendrometers*, a category of diameter measurement tool that also includes the permanent diameter tape, the Finnish parabolic calliper, and the electronic tree measuring fork. *Contact dendrometers* come into physical contact with the tree stem. In contrast, *optical dendrometers* remotely estimate the diameter of trees; these include Wheeler's pentaprism, McClure's pentaprism, and the Barr and Stroud optical dendrometer. Hybrid *optical/contact dendrometers* also exist, including the Biltmore stick, Bitterlich's sector fork, and the Samoan diameter stick. Clark et al. (2000a) comprehensively review many of these tools, describing them in detail and summarising their benefits and drawbacks. They conclude that the diameter tape and calliper provide the greatest accuracy for the lowest cost and that the two tools do not differ in their accuracy.

TREE HEIGHT AND CROWN MEASUREMENTS

Accurately measuring tree height and crown variables (e.g. height to base of live crown, height to live top, crown spread) is relatively challenging, with errors exceeding 30 per cent having been documented (Bragg, 2008). The large size of trees makes direct measurement difficult, though not impossible. Telescopic height rods are useful for directly measuring small trees, but are limited as tree height increases beyond the height of the extended pole. For large trees, direct measurement is much more difficult, but is possible by dropping a fabric tape from the highest point of a tree to the ground; this method is practically limited to those with adequate climbing training (Bragg, 2008). Indirect methods are generally better suited for estimating, rather than measuring, the height of large trees. Hypsometers estimate tree height using trigonometric (e.g. Suunto, Blume-Leiss, Haga, Abney level; see Figure 3.2) or geometric (e.g. Christen, Merritt, Vorkampff-Laue, Chapman) principles.

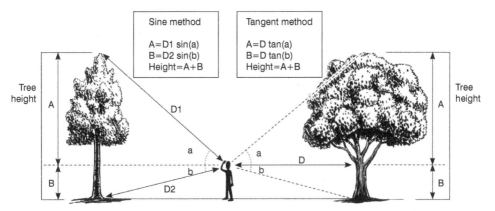

Figure 3.2 How to estimate tree height using trigonometric methods, the sine method (left) and the tangent method (right)

Angles 'a' and 'b' are measured in degrees. Distances D, D1, and D2 are all measured in the same units as desired for tree height

Source: Allan McInnes

Though the term hypsometer is commonly used, some literature refers to hypsometers as clinometers or altimeters.

More recently, multi-purpose instruments using laser or ultrasonic technologies have been developed. Laser dendrometers combine laser range finding and angle measurement for estimating tree structure, while the Vertex (Haglöf Sweden AB, Långsele, Sweden) uses ultrasonic sensing and angle measurement to estimate height or other vertical linear variables. A benefit of these instruments is that height is calculated automatically by the instrument and displayed to the user. In contrast, manual hypsometers require the user to record distances and angles from which height can subsequently be calculated (Figure 3.2).

Technological advance in urban forest measurement

Technological advances have yielded the possibility of measuring tree structure using remote sensing approaches. Though photography has been used to measure tree structure for decades (e.g. Crosby et al., 1983), it suffers from several limitations. Photographic techniques only estimate one-dimensional tree structure variables (e.g. DBH, height) and thus, fail to provide measurements of three-dimensional variables like total tree volume. Moreover, they require fastidious attention to calibration and procedure to ensure accurate estimates. Perhaps for these reasons, photographic techniques have generally only been used in research settings (e.g. Clark et al., 2000b), while contact dendrometers and hypsometers have remained the preferred tools for users outside the research community.

More recently, advances in sensor technology and computer vision approaches have resulted in the potential to remotely measure the structure of urban trees. LiDAR (light detection and ranging) sensors can estimate one-dimensional, two-dimensional and three-dimensional tree variables in natural and planted forests, including urban forests. LiDAR sensors generate three-dimensional point clouds of trees (Figure 3.3), from which any number of structural variables can be estimated. LiDAR sensors can be attached to an aerial platform (e.g. fixed-wing aircraft), a terrestrial platform (tripod), or a mobile platform (e.g. car) for flexibility of use.

In urban settings, mobile and terrestrial LiDAR sensors have been used to estimate a variety of structural tree variables, though this has generally been limited to research settings. Mobile LiDAR scanners have been used to detect street trees and estimate their height, crown diameter, and DBH with a comparable accuracy to field-based measurements (Wu et al., 2013). Terrestrial LiDAR data has also been used to estimate tree height, stem diameter, total volume, and tree location (Holopainen et al., 2013). In contrast to terrestrial and mobile LiDAR acquisition, aerial LiDAR data cannot accurately measure DBH due to its aerial perspective (Figure 3.3). Instead, aerial LiDAR's main uses in urban forestry are for estimating canopy cover, tree numbers, tree height, crown depth, and crown spread. Of the three platforms, aerial LiDAR is most likely to be used operationally for urban forest inventory, because it can be undertaken over large areas. Moreover, the fact that it does not directly measure DBH can be overcome by using height (which is directly measured) to estimate DBH (Saarinen et al., 2014). When combined with hyperspectral imagery, it is possible to identify individual trees, determine their species, and measure common structural variables (Alonzo et al., 2014).

Though LiDAR technology has demonstrated utility for forest assessment, its use for urban forestry inventory is less common. Zhang et al. (2015) suggest three reasons for this, namely the 'complexity of urban areas, the spatial heterogeneity of urban forests, and the diverse structure and shape of urban trees'. We add to that list by suggesting that the high cost of data acquisition, as well as the technically challenging and computationally intensive data analysis, are contributors to the slow uptake of LiDAR-based urban forest inventories.

An alternative to LiDAR is structure-from-motion with multi-view stereo-photogrammetry (SfM-MVS) – an approach that combines stereo-photogrammetry with computer vision. Like LiDAR, SfM-MVS produces spatially accurate three-dimensional models using sets of overlapping two-dimensional digital images. SfM-MVS has not been used in urban forestry outside of specialised research applications (Miller et al., 2015a; Morgenroth and Gomez, 2014) and it is unlikely to be of use over large scales. However, for measuring the size, structure, and form of individual trees using SfM-MVS is intriguing due to its high degree of accuracy and low cost.

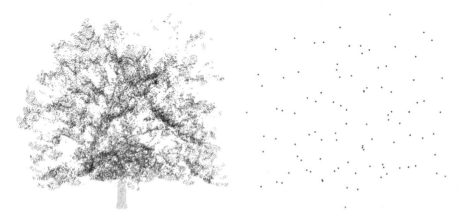

Figure 3.3 Point cloud of a *Fraxinus* spp. tree created with a terrestrial laser scanner (left) and aerial laser scanner (right)

Only basic tree structural data can be measured from aerial laser scanning (e.g. height, crown depth, average crown spread), while terrestrial laser scanning can measure DBH and even crown or woody volume with a high degree of accuracy

General approaches to undertaking urban forest measurement

Presently, the vast majority of urban forest measurement and monitoring is conducted by ground-based field surveys (Nielsen et al., 2014), but other approaches exist including windshield surveys and remote sensing methods. The approach to data collection must be considered in the context of the desired outcomes, as not all approaches can accurately describe all tree and urban forest variables (Table 3.2). Even when an approach to measurement has been demonstrated to accurately measure or estimate a variable, it is important to recognise that the approach may not succeed in all scenarios. For example, a terrestrial LiDAR approach may be useful for describing the structure of an individual street tree, but will fail to do so for trees in an urban woodland due to occlusion effects. Likewise, a ground-based survey of species diversity will be appropriate for a public park, but perhaps not for trees on private property where access restrictions may limit visually assessing trees. Even if an approach is possible, it may not be practical. For example, measuring tree volume by ground-based measurements is possible through laborious destructive methods, but impractical for a large number of trees or trees that cannot be destroyed. However, it is possible to estimate volume reasonably well from ground-based measurements (Nowak et al., 2008a). Alternatively, modern LiDAR and photogrammetric techniques are non-destructive and can be used to efficiently estimate tree volume with a high degree of accuracy (Hackenberg et al., 2014).

Sample, partial, and complete inventories

Two important considerations during urban forest assessment are:

1 how many trees to measure; and
2 which trees to measure.

Depending on the answers to these questions, there are three options to choose from, each with their own benefits and drawbacks: sample inventory, partial inventory, and complete inventory. Sample and partial inventories measure a subset of the whole urban forest, in contrast to a complete inventory in which all trees are measured. Sample inventories rely on selecting a subset of trees from the entire population, whereas a partial inventory measures all trees meeting a particular condition. This could include all street trees or all trees in a park/neighbourhood. Meanwhile a complete inventory measures all trees in the urban forest population and thus requires a high degree of resourcing relative to sample or partial inventories. Practically, complete inventories are often limited to trees on public lands (Miller et al., 2015b), though improvements in remotely sensed data (e.g. satellite imagery, LiDAR) and associated analysis methods allow for complete inventories to incorporate trees on private property (albeit with relatively limited information; see Table 3.2).

The quality of data obtained from sample, partial, and complete inventories can differ markedly. Complete inventories are the most accurate (Nowak et al., 2008b) given they directly measure the attributes of all trees in the urban forest. Partial inventories provide the same level of accuracy, but only for the specific subset they have measured. For example, the city of Milwaukee used a partial inventory (only street trees) to map ash trees (*Fraxinus* spp.) in preparation for an invasion of emerald ash borer. Numerous partial inventories can be strategically implemented over time to yield a complete inventory that is as old as the oldest partial inventory.

Sample inventories use mean and summary data from sampled data to estimate attributes for the whole urban forest (Nowak et al., 2008a). Because not all trees are measured, the

Table 3.2 Different approaches to measuring the urban forest are required depending on the desired outcomes. A ✓ suggests that the approach to measurement is likely to be able to measure or estimate the metric with a high degree of accuracy. A ✗ suggests that the approach to measurement is unlikely to be able to measure or estimate the metric with a high degree of accuracy. The ✓ and ✗ symbols are only a guide; specific examples that contradict the symbols may exist

Ability to accurately measure or estimate	Approach to measurement							
	Low and medium resolution satellite imagery †	High resolution satellite imagery †	Aerial photography †	Terrestrial photography ‡	Aerial LiDAR †	Terrestrial LiDAR ‡	Windshield survey ‡	Ground-based measurement ‡
Tree species	✗	✗	✗	✓	✗	✗	✓	✓
DBH	✗	✗	✗	✓	✗	✓	✗	✓
Total height	✗	✗	✗	✓	✓	✓	✗	✓
Crown depth	✗	✗	✗	✓	✓	✓	✗	✓
Crown spread/area	✗	✓	✓	✓	✓	✓	✗	✓
Tree/crown volume	✗	✗	✗	✓	✓	✓	✗	✓
Leaf area/crown density	✓	✓	✓	✓	✓	✓	✗	✓
Leaf/crown health	✓	✓	✓	✓	✗	✗	✓	✓
Tree damage	✗	✗	✗	✓	✗	✗	✓	✓
Canopy cover	✓	✓	✓	✓	✓	✓	✗	✗

† denotes an approach capable of being employed for trees on public and private land without seeking land owner permission, while ‡ denotes an approach that may require seeking land owner permission.

estimates yielded by sampling approaches have inherent variation (Nowak et al., 2015). But the variation can be minimised by increasing the number of samples (Miller et al., 2015b). Different strategies have been developed to obtain more precise estimates from sampled data that are representative of the urban forest as a whole. These include random, systematic, cluster, and stratified sampling. These strategies are generally applicable to any statistical sampling, but all have been used in urban forest inventory (Miller et al., 2015b).

Who measures urban forests?

A variety of stakeholders engage in urban forest assessment, including researchers, governments, and citizen volunteers. This section presents examples of the ways that different stakeholders measure and monitor urban forests.

Urban forest measurement by researchers

Researchers measure and monitor urban forests to quantify the ecosystem services they provide, including provisioning services (e.g. fuel and food), regulating services (e.g. air quality improvement, stormwater management), cultural services (recreation, aesthetics), and supporting services (e.g. primary production, nutrient cycling). By measuring the structure of trees, researchers can quantify the urban forest's impact on micro-climate (Bowler et al., 2010), carbon sequestration (Nowak et al., 2013), atmospheric and particulate air pollution (Escobedo et al., 2011), and stormwater runoff (Kirnbauer et al., 2013). Qualitatively, a description of the urban forest can be correlated with human well-being (Dallimer et al., 2012), resilience during times of war (Lacan and McBride, 2009) or natural disasters (Morgenroth and Armstrong, 2012), food provision (McLain et al., 2012), property values (Dimke et al., 2013), and energy savings (McPherson and Simpson, 2003).

Quantification of ecosystem services, derived from measurements of urban forest structure, allows for economic valuation of the urban forest resource. For example, Nowak et al. (2002) estimated the total compensatory value for the urban forests of the 48 conterminous states in the United States to be 2.4 trillion USD. While measurement allows for ecosystem service quantification and economic valuation of urban forests, monitoring temporal dynamics allows researchers to describe change in urban forest structure or composition and relate this to anthropogenic, environmental, or climatic factors.

Urban forest measurement by government

Global forest monitoring is reported by the Food and Agriculture Organization of the United Nations in their *Global Forest Resources Assessments* reports. However, urban areas – and hence urban forests – are excluded from these reports. As such, efforts to measure and monitor urban forests at larger scales (e.g. region, country) are rare, though some exist (e.g. Nowak et al., 2001). Instead, monitoring is generally left to local governments. Local government tree inventories are motivated by factors including budgeting, maintaining the urban tree stock, identifying and mitigating high risk trees, and creating informed management plans which can include species distribution and canopy cover goals, or pruning frequencies. To highlight different motivations and different approaches to urban forest inventory by local governments, two case studies are presented, on the Swedish city of Malmö and the metropolitan city of New York, USA.

Malmö, Sweden

Malmö's work with tree inventory and digitisation of its maps began in 1995 when houses, roads, and trees (in parks and streets) were digitised using computer-aided design (CAD). As part of the digitisation, a tree inventory was conducted. The tree inventory was fully undertaken when Dutch elm disease (DED) was found to have reached the city. The inventory data were used to efficiently identify susceptible trees and remove those infected by DED. This reduced the spread and impact of the disease, making it possible for Malmö to focus efforts and resources on replanting trees. The inventory helped Malmö recognise its over-reliance on elms (*Ulmus* spp.), which consequently resulted in the city's decision to replant with a large diversity of species. That decision has resulted in Malmö having greater tree species diversity than many other Scandinavian cities (Sjöman et al., 2012).

The initial inventory (1995–1996) was continually updated with new plantings and removals, however, there was no full re-inventory until 2008. The re-inventory was a consequence of a tree falling and nearly hitting a person sitting on a park bench under the tree. This incident made Malmö realise that they needed to implement a new inventory, including a risk assessment for all trees. The new variables included in the 2008 inventory were: risk classification, damage (root/stem base, trunk, crown), presence of fruiting bodies, DBH, and recommended management action.

Malmö faces several monitoring challenges, including extracting the most value from their inventory data and data accessibility. The tree inventory database's initial use was for paying contractors, whereby contractors are paid a fixed amount annually for regular maintenance and oversight of each tree in the database. To extract further value from the database, it is also used for strategic planning, risk management, as well as planning establishment and maintenance activities like planting, pruning, and removals. Data accessibility was also a problem for city arborists in Malmö. Though new tree variables were collected in 2008, the data were not available to city arborists until 2014 due to a poorly executed software upgrade. By the time city arborists could view the data, much of it was out of date. Further to this accessibility issue, Malmö's tree inventory is not linked to the city's other infrastructure databases. With better linkages, the city would have a better understanding of the spatial overlap between trees and below-ground infrastructure, which could reduce the risk of pipe root intrusion and allow planning for open stormwater systems.

New York City, USA

New York has conducted three street tree inventories (1995, 2005, and 2015). As part of the 2005 tree census, all the parameters necessary to conduct an i-Tree Eco assessment were collected. This provided an overview of the whole urban tree population, but also allowed for estimation of the economic value of the urban tree population, thereby allowing for a cost-benefit analysis to be undertaken. These calculations led to the city's MillionTreesNYC programme, which aimed to plant one million trees around New York (the millionth tree was planted in 2015). This result demonstrates the importance of using an inventory to understand the current resource as well as plan strategically for the future of the urban forest.

The number of variables collected during the inventory was reduced significantly between 2005 and 2015. In 2005, approximately 50 different variables were collected, some of which provided little or no value to urban forest managers. For the 2015 census, the number of collected variables was optimised to include only those with a targeted end use. For example,

Figure 3.4 Planting of annual flowers in a street tree pit are obvious signs of stewardship

Photo courtesy of L. A. Roman

obvious signs of tree stewardship, like flowers planted beneath street trees (Figure 3.4) were noted, because this was evidence of a community group who took an active interest in the urban forest. NYC used this data to catalyse new tree stewardship programmes with community groups. The lesson learned between the 2005 and 2015 inventories was to select inventory variables only after the purpose of the inventory was well defined.

Urban forest measurement by citizens

Volunteer involvement in measuring and monitoring urban forests is a paradigm shift from traditional approaches conducted by trained government staff and professional contractors. But there are numerous potential benefits. Involving citizen scientists directly links a community to its urban forest, thereby increasing public consciousness of the benefits provided by the urban forest, and enhancing support for its stewardship. Cities are not the only beneficiaries of volunteer involvement in urban forest monitoring. Citizens become empowered by partaking in the betterment of their own community and develop an enhanced appreciation of the urban forest (Van Herzele et al., 2005).

Cities are increasingly soliciting volunteers for help with urban forest monitoring – a top-down approach to involving citizens in urban forest measurement and monitoring. The 2015 TreesCount! inventory in New York City involved over 8,000 volunteers who signed up for online training, participated in inventories and became independent mappers. New York achieved their goal of updating their street tree inventory while at the same time engendering a sense of stewardship among the citizen volunteers. London, England, undertook an i-Tree Eco assessment using over 200 volunteers to measure all trees in 476 plots of 0.4 hectares. Volunteers were trained prior to surveying, but also there was a professional arborist or forester in each volunteer surveying team. Ecosystem services and urban forest values were quantified using the measured urban forest structural data as model inputs. The city of Melbourne, Australia, trains volunteers in data collection methods such that they can become citizen urban foresters, helping with street and park tree vegetation mapping, among other activities.

Despite the numerous benefits of using a citizen-science approach to urban forest monitoring, it remains important to consider data quality, cost, and citizen uptake. Training is essential to ensure data quality and the potential for newly collected data to be integrated seamlessly with existing inventory data. A study of an inventory of qualitative characteristics (e.g. species, condition) of street trees in Massachusetts, USA, found that the accuracy of data collected by trained volunteers compared favourably with that collected by certified arborists (Bloniarz and Ryan III, 1996). A previous study in *Quercus garryana* and *Pinus ponderosa* forests in Oregon, USA, showed that student volunteers measured quantitative characteristics (e.g. count, diameter at breast height) as accurately as professionals, but estimates of qualitative characteristics (e.g. crown shape) differed significantly from professional estimates (Galloway et al., 2006). An issue directly related to data quality is the availability of simple standards that are designed for citizen scientists rather than an academic or professional audience (Roman et al., 2013). Another consideration for citizen- or volunteer-led efforts for urban forest measurement is the cost. While volunteers donate their time, costs exist for recruitment, training, mobilisation, and supervision. Twenty years ago, in a study in Massachusetts, USA, these costs were in line with the costs of an urban forest inventory conducted by professional tree care companies (Bloniarz and Ryan III, 1996); it is unclear whether that remains true today.

Perspectives and conclusion

The importance of measuring and monitoring urban forests is being increasingly recognised by a variety of stakeholders, including communities, researchers, and governments. Their motivations range from a desire to quantify the ecosystem services provided by urban forests, to assessing the risks posed by unhealthy or damaged trees, to monitoring long-term urban forest dynamics. Sample, partial, and complete inventories should be undertaken with recognised, standardised approaches, with the provision that these may be modified to meet the specific objectives of any given mensurational undertaking. Though ground-based surveys remain the most common means to measure urban forests, windshield surveys and remote sensing approaches are also undertaken during monitoring. These approaches can be used independently or combined; the key point is to ensure that selected approach is appropriate for measuring the desired urban forest variables, recognising that approaches vary in their accuracy, precision, applicability, and practicality.

References

Alonzo, M., Bookhagen, B. and Roberts, D. A. 2014. Urban tree species mapping using hyperspectral and lidar data fusion. *Remote Sensing of Environment*, 148, 70–83.

Bloniarz, D. V. and Ryan III, H. D. P. 1996. The use of volunteer initiatives in conducting urban forest resource inventories. *Journal of Arboriculture*, 22, 75–82.

Bowler, D. E., Buyung-Ali, L., Knight, T. M. and Pullin, A. S. 2010. Urban greening to cool towns and cities: A systematic review of the empirical evidence. *Landscape and Urban Planning*, 97, 147–155.

Bragg, D. C. 2008. An improved tree height measurement technique tested on mature southern pines. *Southern Journal of Applied Forestry*, 32, 38–43.

Clark, N. A., Wynne, R. H. and Schmoldt, D. L. 2000a. A review of past research on dendrometers. *Forest Science*, 46, 570–576.

Clark, N. A., Wynne, R. H., Schmoldt, D. L. and Winn, M. 2000b. An assessment of the utility of a non-metric digital camera for measuring standing trees. *Computers and Electronics in Agriculture*, 28, 151–169.

Crosby, P., Barrett, J. P. and Bocko, R. 1983. Photo estimates of upper stem diameters (pine trees). *Journal of Forestry*, 81, 795–797.

Dallimer, M., Irvine, K. N., Skinner, A. M. J., Davies, Z. G., Rouquette, J. R., Maltby, L. L., Warren, P. H., Armsworth, P. R. and Gaston, K. J. 2012. Biodiversity and the feel-good factor: Understanding associations between self-reported human well-being and species richness. *BioScience*, 62, 47–55.

Dimke, K. C., Sydnor, T. D. and Gardner, D. S. 2013. The effect of landscape trees on residential property values of six communities in Cincinnati, Ohio. *Arboriculture and Urban Forestry*, 39, 49–55.

Escobedo, F. J., Kroeger, T. and Wagner, J. E. 2011. Urban forests and pollution mitigation: Analyzing ecosystem services and disservices. *Environmental Pollution*, 159, 2078–2087.

Galloway, A. W., Tudor, M. T. and Vander Haegen, W. 2006. The reliability of citizen science: A case study of Oregon white oak stand surveys. *Wildlife Society Bulletin*, 34, 1425–1429.

Hackenberg, J., Morhart, C., Sheppard, J., Spiecker, H. and Disney, M. 2014. Highly accurate tree models derived from terrestrial laser scan data: A method description. *Forests*, 5, 1069–1105.

Holopainen, M., Kankare, V., Vastaranta, M., Liang, X., Lin, Y., Vaaja, M., Yu, X., Hyyppä, J., Hyyppä, H., Kaartinen, H., Kukko, A., Tanhuanpää, T. and Alho, P. 2013. Tree mapping using airborne, terrestrial and mobile laser scanning: A case study in a heterogeneous urban forest. *Urban Forestry and Urban Greening*, 12, 546–553.

i-Tree 2015. i-Tree Eco v. 6: guide to data limitations. www.itreetools.org/resources/manuals/Ecov6_ManualsGuides/Ecov6Guide_DataLimitations.pdf

Jorgensen, E. 1986. Urban forestry in the rearview mirror. *Arboricultural Journal*, 10, 177–190.

Kirnbauer, M. C., Baetz, B. W. and Kenney, W. A. 2013. Estimating the stormwater attenuation benefits derived from planting four monoculture species of deciduous trees on vacant and underutilized urban land parcels. *Urban Forestry and Urban Greening*, 12, 401–407.

Lacan, I. and McBride, J. R. 2009. War and trees: The destruction and replanting of the urban and peri-urban forest of Sarajevo, Bosnia and Herzegovina. *Urban Forestry and Urban Greening*, 8, 133–148.

McLain, R., Poe, M., Hurley, P. T., Lecompte-Mastenbrook, J. and Emery, M. R. 2012. Producing edible landscapes in Seattle's urban forest. *Urban Forestry and Urban Greening*, 11, 187–194.

McPherson, E. G. and Simpson, J. R. 2003. Potential energy savings in buildings by an urban tree planting programme in California. *Urban Forestry and Urban Greening*, 2, 73–86.

McPherson, G., Simpson, J. R., Peper, P. J., Maco, S. E. and Xiao, Q. 2005. Municipal forest benefits and costs in five US cities. *Journal of Forestry*, 103, 411–416.

Miller, J., Morgenroth, J. and Gomez, C. 2015a. 3D modelling of individual trees using a handheld camera: Accuracy of height, diameter and volume estimates. *Urban Forestry and Urban Greening*, 14, 932–940.

Miller, R. W., Hauer, R. J. and Werner, L. P. 2015b. *Urban forestry: Planning and managing urban greenspaces*, Waveland Press, Long Grove, IL.

Morgenroth, J. and Armstrong, T. 2012. The impact of significant earthquakes on Christchurch, New Zealand's urban forest. *Urban Forestry and Urban Greening*, 11, 383–389.

Morgenroth, J. and Gomez, C. 2014. Assessment of tree structure using a 3D image analysis technique-A proof of concept. *Urban Forestry and Urban Greening*, 13, 198–203.

Nielsen, A. B., Östberg, J. and Delshammar, T. 2014. Review of urban tree inventory methods used to collect data at single-tree level. *Arboriculture and Urban Forestry*, 40, 96–111.

Nock, C. A., Paquette, A., Follett, M., Nowak, D. J. and Messier, C. 2013. Effects of urbanization on tree species functional diversity in eastern North America. *Ecosystems*, 16, 1487–1497.

Nowak, D. J., Crane, D. E. and Dwyer, J. F. 2002. Compensatory value of urban trees in the United States. *Journal of Arboriculture*, 28, 194–199.

Nowak, D. J., Greenfield, E. J., Hoehn, R. E. and Lapoint, E. 2013. Carbon storage and sequestration by trees in urban and community areas of the United States. *Environmental Pollution*, 178, 229–236.

Nowak, D. J., Noble, M. H., Sisinni, S. M. and Dwyer, J. F. 2001. People and trees: Assessing the US urban forest resource. *Journal of Forestry*, 99, 37–42.

Nowak, D. J., Walton, J. T., Baldwin, J. and Bond, J. 2015. Simple street tree sampling. *Arboriculture and Urban Forestry*, 41, 346–354.

Nowak, D. J., Crane, D. E., Stevens, J. C., Hoehn, R. E., Walton, J. T. and Bond, J. 2008a. A ground-based method of assessing urban forest structure and ecosystem services. *Arboriculture and Urban Forestry*, 34, 347–358.

Nowak, D. J., Walton, J. T., Stevens, J. C., Crane, D. E. and Hoehn, R. E. 2008b. Effect of plot and sample size on timing and precision of urban forest assessments. *Arboriculture & Urban Forestry* 34(6), 386–390.

Roman, L. A., Mcpherson, E. G., Scharenbroch, B. C. and Bartens, J. 2013. Identifying common practices and challenges for local urban tree monitoring programs across the United States. *Arboriculture and Urban Forestry*, 39, 292–299.

Saarinen, N., Vastaranta, M., Kankare, V., Tanhuanpää, T., Holopainen, M., Hyyppä, J. and Hyyppä, H. 2014. Urban-tree-attribute update using multisource single-tree inventory. *Forests*, 5, 1032–1052.

Sjöman, H., Östberg, J. and Bühler, O. 2012. Diversity and distribution of the urban tree population in ten major Nordic cities. *Urban Forestry and Urban Greening*, 11, 31–39.

Solotaroff, W. 1912. *Shade-trees in towns and cities: Their selection, planting, and care as applied to the art of street decoration; their diseases and remedies; their municipal control and supervision*, J. Wiley & Sons, New York.

UTGL 2014. Urban tree monitoring protocols project overview and data sets summary. Draft. Available at: http://www.urbantreegrowth.org/

Van Herzele, A., Collins, K. and Tyrväinen, L. 2005. Involving people in urban forestry—A discussion of participatory practices throughout Europe. In Konijnendijk, C. C., Nilsson, K., Randrup, T. B. and Schipperijn, J. (eds), *Urban forests and trees*, Springer-Verlag, Berlin.

Wu, B., Yu, B., Yue, W., Shu, S., Tan, W., Hu, C., Huang, Y., Wu, J. and Liu, H. 2013. A voxel-based method for automated identification and morphological parameters estimation of individual street trees from mobile laser scanning data. *Remote Sensing*, 5, 584–611.

Yang, J., La Sorte, F. A., Pyšek, P., Yan, P., Nowak, D. and Mcbride, J. 2015. The compositional similarity of urban forests among the world's cities is scale dependent. *Global Ecology and Biogeography*, 24, 1413–1423.

Zhang, C., Zhou, Y. and Qiu, F. 2015. Individual tree segmentation from LiDAR point clouds for urban forest inventory. *Remote Sensing*, 7, 7892–7913.

PART II

Roles and benefits of urban forests and urban trees

4

ECOSYSTEM SERVICES

Cynnamon Dobbs, Maria Jose
Martinez-Harms and Dave Kendal

The ecosystem service concept and its importance for socio-ecological systems

This section describes and defines the concept of ecosystem services, its application to urban forests and the advantages that the concept has for decision-making and planning in cities. It presents a framework to facilitate comprehensive application of the concept in urban planning. First, the chapter defines the concept of ecosystem services and identifies some classification schemes. Then it outlines how ecosystem services can be valued in an urban context, specifically in relation to the provision of services by urban forests and urban trees, and provides a framework for understanding ecosystem services in urban systems. The chapter then discusses the concept of resilience in relation to ecosystem services, and finishes by discussing the importance of ecosystem services for decision-making.

The concept of ecosystem services refers to the benefits derived from ecological processes that are directly or indirectly provided to humans, and that maintain or improve their well-being (de Groot et al., 2010a; Haines-Young and Potschin, 2010). The publication of the Millennium Ecosystem Assessment report in 2005 established globally, and beyond the scientific world, the importance of nature to human kind. This initiative was followed by The Economics of Ecosystem Services and Biodiversity (TEEB, 2011) which linked the economics of ecosystem services with biodiversity, and by the Intergovernmental Science-Policy Platform on Biodiversity and Ecosystem Services which related scientific information to policy making (Perrings et al., 2011).

The first general classification of ecosystem services, by de Groot et al. (2002), grouped services into four categories: regulating, habitat, production and information functions. Regulating functions included services related to maintaining earth's life support system, such as biogeochemical cycles, chemical composition of the atmosphere and biological diversity. Habitat functions described services that provide space and resources, such as recreation and tourism, while production functions included the provision of resources such as water, food and energy. Finally, information functions referred to services such as cultural enrichment, education and systems of knowledge. This classification was later rearranged to become the basis for the Millennium Ecosystem Assessment's (2005) four categories of ecosystem services:

- *Provisioning services*: products from ecosystems, equivalent to de Groot's et al. (2002) production functions.
- *Regulating services*: benefits from ecological processes, equivalent to regulating functions.
- *Supporting services*: necessary for the production of other ecosystem services, such as soil and oxygen formation, equivalent to a mix of regulating and habitat functions.
- *Cultural services*: the non-material benefits from ecosystems, equivalent to information functions.

This classification has been used as the basis for other frameworks, such as those provided by TEEB (2011) and CICES (Haines-Young and Potschin, 2011; Table 4.1).

An ecosystem service approach can facilitate the understanding and communication of resource use and the impacts of overconsumption and global change on resources across multiple stakeholders (e.g., academic, governmental, NGOs, private sector; Villa et al., 2014). The concept has been incorporated in the United Nations' Sustainable Development Goals (UNDP, 2015), from where it will be integrated into policies, accounting of natural capital, national assessments and planning (Wong et al., 2015).

Different kinds of value can be assigned to ecosystem services, including economic, socio-cultural, ecological and insurance values. The economic value of ecosystem services

Table 4.1 CICES and TEEB classification for ecosystem services

CICES theme	CICES class	TEEB categories
Provisioning	• Nutrition	• Food, water
	• Materials	• Raw materials, genetic resources, medicinal resources, ornamental resources
	• Energy	
Regulating and maintenance	• Regulation of waste	• Air purification, waste treatment
	• Flow regulation	• Disturbance prevention, regulation of water flows, erosion prevention
	• Regulation of physical environment	• Climate regulation, maintaining soil fertility
	• Regulation of biotic environment	• Gene pool protection, lifecycle maintenance, pollination, biological control
Cultural	• Symbolic	• Information of cognitive development
	• Intellectual and experiential	• Aesthetic, inspiration for culture, art and design, spiritual experience, recreation and tourism

Source: adapted from Haines-Young and Potschin (2011)

can be classified either as use or non-use values (de Groot et al., 2010a). Use values are those related to direct consumption and can be assigned to goods from ecosystems, such as food and wood, and use monetary units as indicators (Haase et al., 2014). Non-use values are non-consumptive values that can be indirectly enjoyed by humans, such as aesthetics, climate and other regulating services (de Groot et al. 2010a). They take into account perceptions of ecosystem services and their importance for human well-being (Haase et al., 2014). Avoided costs methods are commonly used to calculate use value, for example to show how the loss of urban trees can increase the cost of energy in summer, or the increased cost for water purification due to the loss of water regulation services (review of the literature in Gómez-Baggethun and Barton, 2012). Other economic methods (hedonic pricing, contingent valuation and stated preference) are commonly used to value ecosystem services in cities. Cultural services are often measured as non-use values, and can be accounted for using monetary (e.g. willingness to pay for a service) or non-monetary metrics, such as qualitative assessments, constructed scales or narrations (Gómez-Baggethun and Barton, 2012). Ecological valuation does not directly include the preferences or desire of humans; they are non-monetary and typically focus on regulating and supporting services such as air pollution removal, cooling effect of trees and pollination (Haase et al., 2014). Insurance values can be used when ecosystem services can be related to the resilience and adaptive capacity of cities; related services include flood regulation, urban cooling, water supply and food production (Lafortezza et al., 2009). A full review of methods for valuing ecosystem services can be found in Farber et al. (2002).

Urban ecosystem services and their assessment

Urban socio-ecological systems are major users of ecosystem services, not only provided by the urban landscape itself, but also by surrounding natural areas. Cities generate a large demand for resources and have been named as major drivers of environmental change. The provision of ecosystem services in urban areas is the result of complex interactions between ecological processes, human activity and social organization. Provision will be dependent on the biophysical condition of the city but also on socio-cultural factors such as property rights, social movements, cultural narratives, social memory and governance (Andersson et al., 2014). Urban areas, if well planned, can be major providers of ecosystem services of local and regional value, through supporting services such as nutrient cycling, provisioning services such as urban food production, and cultural services such as aesthetics and sense of place (Gaston et al., 2013). Services can be delivered by urban forests and other green spaces, wetlands, streams, lakes, green roofs, green walls and gardens, and individual trees (Gómez-Baggethun and Barton, 2012).

Assessments of urban ecosystem services are complex, involving multiple actors and drivers which makes their quantification difficult. Most studies on urban ecosystem services focus on the supply of services more than the demand for them, and are largely empirical, GIS-based or bio-physical (Haase et al., 2014). Within the bio-physical categories several studies explore regulating services such as carbon sequestration and storage from green infrastructure and urban forests. GIS-based models can better model both the supply and demand for services, but their scale of analysis is usually coarser, and the detail needed by decision-making can be lost when analysis is based on land use or land cover information. Quantitative assessments that are transferable, practical and appropriate are needed. Most studies have been case studies, with few examples of comparative studies of urban ecosystem services (i.e. Larondelle and Haase, 2013; Dobbs et al., 2014a).

Recognition of the links between the different components of the urban ecosystem and how these interact within the socio-ecological system can improve the quantification of urban forest ecosystem services. This can generate the information necessary for improving decision-making, management and implementation of policies for developing sustainable cities. The following sections list the ecosystem services derived from urban forest and trees, and propose a framework for recognizing different factors involved in the provision of urban ecosystem services.

Ecosystem services from urban forests and trees

Urban ecosystem services are important for maintaining the well-being of urban dwellers. Therefore, they need to be incorporated in urban planning and development (Niemelä et al., 2010; Cilliers et al., 2012). Urban forests, as common (and often managed) natural components of urban areas, provide a variety of ecosystem services (Dobbs et al., 2011; Escobedo et al., 2011; Pataki et al., 2011). Urban trees can be found in a wide range of settings, including parks, streets, natural areas, housing areas, wetland areas, and along streams and ponds.

Forest and trees in urban landscapes are among the main providers of services including climate amelioration, recreation and air pollution removal, capable of enhancing and maintaining human well-being (Cilliers et al., 2012; Dobbs et al., 2011; de Groot et al., 2010a). Several studies have addressed the positive (services) and negative (disservices) contributions of urban forest and trees, which can affect human well-being at different scales (Nowak and Crane, 2000; Lyytimäki et al., 2008; Dobbs et al., 2011; Escobedo et al., 2011; Cilliers et al., 2012). Other chapters in this book look into urban forest services and disservices in greater detail. Most research on urban forest ecosystem services has occurred in Europe, North America and China, with some examples from Africa and South America. The most commonly assessed services are regulating services, including air pollution removal and carbon storage and sequestration. Fewer studies have focused on cultural services, often limited to recreation and aesthetic services (Haase et al., 2014). Cultural ecosystem services have been poorly defined in comparison with other services, and improved definitions and the integration of existing social research on the human benefits of urban nature into ecosystem service frameworks are required to facilitate their study (Daniel et al., 2012).

A detailed list of services and disservices for urban forest and trees is provided in Tables 4.2 and 4.3. The methodologies and proxies to quantify these services and disservices are explained in the following chapters of the book.

Ecosystem service framework adapted to urban systems

During recent decades, frameworks that link human well-being with the provision of ecosystem services from socio-ecological systems like urban, agricultural, semi-natural or natural systems have been developed (Millennium Ecosystem Assessment, 2003, 2005; de Groot et al., 2010b; Haines-Young and Potschin, 2010; Sukhdev, 2010; Bennett et al., 2015). We adapted the most relevant and recent frameworks (Millennium Ecosystem Assessment, 2003, 2005; de Groot et al., 2010b; van Oudenhoven et al., 2012; Bennett et al., 2015) to the urban realm to show how ecosystem services are provided, what their drivers are, how they are received by beneficiaries and the role of social preferences in the supply of ecosystem services.

Table 4.2 Urban forest ecosystem services and their ecological processes, well-being benefits and scale of importance

Ecosystem service	Ecological process	Benefit for human well-being	Scale of benefit
Global climate regulation	Carbon sequestration	Global climate change	Global
Microclimate regulation	Temperature amelioration by shading	Thermal comfort	Local
Air quality regulation	Pollution removal, absorption leaf biomass	Clean air	Regional
Flood regulation	Rainfall interception	Reduce flooded areas	Local
Storm protection	Waste control	Reduce amount of debris	Local
Maintenance of soil quality	Soil nutrient cycling and fertility	Reduction in the use of fertilizers and vegetation in good condition	Local
Filtering of dust particles	Capture of particulate matter in the tree structure	Reduction in asthma attacks, and respiratory diseases	Regional
Noise reduction	Buffering of traffic noise	Quieter urban environment	Local
Maintenance of genetic and biological diversity	Sustainability of several ecological processes	Resilience and sustainability of the urban forest in cities	Regional, global
Habitat provisioning	Shelter for wildlife	Connectedness to nature	Regional, local
Pollination and seed dispersal	Movements of floral gametes	Biodiversity, honey production	Regional, local
Food security	Fruit production	Urban agriculture	Local
Raw materials	Biomass	Firewood, wood chips	Regional
Recreation	Space for enjoyment	Physical and psychological health	Local
Aesthetic	Scenic beauty	Increase property price	Local
Natural heritage and education	Conservation of native species	Connectedness to nature	Regional
Sense of place	Conservation of green spaces and urban nature	Sense of belonging, neighbourhood, pride	Local
Ecotourism	Conservation of nature in urban areas	Tourism, physical and psychological enjoyment	Local

Source: adapted from Dobbs et al. (2011), Gómez-Baggethun and Barton (2012)

Table 4.3 Urban forest ecosystem disservices and their ecological processes, effects on human well-being and scale of importance (see Chapter 13 for a more detailed exploration)

Ecosystem disservice	Ecological process	Effects for human well-being	Scale of benefit
Emissions of volatile organic compounds	Emissions from photosynthesis	Ozone formation	Global
Allergies	Movements of floral gametes	Allergies	Local
Damage to infrastructure	Wood decomposition	Damage to people or infrastructure	Local, regional
Fear of crime	Dense vegetated areas that create dark and perceived unsafe places	Fear	Local

Source: adapted from Dobbs et al. (2011), Gómez-Baggethun and Barton (2012)

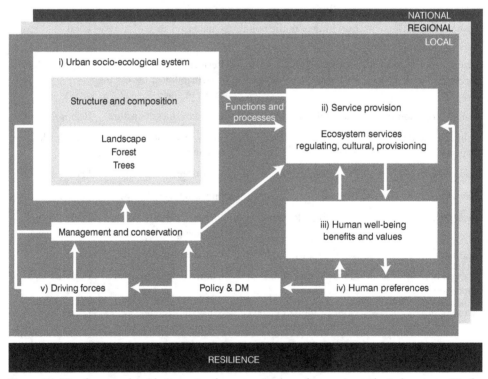

Figure 4.1 The framework with its main elements: (i) the urban socio-ecological system, (ii) the ecosystem services provision mediated by ecosystem processes and functions, (iii) human well-being, (iv) human responses in the form of preferences for ecosystem services and policy decisions developed to manage urban landscapes and (v) multiple driving forces

DM = decision-making

The structure and composition of the ecological features in cities determines the provision of ecosystem services (Dobbs et al., 2014b). Social preferences, used here in the economic sense of a preferred order among a set of alternatives to generate an optimal choice, influence the development of policy decisions to maintain sustainable levels of ecosystem services through appropriate management and planning. Resilience shapes the response of the socio-ecological variations in driving forces impacting the system, such as climate change.

(i) The urban socio-ecological system

The urban socio-ecological system encompasses the set and interplay of biophysical, ecological and social conditions, in which the structure, composition and distribution of the urban forest underpin the supply of ecosystem functions and services (Bennett et al., 2015). The structure and the composition of the urban forest vary according to the scale of analysis. At the landscape scale, the urban forest is composed of a group of vegetation patches that are characterized by their size, distribution and connectedness, while composition is shaped by the dominant vegetation cover of patches (e.g., park, grassland, natural area). A second scale of analysis is the forest level, in areas where groups of trees are found. The structure of these forest areas is characterised by the average tree density, height and canopy cover, while the species occurring in the forest determines composition. Finally, analysis can occur at the individual tree scale, where structure can be characterized by tree size, and composition by species or sub-species. The dynamics of change in this socio-ecological system occur at all three levels of analysis, and will have an effect on the ecological function of the ecosystem at these three scales. For example, changes in the spatial arrangement of natural land cover types as a result of urban expansion can result in large changes to the structure and composition of the urban forest. This can influence ecosystem functioning and the provision of ecosystem services directly and indirectly, as animal movement and behaviour changes in response to the new landscape configuration. At the city scale, factors such as the age of the urban area, species composition, population density and history can shape different carbon storage in cities with a similar climate (Strohbach and Haase, 2012). The level of development of the city and the level of democracy of the country also has been shown to influence the structure of the urban forest; for example, cities with a high Human Development Index tend to have more urban forest per capita, while cities in democratic countries tend to have less urban forest in large, continuous areas that is potentially suitable as habitat for other species (Dobbs et al., 2014b).

(ii) The provision of ecosystem services

Ecosystem services are the processes and conditions derived from ecosystems that sustain and enhance human well-being. Their provision depends on the supply, demand and flow of benefits and value to people (Mitchell et al., 2015). Supply is the full potential of ecological functions or biophysical elements in an ecosystem to provide a given ecosystem service, without consideration of whether humans actually recognize, use or value that function or element (de Groot et al., 2010b; Haines-Young and Potschin, 2010). Examples of ecosystem functions supplied by urban forest patches include net primary productivity, habitat for species and soil stability (Nowak and Crane, 2000; Dobbs et al., 2011). Ecosystem service demand is the level of service provision desired or required by people (Villamagna et al., 2013). Demand is influenced by people's needs and preferences, and is therefore context dependent. For example, the demand for urban trees to provide shade also needs to fulfil the aesthetic preferences of urban inhabitants, and will be influenced by the climate of the city; shade is likely to be more preferred in arid and semi-arid cities rather than continental cities. For the benefits of ecosystem services to be realized, people must interact with ecosystems. This interaction connects ecosystem service supply with demand to produce the delivery of the service to be enjoyed by people through a service flow (Villamagna et al., 2013; Mitchell et al., 2015). Examples of social benefits are health, employment and income (van Oudenhoven et al., 2012). Finally, actors in society can attach a value to these benefits, which is the contribution of ecosystem services to goals, objectives, or conditions that are specified by user.

(iii) Human well-being

Ecosystem service provision contributes directly to human well-being (van Oudenhoven et al., 2012), a multidimensional concept that is measured along a quality-of-life gradient that reflects the spatiotemporal variability of material (food, water, shelter) and nonmaterial needs (good health, social cohesion, security) (Villamagna and Giesecke, 2014). Currently there is little empirical understanding of the diversity of stakeholders and their relationships determining access and distribution of ecosystem services benefits and their implications for livelihoods and human well-being (Bennett et al., 2015). Some efforts have been made in epidemiological studies, testing the relations of nature with psychological, cognitive and physiological health (a list of references can be found in Sandifer et al., 2015). For example, the presence of trees in urban areas has been correlated with decreased depression and stress, improved learning opportunities and quality of life (Kuo and Sullivan, 2001; Song et al., 2012).

(iv) Human responses: preferences, decisions and ecosystem services management

Management of the urban forest must be connected to the socio-ecological dynamics of the urban system they are part of. Diverse stakeholders have different preferences for ecosystem services based on their worldview, cultural background and individual interests. Services that can be more valued by urban dwellers include air pollution removal, microclimate regulation, aesthetics, tourism and environmental education, while services such as food provisioning are commonly perceived as less important in the developed world (Martin-Lopez et al., 2012). Knowing people's preferences enables decision-makers to determine how people value ecosystem services. This social process facilitates the development of policies that maintain and conserve preferred ecosystem services. Policy and decision-making provide preconditions, constraints and incentives for landscape management. The dependence of cities on the surrounding landscape and its biodiversity, as well as the ongoing interactions between processes occurring in urban, peri-urban and rural contexts are essential for sustaining urban ecosystem services and the overall resilience of the urban socio-ecological system.

(v) Driving forces

Driving forces are natural or human-induced factors which can influence the ecosystem, either directly (e.g., through climate change or environmental pollution) or indirectly (e.g., through changes in demography or economy) (Millennium Ecosystem Assessment, 2003, 2005). Analysis of historical and current drivers of change, and the thresholds that distinguish different system states, is needed to identify key social-ecological features that help cope with disturbances. Maintaining these features contributes to making the system resilient while maintaining a diverse and preferred set of ecosystem services for a diversity of beneficiaries (Bennett et al., 2015). The use of scenarios can show possible trends in ecosystem services, the impact of new drivers and the related impacts on the well-being of future generations. Understanding how ecosystem services provision might change over time is a prerequisite for securing a global future that can meet the sustainable provision of multiple ecosystem services, and avoid transitioning to new, less sustainable states (Schröter et al., 2005). Assessing ecosystem service scenarios is also important for determining the resilience of different ecosystem services.

Resilience of ecosystem service provision

Resilience is the capacity of a system to persist and adapt to sudden shocks (e.g., pest outbreaks) and global environmental change (e.g., climate change, urban densification) (Walker and Salt, 2006; McPhearson et al., 2015). There are two key concepts that connect resilience and urban forest ecosystem services; the resilience provided by the urban forest to the social-ecological system, and resilience of the urban forest itself (*sensu* McPhearson et al., 2015).

First, the ecosystem services provided by the urban forest make an important contribution to the resilience of the urban social-ecological system by buffering the system from external shocks and mitigating the effect of, and adapting to, changing external forces. For example, the provision of cooling by the urban forest can buffer the effects of heat-waves by reducing maximum temperatures, and improving people's well-being or reducing negative health impacts. Increasing urban forest canopy cover can help mitigate the long-term effects of urban heat and climate change through shading, which can reduce local temperatures and urban heat island effects.

The second concept is the resilience of the urban forest itself, to ensure the persistent supply of ecosystem services in the face of shocks and global environmental change. Urban forests must be managed in ways to reduce the likelihood of catastrophic loss through sudden pest or disease outbreaks. An example of this is the case of Dutch elm disease (*Ophiostoma ulmi* and *O. novo-ulmi*), which wiped out trees across Europe and North America, resulting in a loss of over 50 per cent of the urban trees in some US cities in the 1960s and 1970s (Karnosky, 1979). This loss of a major part of the mature urban tree population is likely to have had a similarly catastrophic effect on the provision of ecosystem services by the urban forest. Planning for persistence of the urban forest in the face of urban heat and climate change poses an enormous challenge for urban forest managers around the world. It is clear that even relatively small changes in temperature can have a large effect on the species composition of the urban forest (Kendal et al., 2012). It is likely that some, and perhaps many, species that have performed well in the past are likely to fail in future climates, again with potentially catastrophic effects on the provision of ecosystem services (further discussion in Chapter 23).

Two key considerations in managing the resilience of the urban forest and the ecosystem services it provides are redundancy and response diversity (Biggs et al., 2012). Redundancy in the provision of ecosystem services occurs when more than one species is providing the same service. Increased redundancy reduces the effects of catastrophic loss of any particular species on the provision of ecosystem services (Biggs et al., 2012). Redundancy could be increased by the use of morphologically or physiologically similar species that are taxonomically distinct from a dominant species. Response diversity means that different species respond in different ways to the same change in external forces. Increased diversity is an important driver of functional resilience in ecological systems and likely to be an important characteristic of adaptive social-ecological systems with a persistent supply of ecosystem services. Increased diversity increases the resilience of the urban social-ecological system as a diverse urban forest can mitigate a broader range of external forces, and is more likely to be able to mitigate and adapt to unanticipated changes. Increased diversity also limits the vulnerability of the forest to specific threats such as pests and diseases, or a changing climate (Kendal et al., 2014).

Ultimately, diversity is determined by species selection, and taxonomic diversity is a key indicator of diversity. However, age–class diversity, and trait diversity are also important considerations for urban forest management. Age–class diversity is important to ensure that the urban forest does not senesce unevenly, leading to the loss of large proportions of the mature canopy in a short period of time. Traits (characteristics of trees such as size, canopy density and colour) are important indicators of the mechanisms that underpin the provision of ecosystem services. For example, the provision of shade is largely determined by traits such as crown size, leaf area and canopy density. The provision of some cultural ecosystem services is determined by aesthetic traits such as colour and texture. Trait diversity leads to a wider range of services being provided by the forest, and increases the likelihood of those services persisting as external forces change.

There is a danger that simplistic approaches to incorporating ecosystem services into urban forest management will result in simplification of the urban forest itself – for example by favouring a small suite species and cultivars that are good providers of a small set of targeted services. It is crucial that the concepts of resilience and diversity are also applied when an ecosystem services approach is adopted in urban forest management.

Ecosystem services in decision making

Urban ecosystem service assessments are becoming more common (Haase et al., 2014; see also Chapter 11) but it is unclear whether they actually generate information relevant for decision making (Martinez-Harms et al., 2015). The majority of urban ecosystem services assessments have focused on quantifying and valuing the supply of ecosystem services like local climate and air quality regulation, carbon sequestration and physical and mental health (Haase et al., 2014). However, there has been little consideration of the information demanded by decision makers for addressing urban landscape management and urban policies. One rare example of such consideration happened in the city of Vancouver, Canada, where a key concern of stakeholders and managers was to maintain access to shellfish harvest areas (Guerry et al., 2012). After a participatory process with scientists and decision makers, ecosystem services models are now used at the government level to quantify baseline conditions for ecosystem services (e.g. shellfish harvest areas, renewable energy, coastal protection, recreation and water quality) and to develop scenarios representing different land-use policies (Guerry et al., 2012; Ruckelshaus et al., 2015).

Components or processes of the urban social-ecological system only become an ecosystem service when they interact with humans, and they only become realized if someone benefits from them, often involving subjective judgements of what is perceived as a benefit. The urban ecosystem service concept allows for assessing human dependence on the social ecological system through transdisciplinary research, integrating the perspectives and values of different stakeholder groups, and guiding decisions on landscape management (Schröter et al., 2014; Martinez-Harms et al., 2015). The incorporation of urban ecosystem services into landscape management requires a transdisciplinary effort. A useful approach is applying a formal decision-making process. These typically comprise five main steps: the identification of the social ecological context; specification of objectives and performance measures; defining alternative management actions; assessment of trade-offs and prioritization of actions; and implementation of the management action (Martinez-Harms et al., 2015). A management decision occurs when an alternative action is selected from the alternatives, and implemented through its internalization in policy, plans or an institutional arrangement, typically operationalized as some form of regulation or incentive (Martinez-Harms et al., 2015). Examples from the urban realm include urban water management plans (Hogan et al., 2014), the protection of urban natural parks (Ernstson and Sörlin, 2013), the implementation of green infrastructure for urban storm water management, re-vegetation measures for flood mitigation (Pataki et al., 2011), and stream restoration plans for drinking-water treatment (Honey-Rosés et al., 2013), among others. The implementation of management decisions for ecosystem services should be underpinned by the best available science by integrating primary research and systematic reviews. However, ecosystem service assessments typically do not currently follow the steps of a formal decision-making process (Martinez-Harms et al., 2015).

Management should take into account the multiple values and preferences of stakeholders to inform management objectives, identify performance measures and choose management actions. Deliberative and participatory methods can facilitate this, and enable the opportunities for and constraints on effective management to be identified. Experience shows that complex assessments are not necessarily more helpful for decision making (Ruckelshaus et al., 2015). Decision makers do not necessarily need an exhaustive understanding of the social-ecological system, but they need sufficient information to make a choice between management options. Therefore, designing problem-oriented urban ecosystem services assessments, which focus on the information needed by decision makers, can help incorporate urban ecosystem services into management decisions (Honey-Rosés et al., 2013).

Conclusion and perspective

This chapter has highlighted the importance of ecosystem services for improving the well-being of urban dwellers. The forests and trees found in urban areas can deliver a variety of ecosystem services that can fulfil a variety of social needs and meet the preferences of urban inhabitants. The urban forest ecosystem service framework presented in this chapter recognizes the components and driving forces that exist in this socio-ecological system and that occur at different scales. This framework explores the existence of supply and demand of ecosystem services, while recognising provision of services only occurs when people use services to meet their needs.

Governance, policy and planning in urban areas should include ecosystem services in long-term planning, management and conservation of urban forests. It is important to engage with planners, urban designers, landscape architects and park managers to translate urban forestry research into information that is useful for these disciplines. Communication and transdisciplinary work is essential for developing sustainable and resilient cities capable of maintaining the well-being of urban dwellers.

The framework has the advantage of recognizing the elements involved in urban ecosystem service provision. However, its application will depend on capacity to quantify structure, composition, service provision and human values and preferences in a manner that is transferable and easy to implement. This can enable comparisons across different cities, planning schemes and future development scenarios for urban forests, and identify trade-offs that might reduce the well-being of urban inhabitants.

References

Andersson, E., Barthel, S., Borgström, S., Colding, J., Elmqvist, T., Folke, C. and Gren, Å. (2014) 'Reconnecting cities to the biosphere: Stewardship of green infrastructure and urban ecosystem services', *Ambio*, vol. 43, pp. 445–453.

Bennett, E. M., Cramer, W., Begossi, A., Cundill, G., Díaz, S., Egoh, B. N., Geijzendorffer, I. R., Krug, C. B., Lavorel, S. and Lazos, E. (2015) 'Linking biodiversity, ecosystem services, and human well-being: three challenges for designing research for sustainability', *Current Opinion in Environmental Sustainability*, vol. 14, pp. 76–85.

Biggs, R., Schlüter, M., Biggs, D., Bohensky, E. L., BurnSilver, S., Cundill, G. … and West, P. C. (2012) 'Toward principles for enhancing the resilience of ecosystem services', *Annual Review of Environment and Resources*, vol. 37, pp. 421–448.

Cilliers, S., Cilliers, J., Lubbe, R. and Siebert, S. (2012) 'Ecosystem services of urban green spaces in African countries–perspectives and challenges', *Urban Ecosystems*, vol. 16, pp. 681–702.

Daniel, T. C., Muhar, A., Arnberger, A., Aznar, O., Boyd, J. W., Chan, K. M. … and von der Dunk, A. (2012) 'Contributions of cultural services to the ecosystem services agenda', *Proceedings of the National Academy of Sciences of the United States of America*, vol. 109, pp. 8812–8819.

De Groot, R. S., Wilson, M. A. and Boumans, R. M. J. (2002) 'A typology for the classification, description and valuation of ecosystem functions, goods and services', *Ecological Economics* vol. 41, pp. 393–408.

De Groot, R. S., Alkemade, R., Braat, L., Hein, L. and Willemen, L. (2010a) 'Challenges in integrating the concept of ecosystem services and values in landscape planning, management and decision making', *Ecological Complexity*, vol. 7, pp. 260–272.

De Groot, R., Fisher, B., Christie, M., Aronson, J., Braat, L., Haines-Young, R., Gowdy, J., Maltby, E., Neuville, A. and Polasky, S. (2010b) 'Integrating the ecological and economic dimensions in biodiversity and ecosystem service valuation', in Pushpam, K. (ed.), *The Economics of Ecosystems and Biodiversity (TEEB): Ecological and Economic Foundations*, Earthscan, London.

Dobbs, C., Escobedo, F. J. and W. C. Zipperer (2011) 'A framework for developing urban forest ecosystem services and goods indicators', *Landscape and Urban Planning*, vol. 99, pp. 196–206.

Dobbs, C., Kendal, D. and Nitschke, C. R. (2014b) 'Multiple ecosystem services and disservices of the urban forest establishing their connections with landscape structure and sociodemographics', *Ecological Indicators*, vol. 43, pp. 44–55.

Dobbs, C., Nitschke, C. R. and Kendal, D. (2014a) 'Global drivers and tradeoffs of three urban vegetation ecosystem services', *PLoS One*, vol. 9, pp. e113000.

Ernstson, H. and Sörlin, S. (2013) 'Ecosystem services as technology of globalization: On articulating values in urban nature', *Ecological Economics*, vol. 86, pp. 274–284.

Escobedo, F. J., Kroeger, T. and Wagner, J. E. (2011) 'Urban forests and pollution mitigation: Analyzing ecosystem services and disservices', *Environmental Pollution*, vol. 159, pp. 2078–2087.

Farber, S., Costanza, R. and Wilson, M. (2002) 'Economic and ecological concepts for valuing ecosystem services', *Ecological Economics*, vol. 41, pp. 375–392.

Gaston, K.J., Ávila-Jiménez, M.L. and Edmondson, J.L. (2013) 'Managing urban ecosystems for goods and services', *Journal of Applied Ecology*, vol. 50, pp. 830–840.

Gómez-Baggethun, E. and Barton, D. N. (2013) 'Classifying and valuing ecosystem services for urban planning', *Ecological Economics*, vol. 86, pp. 235–245.

Guerry, A. D., Ruckelshaus, M. H., Arkema, K. K., Bernhardt, J. R., Guannel, G. … and Spencer, J. (2012) 'Modeling benefits from nature: Using ecosystem services to inform coastal and marine spatial planning', *International Journal of Biodiversity Science Ecosystem Services & Management*, vol. 8, pp. 107–121.

Haase, D., Larondelle, N., Andersson, E., Artmann, M., Borgström, S., Breuste, J. … and Elmqvist, T. (2014) 'A quantitative review of urban ecosystem service assessments: Concepts, models, and implementation', *Ambio*, vol. 43, pp. 413–433.

Haines-Young, R. and Potschin, M. (2010) 'The links between biodiversity, ecosystem services and human well-being', in Raffaelli, D. and Frid, C. (eds), *Ecosystem Ecology: A New Synthesis*, BES Ecological Reviews Series, Cambridge University Press, Cambridge.

Hogan, D. M., Shapiro, C. D., Karp, D. N. and Wachter, S. M. (2014) *Urban Ecosystem Services and Decision Making for a Green Philadelphia*, US Geological Survey, Washington, DC.

Honey-Rosés, J., Acuña, V., Bardina, M., Brozović, N., Marcé, R., Munné, A. … and Schneider, D. W. (2013) 'Examining the demand for ecosystem services: The value of stream restoration for drinking water treatment managers in the Llobregat River, Spain', *Ecological Economics*, vol. 90, pp. 196–205.

Karnosky, D. F. (1979) 'Dutch elm disease: A review of the history, environmental implications, control, and research needs', *Environmental Conservation,* vol. 6(4), p. 311.

Kendal, D., Dobbs, C. and Lohr, V. I. (2014) 'Global patterns of diversity in the urban forest: is there evidence to support the 10/20/30 rule?', *Urban Forestry & Urban Greening,* vol. 13, pp. 411–417.

Kendal, D., Williams, N. S. G. and Williams, K. J. H. (2012) 'A cultivated environment: Exploring the global distribution of plants in gardens, parks and streetscapes', *Urban Ecosystems,* vol. 15, pp. 637–652.

Kuo, F. E. and Sullivan, W. C. (2001) 'Environment and crime in the inner city: Does vegetation reduce crime?', *Environmental Behaviour* vol. 33, pp. 343–367.

Lafortezza, R., Carrus, G., Sanesi, G. and Davies, C. (2009) 'Benefits and well-being perceived by people visiting green spaces in periods of heat stress', *Urban Forestry & Urban Greening,* vol. 8, pp. 97–108.

Larondelle, N., Haase, D. and Kabisch, N. (2013) 'Mapping the diversity of regulating ecosystem services in European cities', *Global Environmental Change,* vol. 26, pp. 119–129.

Lyytimäki, J., Petersen, L. K., Normander, B. and Bezák, P. (2008) 'Nature as a nuisance? Ecosystem services and disservices to urban lifestyle', *Environmental Science,* vol. 5, pp. 161–172.

Martinez-Harms, M. J., Bryan, B. A., Balvanera, P., Law, E. A., Rhodes, J. R., Possingham, H. P. and Wilson, K. A. (2015) 'Making decisions for managing ecosystem services', *Biological Conservation,* vol. 184, pp. 229–238.

Martín-López, B., Iniesta-Arandia, I. García-Llorente, M., Palomo, I. Casado-Arzuaga, I. García Del Amo, D., Gómez-Baggethun, E. Oteros-Rozas, E., Palacios-Agundez, I. Willaarts, B. González, J. A., Santos-Martín, F., Onaindia, M., López-Santiago, C.A., and Montes, C. (2012) 'Uncovering ecosystem services bundles through social preferences', *PLoS ONE,* 7 (2012), p. e38970 http://dx.doi.org/10.1371/journal.pone.0038970

McPhearson, T., Andersson, E., Elmqvist, T. and Frantzeskaki, N. (2015) 'Resilience of and through urban ecosystem services', *Ecosystem Services,* vol. 12, pp. 152–156.

MEA (2003) 'Ecosystems and human well-being', In *Ecosystems and Human Well-being: A Framework for Assessment,* Island Press, Washington, DC.

Millennium Ecosystem Assessment (2005) *Ecosystems and Human Well-being. Synthesis.* http://www.unep.org/maweb/documents/document.356.aspx.pdf

Mitchell, M. G. E., Suarez-Castro, A. F., Martinez-Harms, M., Maron, M., McAlpine, C., Gaston, K. J., Johansen, K. and Rhodes, J. R. (2015) 'Reframing landscape fragmentation's effects on ecosystem services', *Trends in Ecology & Evolution,* vol. 30, pp. 190–198.

Niemelä, J., Saarela, S., Söderman, T., Kopperoinen, L., Yli-Pelkonen, V., Väre, S. and Kotze, J. (2010) 'Using the ecosystem service approach for better planning and conservation of urban green spaces: a Finland case study', *Biodiversity and Conservation* vol 19, pp. 3225–3243.

Nowak, D. J., and Crane, D. E. (2000) 'The urban forest effects (UFORE) model: quantifying urban forest structure and functions', in Hansen, M., Burk, T. (Eds.), *Proceedings: Integrated Tools for Natural Resources Inventories in the 21st Century.* IUFRO Conference, 16–20 August 1998, Boise, ID. General Technical Report NC-212. US Department of Agriculture, Forest Service, North Central Research Station, St. Paul, MN, pp. 714–720.

Nowak, D. J., Crane, D. E., Stevens, J. C., David, J., Crane, E. and Stevens, J. C. (2006) 'Air pollution removal by urban trees and shrubs in the United States', *Urban Forestry and Urban Greening,* vol. 4, pp. 115–123.

Pataki, D. E., Carreiro, M. M., Cherrier, J., Grulke, N. E., Jennings, V., Pincetl, S., Pouyat, R. V., Whitlow, T. H. and Zipperer, W. C. (2011) 'Coupling biogeochemical cycles in urban environments: Ecosystem services, green solutions, and misconceptions', *Frontiers in Ecology and the Environment,* vol. 9, pp. 27–36.

Perrings, C., Duraiappah, A., Larigauderie, A. and Mooney, H. (2011) 'The biodiversity and ecosystem service science-policy interface', *Science,* vol. 331, pp. 17–19.

Ruckelshaus, M., McKenzie, E., Tallis, H., Guerry, A., Daily, G. … and J. Bernhardt. (2015) 'Notes from the field: Lessons learned from using ecosystem service approaches to inform real-world decisions', *Ecological Economics,* vol. 115, pp. 11–21.

Sandifer, P. A., Sutton-Grier, A. E. and Ward, B. P. (2015) 'Exploring connections among nature, biodiversity, ecosystem services, and human health and well-being: Opportunities to enhance health and biodiversity conservation', *Ecosystem Services,* vol. 12, pp. 1–15.

Schröter, D., Cramer, W., Leemans, R., Prentice, I. C., Araújo, M. B. … and Zierl, B. (2005) 'Ecosystem service supply and vulnerability to global change in Europe', *Science,* vol. 310, pp. 1333–1337.

Song C., Ikei, H., Igarashi, M., Miwa, M., Takagaki, M. and Miyazaki, Y. (2014) 'Physiological and psychological responses of young males during spring-time walks in urban parks', *Journal of Physiological Anthropology*, vol. 33, p. 8.

Strohbach, M. W. and Haase, D. (2012) 'Aboveground carbon storage by urban trees in Leipzig, Germany: Analysis of patterns in a European city', *Landscape and Urban Planning*, vol. 104, pp. 95–104.

Sukhdev, P. (2010) *The Economics of Ecosystems and Biodiversity: Mainstreaming the Economics of Nature: A Synthesis of the Approach, Conclusions and Recommendations of TEEB*, TEEB, Geneva.

TEEB (2011) *TEEB Manual for Cities: Ecosystem Services in Urban Management*, TEEB, Geneva.

UNDP (2015) Sustainable Development Goals. Retrieved from www.un.org/sustainabledevelopment.

van Oudenhoven, A. P. E., Petz, K., Alkemade, R., Hein, L. and de Groot, R. S. (2012) 'Framework for systematic indicator selection to assess effects of land management on ecosystem services', *Ecological Indicators*, vol. 21, pp. 110–122.

Villa, F., Bagstad, K. J., Voigt, B., Johnson, G. W., Portela, R., Honzák, M. and Batker, D. (2014) 'A methodology for adaptable and robust ecosystem services assessment', *PLoS One*, vol. 9, p. e91001.

Villamagna, A. and Giesecke, C. (2014) 'Adapting human well-being frameworks for ecosystem service assessments across diverse landscapes', *Ecology and Society*, vol. 19.

Villamagna, A., Angermeier, P., Bennett, E. (2013) 'Capacity, demand, pressure, and flow: a conceptual framework for analysing ecosystem service provision and delivery', *Ecological Complexity* vol 15: pp. 114–121.

Walker, B. and Salt, D. (2006) *Resilience Thinking: Sustaining Ecosystems and People in a Changing World*, Island Press, Washington, DC.

Wong, C. P., Jiang, B., Kinzig, A. P., Lee, K. N. and Ouyang, Z. (2015) 'Linking ecosystem characteristics to final ecosystem services for public policy', *Ecology Letters*, vol. 18, pp. 108–118.

5

SOCIAL ASPECTS OF URBAN FORESTRY AND METRO NATURE

Kathleen L. Wolf

Introduction

As the world has become more urbanized, planning, social justice, and public health professionals are paying more attention to outdoor place, space, and environments as human habitat. The accelerating pace of research focused on human experiences of nearby nature is remarkable. Studies reveal a complex array of human-with-nature interactions, with many being quite beneficial, and these range in scale from the individual, to household, neighbourhood, and broader community. Physical surroundings in cities, including nature, have received greater attention as important determinants in human relationships and community conditions.

This chapter highlights how city trees and green space serve as a formative, and often supportive backdrop for social benefits ranging from personal development to more liveable, safer communities. Urban nature in all its forms – including urban forests, parks, greenbelts, streetscapes, green space, and green infrastructure – is referred to here as metro nature. Studies about humans and their communities may be less tangible than the science of biophysical effects, as people are highly mobile, and exist in complex situations having multiple influences on outcomes. Nonetheless, social sciences, humanities, and public health research methods reveal underlying responses and correlations, though causal underpinnings of outcomes are more difficult to confirm. Despite the challenges of studying the most complex species on the planet, evidence-based interpretations and inferences point to why all people need metro nature in their lives in order to attain better quality of life, and to engage with other people in more constructive and productive ways.

Economic values

An important and often prominent concern in urban forestry planning and management is the question of value (see Chapter 11). Metro nature provides a range of benefits and services, most of which are not readily bought and sold. Yet communities often invest in public goods that members of society accept as providing value, such as education or emergency response systems. Estimating the values of nature's services helps local governments to weigh costs against returns from development or prioritize payments for green versus grey infrastructure. Without some indicator of economic value, there may be little financial incentive to consider urban nature in land-use decisions, market transactions, and capital investment budgets.

Residential properties

Hedonic pricing has been used extensively for residential property valuation (Wolf, 2007). Property sales prices or assessments are statistically regressed against home and property characteristics to estimate how a difference in a natural feature relates to a change in property value. Mature yard trees can contribute 2–15 percent to home value, and tree canopy in neighbourhoods has been found to increase value 6–9 percent. Tree retention in new development is often claimed to be expensive yet parcels having mature trees are found to have up to 22 percent higher market value. The 'proximate principle' describes how properties near naturalistic open spaces are typically valued at about 8 percent to 20 percent higher than comparable properties, with the effect observable up to a half mile away. Property value gains can be captured by local governments as increased property tax assessments or as excise taxes on sales, with the revenues applied to the annual costs of urban forest or parks programs. A study in Portland, Oregon US, found that the aggregated pricing effect of street trees on all houses yielded a potential boost in annual property tax revenues of USD 15.3 million (Donovan and Butry, 2010).

Figure 5.1 The presence of trees in parks, in streetscapes, and in yards may boost the market value of residential properties

Source: Guy Kramer

Retail and consumer environments

Business owners are often quite influential in communities, and their outlook on green investments can set the tone for tree program support in local government. Merchants are able to tally the direct costs of trees (such as pruning and debris clean up), but may not recognize shoppers' values. Consumer response to urban forests was the focus of a series of studies about central business districts (Wolf, 2005). Visual quality scores were lower for streetscapes without trees and much higher for places with trees. Study respondents also reported that they would be willing to pay from 9 percent (in small cities) to 12 percent (in large cities) more for a set of goods and services in a place having well-maintained trees.

What might contribute to the positive responses to trees? Retail marketers use 'atmospherics' to set the stage for desired shopper behaviours. Colour, light, product placement and other indoor strategies influence purchase choices. Trees are an outdoor atmospheric that sets up favourable impressions, including cues about social interactions. Shoppers buy things to satisfy needs, and also enjoy positive experiences with friends and family. The streetscape can be a welcoming, interesting place, and shape consumer expectations before even entering a store.

Nature and community

Purchases of properties or goods are an important contribution to local economies. But market transactions don't capture the diversity of human experiences and interactions related to metro nature. People are directly dependent on nature for life's necessities, such as clean air and water. Yet the presence of nature sustains people in more subtle, yet very important ways. Studies have spanned the social scale, ranging from individuals, to households, to communities, and include people/plant relationships that are fundamental to basic functioning and quality of life. Trees and the urban forest are a prominent presence in the places where people come together, social interactions occur, and relationships or partnerships take form. Green spaces can be the settings where social cohesion and increased community capital emerge, particularly if people share work on a project or goal.

What is 'community?'

People in cities often gather in public spaces to accomplish goals, relax and play, learn, or enjoy an activity. It is around these multiple interactions that social community takes shape. Early on, sociologists defined 'community' as a group of people living and interacting within a geographical boundary, such as a workplace or a school. The definition has since expanded to include any interrelatedness of individuals, whether physically near each other or not. It includes a social group of any size that shares beliefs, values, interests, intentions, needs, or place and is encouraged by social exchanges. Community may also form around purpose; it can be a collection of people with differing but harmonious views, skills, perceptions, or shared interests who can, with some support (e.g. funding, professional advice, or membership), develop in a cooperative way to achieve valued outcomes. For example, several business communities, both place and interest based, were collaborators in the trees and retail research described earlier. These groups of individuals shared economic interests, and developed programs to support their connections, such as shared tree care.

The concept of community has many expressions (Peters et al., 2010). *Social interactions* are the formal or informal opportunities for people to develop interpersonal relationships. Casual social encounters and interactions in day-to-day life leads to feelings of acceptance and

Figure 5.2 Forested parks are the places in cities where casual social encounters can happen, leading to improved social cohesion

Source: Guy Kramer

sense of belonging. *Social cohesion* is the interdependence between members of a community, experienced as shared values, loyalties, and cooperation. An absence of social cohesion is indicated by feelings of loneliness and not being able to call on others for social support. Public spaces, particularly parks and gardens, offer opportunity for open interactions between people of different backgrounds, helping to develop cohesion and alleviate tensions.

Finally, *social capital* is loosely defined as the 'glue' that holds a community together (Putnam, 2000). It emerges within networks of social relations that are characterized by norms of trust and reciprocity, and a shared give-and-take for mutual benefit. Increased social capital can strengthen and enhance community wellness. Also, individuals can further their own goals by enhancing social capital and relationships. Social capital makes it possible to achieve things that cannot be accomplished solely by individuals, and makes community-oriented goals possible.

Benefits of social ties

Why would community leaders and natural resources managers be interesting in encouraging interaction, cohesion, and increased social capital?

Social relationships are important to individuals in all cultures and at all times throughout their lives. Individuals and groups within communities with strong social cohesion and social capital experience many positive benefits, that are expressed as community wellness (Wood and Giles-Corti, 2008; Elands et al., in press). For example, children and youth in close-knit communities are less likely to participate in health-threatening behaviours such as smoking, drinking, gang involvement, or drug use. Stronger neighbourhood social ties provide more available adult guidance and positive role models. Also, elderly individuals with strong social connections have lower rates of early mortality, reduced suicide rates, less fear of crime, and better physical health. Lack of social interactions and loneliness is associated with a range of human health concerns (Holt-Lunstad et al., 2015). The role of social aspects in health is substantial; social isolation contributes to mortality and illness at the same scale as smoking and other medically-confirmed risk factors.

Strong community relationships may prompt individuals to work together to achieve common goals (e.g. create clean and safe public spaces), to exchange valuable information, and to maintain informal social controls (e.g. discourage crime or other undesirable behaviours), all of which can directly or indirectly influence personal health (Semenza and March, 2008). Communities in which residents experience frequent interactions, high mutual trust, and social reciprocity have been linked with lower homicide rates and less crime (Bellair, 1997). Conversely, neighbourhoods lacking social cohesion and community connectedness have been linked to higher levels of social disorder, leading to personal anxiety, and depression (Ross, 2000).

Nature for community

Metro nature plays an important role in creating the vital neighbourhood spaces that enable more diverse and stronger social ties. It is likely that many of the well-being benefits that are associated with experiences of greenspace are partly mediated through people's local social interactions (de Vries et al., 2013). The associations between social relations and social health

Figure 5.3 The presence of trees and quality landscape can encourage more use of outdoor spaces and the social activity within them

Source: Kathleen Wolf

(as well as individual mental and physical health) are now being more thoroughly explored, including how greenspace can be planned and managed for greater benefit for diverse people and groups (Dinnie et al., 2013).

Early studies of social strengths tended to focus almost exclusively on human-to-human interactions. Yet the physical configuration of place influences the extent and richness of social contacts. People prefer natural over hardscape settings, and preferences may be predictors of the use of environments. The presence of trees, parks, and gardens is related to the rate of use of outdoor spaces, the amount of social activity that takes place within them, and the proportion of social to non-social activities they support. The quality of the space matters, as more beautiful nature is associated with more positive social behaviours (Zhang et al., 2014).

In studies of US public housing (Kuo, 2003), spaces with trees attracted larger groups of people – as well as more mixed groups of youth and adults – than did spaces devoid of natural elements. The presence, number, and location of trees strongly predicted the amount of time that inner-city residents actually spent in outdoor common spaces. Residents reported that they disliked and feared treeless, empty common spaces. Additionally, the more time people spent in this common space, the better they knew their neighbours and the greater their sense of community. Also, inner-city parks that are well-maintained and provide good recreational facilities encourage the social interactions that lead to social ties (Kaźmierczak, 2013).

Such findings span cultures and life-cycle. In the Netherlands, parks have helped generate social cohesion; people of diverse ethnic groups mingle there, forming cross-cultural social ties (Peters et al., 2010). Aging adults report a stronger sense of unity among residents, experience a stronger sense of belonging to the neighbourhood, and feel that neighbours are more supportive of one another compared to individuals who have less exposure to green common spaces (Kweon et al., 1998). As people get older, they are generally less mobile, thus having metro nature in their direct living environments is increasingly important.

Civic environmental stewardship, such as the engagement of volunteers to care for trees and landscapes (see also Chapter 15), can promote social connections. Inner-city residents who spend time in outdoor common spaces caring for flowers, grass, or trees outside of their homes have stronger social networks with their neighbours. A study of beautification projects found increased perceptions of social capital among community members, as people felt more connected with and trusting of their neighbours (Alaimo et al., 2010). Stewardship Mapping ('Stew-Map') is a US Forest Service protocol to inventory stewardship organizations. Results reveal the aggregate activity footprint across a region or metropolitan area, as well as social network linkages (Svendsen et al., 2016). Community greening projects enhance relationships among people, increase community pride, and can serve as a catalyst for broader community improvements.

Place attachment

Connections to place

Certain terms are commonly used in public discussions about urban nature settings. Space, for example, is the physical dimension of the outdoors, represented on maps and in planning documents. It is defined in cities by buildings, paving, or vegetation. Space is transformed into 'place' when humans give it bounds and believe it has value. Place is constructed and reconstructed over time by different groups of people (Tuan, 1990). Perceptions of place, or *sense of place*, are ever-changing, depending on social interactions, context, and events. In cities social communications can make and unmake places, elevating or diminishing the appeal of a site or business. This process has been accelerated by internet communications and crowdsourced inputs, such as online reviews.

Often taken for granted, individuals and groups may have significant relationships to place, expressed in several ways (Manzo and Devine-Wright, 2013). *Place attachment* is a personal bond with a location or landscape on an emotional level as an individual or as a member of a community. It can emerge from certain physical conditions, characteristics of people, or events. *Place identity* is attachment based on emotional or symbolic meanings. The physical landscape or place becomes part of a person's self-identity, perhaps due to childhood or significant personal experiences. *Place dependence* is an attachment based on function. The value of a specific place depends on its ability to satisfy the needs or behavioural goals of an individual or group (such as a community garden).

A similar notion is *topophilia*, the affective bond between people and place or setting. A strong sense of place is often blended with the sense of cultural identity for certain groups, or may simply be an individual's love of certain aspects of a place. As Tuan described, 'diffuse as concept, vivid and concrete as personal experience, the emotional human relationship to landscape is elusive' (Tuan, 1990). The bonds that a person or community feel toward a space may contribute to both market dynamics and social conditions of a community. Interestingly, neuroscience research suggests that response to place constitutes a distinct dimension in mental processing.

Nature attachment

Early research on place attachment and meaning focused on rural, scenic, and residential settings. Yet nature is a part of preferred and meaningful urban places in terms of aesthetics, restorative effects, active use and value, attraction to the familiar, and emotional importance (Kaplan, 1995). Individuals seek natural environments as places to process personal circumstances, and consider goals and priorities. Urban forests and parks can be places of refuge, where one can recover from urban-associated mental fatigue. People become attached to peaceful, restorative green spaces that offer mental and physical respite and may come to depend on them to fulfil health needs, and so incorporate them into their self-identity (Stoner and Rapp, 2008).

Social interactions and shared cultural values of groups are also a part of how attachment forms (Kuo, 2003). The presence of urban nature contributes to greater neighbourhood satisfaction. Increased casual interactions contribute to shared sense of place. Nature's amenities encourage people to spend more time outside, creating stronger social ties and friendships with neighbours through spontaneous face-to-face encounters.

Active engagement outdoors, such as food forest projects or urban foraging, can also strengthen connection to place. Participants may experience a sense of accomplishment, greater community development, and strengthened intergenerational ties. As places gain more social significance, the interdependence between social and physical components increases, binding groups to particular places.

Emotion is central to the formation of place attachment and reinforces relationships between individuals and their environment. Childhood, particularly middle childhood, appears to be a formative time for place attachment. Feelings of connection or belonging initiated at an early age tend to become stronger in later years (Morgan, 2009). Adult remembrance of childhood place can invoke vivid memories and emotional connection, based on experiences of play, adventure, and freedom. Memories from childhood may thus be particularly meaningful.

Sites of loss or tragedy often emerge as places of attachment and self-identity, such as former battlefields or other sites associated with pain or grief. A study of community-based memorials created by victims of the 11 September 2001 terrorist attacks in New York

City ('9/11') found that memorial locations served three core social functions – a place to remember and honour victims, a location for special tribute events, and a sacred space (Svendsen and Campbell, 2010).

Relationships with place can lead to heightened sense of environmental and social responsibility. Stewardship for both public and private lands are acts of caring. Projects within meaningful natural environments can boost volunteer rates for environmental stewardship. People who experience an emotional affinity with nature and perceive natural environments as restorative are more likely to protect natural spaces and engage in pro-environment activities (Zelenski et al., 2015). Frequency of visitation to a natural area can also increase place identity as well as sense of environmental responsibility. At times such connections can be problematic for managers, as people protest change in cherished landscapes, such as the removal of hazard trees.

Crime and safety

Acknowledging potential dis-services is important for effective resource planning and management (see Chapter 12). In some communities metro nature may be implicated as a setting or screen for criminal activity. A limited number of studies addresses the relationship between urban vegetation and crimes, aggressive behaviour, and safety. Findings are not conclusive and even may appear conflicting, yet there are patterns within. Both frequency and type of crime are influenced by the physical characteristics of a space or community. Nature affects the local sociology of security, fear, and crime (Kondo et al., 2017).

Figure 5.4 Civic stewardship of the urban forest can increase forest health, enhance meaning of landscapes, and make communities more secure

Source: Kathleen Wolf

Aggression, violence, and crime

Residents of lower income housing in Chicago reported less graffiti, vandalism, and littering in outdoor spaces containing trees and grass than in more barren spaces. Incidents of social disruption and incivilities, such as the presence of noisy individuals, loitering strangers, and illegal activity, were also less in planted areas. Women who had trees and grass cover outside their apartments reported significantly less aggression against their partners than did those living in un-landscaped areas; for all residents, 25 percent fewer incidents of domestic violence were reported for landscaped homes compared to those having barren surroundings (Kuo, 2003).

Two years of police data on property and violent crimes within the same public housing communities showed fewer total crimes for buildings having greener surroundings (while including other crime predictors in the analysis). Buildings having high levels of vegetation recorded 52 percent fewer total crimes, 48 percent fewer property crimes, and 56 percent fewer violent crimes than buildings with low levels of vegetation (Kuo, 2003).

A study of residential subdivisions in Florida found that the more abundant the vegetation around a house, the less frequently property crimes occurred. Another study of residential properties in Portland, Oregon, found that trees in the public right-of-way were generally associated with a reduction in crime. The effect of trees on crime rates on house lots was mixed; smaller, view-obstructing trees tended to increase crime, whereas larger trees reduced crime (Donovan and Prestemon, 2012).

Fear and visibility

Safety can be judged objectively, as measured by facts and reports, and subjectively as personal perceptions and inferences (Nasar and Jones, 1997). Perceptions influence behaviour and people will avoid places they associate with personal risk. Large shrubs, underbrush, and dense woods can diminish visibility and view distances. Impressions of crime likelihood (irrespective of actual crime rates) can lead people to choose to not enter public spaces, retreat within their homes, and cease on-street socializing.

Managers must consider how to integrate public safety into the planning and management of urban parks, forests, and green spaces. The practices of crime prevention through environmental design (CPTED) rely upon a combination of built, social, and administrative strategies to reduce the circumstances that precede criminal acts. Some law enforcement guidelines focus only on the offender, but CPTED integrates physical environment influences that enable or deter crime.

Vegetation can be retained and managed to reduce perceived and actual risk. A site can support both trees and visibility. A more open understory that provides adequate sight lines increases perceived safety in green spaces. This does not require a landscape devoid of understory, but rather suggests that managers should be sensitive to where they place and how they manage vegetation, in response to personal safety concerns.

Caring for place

Crime behaviour is the result of a complex blend of social, genetic, and environmental factors. Direct interventions (such as more police patrols, or higher arrest rates) are strategies that may first come to mind. Research indicates that metro nature management is another effective strategy (Kondo et al., 2017). Physical features and layout influence patterns of informal contacts among users and residents.

Defensible Space is the notion that vital, well-used spaces aid the development of the neighbourhood social ties that can 'defend' a place against crime (Kuo, 2003). The presence of trees can influence the extent to which residents use and 'take ownership' of residential outdoor spaces, and discourage crime perpetrators. Well-maintained vegetation, including tree canopy, can act as a territorial marker and cue to care, suggesting that inhabitants pay attention to their home territory and that an intruder would be noticed and confronted. Active community garden sites or adopt-a-park projects send similar crime prevention signals.

Neatness counts! Signals of poor maintenance and neglect (such as litter and graffiti) decreases perceived security in urban parks and neighbourhoods. Positive social messages are conveyed by well-tended, orderly settings. A disorderly environment sends messages that no one cares or will challenge crimes within, thus increasing residents' general fear, weakening community controls, and inviting criminal behaviour. Cleaning up vacant lots and overgrown properties, a transformation of neglected space, has positive affects in reducing vandalism, burglaries, and gun crimes.

Mental state and civility

Positive social interactions are dependent in part on the mental health of individuals. Urban lifestyles include many irritants – such as crowding, high temperatures, and high levels of noise. Such conditions may compromise a person's mental capacity to cope and function (Bratman et al., 2012), and are linked to increased aggression and violence.

Attention restoration theory

The information processing demands of everyday life – e.g. traffic, phones and texting, difficult tasks at work, and complex critical decisions – all take their toll on mental capacities. The resulting mental fatigue is characterized by inattentiveness, irritability, and short attention span. Chapter 6 of this handbook introduces attention restoration theory (ART; Kaplan, 1995). The theory describes how nearby nature settings or features can enable attentional recovery and restore cognitive function.

Studies demonstrate links between green spaces and higher performance on attentional tasks in public housing residents, AIDS caregivers, cancer patients, college students, prairie restoration volunteers, and employees of large organizations (Mantler and Logan, 2015). In one study, participants showed a 20 percent improvement on attention performance after a 50-minute walk in nature, versus a highly built downtown setting. Studies of children link access to urban parks and green outdoors spaces with reduced attention-deficit hyperactivity disorder (ADHD) symptoms (Taylor and Kuo, 2006).

Attention fatigue can be experienced when living in a dangerous setting, with a difficult person, or without enough resources to meet one's needs. Mental fatigue may be compounded by health or family concerns. The result is compromised mental health, not only in the clinical sense, but in reduced capacity to avoid impulsive decisions and identify appropriate solutions. If a person is unable to restore their mental coping capacity they may not be patient in their interactions with loved ones, or may be less able to process the important information needed for good choices.

Social interaction effects

Effective information processing plays a central role in managing social situations, especially in avoiding potential conflicts. In problematic social situations, it takes more reasoning and effort to engage in solution-oriented behaviour. Three psychological factors – impairments in cognitive processing, irritability, and impulsivity – have been linked to aggression. Mental fatigue can contribute to outbursts of anger and even violence. Having nearby nature for respite and restoration can contribute to better coping by individuals, and more civil interactions in communities.

Metro nature and equity

The scientific evidence pointing to the health and wellness benefits of exposure to urban green space has expanded greatly in recent years (Wolf, 2016; Wolf and Robbins, 2015). Yet there are racial, ethnic, and socio-economic disparities in access to metro nature in some cities. Discrepancies between socioeconomic groups and the locations of public lands began as early as the nineteenth century in industrialized cities. Political, social, and economic processes over the years have frequently put less affluent groups at a green space disadvantage.

Amenity access

Urban park access generally varies across US cities. In the early 2000s, only 33 percent of residents in LA lived within a quarter of a mile of a park, compared to 97 percent in Boston and 91 percent in New York (Trust for Public Land, 2004). In addition, advances in remote sensing analysis have revealed disparities in the density and quality of urban forest canopy cover associated with neighbourhood income levels, race, and ethnicity. Figure 5.5 shows differences in canopy distribution across communities in one US state (in 2008).

Figure 5.5 Spatial distribution of tree canopy in six cities in the state of Illinois, USA

Source: Reprinted from Zhou and Kim (2013), with permission from the publisher

The study found significantly less canopy in areas populated by a greater number of African Americans in all but one city. And neighborhoods having more residents with college degrees and higher household income were often associated with more tree canopy (Zhou and Kim, 2013).

Larger green spaces provide the greatest health benefits, but large parks also appear to be less prevalent in more deprived neighbourhoods. Having fewer, smaller parks can affect the level of benefit. High neighbourhood density and persons-per-park-acre counts can lead to congestion and overuse, diminishing benefits. In a Baltimore study, areas with higher persons-per-park-acre were more frequently found in predominately African American neighbourhoods versus white neighbourhoods (Boone et al., 2009).

Vacant lots mar some low-income neighbourhoods, and some communities have programs to clean up and repurpose unused properties. One study found that walking near 'greened' vacant lots significantly decreased heart rates, suggesting that remediation of neighbourhood blight may diminish physical stress and improve health (South et al., 2015). Other studies have found reduced crime and increased resident perceptions of security following vacant lot improvements, in addition to increases in market value for adjacent properties.

Health effects

Researchers usually control for socio-economic status when studying nature effects, as there are other health advantages in affluent neighbourhoods (e.g. better nutrition and health care access). Green surroundings may actually reduce socio-economic health inequalities. Testing the relationship, one study of people with low income and high levels of residential greenery had similar mortality rates to people having higher socio-economic status. However, when low income was associated with little surrounding green space, higher mortality rates were found (Mitchell and Popham, 2008).

Nature-based positive effects may actually be amplified in lower-income, inner-city communities. Poverty is related to poorer health across the human life cycle, from children to elders. Crowding, noise pollution, and the increased threat of crime in poorer urban neighbourhoods contribute to chronic mental fatigue. As described earlier, nature's restorative benefits can help people regain mental capacity and be better prepared to cope. Contact with nature by those in deprived urban neighbourhoods has been linked to residents' lower perceived stress and improved physiological stress recovery (Roe et al., 2013).

Displacement concerns

Addressing inequalities by introducing new parks and tree plantings must be done carefully (Wolch et al., 2014). As described earlier, proximity to parks and trees in streetscapes are correlated with increased property values. Neighbourhoods near large tracts of open lands or parks are frequently unaffordable to low-income households, reinforcing the trend of inequity in access to neighbourhood green space. Similarly, newly built parks in lower socio-economic communities may contribute to gentrification by increasing property values and displacing the less affluent as their neighbourhoods become less affordable. Programs designed to offer free or low-cost trees for homeowners in cities around the US may exacerbate inequalities, as low-income residents are less likely to participate in the programs (Donovan and Mills, 2014). Such trends are complex, and with careful planning communities can work successfully to green their neighbourhoods, plant trees, and close the metro nature and health equity gap.

Cultures and communities

Having community gardens and tree plantings contributes to the quality of life of minority ethnic groups, by creating places for culturally significant gardens and planting – strengthening a sense of community and tradition (Shimada and Johnston, 2013). Yet, it is important to recognize that there are potential differences between various ethnic and cultural groups concerning their preferences for nature experiences. Culturally-dominant ideals of nature often are expressed in planning and design, potentially overlooking preferences of minority users. Participatory planning and design strategies are important, including direct outreach to diverse user groups (see also Chapter 17).

Various cultural groups may differ in their preferences for the character of nature in parks and activities during visits (Ho et al., 2005). African Americans may prefer developed recreation facilities over less managed, naturalized settings. In a study of park users in Chicago, Caucasian visitors were twice as likely as Asian visitors to use the park alone, and were also less likely to visit a park with extended family or with an organized group. Large shaded picnic areas, play equipment, water features, sanitary facilities, and open-air vendors or cafes increase attractiveness of parks for Hispanic users.

A green space between racially or socio-economically distinct neighbourhoods can act either as a barrier, creating a boundary, or a magnet and place of social inclusion for different cultural groups. With intentional design and planning, green spaces can further integrate different ethnic and racial groups. Some factors that actively encourage park use by a range of groups include community and neighbourhood group activities, visible trails and play areas on the perimeter to attract people further into the park, a range of facilities to attract diverse users, programs that reflect cultural traditions, and responsive park maintenance and management.

At a more conceptual level, the residuals of tree and nature symbolism may enter the interactions of green space users. Trees can become symbolic representations that trace territory and memory, even conflict (Shimada and Johnston, 2013). Trees embody expressions of age, resilience, and productivity. Forests and trees represent popular stories and meanings in the cultures of many nations, in the east and west. Trees are even present in the discourse of conflict. In the ongoing Israeli–Palestinian conflict, landscapes of pine and cypress grow alongside olive groves, symbolizing counter claims to the land. Forest destruction has often come to symbolize the devastation of war, such as the cleared battlefields of Europe in World War I or the US military's chemical destruction of extensive forests during the Vietnam War. Recognizing imbedded associations is an important part of participatory design for new tree plantings in parks, gardens, and streetscapes.

Conclusion: Nature and resilience

This chapter started with the more straightforward expression of value, the market-based monetization of properties and goods. Yet there are other complicated and meaning-filled values associated with outdoor spaces. Values emerge from direct experience of metro nature, but are also formed during the social interactions that take place within spaces and across time. This chapter offers interpretations of both early and recent notions about the social aspects of nearby nature experiences, including trees and forests.

Some of the most recent insights about social aspects are addressing resilience. The term, similar to sustainability, is defined in many ways, depending on context and spatial scale. Fostering resilience in the face of uncertainty and risk has become the focus of policy and decision makers. Urban areas have become resilience laboratories as they are centres of both impact and innovation. Landscapes and natural resources are increasingly valued as buffers that protect human populations against chronic stress or acute shock – coastal wetlands to mitigate storm surges, green infrastructure to manage water quality, or urban forest canopy to reduce extreme heat effects. But these biophysical processes and outcomes are just one aspect of resilience goals. Community-based responses also contribute to relief and recovery from threats.

Event response

'Urgent biophilia' refers to intentional, purposeful contact with green space that people seek to summon and demonstrate resilience in the face of a crisis (Tidball, 2012). The affinity that humans may have for nature (biophilia), the process of remembering valued landscapes, and the urge to create restorative environments often emerges in the face of challenges. People, as individuals and as communities, often seek engagement with nature, working together to restore the built environment and sustain memory, including tree plantings, parks, and memorials.

Stewardship activity, in particular, can foster community resilience (Svendsen et al., 2014). Local, community-based organizations, such as environmental, community garden, and urban forestry non-profit groups, can activate their established networks and capacities to respond in times of crisis, becoming 'first responders' following a disaster. Such groups have learned to adaptively address multiple needs in their communities. In times of disturbance the persistent, trusted, and networked relationships of community-based organizations can be mobilized to meet immediate community needs. Social resilience depends upon connectedness and innovation, and day-in and day-out nature-based experiences and interaction encourage locally relevant and resourceful response.

Resilient communities

At the individual scale, major disruptions can bring on attention overload. Nature encounters offer the space and cognitive support to review one's circumstances, consider solutions, and make choices that can better the situation of self and loved ones. The nature places where a person has developed a sense of attachment may be particularly supportive as a person seeks to understand and cope. And being within spaces where one feels secure and is not fearful of personal harm also support individual and group activities of recovery.

From casual interactions to organized events, urban green spaces can be the neutral ground where people gather and informal relationships can take form, including opportunities for people from different backgrounds and diverse cultural origins to connect socially. Based on the character of the space and the program of activities that happen within, community building and increased social capital can emerge, particularly if people share work on a project or goal. A sense of connectedness fosters greater trust and cooperation among individuals, potentially leading to greater community resilience in the face of adversity.

It's been said that change is the new normal. In the face of (sometimes rapid) social, political, economic, emotional, and psychological change, green spaces provide respite on many levels. The social cohesion that emerges in both passive nature encounters and stewardship activities enables emergent social systems, that can become the social infrastructure of resilience. As communities face both challenge and stability, trees, forests, and metro nature generally, are ever more important in our urbanizing world. These resources support the social aspects of economic value, social cohesion, place attachment, safety and security, and mental and physical health, and should include attention to diversity and equity.

Figure 5.6 Trees in cities provide many social benefits – only possible if the urban forest is managed and sustained by local policy and programs

Source: Kathleen Wolf

Acknowledgements

Writing support was provided by the USDA Forest Service National Urban and Community Forestry program; the USDA Forest Service, Pacific Northwest Research Station; and the TKF Foundation, Nature Sacred project. Some content of this chapter was adapted from the Green Cities: Good Health web site literature summaries; thanks is extended to the project's author team: Katrina Flora, Mary Ann Rozance, Sarah Krueger. The reference list contains a subset of publications supporting ideas and examples in this chapter; contact the author for a fully cited version of the chapter.

References

Alaimo, K., Reischl, T.M., Allen, J.O. (2010) 'Community gardening, neighborhood meetings, and social capital', *Journal of Community Psychology*, vol. 38, issue 4, pp. 497–514.

Bellair, P.E. (1997) 'Social interaction and community crime: examining the importance of neighbor networks', *Criminology*, vol. 35, issue 4, pp. 677–703.

Boone, C.G., Buckley, G.L., Grove, J.M., Sister, C. (2009) 'Parks and people: an environmental justice inquiry in Baltimore, Maryland', *Annals of the Association of American Geographers*, vol. 99, issue 4, pp. 767–87.

Bratman, G.N., Hamilton, J.P., Daily, G.C. (2012) 'The impacts of nature experience on human cognitive function and mental health', *Annals of the New York Academy of Sciences*, vol. 1249, pp. 118–36.

de Vries, S., van Dillen, S.M., Groenewegen, P.P., Spreeuwenberg, P. (2013) 'Streetscape greenery and health: stress, social cohesion and physical activity as mediators', *Social Science and Medicine*, vol. 94, pp. 26–33.

Dinnie, E., Brown, K.M., Morris, S. (2013) 'Community, cooperation and conflict: negotiating the social well-being benefits of urban greenspace experiences', *Landscape and Urban Planning*, vol. 112, pp. 1–9.

Donovan, G.H., Butry, D.T. (2010) 'Trees in the city: valuing street trees in Portland, Oregon', *Landscape and Urban Planning*, vol. 94, pp. 77–83.

Donovan, G.H., Mills, J. (2014) 'Environmental justice and factors that influence participation in tree planting programs in Portland, Oregon, US', *Arboriculture and Urban Forestry*, vol. 40, issue 2, pp. 70–77.

Donovan, G.H., Prestemon, J.P. (2012) 'The effect of trees on crime in Portland, Oregon', *Environment and Behavior*, vol. 44, issue 1, pp. 3–30.

Elands, B., Peters, K., de Vries, S. (in press) 'Promoting social cohesion and social capital – increasing wellbeing', in M. van den Bosch, W. Bird (eds), *Nature and Public Health*, Oxford University Press, Oxford, UK.

Ho, C., Sasidharan, V., Elmendorf, W., Willits, F.K., Graefe, A., Godbey, G. (2005) 'Gender and ethnic variations in urban park preferences, visitation, and perceived benefits', *Journal of Leisure Research*, vol. 37, issue 3, pp. 281–306.

Holt-Lunstad, J., Smith, T.B., Baker, M., Harris, T., Stephenson, D. (2015) 'Loneliness and social isolation as risk factors for mortality: a meta-analytic review', *Perspectives on Psychological Science*, vol. 10, issue 2, pp. 227–37.

Kaplan, S. (1995) 'The restorative benefits of nature: toward an integrative framework', *Journal of Environmental Psychology*, vol. 15, issue 3, pp. 169–82.

Kaźmierczak, A. (2013) 'The contribution of local parks to neighbourhood social ties', *Landscape and Urban Planning*, vol. 109, issue 1, pp. 31–44.

Kondo, M.C., Han, S., Donovan, G.H., MacDonald, J.M. (2017) 'The association between urban trees and crime: evidence from the spread of the emerald ash borer in Cincinnati', *Landscape and Urban Planning*, vol. 157, pp. 193–9.

Kuo, F.E. (2003) 'The role of arboriculture in a healthy social ecology', *Journal of Arboriculture*, vol. 29, issue 3, pp. 148–55.

Kweon, B.S., Sullivan, W.C., Angel, R. (1998) 'Green common spaces and the social integration of inner-city older adults', *Environment and Behavior*, vol. 30, issue 6, pp. 832–58.

Mantler, A., Logan, A.C. (2015) 'Natural environments and mental health', *Advances in Integrative Medicine*, vol. 2, issue 1, pp. 5–12.

Manzo, L.C., Devine-Wright, P. (2013) *Place Attachment: Advances in Theory, Methods and Applications*, Routledge, Abingdon, UK.

Mitchell, R., Popham, F. (2008) 'Effect of exposure to natural environment on health inequalities: an observational population study', *The Lancet*, vol. 372, issue 9650, pp. 1655–60.

Morgan, P. (2010) 'Towards a developmental theory of place attachment', *Journal of Environmental Psychology*, vol. 30, pp. 11–22.

Nasar, J.L., Jones, K.M. (1997) 'Landscapes of fear and stress', *Environment and Behavior*, vol. 29, issue 3, pp. 291–323.

Peters, K., Elands, B., Buijs, A. (2010) 'Social interactions in urban parks: stimulating social cohesion?', *Urban Forestry and Urban Greening*, vol. 9, issue 2, pp. 93–100.

Putnam, R. (2000) *Bowling Alone: The Collapse and Revival of American Community*, Simon & Shuster, New York.

Roe, J.J., Thompson, C.W., Aspinall, P.A., Brewer, M.J., Duff, E.I., Miller, D., Mitchell, R., Clow, A. (2013) 'Green space and stress: evidence from cortisol measures in deprived urban communities', *International Journal of Environmental Research and Public Health*, vol. 10, issue 9, pp. 4086–103.

Ross, C.E. (2000) 'Neighborhood disadvantage and adult depression', *Journal of Health and Social Behavior*, vol. 2, pp. 177–87.

Semenza, J.C., March, T.L. (2008) 'An urban community-based intervention to advance social interactions', *Environment and Behavior,* vol. 41, issue 1, pp. 22–42.

Shimada, L.D., Johnston, M. (2013) 'Tracing the troubles through the trees: conflict and peace in the urban forest of Belfast, Northern Ireland', *Journal of War and Culture Studies*, vol. 6, issue 1, pp. 40–57.

South, E.C., Kondo, M.C., Cheney, R.A., Branas, C.C. (2015) 'Neighborhood blight, stress, and health: a walking trial of urban greening and ambulatory heart rate', *American Journal of Public Health*, vol. 105, issue 5, pp. 909–13.

Stoner, T., Rapp, C. (2008) *Open Spaces, Sacred Places*, The TKF Foundation, Annapolis, MD.

Svendsen, E.S., Campbell, L.K. (2010) 'Living memorials: understanding the social meanings of community-based memorials to September 11, 2001', *Environment and Behavior*, vol. 42, issue 3, pp. 318–34.

Svendsen, E.S., Baine, G., Northridge, M.E., Campbell, L.K., Metcalf, S.S. (2014) 'Recognizing resilience', *American Journal of Public Health*, vol. 104, issue 4, pp. 581–3.

Svendsen, E.S., Campbell, L.K., Fisher, D.R., Connolly, J.J., Johnson, M.L., Sonti, N.F., Locke, D.H., Westphal, L.M., Fisher, C.L., Grove, M., Romolini, M., Blahna, D.J., Wolf, K.L. (2016) *Stewardship Mapping and Assessment Project: A Framework for Understanding Community-Based Environmental Stewardship*, General Technical Report NRS-156, US Forest Service, Newtown Square, PA.

Taylor, A.F., Kuo, F.E. (2006) 'Is contact with nature important for healthy child development? state of the evidence', in C. Spencer, M. Blades (eds), *Children and Their Environments: Learning, Using and Designing Spaces*, Cambridge University Press, Cambridge UK.

Tidball, K.G. (2012) 'Urgent biophilia: human–nature interactions and biological attractions in disaster resilience', *Ecology and Society*, vol. 17, p. 25.

Trust for Public Land (2004) *No Place To Play: A Comparative Analysis Of Park Access In Seven Major Cities*, The Trust for Public Land, San Francisco, CA.

Tuan, Y.-F. (1990) *Topophilia: A Study of Environmental Perception, Attitudes, and Values*, Columbia University Press, New York.

Wolch, J.R., Byrne, J., Newell, J.P. (2014) 'Urban green space, public health, and environmental justice: the challenge of making cities "just green enough"', *Landscape and Urban Planning*, vol. 125, pp. 234–44.

Wolf, K.L. (2005) 'Business district streetscapes, trees, and consumer response', *Journal of Forestry*, vol. 103, issue 8, pp. 396–400.

Wolf, K.L. (2007) 'City trees and property values', *Arborist News*, vol. 16, issue 4, pp. 34–6.

Wolf, K.L. (2016) 'Green Cities: Good Health', www.greenhealth.washington.edu, accessed 23 June 2016.

Wolf, K.L., Robbins, A.S.T. (2015) 'Metro nature, environmental health, and economic value', *Environmental Health Perspectives,* vol. 123, issue 5, pp. 90-8.

Wood, L., Giles-Corti, B. (2008) 'Is there a place for social capital in the psychology of health and place?', *Journal of Environmental Psychology*, vol. 28, issue 2, pp. 154–63.

Zelenski, J.M., Dopko, R.L., Capaldi, C.A. (2015) 'Cooperation is in our nature: nature exposure may promote cooperative and environmentally sustainable behavior', *Journal of Environmental Psychology*, vol. 42, pp. 24–31.

Zhang, J.W., Piff, P.K., Lyer, R., Koleva, S., Keltner, D. (2014) 'An occasion for unselfing: beautiful nature leads to prosociality', *Journal of Environmental Psychology*, vol. 37, pp. 61–72.

Zhou, X., Kim, J. (2013) 'Social disparities in tree canopy and park accessibility: a case study of six cities in Illinois using GIS and remote sensing', *Urban Forestry and Urban Greening*, vol. 12, issue 1, pp. 88–97.

6

IMPACTS OF URBAN FORESTS ON PHYSICAL AND MENTAL HEALTH AND WELLBEING

Matilda van den Bosch

Introduction: A changing disease scenario in a changing environment

'Health is a state of complete physical, mental and social well-being and not merely the absence of disease or infirmity', according to the World Health Organization (WHO, 1948). This holistic interpretation of human health is particularly important to be aware of in a world where our health and wellbeing is to an increasing extent determined by environmental, social, and contextual factors. The last century's focus on bio-pathological aspects of health has indeed contributed to substantial progress in medical science and significant improvements in treatment of many infectious diseases. On the other hand, both human beings and the environment are now paying a price for this development. The environment has become polluted by pharmaceuticals, causing imbalance and disruption in several ecosystems, and by overuse of antibiotics we are today facing an escalating challenge of antimicrobial resistance. As we depend on healthy ecosystems for our survival and wellbeing this environmental degradation threatens our own health.

Moreover, the globally changing circumstances, including urbanisation and climate change, have altered the global disease burden. According to the latest update of the *Global Burden of Disease* (GBD) study, apart from climate change and environmental threats, the main health risks today in terms of disability-adjusted life years (DALYs) are high systolic blood pressure and smoking, rather than infectious diseases which were previously dominating. Other risk factors that have increased substantially are high body mass index (BMI) and high fasting plasma glucose. Neither of these risks are well treated with pharmaceutical interventions, but must be approached from a contextual perspective, simply speaking – by making it easier for people to live healthy lives. This can be supported by healthy environments.

The majority of disease burden is now also attributable to so called *non-communicable diseases* (NCDs) (i.e. chronic diseases such as cardiovascular diseases, diabetes, obesity, cancer, and mental disorders). For example, in Europe unipolar depression is today the leading chronic disease. NCDs disproportionally affect low- and middle-income countries, due to unequal distribution of healthy conditions and environments as well as health resources (Vos et al., 2015).

Common risk factors for many NCDs are chronic stress, physical inactivity, and social isolation. Thus, by preventing these risk factors, better public health and increased wellbeing can be achieved.

This chapter starts with outlining the most important theories regarding associations between natural environments and health, often attributed to restoration from mental fatigue and stress recovery. Secondly, the importance of urban green spaces for physical activity and social interactions is explained. Following a section on how biodiverse urban greenery can promote healthy immune system development, the chapter summarises how urban ecosystem services contribute to human health. Finally, the chapter concludes that the relations between urban green and human health are multiple, interrelated, and complex and that in order to optimally plan for healthy and liveable cities we need to find novel ways of collaborating across science, policy, and practice.

How nature can reduce stress and mental fatigue – the theories

The initial theories linking natural environments (such as urban forests) and health focused much on nature's potential for recovery from stress or attentional fatigue. The leading theoretical frameworks for research on restorative environments are the *stress reduction theory* (SRT) and *attention restoration theory* (ART).

SRT postulates that natural features and patterns elicit rapid, affective reactions which occur without conscious processing (Ulrich, 1983). These reactions initiate physiological mobilisation and subsequent stress recovery. The theory is supported by empirical studies, which have confirmed that viewing natural settings can provide more efficient stress recovery than viewing built scenes, as indicated by, for example, a stronger reduction in muscle tension, skin conduction, saliva cortisol, blood pressure, and heart rate, and more positive changes in self-reported affect (Kuo, 2015).

ART, on the other hand, emphasises the importance of cognitive mechanisms. A core assumption of ART is that people only have a limited capacity to direct their attention to something that is not in itself captivating. The mechanism necessary to focus on things that require cognitive effort ('directed attention') becomes depleted with prolonged or intensive use, according to the theory (Kaplan and Berman, 2010). Entering a situation that does not require directed attention permits a fatigued person to rest and replenish central executive function.

Natural environments tend to provide settings which are fascinating but do not require directed attention and therefore contribute to rest and recovery. The theory is supported by findings demonstrating that people who are in particular need of attentional restoration, rather than emotionally stressed, have a preference for natural over urban settings (Staats et al., 2003).

Later theories claim that the visual input of natural scenes have a specific effect on our wellbeing, mediated by a calming effect on the brain. Swedish scholar Caroline Hägerhäll investigated the occurrence of *fractal patterns* (repetitive patterns over different scales, for example in ferns, tree silhouettes, or plants) in nature (Figure 6.1). She could demonstrate with Electroencephalography (EEG) that by watching natural fractal patterns, which display a certain repetitive dimension, the brain enters an awake but resting mode (Hägerhäll et al., 2008).

The urban environment, stress, and green spaces

Although cities are centres of wealth, culture, and innovation they also have specific health challenges by adverse environmental exposures, such as high pollution levels, noise and urban heat islands. Additionally, mental disorders are more common in urban than in rural areas (Peen et al., 2010), partly due to a stressful urban environment and lifestyle (Figure 6.2).

Figure 6.1 Repeating fractal patterns in a cauliflower

Source: ElbtheProf, CC BY 2.0

Figure 6.2 Many people perceive the city as a stressful environment

Source: Abbilder, CC BY 2.0

Particularly, mental disorders are associated with *chronic stress*. Stress is a natural, physiological response to any challenging condition, to which the body needs to adapt. It works through an intricate system of neuro-sensory reactions initiating hormone cascades and feedback loops. These cascades and loops affect the bodily organs to turn into an 'alarm mode', adequate for dealing with a physical stressor. The alarm mode includes increased heart rate and blood pressure, decreased salivation, and direction of blood flow from the brain to the muscular system. However, if the challenge is not conquered or there is no option for recovery, the stress reactions are maintained and we enter a chronic stress state. This implies a wear-and-tear of the body, eventually leading to exhaustion and mental disease.

From this it is clear that by preventing stress and promoting less stressful lifestyles, much can be won for mental and public health.

It is suggested that the disconnection from nature is one of the reasons for the higher prevalence of stress and mental disorders in cities (Peen et al., 2010). Urban green spaces have the potential to reduce stress, by offering an escape from daily hectic. Entering a park can give a feeling of being in another world, a world without the noise and overload of demanding sensory input associated with city life. Being in a natural environment can provide a connection to our inner selves and our inherent connection to nature. This is often perceived as soothing for the mind, and hassles of work or other earthly demands fade.

Having access to and visiting urban parks provide opportunities for recreation and stress relief, but also views of nature from, for example, an office window may improve wellbeing by offering restorative visual input (Brown et al., 2013). Also sounds of nature are stress-reducing (Annerstedt et al., 2013).

Intriguing recent research findings have shown that exposure to nature may reduce depression symptoms by direct functional effects on the brain (Bratman et al., 2015). We also know that the brain displays functional and structural differences between urban and rural populations, something that can explain the increased vulnerability to social stress and higher prevalence of depression and schizophrenia among people living in cities (Lederbogen et al., 2013). By moving to greener areas you also seem to reduce your risk for developing mental disorders (Annerstedt et al., 2015).

Providing large, accessible green spaces and an urban green infrastructure can, as a consequence of the above described findings, be a cost-efficient strategy to prevent chronic stress reactions and mental disorders on a population level, with a substantial impact on the burden of disease. Thus urban planning turns into an active public health intervention.

Physical inactivity is a risk factor that may be counteracted by urban green spaces

Apart from the stress-reducing effect, another suggested pathway between urban green and health is increased *physical activity*.

Physical inactivity is a major obstacle to population health. Globally, around 31 per cent of adults aged 15 and over were insufficiently active in 2008 and the issue is escalating with increasing urbanisation. Approximately 3.2 million deaths each year are attributable to insufficient physical activity (WHO, 2010). Several environmental factors are recognised as contributing to physical inactivity in cities, such as high-density traffic and lack of parks and sidewalks.

Human beings are biologically adapted to a physically active lifestyle, only occasionally remaining seated (Figure 6.3). Our current sedentary lifestyles contribute to an epidemic of obesity, also among children, high prevalence of cardiovascular morbidity, and many other serious diseases, such as diabetes and cancer.

During the twenty-first century, renewed interest in accessible urban open spaces has been sparked by its role in encouraging active living. For example, the World Health Organization (WHO) recommends that urban citizens should have access to green spaces of at least 1 ha size within 300m distance, considered as an *urban green space indicator* (UGSI) for public health (Annerstedt van den Bosch et al., 2016).

Figure 6.3 Human beings are not biologically adapted to urban lifestyles with high consumerism, cheap food, and sedentary lifestyles

Source: Christopher Dombres, CC BY 2.0

It is easy to imagine that open, green spaces would encourage physical activity in a city. In an urban woodland or a park the air quality is better, the thermal conditions more comfortable, and traffic is less of an issue. All these factors promote active transportation and contribute to a better effect of the physical performance. It is claimed that '*green exercise*' may be even more beneficial than indoor physical activity (Mitchell, 2013).

Particularly for children and adolescents, access to urban green spaces encourages physical activity and decreases levels of overweight and obesity (Potwarka et al., 2008). Green spaces also facilitate 'active play', which is an important source for children's development. Natural features (i.e., trees, rocks, and water) in parks attract children and youth of all ages (Fjørtoft and Sageie, 2000).

Also adults living closer to urban green spaces or whose closest green space is larger, report more physical activity (Astell-Burt et al., 2013). Causality has been demonstrated in a couple of longitudinal studies (Giles-Corti et al., 2013; Sugiyama et al., 2013).

Obviously not any green space encourages physical activity. The quality and amenities offered play a certain role, where for example walking paths, running tracks, and streetlights can be important for use of the park (Cohen et al., 2006).

Although research as well as common sense tell us that people become more active by having access to urban green spaces, the empirical evidence is somewhat inconsistent (Foster et al., 2009). Several factors are interplaying, and while accessibility can encourage parts of the population it may not be true for everyone.

Social capital growth in urban green spaces

Social isolation is a risk factor for mortality of the same magnitude as smoking (Pantell et al., 2013) (Figure 6.4). Some predictors of social isolation may be obvious, such as being un-married or unemployed, while others may be less apparent, such as environmental influence.

Figure 6.4 Social isolation and feeling unconnected to the neighbourhood or other parts of society is a major risk factor for disease development

Source: KellyB, CC BY 2.0

When discussing social cohesion in an environmental context the terms 'sense of community' and 'social ties' are often used as equivalents. Neighbourhood open spaces, particularly green spaces, provide opportunities for people to meet other people and also interact interculturally. In general, green spaces and public parks seem to strengthen social ties of the neighbourhood (Sugiyama et al., 2008). It is possible that a non-demanding natural environment, to which all inhabitants in a neighbourhood can develop an attachment, the obstacles for social interactions across common barriers are diminished. By developing a sense of community, social ties can be bound around a mutual, but external interest.

The topic of urban green spaces' relevance for social interactions is further elaborated on in Chapter 5.

The value of urban green spaces for development and learning

The particular value of natural environments for children has already been elaborated on in the context of play and physical activity. However, connection to nature is also important for behavioural and cognitive development.

Already from a foetal and infancy state, children's development seems to benefit from green exposure. Several studies have shown improved pregnancy outcomes and increased birth weight of children borne by mothers in greener urban environments (Donovan et al., 2011). Like with many other effects, this benefit seems to be most pronounced in less wealthy populations (Dadvand et al., 2012a).

Children growing up in the countryside with access to animals and natural features become aware of ecosystem functions and often display a higher environmental connectedness and awareness as adults (Hacking et al., 2007). By realising their own part in the ecological system, an evident relation to nature can be created (Louv and Hogan, 2005).

The author Richard Louv coined the expression '*nature-deficit disorder*', which may be part of the explanation of the high prevalence of behavioural disorders among children, such as attention deficit hyperactivity disorder (ADHD) and other neurodevelopmental diseases (Louv and Hogan, 2005). Experiences from parents and research show that exposure to nature seems to improve behavioural development and relieve symptoms of ADHD (Amoly et al., 2014). This is particularly important in cities, where the disconnection to nature is obvious and the stress levels are high, exposing already vulnerable children to a risk of developmental disturbances.

Harmful environmental exposure during childhood is detrimental for later life health development (Gluckman et al., 2007). Similarly, healthy environments during childhood could support a healthy development. There is clear evidence that children's mental development and cognitive capacity is supported by access and exposure to urban green environments (Dadvand et al., 2015; Amoly et al., 2014). Therefore, investments in green neighbourhoods and urban green infrastructure where children dwell and go to school is unreservedly important in any urban policy making. This will have a long-term impact throughout the life course on health, wellbeing, and economy.

Immune system development is boosted by contact with nature

For a normal development of physiological functions, we require exposure to the natural environment because it is a crucial source of *microbial biodiversity* (Rook, 2013). Particularly the human immune system is dependent on biodiverse exposure in early life for providing the immunoregulatory 'software' with data, derived from the microbial environment.

Modern urban lifestyles deprive us from exposure to necessary microorganisms. Antibacterial and antifungal agents in many cleaning materials in combination with living in air-conditioned buildings with little exposure to natural organisms disrupt normal immune system development and the risk of several types of inflammatory disorder increases. In high-income countries, especially in urban environments, the incidence and prevalence of disorders that are at least partly attributable to failure of immunoregulation are escalating (Bach, 2002). Examples of such disorders are inflammatory bowel diseases (IBDs), multiple sclerosis, diabetes (type 1), and allergies.

Interacting with biodiverse natural environments increases the exposure to diverse microorganisms. For example, living on farms protects children against allergies and juvenile forms of IBDs (Radon et al., 2007); and contact with animals, in for example urban parks, provides important microbiota (Mcdade et al., 2012).

The commonly observed tendency among human infants to ingest soil (geophagy) is probably an evolutionary relic of our inherent drive to obtain large parts of our microbiota from the external, natural environment (Mulder et al., 2009).

Although biodiversity is already decreasing, also on a microorganism level, the health benefits to be derived by boosting our immune systems are substantial through contact with green environments. Human babies of today, being reared in high-income settings with minimal contact with environmental biodiversity, have a significantly increased risk of developing food allergies and other immunoregulatory abnormalities. Especially in cities, it is a duty to our children and coming generations to maximise the potential for exposure to microbiota by providing accessible and biodiverse green spaces.

Urban green spaces promote pro-environmental and pro-social behaviour

We know that climate change will have detrimental effects on both human health and the environment. The reason for the changing climate is human behaviour, the anthropogenic impact on ecosystems and climate. An up-stream solution to the issue is to apply methods of *behaviour change* in order to prompt pro-environmentalism and thereby preventing further climate change.

Pro-environmental behaviour can be defined as the propensity to take actions and decisions with environmentally-friendly impact and is commonly understood to be a consequence of attitudes and concerns related to ecosystem destruction, climate change, and other adverse ecological impacts of human activity (Figure 6.5). Environmentally aware actions by people can substantially reduce carbon emissions.

Like many social behaviours, pro- environmental behaviour can be automatically induced by external stimuli, particularly by experiencing natural environments (Zelenski et al., 2015).

In urban settings where the incitements for unsustainable living can be many – for example car-dependent transportation, consumerism, and easily accessible fast food from unsustainable food systems – providing access to nature is important both to increase the connection to nature and to promote awareness of ecosystems and the importance of sustainable living. Childhood experiences in nature appear to enhance adult environmentalism (Wells and Lekies, 2006), a perturbing fact as children of today spend less and less time outdoors. Planning for general green space accessibility could eventually prove to be a cost-efficient nature-based solution with positive impact on both urban environments and public health.

Figure 6.5 Recycling is an example of pro-environmental behaviour. This woman removes cardboard from a Temple in Hanoi, Vietnam

Source: Dennis Jarvis, CC BY-SA 2.0

Urban ecosystem services and human health

As mentioned above, climate change and ecosystem degradation are major threats to global health. A puzzling, but seemingly general trait of humankind, is our propensity not to realise what we have until it's lost. Unfortunately, we are now facing the results of our inability to realise what we had in terms of healthy ecosystems which provided life-sustaining services.

Chapter 4 of this book outlines the concept of ESS and the various benefits to be derived from healthy ecosystems. This section will highlight the health gains we can make by increasing the awareness of urban ESS and exploiting those in a resilient manner.

Heat-related disorders and increased mortality during heat waves are becoming increasing threats (Basagaña et al., 2011), especially in cities, due to the *urban heat island* (UHI) effect (Oke, 1973). The urban heat island is a phenomenon based on the use of impervious materials that absorb and store solar heat radiation and the relative lack of vegetated land (Taha, 1997).

Trees and vegetation help cool the urban environment through both shading and evapotranspiration (Taha, 1997). There is increasing evidence that urban greening can be an effective strategy for preventing UHI and adapt cities to global warming, thereby constituting an ESS of substantial relevance for public health (Bowler et al., 2010; see also Chapter 7 of this volume for more about urban forestry and climate).

Another major health threat is *urban air pollution*, with an estimated 524,000 premature deaths annually in Europe alone ($PM_{2.5}$, NO_2, ozone) (EEA, 2015) (Figure 6.6).

The evidence around green spaces' potential to reduce air pollution levels in cities is inconsistent, with some studies showing significant effects (Nowak et al., 2013), while others show no effect (Setälä et al., 2013) or even worsened pollution levels under street tree canopies (Jin et al., 2014). An important aspect in this matter is careful planning of street trees, where

Figure 6.6 Air pollution is a global threat to human health, with particularly negative impact in cities

Source: Jean-Etienne Minh-Duy, CC BY-SA 2.0

pollution may otherwise be trapped on the street level. By optimising the relation between tree canopy density and leaf area index, strategies for planting and using trees to decrease air pollution can potentially be achieved, while simultaneously providing multiple other services. Read more about urban forestry and pollution mitigation in Chapter 8.

Noise reduction is another ESS. The exposure to noise is increasing as a result of continuing urbanisation, rising traffic volumes, industries, and a decreasing availability of quiet places in cities. The range of disease burden from noise pollution is estimated to 1.0–1.6 million DALYs in the European Region (WHO, 2011). Urban green spaces may ameliorate the negative perception of noise, though the evidence is inconsistent (Dzhambov and Dimitrova, 2014). However, a well-designed urban green space, of a certain size and tree canopy cover, offers a natural soundscape environment with an opportunity to rest from the negative influence of city noise.

Altogether, millions of lives, particularly in vulnerable populations, can be saved now and in future generations by preserving urban ecosystems and maintaining the necessary services they provide.

Inequalities in health and access to green

Referring to the particular importance of green spaces for vulnerable populations, equal distribution of high-quality green spaces is important. Unfortunately, a gentrification effect has been observed, as informed people with already healthy lifestyles and more resources tend to live in greener areas, something which may also bias some of the research results. This inequality aspect may be due to segregation and intra-urban *environmental injustice* (Wolch et al., 2014). The inequity in access to high-quality green spaces in cities requires affirmative action among urban planners in collaboration with public health workers. Green spaces' potential of being so-called *equigenic environments* – disrupting the usual conversion of socioeconomic inequality to health inequality – is an important topic for urgent exploration and societal implementation.

The complex web of urban health, green spaces, science, and policy making

Modern public health has its roots in urban planning, and urban planning can partly be considered a public health tool. However, this interdependence is not always fully recognised in legislation or among urban planners, public health workers, and decision makers. Stronger acknowledgement of this relation would support the goal of healthier cities for both the environment and humankind.

This chapter has tried to introduce and categorise the various health effects of urban forests and other urban green spaces, outlining pathways like stress recovery, physical activity, social cohesion, as well as mechanistic, biological and ecosystem explanations and services. However, this must actually be considered a simplified and even flawed way of approaching the topic. Human health and interactions with natural environments are inherently complex, dynamic, and uncertain, and cannot be studied or approached within silos or separate categories.

Cross-collaborating

For urban ecosystems and populations to survive, novel methods and transdisciplinary efforts are required. Within such efforts scientists must no longer rely on reductionism and single risk factor modelling, but systems science is required, in which policy makers, practitioners, and stakeholders are involved from problem formulation to solution. Environmental exposures, be they pathological or healthy, lack the stability, clarity, and immediate social impact that attach to individual-level 'risk or health factors'. Similarly, the currently prevailing medical models of 'health and disease' do not extend to understanding the long-term foundations of and the complex ecosystem services' support to human health. Within this context lies the realisation of how urban green spaces promote and protect health and prevent disease in a life course perspective, through multiple interactive chains, networks, synergies, trade-offs, and feed-back loops.

Health arguments for green space investments

From recent research projects a large number of studies around green spaces and health has emerged. Those demonstrate that access to urban green spaces reduces mortality (Gascon et al., 2016), cardiovascular morbidity (Donovan et al., 2015), health inequalities (Dadvand et al., 2012a), blood pressure (Grazuleviciene et al., 2014), and depressive symptoms (Reklaitiene et al., 2014), while improving pregnancy outcomes (Dadvand et al., 2012b), general physical and mental health (Triguero-Mas et al., 2015), and behavioural and cognitive development (Dadvand et al., 2015). These are all intriguing findings, based on high-quality research, and are good arguments for investing in, establishing, and maintaining urban green spaces.

Scientific simplifications

The problem, though, is that most of these studies take into account only one or two health outcomes, applying objectively measurable green variables (e.g. Normalised Difference Vegetation Index, NDVI), studying risk factors with sophisticated epidemiological methods, originally developed for much less dynamic and integrated factors. Unfortunately, this implies a risk for over-simplification where synergies and trade-offs are neglected. Eventually, this may create a sound evidence-base on which to build policies and accurate decisions.

The same risk applies for the segmentation in ecosystem services and disservices, echoing the inadequate dichotomy between health and disease. Human health and wellbeing, as well as ecosystem resilience, are not either/or, but contingent, dynamic, irregular scales of features and events. Therefore, we cannot disentangle the so-called ecosystem disservices from the services. For example, aeroallergens, like pollen from urban trees, are not health risks as such, but part of larger ecosystems. Human behaviour, our interference with nature and ecosystems, urbanisation, overexploitation, and disrupted immunoregulatory mechanisms are part of the intricate web that seemingly turns pollen emissions into a 'disservice', although the issue may have been initiated by poor management and creating conditions favourable for invasive species.

Therefore, we need to apply a different perspective in order to protect humans from the 'disservices', using systems science methods that can solve the root to the problems rather than the symptoms. The issue of pollen emissions, like issues of fear of urban green spaces or falling branches, should be approached by transdisciplinary methods where *human behaviour* is at the core of the problem to be solved.

Responding ethically to the environmental demands of cities

The challenge for scientific exploration of health and environment has indeed entered a new era in which classical reductionist research on specific physical, chemical, or epidemiological relationships must be replaced with a systems science view, incorporating the *precautionary principle* and the use of dynamic systems modelling, estimates, and scenarios. The application and transfer of science across borders of policies, decision making, and practices is as challenging as it is ethically compelling. Only then can we achieve the credibility required to establish long-term policies and legacies for creating and maintaining urban green spaces on a broader scale, adapted to prevailing climates and conditions. Only then can it become an evident, wide-spread action of 'science, precaution, and innovation' aiming at providing green, healthy cities for our children and future generations, in which to live, work, and play.

References

Amoly, E., Dadvand, P., Forns, J., López-Vicente, M., Basagaña, X., Julvez, J., Alvarez-Pedrerol, M., Nieuwenhuijsen, M. J., Sunyer, J. 2014. 'Green and blue spaces and behavioral development in Barcelona schoolchildren: The BREATHE Project'. *Environmental Health Perspectives*, vol 122, 1351–1358.

Annerstedt, M., Östergren, P.-O., Grahn, P., Skärbäck, E., Währborg, P. 2015. 'Moving to serene nature may prevent poor mental health—results from a Swedish longitudinal cohort study'. *International Journal of Environmental Research and Public Health*, vol 12, 7974–7989.

Annerstedt, M., Jönsson, P., Wallergård, M., Johansson, G., Karlson, B., Grahn, P., Hansen, Å. M., Währborg, P. 2013. 'Inducing physiological stress recovery with sounds of nature in a virtual reality forest: Results from a pilot study'. *Physiology and Behavior*, vol 118, 240–250.

Annerstedt van den Bosch, M., Mudu, P., Uscila, V., Barrdahl, M., Kulinkina, A., Staatsen, B., Swart, W., Kruize, H., Zurlyte, I, Egorov, E. 2016. 'Development of an urban green space indicator and the public health rationale'. *Scandinavian Journal of Public Health*, vol 44, no 2, 159–167.

Astell-Burt, T., Feng, X., Kolt, G. S. 2013. 'Green space is associated with walking and moderate-to-vigorous physical activity (MVPA) in middle-to-older-aged adults: Findings from 203 883 Australians in the 45 and Up Study'. *British Journal of Sports Medicine*, vol 48, no 5, pp. 404–406.

Bach, J.-F. 2002. 'The effect of infections on susceptibility to autoimmune and allergic diseases'. *New England Journal of Medicine*, vol 347, 911–920.

Basagaña, X. S. C., Barrera-Gómez, J., Dadvand, P., Cunillera, J., Ostro, B., Sunyer, J., Medina-Ramòn, M. 2011. 'Heat waves and cause-specific mortality at all ages'. *Epidemiology*, vol 22, 765–772.

Bowler, D. E., Buyung-Ali, L., Knight, T. M., Pullin, A. S. 2010. 'Urban greening to cool towns and cities: A systematic review of the empirical evidence'. *Landscape and Urban Planning*, vol 97, 147–155.

Bratman, G. N., Hamilton, J. P., Hahn, K. S., Daily, G. C., Gross, J. J. 2015. 'Nature experience reduces rumination and subgenual prefrontal cortex activation'. *Proceedings of the National Academy of Sciences*, vol 112, 8567–8572.

Brown, D., Barton, J., Gladwell, V. 2013. 'Viewing nature scenes positively affects recovery of autonomic function following acute mental stress'. *Environmental Science and Technology*, vol 47, 5562–5569.

Cohen, D. A., Ashwood, S., Scott, M., Overton, A., Evenson, K. R., Voorhees, C. C., Bedimo-Rung, A., Mckenzie, T. L. 2006. 'Proximity to school and physical activity among middle school girls: The Trial of Activity for Adolescent Girls Study'. *Pediatrics*, vol 118, e1381–e1389.

Dadvand, P., De Nazelle, A., Figueras, F., Basagaña, X., Su, J., Amoly, E., Jerrett, M., Vrijheid, M., Sunyer, J., Nieuwenhuijsen, M. J. 2012a. 'Green space, health inequality and pregnancy'. *Environment International*, vol 40, 110–115.

Dadvand, P., Sunyer, J., Basagana, X., Ballester, F., Lertxundi, A., Fernandez-Somoano, A., Estarlich, M., Garcia-Esteban, R., Mendez, M., Nieuwenhuijsen, M. 2012b. 'Surrounding greenness and pregnancy outcomes in four Spanish birth cohorts'. *Environmental Health Perspectives*, vol 120, 1481–1487.

Dadvand, P., Nieuwenhuijsen, M. J., Esnaola, M., Forns, J., Basagaña, X., Alvarez-Pedrerol, M., Rivas, I., López-Vicente, M., De Castro Pascual, M., Su, J., Jerrett, M., Querol, X., Sunyer, J. 2015. 'Green spaces and cognitive development in primary schoolchildren'. *Proceedings of the National Academy of Sciences*, vol 112, 7937–7942.

Donovan, G. H., Michael, Y. L., Butry, D. T., Sullivan, A. D., Chase, J. M. 2011. 'Urban trees and the risk of poor birth outcomes'. *Health and Place*, vol 17, 390–393.

Donovan, G. H., Michael, Y. L., Gatziolis, D., Prestemon, J. P., Whitsel, E. A. 2015. 'Is tree loss associated with cardiovascular-disease risk in the Women's Health Initiative? A natural experiment'. *Health and Place*, vol 36, 1–7.

Dzhambov, A., Dimitrova, D. 2014. 'Urban green spaces effectiveness as a psychological buffer for the negative health impact of noise pollution: A systematic review'. *Noise and Health*, vol 16, 157–165.

EEA 2015. *Air Quality in Europe: 2015 Report*. Publications Office of the European Union, Luxembourg.

Fjørtoft, I., Sageie, J. 2000. 'The natural environment as a playground for children: Landscape description and analyses of a natural playscape'. *Landscape and Urban Planning*, vol 48, 83–97.

Foster, C., Hillsdon, M., Jones, A., Grundy, C., Wilkinson, P., White, M., Sheehan, B., Wareham, N., Thorogood, M. 2009. 'Objective measures of the environment and physical activity: Results of the Environment and Physical Activity Study in English adults'. *Journal of Physical Activity and Health*, vol 6, S70–S80.

Gascon, M., Triguero-Mas, M., Martínez, D., Dadvand, P., Rojas-Rueda, D., Plasència, A., Nieuwenhuijsen, M. J. 2016. 'Residential green spaces and mortality: A systematic review'. *Environment International*, vol 86, 60–67.

Giles-Corti, B., Bull, F., Knuiman, M., Mccormack, G., Van Niel, K., Timperio, A., Christian, H., Foster, S., Divitini, M., Middleton, N., Boruff, B. 2013. 'The influence of urban design on neighbourhood walking following residential relocation: Longitudinal results from the RESIDE study'. *Social Science and Medicine*, vol 77, 20–30.

Gluckman, P. D., Hanson, M. A., Beedle, A. S. 2007. 'Early life events and their consequences for later disease: A life history and evolutionary perspective'. *American Journal of Human Biology*, vol 19, 1–19.

Grazuleviciene, R., Dedele, A., Danileviciute, A., Vencloviene, J., Grazulevicius, T., Andrusaityte, S., Uzdanaviciute, I., Nieuwenhuijsen, M. J. 2014. 'The influence of proximity to city parks on blood pressure in early pregnancy'. *International Journal of Environmental Research and Public Health*, vol 11, 2958–2972.

Hacking, E. B., Barratt, R., Scott, W. 2007. 'Engaging children: Research issues around participation and environmental learning'. *Environmental Education Research*, vol 13, 529–544.

Hägerhäll, C. M., Laike, T., Taylor, R. P., Küller, M., Küller, R., Martin, T. P. 2008. 'Investigations of human EEG response to viewing fractal patterns'. *Perception*, vol 37, 1488–1494.

Jin, S., Guo, J., Wheeler, S., Kan, L., Che, S. 2014. 'Evaluation of impacts of trees on PM2.5 dispersion in urban streets'. *Atmospheric Environment*, vol 99, 277–287.

Kaplan, S., Berman, M. G. 2010. 'Directed attention as a common resource for executive functioning and self-regulation'. *Perspectives on Psychological Science*, vol 5, 43–57.

Kuo, M. 2015. 'How might contact with nature promote human health? Exploring promising mechanisms and a possible central pathway'. *Frontiers in Psychology*, vol 6.

Lederbogen, F., Haddad, L., Meyer-Lindenberg, A. 2013. 'Urban social stress – Risk factor for mental disorders. The case of schizophrenia'. *Environmental Pollution*, vol 183, 2–6.

Louv, R., Hogan J. 2005. *Last Child in the Woods: Saving Our Children from Nature-Deficit Disorder*, Algonquin Books of Chapel Hill, Chapel Hill, NC.

Mcdade, T. W., Tallman, P. S., Madimenos, F. C., Liebert, M. A., Cepon, T. J., Sugiyama, L. S., Snodgrass, J. J. 2012. 'Analysis of variability of high sensitivity C-reactive protein in lowland Ecuador reveals no evidence of chronic low-grade inflammation'. *American Journal of Human Biology*, vol 24, 675–681.

Mitchell, R. 2013. 'Is physical activity in natural environments better for mental health than physical activity in other environments?'. *Social Science and Medicine*, vol 91, 130–134.

Mulder, I. E., Schmidt, B., Stokes, C. R., Lewis, M., Bailey, M., Aminov, R. I., Prosser, J. I., Gill, B. P., Pluske, J. R., Mayer, C.-D., Musk, C. C., Kelly, D. 2009. 'Environmentally-acquired bacteria influence microbial diversity and natural innate immune responses at gut surfaces'. *BMC Biology*, vol 7, 1–20.

Nowak, D. J., Hirabayashi, S., Bodine, A., Hoehn, R. 2013. 'Modeled PM2.5 removal by trees in ten U.S. cities and associated health effects'. *Environmental Pollution*, vol 178, 395–402.

Oke, T. R. 1973. 'City size and the urban heat island'. *Atmospheric Environment (1967)*, vol 7, 769–779.

Pantell, M., Rehkopf, D., Jutte, D., Syme, S. L., Balmes, J., Adler, N. 2013. 'Social isolation: A predictor of mortality comparable to traditional clinical risk factors'. *Am J Public Health*, vol 103, 2056–2062.

Peen, J., Schoevers, R., Beekman, A., Dekker, J. 2010. 'The current status of urban-rural differences in psychiatric disorders'. *Acta Psychiatrica Scandinavica*, vol 121, 84–93.

Potwarka, L. R., Kaczynski, A. T., Flack, A. L. 2008. 'Places to play: Association of park space and facilities with healthy weight status among children'. *Journal of Community Health*, vol 33, 344–350.

Radon, K., Windstetter, D., Poluda, A. L., Mueller, B., Mutius, E. V., Koletzko, S., Group, C. a. U. K. Z. T. S. 2007. 'Contact with farm animals in early life and juvenile inflammatory bowel disease: A case-control study'. *Pediatrics*, vol 120, 354–361.

Reklaitiene, R., Grazuleviciene, R., Dedele, A., Virviciute, D., Vensloviene, J., Tamosiunas, A., Baceviciene, M., Luksiene, D., Sapranaviciute-Zabazlajeva, L., Radisauskas, R., Bernotiene, G.,

Bobak, M., Nieuwenhuijsen, M. J. 2014. 'The relationship of green space, depressive symptoms and perceived general health in urban population'. *Scandinavian Journal of Public Health*, vol 42, 669–676.

Rook, G. A. 2013. 'Regulation of the immune system by biodiversity from the natural environment: An ecosystem service essential to health'. *Proceedings of the National Academy of Sciences*, vol 110, 18360–18367.

Setälä, H., Viippola, V., Rantalainen, A. L., Pennanen, A., Yli-Pelkonen, V. 2013. 'Does urban vegetation mitigate air pollution in northern conditions?'. *Environmental Pollution (Barking, Essex: 1987)*, vol 183, 104–112.

Staats, H., Kieviet, A., Hartig, T. 2003. 'Where to recover from attentional fatigue: An expectancy-value analysis of environmental preference'. *Journal of Environmental Psychology*, vol 23, 147–157.

Sugiyama, T., Leslie, E., Giles-Corti, B., Owen, N. 2008. 'Associations of neighbourhood greenness with physical and mental health: Do walking, social coherence and local social interaction explain the relationships?'. *Journal of Epidemiology and Community Health*, vol 62.

Sugiyama, T., Giles-Corti, B., Summers, J., Du Toit, L., Leslie, E., Owen, N. 2013. 'Initiating and maintaining recreational walking: A longitudinal study on the influence of neighborhood green space'. *Preventive Medicine*, vol 57, 178–182.

Taha, H. 1997. 'Urban climates and heat islands: Albedo, evapotranspiration, and anthropogenic heat'. *Energy and Buildings*, vol 25, 99–103.

Triguero-Mas, M., Dadvand, P., Cirach, M., Martínez, D., Medina, A., Mompart, A., Basagaña, X., Gražulevičienė, R., Nieuwenhuijsen, M. J. 2015. 'Natural outdoor environments and mental and physical health: Relationships and mechanisms'. *Environment International*, vol 77, 35–41.

Ulrich, R. 1983. 'Aesthetic and affective response to natural environment'. *Human Behavior and Environment: Advances in Theory and Research*, vol 6, 85–125.

Vos, T., Barber, R. M., Bell, B., Bertozzi-Villa, A., Biryukov, S. et al. 2015. 'Global, regional, and national incidence, prevalence, and years lived with disability for 301 acute and chronic diseases and injuries in 188 countries, 1990–2013: A systematic analysis for the Global Burden of Disease Study 2013'. *The Lancet*, vol 386, 743–800.

Wells, N. M., Lekies, K. S. 2006. 'Nature and the life course: Pathways from childhood nature experiences to adult environmentalism'. *Children, Youth and Environments*, vol 16, 1–24.

WHO 1948. 'Preamble to the Constitution of the World Health Organization as adopted by the International Health Conference, New York, 19–22 June, 1946; signed on 22 July 1946 by the representatives of 61 States; entered into force on 7 April 1948.' *Official Records of the World Health Organization*, no. 2, p. 100. World Health Organization, New York.

WHO 2010. *Global Recommendations on Physical Activity for Health*. World Health Organization, Geneva, Switzerland.

WHO 2011. 'Burden of disease from environmental noise: Quantification of healthy life years lost in Europe'. *The World Health Organization European Centre for Environment and Health, Bonn* vol.

Wolch, J. R., Byrne, J., Newell, J. P. 2014. 'Urban green space, public health, and environmental justice: The challenge of making cities "just green enough"'. *Landscape and Urban Planning*, vol 125, 234–244.

Zelenski, J. M., Dopko, R. L., Capaldi, C. A. 2015. 'Cooperation is in our nature: Nature exposure may promote cooperative and environmentally sustainable behavior'. *Journal of Environmental Psychology*, vol 42, 24–31.

7

URBAN FORESTRY
AND MICROCLIMATE

Simone Orlandini, Jennifer K. Vanos,
Andreas Matzarakis, Luciano Massetti and Martina Petralli

Introduction

The urban climate is a very complex system because of the large number of characteristics that affect weather variables in the urban environment and the different kind of morphologies and materials used in cities all over the world. Urban areas have higher air temperature values than the surrounding rural areas and this phenomenon is called the urban heat island (UHI) effect. The variations in heat storage, radiation absorption, and sensible and latent heat produced by urban land-use types determine large intra-urban air temperature differences and therefore can cause different UHI values within the urban area (Lowry, 1967; Oke, 1982).

Impervious surfaces are the major cause of the UHI because they alter the surface energy budget by slowing heat release and lowering evapotranspiration due to the reduced percentage of vegetation. The resulting energy budget change causes the UHI effect that is particularly intense in summer and winter, and at night due to the release of heat stored during the day. Moreover, higher temperature and reduced evapotranspiration cause lower humidity levels than rural areas. The UHI and temperature variability have been shown to cause decreased health and thermal comfort of people living in cities in the summer within an urban area (Brandani et al., 2016; Petralli et al., 2012; Vanos et al., 2015). These effects are measured by air temperature indices or by complex indices that take into account the combined effect of temperature, relative humidity, radiation and wind to estimate the comfort level of a person (Matzarakis et al., 1999). The consequences of the UHI can be reduced by appropriate design of urban areas, with vegetation playing an important role in mitigating local heating: urban trees and forest modify the energy balance of the surfaces, first and foremost providing shade and evapotranspiration. The presence of urban trees and forest affects all atmospheric variables; those that show the greatest evidence are the air temperature, surface temperatures and mean radiant temperature (Streiling and Matzarakis, 2003).

This chapter presents knowledge concerning the effects of urban trees and forests on urban microclimate variables and human thermal comfort. Benefits of urban vegetation are evidenced by the growing interest on urban tree planting as an energy use control during summer: the number of papers published on this argument is growing in journals related to urban planning and sustainable urban design, building energy, climatology/meteorology and related fields (Abdel-Aziz et al., 2015). The chapter is divided into three sections. The first is dedicated to the effect of urban trees and forest on air temperature and its variability. The second assesses the effect of urban tree and forest on energy exchange and thus on surface temperature and mean radiant temperature. The last section presents an analysis of the effect of microclimate interventions with trees on weather parameters and human thermal comfort.

Air temperature

Vegetation is commonly considered a factor in climate control in the urban environment. During the cold season, trees represent shelters from cold winds, while during the hot season they provide shade for people, streets and buildings. In particular, trees are defined as 'the most effective vegetation element for reducing overheating in urban areas' (Konijnendijk et al., 2005). The overheating of urban areas, resulting in the UHI phenomenon, is the most evident effect strictly associated with the built environment and represents increasing temperatures in urban areas with respect to near rural areas (Lowry, 1967; Oke, 1982). It is typically higher at night, under clear and calm conditions, and may have a large impact on public health and well-being during the hot season. Air temperature (T_a) is not distributed evenly within a single urban area. UHI intensity and intra-urban temperature distribution are linked to various factors, such as building density and height, street cover and vegetation cover ratio (Gosling et al., 2014; Petralli et al., 2014; Massetti et al., 2014). The cooling effect of vegetation has been well-documented worldwide, and it is primary due to the combined effect of evapotranspiration and shade. The cooling effect of trees is more significant for large urban parks than for single trees. The park cool island (PCI; Spronken-Smith and Oke, 1998) is the term used to define the cooling effect of an urban park as compared to the surrounding built environment. It can reach values of 4–6°C, but in mid-latitude cities it averages 0.5–3°C depending on park design and magnitude (Skoulika et al., 2014). A significant positive correlation between PCI intensity and urban park size was found by many authors, showing that larger parks had a stronger PCI effect than the smaller ones (Ren et al., 2013; Zhang et al., 2009). The park design and the presence of forested areas has a strong effect on T_a reduction in urban areas. This reduction was recently confirmed by a study conducted in Leipzig (Germany) analysing T_a data from 62 urban parks and forests, where cooling increased with increasing size, and the cooling effects were greater in urban forests than in non-forested parks (Jaganmohan et al., 2015). In Florence during the hot season, thoroughly forested urban areas, expressed as a tree cover ratio (TCR; percentage of the urban area covered by trees against the total area), was related to a reduction of maximum T_a within 50m, also reducing minimum T_a at distances of more than 200m (Petralli et al., 2014).

The cooling effect of boulevard or single street trees is more evident on maximum T_a and is linked to several tree and street characteristics. According to tree species, the transpiration rates, canopy architecture, and leaf size are important. For the street, urban geometry, street width and orientation, and the height of buildings influence the wind direction and intensity, and influence the duration of solar radiation interception by tree crowns (Loughner et al., 2012).

According to the evidence that urban trees can reduce T_a and the UHI, some cities have already included the plantation of millions of trees in their city plan, while others aim to increase their tree canopy cover to mitigate the impacts of the UHI in the coming years. Nowadays, Chicago achieves a city-wide tree count of 4.1 million trees and plans to plant many other trees, almost one million, by 2020 (Slavin, 2011). Similarly in Australia, cities such as Melbourne and Sidney aim to increase their tree canopy cover from 22 to 40 per cent by 2032 and from 15.5 to 23.3 per cent by 2030, respectively (Sanusi et al., 2016).

Solar radiation, surface temperature and mean radiant temperature

The main difference between an open area and one with a forest or trees is the energy balance of the surfaces. The structural and morphological characteristics of a tree or a forest modify the energy fluxes, especially the radiation fluxes, resulting in a unique microclimate (Bonan, 2002). Besides the modified radiation fluxes, the urban forest microclimate includes specific features such as a weaker diurnal cycle of T_a, increased relative humidity (RH), and reduced wind speed (v), which can be important for several issues and applications. Because of the three-dimensionality of trees and forests, the main energy exchanging surface is to the surface of the tree or trees (the top of the canopy), rather than the ground. Inside the canopy, the main energy gain – the short wave radiation from the upper hemisphere – is reduced and mostly diffuse radiation is received. This has implications not only on the amount of solar radiation, but also on the surface temperature (T_s) inside the canopy. Surface temperature is directly connected to the absorbed radiation and affects the long-wave emission of a given surface. A dry surface partitions more energy into sensible heat, which increases T_s (such as tree or ground), influencing the surrounding microclimate.

The mean radiant temperature (T_{mrt}) is defined as the uniform temperature of an imaginary enclosure in which the radiant heat transfer from the human body is equal to the radiant heat transfer in the actual non-uniform enclosure (Matzarakis et al., 2007, 2010; ISO, 1998). It is based on the fact that the exchange of radiant fluxes between two objects is approximately proportional to their temperature difference multiplied by their ability to emit and absorb heat (emissivity) from all objects surrounding the body of interest. This is valid as long as the absolute temperatures of objects in question are large compared to the temperature differences, allowing linearization of the Stefan-Boltzmann law in the relevant temperature range.

T_{mrt} cannot be measured directly but is estimated based on the three-dimensional integration of the radiation fluxes (VDI, 1998). For the estimation of T_{mrt}, the following radiation flux densities are required (Matzarakis et al., 2010; VDI, 1998):

- direct solar radiation;
- diffuse solar radiation;
- reflected short-wave radiation;
- atmospheric radiation (longwave) from the open sky; and
- long-wave radiation from solid surfaces (lower hemisphere and horizon limitation).

The following parameters describing the surroundings of the human body must be known:

- sky view factor – because of the limitation of the horizon and the influence of short- and long-wave radiation flux densities;
- view factor of the different solid surfaces – because of the modification of the reflected short-wave radiation;
- albedo of the different solid surfaces – because of the influence of the incoming short-wave radiation; and
- emissivity of the different solid surfaces – because of the influence on the T_s.

To calculate T_{mrt}, the entire surroundings of the human body are divided into n isothermal surfaces with the temperatures T_i ($i = 1$ to n) and emissivities ε_i, for which the solid angle portions ('angle factors'), F_i, are used as weighting factors. Long-wave radiation ($E_i = \varepsilon_i \sigma T_i^4$) and diffuse short-wave radiation, D_i, are emitted from each of the n surfaces of the surroundings (Jendritzky et al., 1990; Matzarakis et al., 2010).

$$T_{mrt} = [\frac{1}{\sigma} \sum_{i=1}^{n} (E_i + a_k \frac{D_i}{\varepsilon_p}) F_i]^{0.25} \tag{7.1}$$

where σ is the Stefan-Boltzmann constant [5.67×10^{-8} W/(m²K⁴)] and ε_p is the emission coefficient of the human body (standard value 0.97). D_i comprises the diffuse solar radiation and the diffusely reflected global radiation, whereas a_k is the absorption coefficient of the irradiated body surface area of short-wave radiation (standard value 0.7). T_{mrt} is incremented to T^*_{mrt}, if there is also direct solar radiation (Matzarakis et al., 2010):

$$T^*_{mrt} = \left[T_{mrt}^4 + \frac{f_p a_k I^*}{\varepsilon_p \sigma} \right]^{0.25} \tag{7.2}$$

In this case, I is the radiation intensity of the sun on a surface perpendicular to the incident radiation direction. The surface projection factor is f_p. For applications in human-biometeorology, it is generally sufficient to determine f_p for a rotationally symmetrical person standing or walking (Jendritzky et al., 1990; VDI, 1998; Matzarakis et al., 2007, 2010). For complex environments like forests and urban areas, the sky view factor affects the T_{mrt} the most. In most cases, combined methods for the estimation of T_{mrt} should be used in order to increase precision (Krüger et al., 2014).

Mean radiant temperature is one of the most influencing parameters for the quantification of thermal comfort or heat stress by thermal indices, such as the physiologically equivalent temperature (PET) or Universal Thermal Climate Index (UTCI), which are based on and derived from the human energy balance (Höppe, 1999; Matzarakis et al., 1999; Jendritzky et al., 2012). The indices require T_a, RH, v and T_{mrt} meteorological input parameters and in addition thermo-physiological factors (human metabolism, activity and heat resistance of clothing). Free available micro scale models allow the calculation of the thermal indices (Matzarakis et al., 2007, 2010).

Urban forestry and greenery require different considerations for design (e.g. architectural, functional, climate and aesthetical aspects).

The following figures show the temporal pattern of several parameters in urban areas of Freiburg, Germany. The investigation was conducted in a location with chestnut trees (Streiling and Matzarakis, 2003) in the northern downtown of Freiburg (270 m a.s.l.). Five sites (Table 7.1 and Figure 7.1) arranged in a circle with a diameter of less than 100 m were selected, which differ in their structure and dimension of trees and buildings. The first site was under a tree. The second was located east of a clump of trees, the third under a few trees, the fourth on the south side of a street canyon, and the fifth on the north side of the street canyon (Table 7.1 and Figure 7.1). The sky view factors (SVF) are listed in Table 7.1.

For the selected site it can be seen that the differences in the mean T_a are very small (less than 1°C) between the different locations. For surface temperatures, the differences depend on the site characteristics (surface conditions and SVF) with the lowest T_s found in locations where the global radiation is reduced, and the highest in locations with high exposure to solar radiation during the day. This observation can be seen in the mean conditions of the short-wave radiation from the upper hemisphere. The same pattern can be seen also in the T_{mrt} and the differences between T_{mrt} and T_a, which are driven mostly by T_s and the short-wave radiation.

Table 7.1 Sky view factor (SVF), mean daytime air temperature (T_a), surface temperature (T_s), short-wave radiation (K) from upper hemisphere, mean radiant temperature (T_{mrt}), and deviation of T_{mrt} and T_a at different sites in Freiburg

Sites	SVF	T_a (°C)	T_s (°C)	K (W m⁻²)	T_{mrt} (°C)	T_{mrt}–T_a (°C)
Below a tree canopy	0.25	20.4	23.3	68.6	35.4	11.0
East of a clump of trees	0.47	20.7	24.9	201.9	45.1	24.4
Below some tree canopies	0.07	20.3	19.0	18.7	32.1	11.8
Street canyon, south	0.42	20.6	15.3	89.7	34.2	13.6
Street canyon, north	0.61	21.3	32.6	322.0	51.4	30.1

Source: Mayer and Matzarakis (2006)

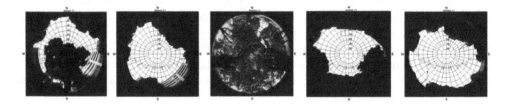

Figure 7.1 Fish-eye picture of sites described in Table 7.1

The daily changes and pattern of the K, T_s, T_{mrt} and PET are shown in Figures 7.2–7.5. The maximum T_a difference between locations was 2.2°C and was calculated comparing areas with trees and areas without trees. The same pattern is also visible in T_s. The maxima T_s measured was about 25°C. The highest value in short-wave radiation from the upper hemisphere was 660.4 W m⁻² and the lowest maximum, recorded below the trees, only reaching 34.3 W/m². For T_{mrt} the lowest values were recorded in the morning with few differences between the sites. The highest values midday depended strongly on direct exposure to the sun. T_{mrt} maxima ranged from around 24.9°C in the shade to 63.3°C in the urban canyon, which means a difference in T_{mrt} of about 35°C. From the pattern of PET (Figure 7.5), it is shown that PET mostly follows T_{mrt}, with maximum differences of 10°C between T_a and the PET in the sun.

In general, it can be summarized, that urban trees and forest have the highest influence on T_{mrt}, PET and T_s, and less on T_a.

Thermal comfort

The urban forest can have an important influence on human thermal comfort, with proper design strategies allowing people to stay warm and protected from the cold winter winds, or helping people stay cool and protected from the hot summer sun. Thermal comfort is defined as the condition of the mind that expresses satisfaction with the thermal environment and is determined by subjective evaluation. This decision is based on six primary factors from two categories: *personal factors* (metabolism and clothing) and *environmental factors* (air temperature, wind speed, relative humidity, and short- and long-wave radiation). On average, when a person feels no desire to alter the environment or activity/clothing, they are said to be

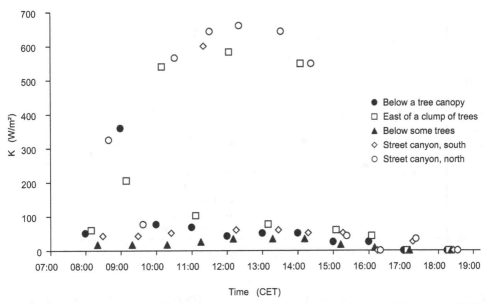

Figure 7.2 Short-wave radiation (K) from the upper hemisphere on a typical day (19 September 2000) at different sites in Freiburg

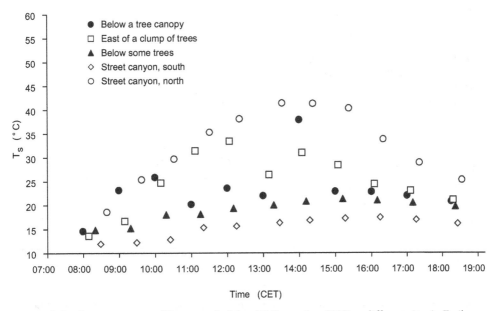

Figure 7.3 Surface temperature (T_s) on a typical day (19 September 2000) at different sites in Freiburg

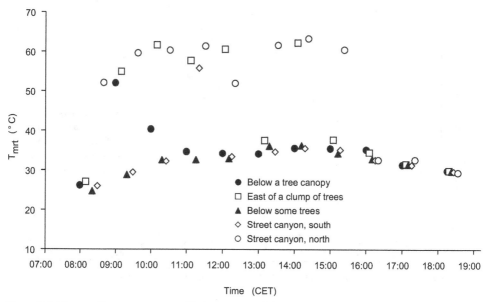

Figure 7.4 Mean radiant temperature (T_{mrt}) on a typical day (19 September 2000) at different sites in Freiburg

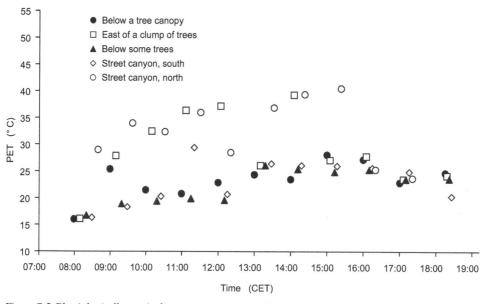

Figure 7.5 Physiologically equivalent temperature (PET) on a typical day (19 September 2000) at different sites in Freiburg

comfortable and near ($+/-$ 50 W m^{-2}) of energy balance (Brown and Gillespie, 1986). The energy balance of a human body is stated as (Vanos et al., 2010):

$$0 = S + M + K + R - C - E \qquad (7.3)$$

where S is the energy sum, which is 0 when in perfect balance, M is the metabolic heat flux, K is conductive heat flux, R is the net radiation experienced by a human, C is convective heat flux, and E is evaporative heat flux. As these are fluxes, units are expressed in watts per metre squared of a human's surface area (W m^{-2}), where the average human has a surface area of 1.8 m^2 (DuBois and DuBois, 1916).

Personal factors

The metabolic heat flux is calculated as the individual metabolic activity (M_a) minus respiratory heat loss (f):

$$M = (1 - f) M_a \qquad (7.4)$$

The metabolic activity is often expressed as a metabolic equivalent (MET), where 1 MET $= 58.2$ W m^{-2} (18.4 Btu/h·ft^2), which is the M_a of a person at rest. MET values for common activities include 1.2–1.4 METs for light activity/standing, and 2.0 METs or greater for activities that involve movement, such as walking, working, or lifting heavy objects (ASHRAE, 2013). Metabolic equivalences can be found using activity codes from Ainsworth et al. (2000).

In thermal comfort studies, heat fluxes are often based on Ohm's law analogy, where the flux rate is equal to a difference (temperature or concentration) between 'surfaces' (skin and air), divided by the sum of the resistances to heat transfer in s m^{-1} (e.g. tissue, clothing, aerodynamic resistance) (Campbell and Norman, 1998). For clothing resistance – the second personal factor – the resistance of garments or clothing ensembles can be found in international standards documents, such as ISO 9920 (ISO, 2007). For example, the clothing ensemble of walking shorts and short-sleeved shirt has a clothing resistance (r_c) of 56 s m^{-1}, and recent studies have incorporated area factors of the clothed body to gain a more accurate estimation of energy fluxes from the given body segment (e.g. unclothed arm versus clothed torso; Vanos et al., 2012a).

Environmental factors

Conditions that are very hot or cold are two important considerations for microclimatic design. Heat or cold stress will occur when the body is no longer able to dissipate or create/ maintain enough energy, respectively, to stay at an equilibrium body temperature. This balance is controlled by simultaneous transfers of heat to and from the body.

Because the urban landscape is complex, the environmental variables that affect an individual's thermal comfort can change rapidly through space and time. For example, moving from a sunny asphalt street into a treed and grassy park will result in the following changes in the energy balance equation (while holding personal factors constant):

1 Tree shade blocking short- and long-wave radiation results in a net radiation (R) decrease. A human is exposed to less incoming radiation, but also less reflected short-wave radiation and less emitted long-wave radiation from the surface;

2 Energy is partitioned into evapotranspiration from trees and moist soil, decreasing T_a. Evaporative heat loss (E) may be inhibited or remain constant with higher vapour pressure, and is highly dependent on wind speed;

3 Convective heat loss (C) may slightly increase due to a drop in air temperature from the occurrence of (1) and (2).

Here, we address each component of the energy budget (Equation 7.3) that is impacted by the above 'asphalt-to-greenspace' example, and demonstrate how each component is controlled by the microclimate variables of interest with data from an urban asphalt street and green park in Toronto, Ontario, Canada (Vanos et al., 2012b). Table 7.2 displays the environmental conditions from data collected on an asphalt street and into an urban park in Toronto, Canada, and the corresponding energy streams in each situation for a person walking, all of which utilize in part Equations 7.3–7.15.

Radiation

As discussed earlier, the presence of urban trees and forests impacts the net radiation in the urban microclimate, and consequently the T_{mrt} that a person is exposed to. The combined fluxes of long- and short-wave radiation to and from the body, as well as the geometry and surface colour of an individual, control the amount of radiation absorbed by a human (R_{abs}, W m^{-2}).

Humans are generally cylindrical in shape, and we therefore estimate the intercepted radiation (and hence radiative energy exchange at the human surface) by addressing each

Table 7.2 Example of energy streams and final energy budget values calculated using meteorological data from an open street (asphalt) and greenspace (grass, trees) collected on a moderately warm day in Toronto, Canada (15 September 2009). Average meteorological data are shown in rows 1–4, with calculated energy streams shown in rows 5–9, summing to the final energy budgets. Corresponding equations are listed (+/− 50 W m–2 is considered comfortable)

Meteorological variable/energy stream with corresponding equation	Greenspace (trees and grass)	Open street (asphalt)
Mean air temperature (T_a)	24.5°C	19.5°C
Activity speed (v_a)	1.3 m s^{-1}	1.3 m s^{-1}
Windspeed (v)	3.0 m s^{-1}	3.0 m s^{-1}
Relative humidity (RH)	63%	55%
Metabolic heat (M) (Eq. 7.4)	200 W m^{-2}	202 W m^{-2}
Absorbed radiation (R_{abs}) (Eq.7.10)	298 W m^{-2}	350 W m^{-2}
Emitted long-wave radiation (L_{emit}) (Eq. 7.11)	−341 W m^{-2}	−354 W m^{-2}
Convective heat loss (C) (Eq. 7.12)	−126 W m^{-2}	−89 W m^{-2}
Evaporative heat loss (E) (Eq. 7.13)	−68 W m^{-2}	−69 W m^{-2}
Final energy budget (B) (Eq. 7.3)	−37 W m^{-2}	40 W m^{-2}

Source: Vanos et al. (2012b), Slater (2010)

component of radiation and its interaction with a cylindrical human surface area (A_{cyl}) to determine a R_{abs} from the following steps (also see Kenny et al., 2008).

1 **Diffuse short-wave radiation** (K_d), originates from the scattered shortwave radiation and is absorbed by humans as:

$$K_{d(abs)} = 0.5(1-\alpha_h)K_dA_{cyl}$$ (7.5)

2 **Direct (or beam) short-wave radiation** (K_b) is absorbed by the sunlit portion of the body as:

$$K_{b(abs)} = (1-\alpha_h)K_p\sin(\psi)A_{cyl}$$ (7.6)

where α_h is the average skin and clothing albedo of a human (~0.37) (which can range from 0.4 to 0.45 for light skin tones and from 0.2 to 0.25 for darker skin tones (Monteith and Unsworth, 2008), K_p is the incoming direct perpendicular irradiance (Campbell and Norman, 1998), and ψ is the zenith angle. The sum of $K_{d(abs)}$ and $K_{b(abs)}$ give $K_{in(abs)}$.

3 **Reflected short-wave radiation** (K_{up}), reflected from the ground surface is absorbed by a human as:

$$K_{up(abs)} = 0.5(1-\alpha_h)K_{up}A_{cyl}$$ (7.7)

4 **Longwave radiation** from the ground hemisphere (L_g) and sky hemisphere (L_a), where absorbed longwave radiation is calculated, respectively, as follows:

$$L_{g(abs)} = 0.5\varepsilon_hL_gA_{cyl}$$ (7.8)

$$L_{a(abs)} = 0.5\varepsilon_hL_aA_{cyl}$$ (7.9)

where ε is the emissivity (~0.95 for humans). In Equations 7.5 and 7.7–7.9, the view factor of the cylinder is 0.5 since any point on the cylinder can only 'see' one hemisphere (sky or ground). Equations 7.5–7.9 are integrated to determine the absorbed radiation by a human as follows:

$$R_{abs} = A_{eff}\left(\frac{K_{in(abs)} + K_{up(abs)} + L_{a(abs)} + L_{g(abs)}}{A_{cyl}}\right)$$ (7.10)

where A_{eff} is the effective radiative area fraction of a human (0.78 for standing, 0.70 for sitting). The components of K_{in}, K_{up}, L_g, and L_a can be measured with a net radiometer, a pyranometer and pyrgeometer side by side, or using theoretical estimations.

Concurrently, a person will emit terrestrial, or long-wave, radiation from their body, based on Boltzmann's law:

$$L_{emit} = A_{eff}\varepsilon\sigma(T_{sf} + 273.15)^4$$ (7.11)

where σ is Bolztmann's constant (5.67×10^{-8} W m^{-2} K^{-4}), and T_{sf} is the surface temperature of a human, emitting heat from the skin surface to the ambient environment (Campbell and Norman, 1998). Therefore, the net radiation (R) in the energy budget equation 7.3 is the difference of $R_{abs} - L_{emit}$.

Convection

The convective (or sensible) heat exchange is the result of a gradient in temperature from the skin to the air, whereby resistances can moderate this heat flow:

$$C = \rho C_p \left(\frac{T_{sk} - T_a}{r_H} \right) \tag{7.12}$$

where r_H is the sum of resistance to heat transfer, and T_{sk} is skin temperature. The T_a is generally lower in a vegetated area, and thus the difference between T_{sk} and T_a will increase, increasing heat loss from the body. For a sedentary person in low air flow (<0.1m s^{-1}), free convection is dominant, and r_H is relatively high. In this situation it is difficult to cool in warm conditions, and easier to stay warm in cold conditions. When forced convection occurs due to increasing airflow, the layer of air around the body and skin is disturbed, lowering the aerodynamic and clothing resistance, and increasing the convective heat loss.

In hot conditions where $T_a > T_{sk}$, the convective heat flux becomes positive (or towards) the human body, thus heating the skin and core. Heat stress is a concern under such environmental conditions, and the main mechanism for heat loss becomes evaporative heat loss.

Evaporation

The act of sweating is an effective heat loss mechanism that depends on the evaporation of water from the skin. However, the ambient environment tightly controls this evaporative (or latent) heat flux, whereby the maximum exchange of latent heat (E_{max}) that is possible in a given environment is the result of a specific humidity gradient between the skin (q_{sk}) and the air (q_a):

$$E_{max} = \rho L_v \left(\frac{q_{sk} - q_a}{r_v} \right) \tag{7.13}$$

where q_{sk} and q_a is the specific humidity at T_{sk} and T_a, respectively, r_v is the resistance to vapour transfer, p is the density of air, and L_v is the latent heat of vaporization. At a high humidity, the difference between q_a and q_{sk} will be low; hence evaporation, and therefore heat loss, is decreased.

When considering our example of moving from a sunny, asphalt surface into a grassy forest, the q_a will increase, which will decrease the specific humidity gradient and thus decrease the E_{max}. As E_{max} is the maximum amount of latent heat loss possible given the environmental conditions, if conditions with a high q_a are present, a person can only exchange latent heat based on the sum of insensible (E_i) and sensible evaporation (E_s). The sensible evaporation (heat loss through perspiration) can be calculated for thermal comfort modelling based on metabolic rate:

$$E_s = 0.42\,(M - 58) \tag{7.14}$$

The final evaporative heat exchange value, E, cannot be higher than E_{max}, as it would be physically unreasonable for a human to sweat more than is allowed by the environment (or estimated by E_{max}). Within the context of urban forestry, the presence of moisture in an environment has two effects. First, the evaporation of moisture allows for cooling of the air, which reduces T_a; however, the second effect results in a higher vapour pressure, which limits the body's ability to lose heat through evaporation, as demonstrated in Equation 7.13. In such cases with higher specific humidities leading to lower evaporative exchanges at the skin's surface, the presence of wind helps to reduce aerodynamic resistance, and thus increases evaporation.

Wind speed

Wind speed is an important factor for calculating the aerodynamic resistance (r_a) that affects convection, skin temperature, and evaporation. A higher wind speed also reduces clothing resistance. In thermal comfort modelling, the wind speed (v) is integrated with the activity velocity (v_a) of an individual to determine the relative velocity (v_r), as follows (ISO, 2007):

$$v_r = \sqrt{\left(v_a - v_w \cos(\alpha)\right)^2 + \left(v_w \sin(\alpha)\right)^2} \tag{7.15}$$

where α is the degree angle between the wind direction and body movement ($0°$ if same direction). In a complex urban area, accounting for the wind direction can be challenging given the turbulent nature of the winds, in which case the square root of the sum of squares of v and v_a will suffice.

In summary, each component of the energy budget equation influences human thermal comfort, where personal factors are controlled by behaviour, and environmental factors are based on the microclimate. The microclimate can be controlled by urban landscape design, and by calculating the individual energy exchanges based on meteorological variables, we can pinpoint those variables that are problematic for heat gain or loss.

Design with urban forestry for thermal comfort

The energy budget is a fundamental concept that lies at the core of microclimatology (Brown and Gillespie, 1995). When we design for thermal comfort, we want to design in a way that will allow the body to maintain an energy balance for the *given activity use of the space* (e.g. sitting versus running). The design and composition of urban forests can dramatically influence specific parameters, as shown in the example in Table 7.2. For example, deciduous trees can block shade in summer and allow sunlight in winter, while coniferous trees can block the harsh winter winds if positioned to block a city's predominant wind direction. The parameters that are easily changed through an urban design intervention – such as the addition of trees – are wind speed, radiation, and energy partitioning (Brown and Gillespie, 1995). The T_a and RH are more difficult to manipulate due to the efficiency of wind at mixing heat and moisture.

Table 7.3 presents how a group of trees within an urban area will affect the variables controlling the human energy budget, addressing factors by season. Because the thermal comfort needs of humans vary based on the climate of interest (temperate versus tropical), designers should look to methods that design with the climate in mind for thermal comfort, an approach entitled 'bioclimatic design'. This approach is even more important as urban areas grow in size and population, and vegetation is often replaced by impermeable surfaces that retain heat, do not cool by evaporation, and are not shaded. Although we cannot do anything to change the prevailing atmospheric conditions, we can modify how they are experienced at the microscale with landscape elements such as trees.

Table 7.3 Analysing the effect of design interventions with trees on the energy flows in the urban landscape impacting thermal comfort during the daytime (assuming no change in surface type)

Variable	Season	
	A: Summer	B: Winter
Air temperature (T_a)	Full leaf deciduous or coniferous trees with low transmissivity limit solar gain to surfaces by approximately 75% on average, thus limiting the emission of terrestrial radiation to heat the air. Evapotranspiration by the leaves use available energy. Both processes effectively lower T_a.	No effect from deciduous trees. Shade supplied by a stand of coniferous trees[3] will block sunlight and the ability of the ground to heat and re-emit warming terrestrial radiation.
	A decrease in T_a will (1) increase convective heat loss from the body, and (2) decrease long-wave radiation exposure and absorption by the human. Net effect is a decrease in the energy budget. (+)	*Conifers shading the ground will decrease T_a and increase convective heat loss. Net effect is a decrease in energy budget from conifers (−), and no change from deciduous. (|)*
Vapour pressure	Increased evapotranspiration will increase vapour pressure.	Deciduous trees lacking leaves will not transpire, while coniferous trees have very little transpiration in winter, thus the presence of either will not change the RH.
	Vapour gradient from skin to air will decrease and effectively decrease the amount of evaporative heat loss.[1] (−)	*No effect on energy budget. (|)*
Wind speed (v)	If trees are positioned to decrease wind speed, low winds will increase the vapour gradient from skin to air, and limit convective mixing.[1]	Winter winds can be deflected by coniferous windbreaks.[4]
	Lowering the wind speed will lower evaporative heat loss. Convective heat loss will lessen with lower winds, and as T_a approaches T_{sk}, convective heat exchange will be limited.[2] Energy budget is overall increased by lowering wind speeds in summer. (−)	*Lowering the wind speed will decrease the amount of heat loss through convection, increasing the energy budget. (+)*
Solar radiation (K)	Full leaf deciduous or coniferous trees will limit the amount of solar radiation research a human.	Winter sunlight (low sun angles) can penetrate through a zero leaf canopy of deciduous trees to warm location, yet will be blocked by a stand of conifers.[3]
	Absorbed radiation due to solar radiation gain will be decreased, thus lowering the energy budget. (+)	*Increased solar insolation will increase the amount of energy gained from the sun, increasing the human energy budget (+); equal to that of open space.*

Variable	Season	
	A: Summer	B: Winter
Terrestrial radiation (L_{in}, L_{up})	A decrease is the surface temperature (T_s) due to shading of the surface decreases emission of terrestrial radiation (L_{up}) towards the body. Shade will block Lin.	When providing shade, conifers block incoming terrestrial radiation and lower T_s and thus T_a. A leafless deciduous tree allows penetrations of L_{in} and K_{in} to warm the surface and increase T_a.
	Absorbed longwave radiation will decrease, lowering the energy budget. (+)	*Shade-providing conifers will lower energy budget (–); bare trees will show no effect on energy budget as compared to open space. (	)*

Bioclimatic design will depend on the extremes of hot, cold, wind, and radiation for a given locale and season. Recommendations are provided for design interventions with trees for a mid-latitude city (23.5–66.5°N or S). In a high latitude city (>66°N or S), one would try to incorporate strategies from column B for thermal comfort design interventions with trees, while in a city in the tropics (23.5°S–23.5°N), one would focus on incorporating strategies from column A. However, a combination of strategies can be employed that accounts for seasonal climates in a given location (e.g. wind break in winter, open in summer). Symbols (+), (–), and (|) denote a beneficial, negative, or neutral change to the human energy budget due to design intervention with trees for the given season, respectively, assuming all other variables held constant. Column A can also be directly related to Table 7.2.

[1]If we want convection to help remove energy from the site, design should not block winds.
[2]Assuming low winds.
[3]At the correct tree and sun orientation to supply shade; if trees do not supply shade then no change.
[4]Use of location-specific wind increase can guide the direction to orient trees.

Conclusions

The knowledge of urban climate clearly demonstrates that vegetation can play a key role in mitigating UHI and thermal discomfort in densely built areas. There are several studies that evaluate in detail these effects of vegetation and trees both at the urban and microclimatic scale, so much that climatic variations induced by urbanization or by greening can be estimated with a certain level of accuracy. Furthermore, such relationships were also found with indices that have been defined to assess consequences on human health and thermal comfort (Petralli et al., 2015). This knowledge is well documented and diffused within the scientific community. Nevertheless, there remains a need to translate this knowledge applied by policymakers and urban planners. With this aim, it is necessary that cause and effect relationships are clearly identified and applied to different contexts. So far, most studies have focused on the relationship between greening and thermal comforts, such as in specific contexts, a canyon, a square, a park or a city. There is a need to move forward and perform studies that aim to combine and compare various approaches in different areas, and to find relationships that can be applied depending on the urban structure and climatic zone. Technological development creates available data and tools, such as meteorological networks, land use data, remote sensing, geographic information systems, models and indices, etc. The next challenge is to integrate such data to support urban planners in developing sustainable cities, and creating living environments with thermally comfortable conditions, that limit bioclimatic stresses associated with negative health and well-being.

References

Abdel-Aziz D.M., Al Shboul A., Al-Kurdi N.Y. (2015) Effects of tree shading on building's energy consumption: The case of residential buildings in a Mediterranean climate. *American Journal of Environmental Engineering* 5(5), 131–140. doi:10.5923/j.ajee.20150505.01

ASHRAE (2013) *Standard 55-2013: Thermal Environmental Conditions for Human Occupancy*. New York: American Society of Heating and Refrigeration Engineers.

Bonan, G. (2002) *Ecological Climatology: Concepts and Applications*. Cambridge: Cambridge University Press.

Brandani G., Napoli M, Massetti L., Petralli M., Orlandini S. (2016) Urban soil: Assessing ground cover impact on surface temperature and thermal comfort. *Journal of Environmental Quality* 45(1), 90–97.

Brown R.D., Gillespie T.J. (1986) Estimating outdoor thermal comfort using a cylindrical radiation thermometer and an energy budget model. *International Journal of Biometeorology* 30, 43–52.

Brown R.D., Gillespie T.J. (1995). *Microclimate Landscape Design: Creating Thermal Comfort and Energy Efficiency*. Chichester: John Wiley & Sons.

Campbell G., Norman J. (1998) *An Introduction to Environmental Biophysics*, 2nd edn. New York: Springer.

DuBois D., DuBois E.F. (1916) Clinical calorimetry: a formula to estimate the appropriate surface area if height and weight be known. *Arch. Intern. Med.* 17, 863–871.

Fanger P.O. (1972) *Thermal Comfort*. New York: McGraw-Hill.

Gosling S.N., Bryce E.K., Dixon P.G., Gabriel K.M.A., Gosling E.Y., Hanes J.M., Hondula, D.M., Liang L., Bustos MacLean P.A., Muthers S., Nascimento S.T., Petralli M., Vanos J.K., Wanka E.R. (2014) A glossary for biometeorology. *International Journal of Biometeorology* 58(2), 277–308.

Höppe P. (1999) The physiological equivalent temperature: A universal index for the biometeorological assessment of the thermal environment. *International Journal of Biometeorology* 43, 71–75.

ISO (1998) *ISO 7726: Ergonomics of the Thermal Environment: Instrument for Measuring Physical Quantities*. Geneva, Switzerland: International Organization for Standardization.

ISO (2007) *ISO 9920: Ergonomics of the Thermal Environment: Estimation of the Thermal Insulation and Water Vapour Resistance of a Clothing Ensemble*. Geneva, Switzerland: International Organization for Standardization.

Jaganmohan M., Knapp S., Buchmann C.M., Schwarz N. (2015) The bigger, the better? The influence of urban green space design on cooling effects for residential areas. *Journal of Environmental Quality* 45(1), 134–145.

Jendritzky G., Dear R., Havenith G. (2012) UTCI – Why another thermal index? *International Journal of Biometeorology* 56, 421–428.

Jendritzky G., Menz H., Schirmer H., Schmidt-Kessen W. (1990) Methodik zur raumbezogenen Bewertung der thermischen Komponente im Bioklima des Menschen (Fortgeschriebenes Klima-Michel-Modell). Beitr Akad Raumforsch Landesplan, No. 114.

Kenny N. A., Warland J. S., Brown R. D., Gillespie T. G. (2008). Estimating the radiation absorbed by a human. *International Journal of Biometeorology* 52(6), 491–503.

Konijnendijk C.C., Nilsson K., Randrup T.B., Schipperijn J. (2005) *Urban Forests and Trees: A Reference Book*. Berlin: Springer.

Krüger E.L, Minella F.O., Matzarakis A. (2014) Comparison of different methods of estimating the Mean Radiant Temperature in outdoor thermal comfort studies. *International Journal of Biometeorology* 58, 1727–1737.

Loughner C.P., Allen D.J., Zhang D.L., Pickering K.E., Dickerson R.R., Landry L. (2012) Roles of urban tree canopy and buildings in urban heat island effects: Parameterization and preliminary results. *Journal of Applied Meteorology and Climatology* 51, 1775–1793. doi: 10.1175/JAMC-D-11-0228.1

Lowry W.P. (1967) The climate of cities. *Scientific American* 217(2), 15–24.

Massetti L., Petralli M., Orlandini S. (2015) The effect of urban morphology on Tilia × europaea flowering. *Urban Forestry & Urban Greening* 14(1), 187–193. doi:10.1016/j.envpol.2014.04.026

Matzarakis A., Mayer H., Iziomon M.G. (1999) Applications of a universal thermal index: Physiological equivalent temperature. *International Journal of Biometeorology* 43, 76–84.

Matzarakis A., Rutz F., Mayer H. (2007) Modelling radiation fluxes in simple and complex environments: Application of the RayMan model. *International Journal of Biometeorology* 51, 323–334.

Matzarakis A., Rutz F., Mayer H. (2010) Modelling radiation fluxes in simple and complex environments: Basics of the RayMan model. *International Journal of Biometeorology* 54, 131–139.

Monteith J., Unsworth M. (2008) *Principles of Environmental Physics*, 3rd edn. Burlington, MA: Elsevier.

Oke T.R. (1982) The energetic basis of the urban heat island. *Q J R Meteorol Soc* 108, 1–24.

Petralli M., Massetti L., Brandani G., Orlandini S. (2014) Urban planning indicators: Useful tools to measure the effect of urbanization and vegetation on summer air temperatures. *International Journal of Climatology* 34(4), 1236–1244.

Petralli M., Brandani G., Napoli M., Messeri A., Massetti L. (2015) Thermal comfort and green areas in Florence. *Italian Journal of Agrometeorology* 20, 39–48.

Petralli M., Morabito M., Cecchi L., Crisci A., Orlandini S. (2012) Urban morbidity in summer: ambulance dispatch data, periodicity and weather. *Central European Journal of Medicine* 7(6), 775–782.

Ren Z., He X., Zheng H., Zhang D., Yu X., Shen G., Guo R. (2013) Estimation of the relationship between urban park characteristics and park cool island intensity by remote sensing data and field measurement. *Forests* 4, 868–886. doi:10.3390/f4040868

Sanusi R., Johnstone D., May P., Livesley S.J. (2016) Street orientation and side of the street greatly influence the microclimatic benefits street trees can provide in summer. *Journal of Environmental Quality* 45, 167–174. doi:10.2134/jeq2015.01.0039

Skoulika F., Santamouris M., Kolokotsa D., Boemi N. (2014) On the thermal characteristics and the mitigation potential of a medium size urban park in Athens, Greece. *Landscape and Urban Planning* 123, 73–86.

Slater G. (2010) The cooling ability of urban parks. MS thesis, School of Environmental and Rural Design, University of Guelph.

Slavin M.I. (ed.) (2011) *Sustainability in America's Cities: Creating the Green Metropolis.* Washington, DC: Island Press.

Sponken-Smith R.A., Oke T.R. (1998) The thermal regime of urban parks in two cities with different summer climates. *Int. J. Remote Sensing* 19, 2084–2104.

Streiling S., Matzarakis A. (2003) Influence of singular trees and small clusters of trees on the bioclimate of a city: A case study. *Journal of Arboriculture* 29, 309–316.

Vanos J.K. (2015) Children's health and vulnerability in outdoor microclimates: A comprehensive review. *Environment International* 76, 1–15.

Vanos J.K., Warland J.S., Gillespie T.J., Kenny N.A. (2010). Review of the physiology of human thermal comfort while exercising in urban landscapes and implications for bioclimatic design. *International journal of biometeorology* 54(4), 319–334.

Vanos J.K., Warland J.S., Gillespie T.J., Kenny N.A. (2012a). Thermal comfort modelling of body temperature and psychological variations of a human exercising in an outdoor environment. *International Journal of Biometeorology* 56(1), 21–32.

Vanos J.K., Warland J.S., Gillespie T.J., Slater G.A., Brown R.D., Kenny N.A. (2012b). Human energy budget modeling in urban parks in Toronto, ON. and applications to emergency heat stress preparedness. *Journal of Applied Meteorology and Climatology* 51(9), 1639–1653.

VDI (1998) *VDI 3787: Environmental Meteorology, Methods for the Human Biometeorological Evaluation of Climate and Air Quality for the Urban and Regional Planning at Regional Level. Part I: Climate.* Berlin: Beuth.

Zhang X., Zhong T., Feng X., Wang K. (2009) Estimation of the relationship between forest patches and urban land surface temperature with remote sensing. *International Journal of Remote Sensing* 30, 2105–2118.

8

URBAN FORESTRY AND POLLUTION MITIGATION

Arne Sæbø, Sara Janhäll,
Stanislaw W. Gawronski and Hans Martin Hanslin

Introduction

In this chapter we discuss how urban trees can contribute to mitigation of pollutants in highly urbanised areas. Reducing pollutant emissions is the main change required to improve environmental quality in urban areas, but hard surfaces and all activities in urban settings, especially transport, contribute to generation of pollutants, and will continue to do so in the future low-emitting cities. Mitigation using urban forestry could address part of the problem, with one of the most important ecosystem functions of vegetation in densely populated areas and along roads being to help protect people from pollution. Special attention must be paid to urban green elements, and dedicated designs are needed for mitigation of contaminants, which are most often of a diffuse nature in urban areas. However, the variation in pollution level between sites within cities can be larger than between so-called polluted and clean cities and thus the local situation should be the main focus when considering amelioration measures.

Pollution in urban environments: current trends and expected developments

Air pollution

Urban air pollution comprises a mixture of gaseous compounds and particulate matter (PM), depending on its source within the traffic, industry or energy generation sector, for example (EEA, 2015). The relative importance of each source differs depending on factors such as city size, level of economic development, geographical region, climate and local weather conditions. It is estimated that the average life span of Europeans is decreased by approximately nine months due to air pollution (WHO, 2013), while in heavily air-polluted areas of Europe the average shortening of life can be nearly three years (EEA, 2007). Children are among the groups most susceptible to the effects of air pollution. Reducing the concentration of PM in the ambient air, especially near busy roads, may be one of the most significant ways to decrease health risks. Although industrial pollution has decreased in most places in the Western world during recent decades, urban air quality has deteriorated dramatically in China and India, for example, due to increasing demand for energy. The rapid urbanisation occurring in many regions worldwide may thus contribute to air pollution still being one of the major causes of mortality in the world by 2050 (OECD, 2012). According to the European Environmental Agency (EEA, 2015), PM and associated pollutants are currently among the most health-threatening factors in urban areas.

Natural particles originate from sea spray, volcanoes, fires, dust storms, soil erosion, fungal mycelium, spores and pollen. Particulate matter consists of particles with aerodynamic diameter in the range 0.001–100 μm (AQEG, 2005). Particulate matter of certain sizes can be suspended in the air for long periods and is transported over long distances. Car exhaust fumes and other combustion sources contain many toxic compounds, with nitrogen oxides (NO_x), soot, organic compounds and heavy metals being among the most harmful pollutants. Volatile organic compounds (VOCs), which originate mainly from fuels and detergents, interact with atmospheric processes and contribute to ozone formation. Heating of houses using wood and gas burners also causes a serious decrease in air quality during winter, especially under atmospheric inversion conditions.

Depending on the size, PM may end up in different parts of the human respiration system; particles smaller than 10 μm (PM_{10}) are small enough to enter human airways and coarse particles (2.5–10 μm) are usually deposited in the upper respiratory tract, whereas fine ($\leq$2.5 μm; $PM_{2.5}$) and ultrafine ($\leq$0.1 μm) particles can reach the lungs and the alveolar regions, or even the blood. Thus, PM can reach various organs, causing major health effects (Nemmar et al., 2002). Particles larger than 10 μm are usually not considered to affect human health directly, but can reduce atmospheric visibility, deposit on soil surfaces and can carry constituents that can be toxic.

Pollution in soil and water

Urban soils and waters are affected by centuries of contamination from industry and traffic, both directly and through air pollution and often show elevated levels of a wide range of organic compounds and elements. Nearly 30 per cent of European soils may have elevated levels of one or more heavy metals (Tóth et al., 2016). Although urbanisation leads to serious negative effects on soils over large areas (FAO and ITPS, 2015), urban soils are characterised by very great spatial variability in their physical, chemical and biological properties both at small and larger scales. This also applies to levels of contaminants in soils, as a result of integrating point sources, diffuse contamination, and the redistribution, mixing and amendment of the soil matrix by construction work and urban greening. Pollution levels and trends and the low quality of urban soils due to sealing, compaction, mixing and adverse abiotic conditions have been well documented (Thornton, 1991; FAO and ITPS, 2015).

Many pollutants, for example metals, organic contaminants, polycyclic aromatic hydrocarbons (PAH) and dioxins, are adsorbed to various extents by sorption complexes. These complexes store pollutants, with some of them being held very strongly (stabilisation), as in the case of mercury (Hg) and lead (Pb), and thus rarely threatening groundwater quality. However, these processes depend on the presence of viable soil organisms, a prerequisite that is not always fulfilled, especially in areas with mostly sealed soils. Metal bioavailability also depends on sorption and solubility in water, with the latter being strongly affected by the pH of the soil solution. All metals except molybdenum (Mo) are soluble at soil pH values of about 6 to 7.

The urban contaminant matrix also changes as emissions change owing to technological development, such as the introduction of catalysts and changes in fuels, and owing to expanded and elevated urban metal metabolism and management. The use of composts and biosolids in urban greening also increases the cycling and bioavailability of contaminants, resulting in 'chemicals of emerging concern' (FAO and ITPS, 2015), for example pharmaceuticals, flame retardants and many other compounds of anthropogenic origin.

Managing urban soil quality and restoration is a pressing concern with consequences for urban forestry. Human exposure to contaminants in soil occurs mainly through ingestion of soil or plant parts grown in contaminated soil, but can also take place through inhalation of dust or volatile compounds. This topic is of high importance in urban gardening and agriculture, but since it is of less importance for urban forestry, it is not discussed further in this chapter.

Urban waters not entering wastewater systems receive a complex mixture of contaminants via precipitation runoff, leaching and accidental spills from industry, roads, gardens and public spaces, combined with runoff and leaching of historical contaminants from industrial areas and landfills. Sediments often contain polluted layers and some of the exposed contaminants may be redistributed when urban waterways are reopened and restored.

Dispersion and deposition of air pollutants in the urban environment

Dispersion of traffic-related PM emissions is used as an example in the following discussion, but the principles also apply to other local emissions, since traffic exhaust fumes are rather similar to other combustion effluents and eroded particles behave in many ways like dust. However, the specific compounds emitted by traffic may behave rather differently, depending on particle size. Most eroded particles are coarse, while exhaust particles are in the ultrafine size range. Larger particles are more frequently deposited at sites where the airstream loses speed and changes direction, for example at corners. Particle deposition on objects in the air stream, for example on fibres, is also usual. Ultrafine particles, which are often related to exhaust fumes, settle through diffusion and thus increased deposition occurs when exhaust gases remain close to a deposition surface for some time. Even if particles larger than PM_{10} are less relevant for health, these particles reduce visibility and are deposited on surfaces, causing soiling and changing their properties. They are also frequently included in measurements of particle deposition as they have a comparatively large mass, which may mask the presence of smaller (coarse, fine and ultrafine) particles that are too small to affect the results, but are more damaging to human health.

Urban heat islands affect the wind patterns in cities. Human activities release heat, street canyons reduce cooling during the night, and low-albedo materials with large heat storage capacity lead to greater accumulation of heat in cities than in rural and green areas (Oke, 1987). The warmer urban air lifts during the night, drawing air from the rural surroundings into the city and creating new wind systems. On the other hand, wind patterns caused by urban heat island in cities can mitigate inversions and increase the dilution of urban effluents. If the amount of vegetation in the city increases, the difference between urban and rural areas decreases and the urban heat island does not form as easily. Large parks within the urban area are cooler and can contribute to diluting air pollution by mixing polluted air in the city with the cleaner air in the park (Oke, 1987).

Buildings and other obstacles affect the wind in urban areas. The wind speed. and direction can change totally within local areas compared with regional wind systems. A common example is the street canyon, i.e. the space between buildings where traffic flows.

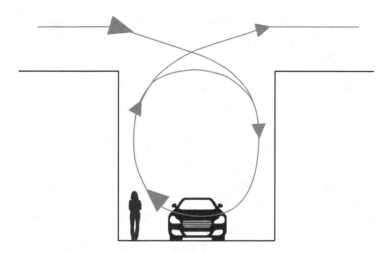

Figure 8.1 Air moving from left to right above the roof tops forms a vortex between the houses in the street canyon, moving the traffic exhaust fumes towards the pavement on the left. If the canyon height is much greater than the width, two vortices may be stacked on top of each other. (Adapted from openclipart.org)

In street canyons, two different systems compete: wind direction along or across the street. Longitudinal winds are normally slightly faster than regional winds, since all air needs to be compressed into the street canyons as the buildings block the other paths (the tunnel effect). For winds perpendicular to the street axis, vortices are formed within street canyons (Figure 8.1) and wind direction at ground level is then opposite to that above roof level. The polluted air on the ground may thus move in circles inside the canyon, reducing the degree of mixing with the cleaner air above the rooftops. In narrow street canyons (large height–width ratio; Figure 8.1), more vortices may form on top of each other, further decreasing the dispersion. When vegetation is introduced into these systems, the intensity of the vortices decreases (Salmond et al., 2013).

Erecting a barrier between the traffic and pedestrians can be one way to decrease their exposure to air pollutants without changing the amount of pollutants in the air. The barrier increases the distance between pedestrians and traffic, directly decreasing exposure, and may

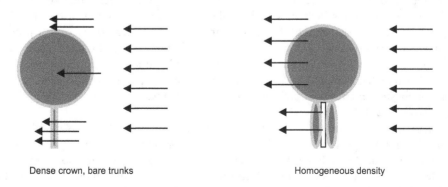

Dense crown, bare trunks   Homogeneous density

Figure 8.2 Air streams go where the resistance is smallest. Homogeneous density is important if the air is to be filtered instead of being deflected to open spaces, for example under a dense crown canopy of street trees

also move pollutants in different ways. Vegetation barriers normally behave differently from impermeable barriers, allowing some of the air to pass through and some of the pollutants to be deposited on the leaves and branches. If there are gaps in the barrier, the air will flow through the gap and only the deposition surface area close to the gap will be used as a filter for the air (Figure 8.2). Downwind from the vegetation the pollution concentrations are lower, because of both dilution and deposition.

Role of vegetation in mitigating air pollution

The effect of vegetation on air quality depends on how it interferes with deposition and dispersion of pollutants (Janhäll, 2015). Vegetation can reduce wind speed, direct air flow and intercept particles and gases, depending on physical attributes such as height and canopy structure, the position in the urban landscape and the distance to pollution sources. Hence, the effects will differ between green walls, hedges, single trees, rows of trees and a dense tree cover in parks. In many regions of the world, the effects also vary throughout the year with changes in canopy cover.

Effects on dispersion

Dispersion of pollutants out of street canyons is essential to prevent accumulation of pollutants and negative health effects. Unless carefully selected and designed, vegetation may hinder the air circulation and decrease the already limited dilution of emissions in cities. The main problem is the use of street trees that trap pollutants below their canopies. One way to handle this is to use wall and roof vegetation, shrubs, grasses and low hedges (Vos et al., 2012) instead of large trees and screens that hinder air circulation. Thus, the aim is to increase leaf area to intercept pollutants without affecting air circulation and dispersion. The effect of plant canopies on air flow depends on their aerodynamic properties and permeability. It is crucial to know how the air will be funnelled through the vegetation at different sites, with different designs and, most importantly, with different densities.

Introducing vegetation further decreases the ventilation and increases the concentration of air pollutants in the street at pedestrian level, as shown using the ENVIMET model, for example (Wania et al., 2012). Those authors suggested using lower vegetation and sufficiently sparse planting to retain good ventilation and prevent accumulation of pollutants at street level. Gromke et al. (2016) found in a wind tunnel experiment and a street canyon model that a continuous hedgerow of high density (low permeability) in the middle of the street could decrease pollution at pedestrian head level by 30–60 per cent if properly designed, but that other designs could increase the pollution levels. This effect was because of changes caused by the hedgerow in the dispersion of pollution. Pollution deposition was not examined in that study, but the importance of the type and placement of the vegetation in obtaining maximum effects on pollution through the ventilation of streets was emphasised (Gromke et al., 2016).

Increasing the vegetation leaf area in pollution hotspots without decreasing dilution of pollution can still improve the positive effect of vegetation on urban air quality by increasing deposition (Janhäll, 2015). However, King et al. (2014) found that the emissions of nitrogen dioxide (NO_2) in New York were much larger than deposition. In that study, the modelled air quality effects of different 900 m^2 square grids were attributed to absence of emission sources in areas with the best air quality, which coincided with the largest tree canopy. Emphasis must be on the correct placement of efficient filtering vegetation close to emission sources. In many places, the air circulation is relatively easy to predict. For example in many suburban areas and along heavily trafficked roads, vegetation can separate the polluted road from vulnerable areas near it without having negative effects on air circulation and pollution dilution.

Effects on deposition

As air moves through the canopy, leaves intercept air pollution, reduce air velocity and act as a passive sink of particulate pollution. Trees are considered to be most efficient in accumulating particulate matter (Nowak, 1994; McDonald et al., 2007) because of their size, large leaf area and complex crown architecture giving turbulent air movements that increase the deposition of air pollution on leaves (Fowler et al., 1989). The interception of pollution is affected by processes at several scales, from the leaf surface to the canopy structure. Leaf surface morphology, such as presence of hairs and trichomes, leaf shape and size, branching patterns and canopy density, are vegetation factors that determine the dry deposition velocity of PM. These attributes differ considerably between plant species and planting designs. Large differences between species in PM accumulation have been observed in several studies (Dzierżanowski et al., 2011; Sæbø et al., 2012). This effect also depends on particle size and many of the relevant studies focus on coarse particles or even particles larger than 10 μm, which have limited health impacts on humans. The differences between species reported in the literature are in the order of 10- to 15-fold, with typical PM accumulation rates of 10-70 μg cm^{-2} leaf area. Pine species have repeatedly been shown to be efficient in capturing PM (Beckett et al., 1998, 2000; Sæbø et al., 2012) and should be used more widely in urban areas, particularly because they are efficient during winter when pollution concentrations are highest. However, conifers should not be placed in the front line, for example, at road edges, where the salt and pollution concentrations are highest. Also, they should be planted only where there is a sufficient space for canopy growth. Several studies have documented positive correlations between vegetation cover and air quality in parts of cities, illustrating the combined effect of vegetation on air pollution and on lowering emissions per unit area. The same trends can be found for air quality from the edge to the interior of urban forests.

Rain can wash off PM deposited on surfaces, leaving the leaf recharged for accumulation of more PM, or pollutants may become tightly bound within the wax layer of the leaf. Some of the PM deposited is re-suspended by wind. The fate of the PM washed off the leaves should be considered in planning processes. If these pollutants end up on hard surfaces, the danger of re-suspension increases. However, if rainwater carries the PM to an infiltration zone, the particles can be immobilised in the soil, for example by phosphate groups (Hafsteinsdottir et al., 2015) or by organic matter (Cui et al., 2016). Soil amendments can increase immobilisation processes (Tica et al., 2011). Gaseous contaminants enter the leaf tissues through the stomata and may be metabolised by plants (NO_2), but can also cause damage to the tissue if absorbed in large quantities (NO_2; O_3; and sulphur dioxide, SO_2). Plants have developed defence mechanisms related to stomatal conductance, resistance in mesophyll gas exchange and action of antioxidants naturally occurring in the plant tissue (Teklemariam and Sparks, 2006). Any NO_x compounds taken up and metabolised by plants will thus be taken out of the system. Some SO_2 and O_3 may also be absorbed by plant tissues, but those gases can easily harm the tissues. Particulate matter carries many pollutants (e.g. PAH, polychlorinated dibenzo-p-dioxins/dibenzofurans, polychlorinated biphenyls, heavy metals) that can penetrate deeper into plant tissues and interfere with photosynthesis and growth (Vardaka et al., 1995). As PM accumulates on the surface, some particles may clog stomata and disrupt their function, while others may cause erosion of the wax layer. Such changes have consequences for photosynthetic efficiency and water use. Thus, the ability to capture PM and an ability to tolerate or handle the associated stress are important traits in selection of optimal vegetation for urban and suburban areas.

Plants may also release pollutants

The release of biogenic volatile organic compounds (BVOC) from vegetation can have significant air quality implications in polluted urban areas, since they are involved in ozone formation or removal (Calfapietra et al., 2013). Since the quality and quantity of BVOC emissions is species-specific (Ciccioli et al., 2004), this aspect must be considered when determining the appropriate choice of vegetation for large-scale urban tree planting programmes.

Scaling up

When scaling up from the individual tree or planting bed to larger areas, much of the efficiency of urban greening to mitigate air pollution has to be estimated using models. In the much used iTree model (see e.g. Morani et al., 2014), the vegetation is represented by a 'big leaf' factor, but the model description does not specify which species comprises the largest proportion of the total canopy. Such models assume homogeneous conditions over large areas, based on point measurements, and do not consider species-specific differences nor spatial differences in accumulation potential. Model runs provide estimates of PM removal from the air, but the values presented do not always correlate well with measured levels of pollutants in air. The predicted effect of vegetation on modelled and measured levels of air pollution is often moderate. One exception is the effects found in large parks, where the air pollution shows large drops but at correspondingly large distances from the source. There is still a considerable level of scientific uncertainty on the effect of vegetation on air quality and models should be further validated by means of *in situ* measurements on order to develop the knowledge needed to make informed choices in urban design and greening.

Vegetation designs for improved effects

Trees interact with soil and water contaminants in several ways. First, they require a certain tolerance to soil contaminants such as heavy metals and salinity for long-term quality and function. They may also play a role in pollution mitigation, since under certain conditions they contribute to degradation of organic contaminants and interfere with the mobility and bioavailability of metal and metalloid contaminants. Only a limited number of organic contaminants can be remediated using trees. The interaction with soil microorganisms is important and can be stimulated with site-specific soil management and choice of vegetation. There are two main approaches to mitigate metal and metalloid air pollution using vegetation, which involve either using the vegetation to increase availability for plant uptake and translocation to aerial parts for harvest, or using the vegetation to bind and stabilise contaminants in the soil (Pulford and Watson, 2003). Short-rotation coppice with *Salix* can lower the levels in soil of mobile elements such as cadmium (Cd) and zinc (Zn). However, this phytoextraction process takes time and is feasible only for mobile elements at low to moderate contamination levels and in locations where aboveground structures, including leaves, can be harvested at regular intervals. Phytostabilisation, on the other hand, prevents the uptake and/ or translocation of contaminants to shoots and leaves and is probably the most feasible method for urban forestry in general. There are large differences between tree species in their ability to immobilise elements in the root zone, accumulate them in roots and translocate them to stem and leaves. Solutions for both phytoextraction and phytostabilisation must be highly site-specific and have to be tailored to soil conditions, climate, soil management, additional biological and chemical treatments, and available plant material.

Design for best effects

Vegetation barriers between polluted areas and vulnerable areas (for example playgrounds, schools, hospitals and residential areas) should be further explored for their environmental effects. However, urban vegetation has the potential to improve environmental quality and reduce human exposure to pollutants only if this vegetation use is strictly targeted. Careful consideration of the design of urban vegetation elements and their relationship to the built environment and to micrometeorology is the key to success (Janhäll, 2015). The approach must be based on the following high priority principles:

1 Plants must be used in a way that does not hinder air circulation, but rather increases ventilation of city streets.
2 Maximising leaf area for pollutant deposition must be part of the strategy.
3 Systems must be put in place to ensure that deposited pollutants are either taken out of the environment or immobilised.

Balancing different processes: air circulation, dispersion and deposition

The placement and type of screens are crucial to the ventilation effect and thus to mitigation of air pollution near roads. Such measures could both re-direct air streams and filter the air, thus shielding vulnerable areas from the highest concentrations of pollutants. However, vegetation can also prevent air movements in street canyons, causing entrapment of contaminants in the breathing zone of city dwellers. Tong et al. (2016) recommend either a wide vegetation belt comprising large leaf area downwind of roads, or a combination of a solid barrier and vegetation. Tall screens can exacerbate the pollution concentrations at the road level. Several of the studies referred to above show that vegetation must be placed so that air exchange is not affected in a negative way. Best practice guidelines for planning urban green structures need to include this point.

Maximising leaf area

Ground-covering grasses or herbs, shrubs, small trees, climbers and large trees can together make an immense total leaf area in urban environments. For example, Sternberg et al. (2010) reported that ivy, when grown on urban walls, can act as a 'particle sink', adsorbing PM.

Mori et al. (2015), showed that the deposition of elements and PM seems to be linked more to the growth parameters of plants rather than leaf anatomy. Diversity makes it possible to use much more of available areas. Road and street verges, parks, walls and roofs should be harnessed for vegetation cover, with walls and roofs still being little utilised in most cities. However, technically sound and secure designs must be employed for climbers and roof vegetation, or else damage to buildings will easily prevent implementation of this measure. Introduction of large new areas of vegetation would not only benefit pollution mitigation but would also improve stormwater management, reduce the heat island effect and increase aesthetic values and residents' wellbeing, issues which have great scope for improvement if available areas were better exploited than they are at present.

Removing contaminants from the environment

The PM deposited in cities often ends up on hard surfaces, from which it is later re-suspended by wind or washed off by rain. After wash-off, the pollutants may reach street level, where they either end up in the drainage system or are re-suspended again. This is a circle of events that preserves too much of the pollutant load in the human environment. It is essential to find ways to immobilise the contaminants shortly after they are deposited. Soft surfaces (i.e. with infiltration capability) or drainage systems that transport the polluted wash-off water from plants will be important for transporting pollutants out of the cycle of deposition and re-suspension. However, if the polluted water is diverted to a drainage system, the pollutants will still end up in other parts of the human environment. The best option would therefore be to funnel the water for sedimentation and decontamination. Systems can be introduced for this purpose, but need careful, thorough planning.

Dialogue and cooperation between different professionals, including policy makers, planners, architects, landscape architects and managers of the urban environment, must be established in this very important process in order to succeed in improving the quality of the urban humanosphere.

Summary

The current level of pollution in cities world-wide shows that there is still a great need to decrease emissions and mitigate pollution. The potential of plants for PM accumulation is still incompletely documented and insufficiently exploited for the benefit of urban dwellers. More good data are needed on total pollutant accumulation on leaves, runoff with rain and wind removal during different seasons. Modelling work performed to date needs to be validated with *in situ* measurements of air quality. The most demanding challenge is probably to gain a good knowledge about the design and appropriate placement of filtering screens consisting of suitable plants, in order to maximise pollutant deposition without negatively affecting air exchange in urban areas.

References

AQEG. 2005. Particulate matter in the United Kingdom. Retrieved from www.defra.gov.uk/environment/airquality/aqeg.

Beckett, K.P., Freer-Smith, P., Taylor, G. 1998. Urban woodlands: Their role in reducing the effects of particulate pollution. *Environmental Pollution* 99(3): 347–360.

Beckett, K.P., Freer-Smith, P., Taylor, G. 2000. Effective tree species for local air quality management. *Journal of Arboriculture* 26(1): 12–19.

Bell, J.N.B., Honour, S.L., Power, S.A. 2011a. Effects of vehicle exhaust emissions on urban wild plant species. *Environmental Pollution* 159: 1984–1990.

Bell, M.L., Morgenstern, R.D., Harrington, W. 2011b. Quantifying the human health benefits of air pollution policies: Review of recent studies and new directions in accountability research. *Environmental Science and Policy* 14: 357–368.

Bergbäck, B., Johansson, K., Mohlander, U. 2001.Urban metal flows: A case study of Stockholm. Review and conclusions. *Water, Air and Soil Pollution* 1: 3–24.

Calfapietra, C., Fares, S., Manes, F., Morani, A., Sgrigna, G., Loreto, F. 2013. Role of biogenic volatile organic compounds (BVOC) emitted by urban trees on ozone concentration in cities: A review. *Environ Pollut* 183: 71–80.

Caricchia, A.M., Chiavarini, S., Pezza, M. 1999. Polycyclic aromatic hydrocarbons in the urban atmospheric particulate matter in the city of Naples (Italy). *Atmospheric Environment* 33(23): 3731–3738.

Chen, L., Liu, C., Zou, R., Yang, M., Zhang, Z. 2016. Experimental examination of effectiveness of vegetation as bio-filter of particulate matters in the urban environment. *Environmental Pollution* 208(A): 198–208.

Ciccioli, P., Baraldi, R., Mannozzi, M., Rapparini, F., Nardino, M., Maglietta, F., Brancaleoni, E., Frattoni, M. 2004. Emission and flux of terpenoids released from the terrestrial ecosystem present in the Pianosa island. *Journal of Mediterranean Ecology* 5(1): 41–51.

Cui, L., Pan, G., Li. L., Bian, R., Liu, X., Yan J., Quan, G., Ding, C., Chen, T., Liu, Y., Liu, Y., Yin, C., Wei, C., Yang, Y., Hussain, Q. 2016. Continuous immobilization of cadmium and lead in biochar amended contaminated paddy soil: A five-year field experiment. *Ecological Engineering* 93: 1–8.

Dzierżanowski, K., Popek, R., Gawrońska, H., Sæbø, A., Gawroński, S.W. 2011. Deposition of particulate matter of different size fractions on leaf surfaces and in waxes of urban forest species. *International Journal of Phytoremediation* 13: 1037–1046.

EEA. 2007. *Air Pollution in Europe 1990–2004*. Report No. 2/2007. Office for Official Publications of the European Communities, Copenhagen.

EEA. 2015. *Air Quality in Europe: 2015 Report*. Office for Official Publications of the European Communities, Copenhagen. doi:10.2800/62459.

FAO and ITPS. 2015. *Status of the World's Soil Resources (SWSR) – Main Report*. Food and Agriculture Organization of the United Nations and Intergovernmental Technical Panel on Soils, Rome, Italy.

Fowler, D., Cape, J.N., Unsworth, M.H. 1989. Deposition of atmospheric pollutants on forests. *Philosophical Transactions of the Royal Society B* 324: 247–265.

Gavett, S.H., Madison, S.L., Dreher, K.L., Winsett, D.W., McGee, J.K., Costa, D.L. 1997. Metal and sulphate composition of residual oil fly ash determines airway hyperactivity and lung injury in rats. *Environmental Research* 72: 162–172.

Ghio, A., Devlin, R. 2001. Inflammatory lung injury after bronchial instillation of air pollution particles. *American Journal of Respiratory and Critical Care Medicine* 164: 704–708.

Gromke, C., Jamarkattel, N., Ruc, B. 2016. Influence of roadside hedgerows on air quality in urban street canyons. *Atmospheric Environment* 139: 75–86.

Hafsteinsdottir, E.G., Camenzuli, D., Rocavert, A.L., Walworth, J., Gore, D.B. 2015. Chemical immobilization of metals and metalloids by phosphates. *Applied Geochemistry* 59: 47–62.

Janhäll, S. 2015. Review on urban vegetation and particle air pollution: Deposition and dispersion. *Atmospheric Environment* 105: 130–137, doi: 10.106/j.atmosenv.2015.01.052.

King, K.K., Johnson, S., Kheirbek, I., Lu, J.W.T., Matte, T. 2014. Differences in magnitude and spatial distribution of urban forest pollution deposition rates, air pollution emissions, and ambient neighborhood air quality in New York City. *Landscape and Urban Planning* 128: 14–22.

Laden, F., Neas, L.M., Dockery, D.W., Schwatz, J. 2000. Association of fine particulate matter from different sources with daily mortality in six U.S. cities. *Environ Health Persp* 108(10): 941–947.

McDonald, A.G., Bealey, W.J., Fowler, D., Dragosits, U., Skiba, U., Smith, R.I., Donovan, R.G., Brett, H.E., Hewitt, C.N., Nemitz, E. 2007. Quantifying the effect of urban tree planting on concentrations and depositions of PM10 in two UK conurbations. *Atmos Environ* 41(38): 8455–8467.

Morani, A., Nowak, D.J., Hirabayashi, S., Calfapietra, C. 2014. How to select the best tree planting locations to enhance air pollution removal in the MillionTreesNYC initiative. *Environ Pollut* 159: 1040–1047.

Mori, J., Sæbø, A., Hanslin, H.M., Teani, A., Ferrini, F., Fini, A., Burchi, G. 2015. Deposition of traffic-related air pollutants on leaves of six evergreen shrub species during a Mediterranean summer season. *Urban Forestry and Urban Greening* 14(2), 264–273.

Nemmar, A., Hoet, P.H.M., Vanquickenborne, B., Dinsdale, D., Thomeer, M., Hoylaerts, M.F., Vanbilloen, H., Mortelmans, L., Nemery, B. 2002. Passage of inhaled particles into the blood circulation in humans. *Circulation* 105: 411–414.

Nowak, D.J. 1994. Air pollution removal by Chicago's urban forest. In G.E. McPherson, D.J. Nowak, R.A. Rowntree (eds), *Chicago's Urban Forest Ecosystem: Results of the Chicago Urban Forest Climate Project*. General Technical Report NE-186: 63-81. USDA, Washington, DC.

OECD. 2012. *OECD Environmental Outlook to 2050: The Consequences of Inaction*. OECD, Paris. doi: 10.1787/9789264122246-en

Oke, T.R. 1987. *Boundary Layer Climates*. Routledge, Abingdon.

Pulford, I.D., Watson, C. 2003. Phytoremediation of heavy metal-contaminated land by trees: A review. *Environment International* 29(4): 529–540.

Sæbø, A., Popek, R., Nawrot, B., Hanslin, H.M., Gawronska, H., Gawronski, S.W. 2012. Plant species differences in particulate matter accumulation on leaf surfaces. *Science of the Total Environment* 427–428: 347–354.

Salmond, J.A., Williams, D.E., Laing, G., Kingham, S., Dirks, K., Longley, I., Henshaw, G.S. 2013. The influence of vegetation on the horizontal and vertical distribution of pollutants in a street canyon. *Sciences of the Total Environment* 443: 287–298.

Sternberg, T., Viles, H., Cathersides, A., Edwards, M. 2010. Dust particulate absorption by ivy (*Hedera helix* L) on historic walls in urban environments. *Science of the Total Environment* 409(1): 162–168.

Teklemariam, A.T., Sparks, J.P. 2006. Leaf fluxes of NO and NO2 in four herbaceous plant species: The role of ascorbic acid. *Atmospheric Environment* 40: 2235–2244.

Thornton, I. 1991. Metal contamination of soils in urban areas. In P. Bullock and P. J. Gregory (eds), *Soils in the Urban Environment*, Blackwell Publishing, Oxford, ch. 4. doi: 10.1002/9781444310603

Tica, D., Udovic, M., Lestan, L. 2011. Immobilization of potentially toxic metals using different soil amendments. *Chemosphere* 85: 577–583.

Tong, Z., Baldauf, R.W., Isakov, V., Deshmukh, P., Zhang, K.M. 2016. Roadside vegetation barrier designs to mitigate near-road air pollution impacts. *Science of the Total Environment* 541: 920–927.

Tóth, G., Hermann, T., Szatmári, G., Pásztor, L. 2016. Maps of heavy metals in the soils of the European Union and proposed priority areas for detailed assessment. *Science of the Total Environment* 565: 1054–1062.

Vardaka, E., Cook, C.M., Lanaras, T., Sgardelis, S.P., Pantis, J.D. 1995. Effect of dust from a limestone quarry on the photosynthesis of *Quercus coccifera,* an evergreen schlerophyllous shrub. *Environmental Contamination and Toxicology* 54: 414–419.

Vos, P.E.J., Maiheu, B., Vankerkom, J., Janssen, S. 2012. Improving local air quality in cities: To tree or not to tree? *Environm Poll* xxx: 1–10.

Voutsa, D., Samara, C. 2002. Labile and bioaccessible fractions of heavy metals in the airborne particulate matter from urban and industrial areas. *Atmospheric Environment* 36: 3583–3590.

Wania, A., Bruse, M., Blond, N., Weber, C. 2012. Analysing the influence of different street vegetation on traffic-induced particle dispersion using microscale simulations. *Journal of Environmental Management* 94: 91–101.

WHO. 2013. *Review of Evidence on Health Aspects of Air Pollution: REVIHAAP Project*. Technical report. World Health Organization, Geneva. Retrieved from www.euro.who.int/pubrequest.

Yang, J., McBride, J., Zhou, J., Sun, Z. 2005. The urban forest in Beijing and its role in air pollution reduction. *Urban For and Urban Green* 3(2): 65–68.

9

URBAN FORESTS AND BIODIVERSITY

Emilio Padoa-Schioppa and Claudia Canedoli

Introduction

Traditionally, urban areas have been regarded as locations of low biodiversity that are dominated by non-native species (Alvey, 2006). Many studies have provided evidence to show that urban and suburban areas can rather contain relatively high levels of biodiversity (Jim and Liu, 2001; Araújo, 2003; Godefroid and Koedam, 2003; Cornelis and Hermy, 2004). Urban forests can play a crucial role for the conservation of many species of flora and fauna and they can therefore be actively managed by foresters and city planners to preserve that diversity. Moreover, urban forests and in general vegetation occurring in urban areas (like parks, patches of woodland, residential areas or other components of a green infrastructure) can comprise a significant percentage of a nation's tree canopy, leading their management to have important effects not only to a local but also to a national scale (Florgård, 2010; Baffetta et al., 2011). Biodiversity in forests can be preserved at high levels by maintaining large and undisturbed natural habitats. However, in urban and suburban areas this may be not feasible because of little natural areas remaining (Alvey, 2006) and negative pressures coming from the urban system. Urban habitats may provide refuges for species whose native habitats have been greatly diminished (Gavareski, 1976; Mills et al., 1989). Unfortunately, only small and isolated natural patches persist in urban areas and they attract people living in cities, being therefore affected by multiple stressors (Ficetola et al., 2007). Due to their fragility in representing suitable natural habitats and the direct disturbance caused by large numbers of visitors, it is therefore urgent to protect biodiversity of residual forested patches, because for many species they represent the only available habitat (and this can be particularly important for rare or endemic species) (Ficetola et al., 2007). Undertaking good practices of management of vegetation in the urban ecosystem is therefore a great strategy for the sustainability of human development and for cities' resilience (Kraxner et al., 2016).

In this chapter, we will discuss why it is of fundamental importance to maintain biodiversity in urban forests, which are the most common taxa that can occur and how the structure of the urban forests can influence the presence of species. Case studies are used as examples to illustrate which actions are needed to maintain and preserve biodiversity in urban environment.

Levels of biodiversity

Usually, when talking about biodiversity it is common to refer to the diversity of species of flora and fauna. However, biodiversity concept involves different levels of organizations, from genes to species, habitats/ecosystems or landscapes (Primack, 2000). *Genetic diversity* is the diversity of genetic materials within species. Individuals of a certain species are carriers of genes and their variability is responsible for the different traits in that species. It is of fundamental importance to maintain a genetic pool in order to allow species to better respond to environmental changes. *Species diversity* includes the variety (number of species) of the living organisms that exist like plants, animals, fungi and microbes. *Habitats/ecosystems diversity* refer to a larger scale, including not only the species that exist at a place but also the interactions between them and the physical environment. An ecosystem is composed of a community of organisms which lives within the physical environment; and the interactions which occur between them and all these features can be expressed at different scales, from microsites (e.g. a dead tree trunk or a pond) to larger habitats (e.g. a forest or a mountain). At a major scale, *landscape diversity* represents the set of landscapes existing in a region, a country or within a certain area of interest. In an urban environment, landscape is profoundly shaped by human imprint and actions are constantly undertaken to maintain that territorial organization and dynamic. Quality and diversity of landscapes evolve under the pressure of human culture over time. Even though the most common actions undertaken to conserve or promote the diversity of the biodiversity interest species', one should be aware that this represents only one level of biodiversity which is needed. Through actions taken at one of these levels, other levels might be influenced (e.g. maintaining a healthy population of woodpeckers in an urban forest, can imply having more individuals of that species to carry genetic diversity or enhance the functioning of the whole ecosystem through their role in the food chain).

Why is it important to preserve biodiversity in urban forests?

Urban areas are responsible for threats to the survival of many species. A study from McDonald et al. (2008) estimate that, only considering vertebrate species, urbanization is implicated in the listing of around 8 per cent of the IUCN (International Union for Conservation of Nature). Protected areas cannot protect biodiversity by itself (Rodrigues et al., 2004), therefore it is of fundamental importance to promote conservation actions also in non-protected areas. At the same time urban areas can bring many species because of geographical reasons or for the concentration of many organisms of different origins. A study from Araújo (2003) showed that urban areas across all of Europe seem to contain higher levels of biodiversity than unpopulated areas (this association was true for widespread, narrowly endemic, and threatened species). Moreover, areas deemed suitable for urban development might often coincide with those areas that also support high native species richness and endemism (Lugo, 2002). These results further emphasize the importance of biodiversity conservation in and around dense human settlement (Alvey, 2006).

Biodiversity is valuable for many reasons. Humankind derives tremendous benefits from the wide diversity of species, genes, ecosystems and landscapes on Earth. The awareness of the importance of nature's services has brought the scientific world to approach ecosystems in terms of services they provide to humans. Biodiversity fits the concepts of ecosystems services in many ways. It is thus possible to value biodiversity as a regulator of ecosystem services, a final ecosystem service, or a good itself (Mace et al., 2012). The regulatory role of biodiversity is expressed in the control made by biodiversity in the ecosystem processes that underpin ecosystem services (e.g. the dynamics of many soil-nutrient cycles are determined by the composition of biological communities in the soil; see Bradford et al., 2002). Biodiversity can be also seen as a final ecosystem service, because biological diversity (at the level of genes and species) contributes directly to some goods and their values (think for example at the importance of wild crop relatives for the improvement of crop strains). Therefore, both genetic and species diversity are final ecosystem services directly contributing to goods (Mace et al., 2012). But biodiversity is also a good in itself with a distinct value. It has a cultural value, that encompasses appreciation of wildlife and scenic places and spiritual, educational, religious and recreational values (Bennett and Hassan, 2003). The 'biophilia hypothesis' (Kellert and Wilson, 1993) may explain also the value of urban forests for human health (Wilson, 2002). In the case of final ecosystem services, ecosystems could be specifically managed for the diversity of the desired biodiversity components (Mace et al., 2012). The loss of biodiversity is of critical concern, given that diversity plays an important role in long-term ecosystem functioning. Interactions among biotic and abiotic components of ecosystems involve ecological and evolutionary processes, and ultimately lead to the stocks and flows that underpin the final ecosystem services (Mace et al., 2012). Regarding their importance, ecosystem services derived from urban forests improve resilience and quality of life in cities (Gómez-Baggethun et al., 2013) and social as well as ecological benefits will be gained through biodiversity protection.

The presence of certain species of flora and fauna might be used as ecological indicator to determine the quality of an environment (Padoa-Schioppa et al., 2006; Barrico et al., 2012). By monitoring some taxa we can gain information about their status (presence, abundance and distribution) but they can also give us precious information about the quality of the environmental conditions. This is particularly of concern in high-altered environment where disturbing factors are several (Larsen et al., 2010) and sometimes of difficult identification or measurement. Conducting a census on species and implementing a monitoring programme allows us to determine the presence and status of species of conservation interest. This is of fundamental importance because the presence of endangered species might determine the implementation of specific actions arising from protection laws.

Which flora and fauna occur in urban forests?

Urban forests have been shown to harbour several taxa of flora and fauna, ranging from native species (some of them endangered or of high conservation value) to exotic and invasive species. A central concern in urban biodiversity is biotic homogenization (Alvey, 2006). It is defined as 'the process of replacing localized native species with increasingly widespread non-native species' (McKinney, 2008). Examples of widespread established species common to urban ecosystems throughout the globe are the rock dove (*Columbia livia*), house mouse (*Mus musculus*) and feral house cat (*Felis catus*) and they have been shown to have a direct impact on the loss of native species (Alvey, 2006). The consequences of biotic homogenization act at a local scale as well as a global scale by leading to local extinctions and to the decrease of the overall global biodiversity. According to this, studies predict that global biodiversity is expected to continue to decline for at least the next few centuries, and urban areas play a crucial role in this process.

Biodiversity in urban environment is also a question of scale. In fact, alpha diversity (species richness in single habitat patches) has often been found to be high in urban habitats (Niemelä, 1999). This is due to the introduction of new organisms by people and to the fact that many species of different origins (from native to exotic) might find suitable conditions in the anthropogenic habitats (where some resources such as food are concentrated). However, high alfa diversity doesn't mean necessarily a healthy environment in ecological terms. As we have previously highlighted, the widespread establishment of some species particularly adaptable at the urban environment can from one side increase the number of species occurring at local scale (when their presence doesn't lead to local extinctions) but at the same time will contribute to a lower biodiversity at larger scale (we can find the same species everywhere around the globe). Moreover, apart from the mere number of species occurring in a habitat, it is fundamental to consider the ecological role of each of these species in the ecosystem functioning and the inter-relationships occurring between them. In fact, simply counting the number of species itself provides little information about specific effects of urbanization. Species richness used as an indication of biological conservation value may be misleading because disturbances may favour certain species (widespread and abundant generalists) leading to an increase in species richness (Niemelä, 2001). Community assemblage or the presence of rare or threatened species can be used (alongside species richness) to better provide an indication of conservation value of the environment.

Birds

Birds are probably the most studied taxon in urban environment. Compared to other vertebrates, they are easy to monitor by skilled observers and provide a mechanism to explore urban effects and responses to different urban designs (Chace and Walsh, 2006). Native avian communities change as urban development rises. There is a general tendency for bird communities to shift towards ones more distinct from the native communities but more similar between other urban areas (Blair, 1996). It has been shown how urbanization tends in general to select for omnivorous, granivorous, and cavity nesting species (see Chace and Walsh 2006 for a review of urban effects on native avifauna). Generally as degrees of urbanization increase, avian biomass tends to increase but at the expense of a reduction of species richness. Chance and Walsh (2006) found that in closed temperate canopy forests, urbanization favours a certain typology of birds (seed eaters, omnivores and ground foragers) while selecting against others (high canopy and foliage foragers, insectivores, bark gleaners and drillers). Specifically, *Columba livia* (rock dove), *Zenaida macroura* (mourning dove), *Chaetura pelagica* (chimney swift), *Turdus migratorius* (American robin), *Sturnus vulgaris* (European starling), *Passer domesticus* (house sparrow) and *Quiscalus quiscula* (common grackle) are species that have been observed to respond positively to urbanization; flycatchers, *Vireo olivaceus* (red-eyed vireo), *Setophaga cerulea* (cerulean warbler) and most woodpeckers – except *Colaptes auratus* (northern flicker) and *Picoides pubescens* (downy woodpecker) – showed instead a generally negative response to urbanization (Chace and Walsh, 2006). As a result of the selection created by urbanization processes against some birds species, the composition of avian communities in urban areas might be significantly different from the natural occurring communities (Beissinger and Osborne, 1982; Mills et al., 1989). It has been also found that for avifauna, native birds respond positively to native vegetation density, while non-native species respond positively to exotic plant biomass (Mills et al., 1989). Some species are among the majority associated with extensive urbanization (e.g. rock dove, starling, robin, house sparrow) and some of them have been found to be more associated with highly managed urban woodland than to unmanaged ones (Burr and Jones, 1968). These species have the ability to exploit anthropogenic structures as nesting sites and to be less subject to stress, and this has undoubtedly contributed to their abundance in urban areas (Gavareski, 1976).

Among birds, nocturnal raptors include species of high conservation interest. Unfortunately, due to their nocturnal and secretive behaviour, their ecology in urban environment is still poorly described in scientific literature. Nocturnal raptors can occupy a wide home range, which in some cases extend beyond city borders. In particular, peripheral forest patches and vacant buildings can be favourable habitats for them. Urban environments may play an important role in terms of refuge for these creatures because there are adequate nest sites and resource availability (their diet includes small rodents, small birds, insects and frogs among others). Trees in urban forests are used as nesting sites, and some species of nocturnal raptors can gather together in groups (known as 'roosts'). Communal roosting is common in areas of high human activity. In fact, supplementary food resources in cities could be augmenting the birds' foraging opportunities, making urban communal roosting advantageous (Jaggard et al., 2015). In presence of birds' roosts it is important to pay attention during pruning operations and parks and gardens maintenance. In fact, due to the large number of birds that comprise the roosts, and their activity (nesting), they represent important sites deserving special management. Even slight alteration to the local environment (i.e. pruning of the trees they are using) might turn into serious damage with implications for the whole local population.

Reptiles and amphibians

Reptiles and amphibians frequently occur around urban areas, although for some reasons they can be considered more susceptible to threats from humans than from other taxa (i.e. they are less likely to be respected by humans because are perceived as scary or dangerous animals). For reptiles, habitat structure can play an important role in providing suitable basking, shelter and foraging opportunities. Some attributes such as the presence of fallen woody material, ground cover/leaf litter, moderate weed cover and bush-rocks has been found positively associated with terrestrial reptiles species because they can facilitate thermoregulation and provide safe refuges from predators while foraging and dispersing (Garden et al., 2007). Weeds provide suitable lower-stratum vegetation cover also because they support the presence of important reptile prey species (Fischer et al., 2003). The positive response to weediness appears true, however, as far as weeds are not associated with increased disturbance of habitat because of the sensitivity of certain reptiles to habitat disturbance. Amphibians are a taxon with conservation concern worldwide. They have special habitat needs that involve presence of water bodies as well as suitable terrestrial sites. Alongside the maintenance of high quality environment, urban forests and other suitable habitat patches have to maintain connection between each other and avoid isolation. In fact, in low-mobility animals, isolation can hinder the flow of individuals between populations and expose them to a higher risk of extinction caused by environmental, demographic and genetic stochasticity (Ficetola and De Bernardi, 2004). Also among amphibians, dispersing ability is different according to species: pool frogs are able to move long distances using water bodies (when available) while tree frogs during the post-breeding season live on trees and shrubs, and use canopies (such as hedgerow networks) to move in the anthropic matrix; newts and toads are more dependent on terrestrial habitats for dispersal, and very sensitive to the presence of barriers (Ficetola and De Bernardi, 2004). Thus, amphibians' persistence in a human-dominated system is dependent on the adoption of adequate maintenance of both high quality habitats and ecological connections. The isolation effect has to be taken into account in the management of every taxa because threats derived from isolated populations are common to different species, even if this is likely to be more damaging for animals that have lower mobility ability (e.g. that don't disperse by flying, such as birds, bats or some arthropods).

Mammals

Mammals occurring in urban forests are mainly species of small or medium size like bats, squirrels, voles, mice, rabbits, foxes or badgers. Presence of forest and shrub in urban areas are important for small mammals, and woodland environment provides unique suitable habitat especially for medium ones, such as many of the carnivores. Mammals particularly studied among urban environment are bats. On a small scale, wooded areas, water bodies and artificial lighting are important for certain species of bats (Avila-Flores and Fenton, 2005). Benefits derived from trees depend also on the fact that in urban environments insect productivity increases with vegetation coverage. Moreover, forests furnish a suitable foraging, roosting habitat and an escape cover from predators and stressors. Bats have been found to use forest edges as foraging areas or as travel corridors in an urban landscape (Gehrt and Chelsvig, 2003). Human-shaped environment might have negative consequences for most species of insectivorous bats while demonstrating an increase in the abundance of a few opportunistic species. Among bats, species that seem to tolerate heavy levels of urbanization are those with flexible roost requirements whose flight abilities allow them to reach scarce but predictable sources of food (Avila-Flores and Fenton, 2005). Due to their high mobility, they are also very sensitive to alteration in the surrounding landscape.

Insects and other invertebrates

Insects and other invertebrates constitute a big portion of species richness in a forest. Carabid beetles are a well-known taxa: they are abundant, speciose, and taxonomically and ecologically well studied.

Generally, in order to preserve the integrity of forest arthropod fauna in urban areas it is essential to leave large, continuous forest tracts untouched to preserve specialist species and forest species assemblage (Niemelä and Halme, 1998). Also shape of the forest patch may affect species richness because higher edge to area ratios usually contain more species because of high invasion rate from the surrounding matrix. Effects of fragmentation have been observed at species level, with some forest species absent from the smallest fragments. At the same time, forest remnants might provide valuable habitats also for some open-land species of ground beetles (Fujita et al., 2008). As happens for other taxa, fragmented patches of forest can result in an increase in total species richness because of invasion from surrounding open habitats, leading to changes in invertebrate species composition of forests. Changes in community composition might have ecological consequences in urban forests (e.g. impoverishment for the specific ecological function that organisms provide), although this aspect is rarely investigated. It is thus important to maintain functional relationships among species for the overall health of the ecosystem. Another factor of concern is that there is a tendency of increasing similarity of habitat patches that can homogenise species assemblages among forest patches. To contrast fragmentation, connecting forest stands with adequate corridors (hedges and wooded corridors) can contribute to preserve forest species among the urban matrix. Some authors suggest how hedges may have surprisingly rapid effects on local carabid distribution (Niemelä, 2001). Tree height and tree species composition may affect animal dispersal. Specifically, hedgerows consisting of native tree species appear to be able to maintain higher abundance of forest species than hedgerows of exotic species. Trying to provide practical management recommendations, some authors suggest maintaining a minimum size of at least tens of hectares of forest patch to guarantee an intact assemblage of interior specialist species of carabids (Niemelä, 2001) (but these indications may vary a lot in tropical habitats).

Plants

Plant diversity in urban parks may vary. Often all the plants (trees, shrubs and herbs) are planted by humans, with a large number of exotic species, and generally a homogenization of floristic and genetic diversity occurs. In some case, in particular in suburban parks, forest patches are remnant patches, and in this situation floristic composition may have a higher conservation value.

Digiovinazzo et al. (2010) studied forests patches in suburbs of Milan, classifying all the herbs in groups of floristic interest, according to Raunkier life forms, Ellemberg indicators for Italian flora (Pignatti, 2005) and phytosociological taxonomy. The result is that in larger forest patches there are more forest indicators species. There are two threshold values: forest patches in 1.5 ha are very poor of 'forests species', while fragments bigger than 35/40 ha do not further increase the number of forest species.

Exotic species

Exotic species are another major threat to biodiversity and urban forests are habitats particularly subject to invasion of non-native organisms. This is due to the fact that they experienced an intense flow of people, which is an agent of dispersion (accidental or intentional) of exotic species (Ficetola et al., 2009). The most effective tool to contrast this problem is the prevention of introductions (Keller et al., 2008) because once established, the management of exotic species can be extremely difficult. Threats derived from alien species are due to the competition for resources that can arise with native species. Each species occupies its own ecological niche in the habitats and when a new species is introduced it is faced with the problem to find resources (food, sites, etc.) to successfully survive. Because of its extraneousness to the environment, it might have no natural enemies and therefore take an advantage at the expense of naturally occurring species. Processes of competitive exclusion might take place with the result of exotic organisms replacing native ones and determining cascade effects on the entire ecosystem. Moreover, we have to remember that impacts of human activities can strongly alter the relative competitive ability between co-occurring species, for example by altering the physical environment (McKinney, 2008). Thus, when managing forests by controling tree density or species composition, we might favour or contrast the presence of native organisms with relevant ecological implications.

How the structure of urban forests influences biodiversity

Vegetation in urban environments is generally highly fragmented, there is less coverage at mid- and upper-canopy levels, and there is more ground cover than in nearby wild sites. Exotic plants often replace native ones and the entire ecological functioning experiences the direct management of human actions (e.g. removal of stumps and logs). Urban forests can and have to be managed to preserve biodiversity and species of conservation interest taking into account some aspects of their structure that can have important implications. Despite the primary importance played by vegetation in harvest species, it is important to remember how the complexity of urban systems need to understand the relative importance of each factor that acts locally. For example, Friesen et al. (1995) observed that forest fragments of similar size and vegetative structure may not be ecologically equal because of differences in their surroundings. They suggest that external effects severely deflate the ecological value of adjacent forests and that anthropogenic stress might act in the same way both on small and large forests. Forest structure has severe effects also on internal processes driving vegetation ecology. Getzin et al. (2008) studied western hemlock in two old-growth stands (Vancouver Island) and proved that successional dynamics are intensified in heterogeneous forest stands with strong spatial structures compared to homogeneous forest stands. Thus, spatial heterogeneity can play a determinant role in affecting plant population dynamics and pattern formation. This is particularly important if we consider cascading feedback effects that can arise where plant demographics interact with spatial structure (Getzin et al., 2008).

Forest size

Maintaining large woodlot size is considered an option to promote urban biodiversity – the larger the size, the greater the species diversity (Alvey, 2006). Several studies demonstrated how the size of green spaces in urban areas was the main factor explaining the variation in biodiversity indicators (Cornelis and Hermy, 2004). For example, Godefroid and Koedam

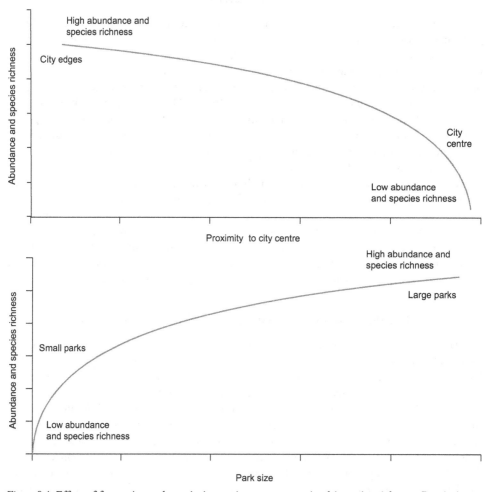

Figure 9.1 Effect of forest size and proximity to city centre on animals' species richness. Proximity to city centre has a negative effect on both species richness and abundance (above); on the contrary, large forests can harbour a higher number of species and higher abundance (below)

(2003) found that one very large woodlot (1,666 ha) in Brussels had greater species richness than 11 small woodlots (2–123 ha). Also fragmentation of forests in small patches contribute to biodiversity loss. Fragmentation dynamics cannot be considered only within the context of habitat size (Friesen et al., 1995). In fact, remnant patches of forests are subject to many adverse effects and in small patches the pressures arising from external forces might be predominant to the internal forces in driven ecosystem dynamics (Saunders et al., 1991). Not all the species are affected in the same way from reduction in forests' patch size. For example, forest interior species might be the main disadvantaged by this situation. Also distance from the city centre is generally correlated with highest species richness (the higher the distance, the higher the diversity). In Figure 9.1 it is illustrated how species richness changes according to the size of the green area and distance from the city core. Both effects of size and position are well explained according to the theory of island biogeography (MacArthur and Wilson, 1967).

Forest composition and tree diameter

The maintenance of forest-stand structural complexity is critical for forest biodiversity conservation (Lindenmayer et al., 2006). It has been shown that structural simplification of vegetation has negative impacts on biodiversity (McKinney, 2008). In fact, structural complexity can provide the within-stand variation in habitat conditions required by some taxa (a 'habitat heterogeneity' function) (Lindenmayer et al., 2006). This allows organisms of different ecological requirements to persist in an area which they wouldn't otherwise survive. Moreover, maintaining structured complexity forests may facilitate a more rapid return of logged and regenerated stands to suitable habitat for species that have been displaced (Lindenmayer et al., 2006).

Trees' diameter is a parameter that may be considered when we want to look at the relationship between a forest's structure and biodiversity. Many studies have revealed the relation between trees' diameter (or basal area) values and species abundance (Sanesi et. al., 2009; Lugo and Helmer, 2004; Pearman, 1997). By studying target species of birds in relation to DBH (diameter breast height), Sanesi et al. (2009) provided evidence on the response of bird species to the presence of mature and heterogeneous forest stands. In Figure 9.2 is illustrated the result of a study conducted in the parks of Milan (Sanesi et al., 2009) describing the effect of structural attributes of forests and bird species abundance, where it is possible to see an 'optimal', a 'medium' and a 'minimal' situation. The assumption is that a larger abundance of a species may be considered better than a lower abundance (see Padoa-Schioppa et al., 2006, for a more detailed explanation of this assumption). There is evidence that the structural heterogeneity of forest trees within greenspaces is a fundamental aspect of supporting high levels of species abundance. The conservation of species diversity in urban areas should be based on the knowledge of one or more indicators species having different habitat requirements in terms of vegetation and trees' structure (Ficetola et al., 2007).

Alongside the need for structural and compositional heterogeneity, control of exotic species (that can in some cases represent a big proportion of tree species composition) is a matter of high concern in urban environments. Urban and suburban areas are subjected to replacement of native plant communities with managed systems of altered structure, a process that influences ecological and environmental relationships. For example it has been observed that bird community composition shifts in response to exotic plantings (Mills et al., 1989). Strong positive correlation has been found between native vegetation (volume and structure) and native bird species richness, as well as non-native species diversity correlated with exotic vegetation (Mills et al., 1989). Maintenance of native vegetation might be very

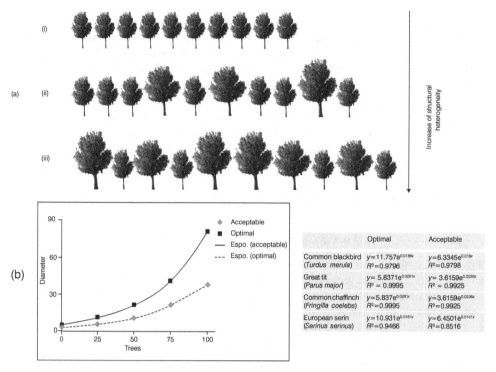

Figure 9.2 (a) Three different stages of structural heterogeneity of forest trees in relation to biodiversity trend. Moving from the simplest stage (i) towards more complex stages (ii and iii), trees' structure composition increase in heterogeneity and this leads to the persistence of trees having different ages and size and to constitute different ecological niches for different species of animals. (b) Optimal model of the DBH distribution and equations describing the model for some target species studied

Source: panel (b) adapted from Sanesi et al. (2009)

positive, as was found for certain native birds in Arizona (Emlen, 1974) which responded positively to urbanization in a city that maintains a high proportion of native vegetation (see also Marzluff et al., 2001).

As much as possible in forested patches, the maintenance of natural vegetation in the shrub layer would provide an increased number of niches exploited by many different species. In Seattle (Washington) the presence of adequate natural brush cover was evidently a habitat requirement for several species of birds like the *Psaltriparus minimus* (bushtit), *Troglodytes hiemalis* (winter wren), *Thryomanes bewickii* (Bewick's wren), *Catharus ustulatus* (Swainson's thrush), *Setophaga nigrescens* (black-throated gray warbler), *Pipilo erythrophthalmus* (rufous-sided towhee) and *Melospiza melodia* (song sparrow) (Gavareski, 1976). Similarly, the presence of other microhabitats such as small-scattered openings and some form of water in or adjacent to the woods, can provide nesting and feeding sites for a variety of wild fauna.

Although a traditional attitude would consider all green spaces associated with the urban environment mostly designed to be exploited by citizens for recreational activities, urban forests can (and must) at the same time provide suitable habitats for wildlife fauna. Undertaking consistent conservation actions would have the double result to preserve biodiversity and to provide recreational opportunity for residents to experience nature and its diversity. As we have shown, actions are required on several spatial scales simultaneously (at local as well as at landscape level).

Although for some taxa (such as birds) the knowledge about their ecological interactions and requirements within the urban environment is well known, for many other taxa it is still lacking. From one side the scarcity of urban-based ecological research (Garden et al., 2006) and from the other the discrepancies in some research findings limit the possibility to make generalization about habitat requirements for the less common taxa (Garden et al., 2007). This in turn makes it difficult to provide effective management recommendations that are not supported by specific addressed studies. There is also to consider the large variation in habitat requirements among species (even within the same taxa) and this, coupled with the intrinsic complexity of urban systems (where disturbances strongly differentiate in type, rate and intensity), contribute to make difficult the generalization of habitat requirements for many taxa in urban forests. An important recommendation is therefore to integrate urban forests' management with the indications provided by studies on species' responses to local-level habitat attributes (when available) addressing environmental needs of local flora and fauna.

References

Alvey, A. A. 2006. Promoting and preserving biodiversity in the urban forest. *Urban Forestry and Urban Greening* 5(4), 195–201.

Araújo, M. B. 2003. The coincidence of people and biodiversity in Europe. *Global Ecology and Biogeography* 12(1), 5–12.

Avila-Flores, R. and Fenton, M. B. 2005. Use of spatial features by foraging insectivorous bats in a large urban landscape. *Journal of Mammalogy* 86(6), 1193–1204.

Baffetta, F., Fattorini, L. and Corona, P. 2011. Estimation of small woodlot and tree row attributes in large-scale forest inventories. *Environmental and Ecological Statistics* 18, 147–167.

Barrico, L., Azul, A.M., Morais, M.C., Pereira Coutinho, A., Freitas, H. and Castro, P. 2012. Biodiversity in urban ecosystems: Plants and macromycetes as indicators for conservation planning in the city of Coimbra (Portugal). *Landscape and Urban Planning* 106, 88–102.

Beissinger, S. R. and Osborne, D. R. 1982. Effects of urbanization on avian community organization. *Condor* 84, 75–83.

Bennett, E. and Hassan, R. M. 2003. *Ecosystems and Human Well-Being: A Framework for Assessment.* Island Press, Washington, DC.

Blair, R. B. 1996. Land use and avian species diversity along an urban gradient. *Ecol Appl* 6, 506–519.

Bradford, M. A. et al. 2002. Impacts of soil faunal community composition on model grassland ecosystems. *Science* 298, 615–618.

Burr, R. M. and Jones, R. E.,1968. The influence of parkland habitat management on birds in Delaware. *Transactions of the North American Wildlife and Natural Resource Conference* 33, 299–306.

Chace, J. F. and Walsh, J. J. 2006. Urban effects on native avifauna: A review. *Landscape and Urban Planning* 74(1), 46–69.

Cornelis, J. and Hermy, M. 2004. Biodiversity relationships in urban and suburban parks in Flanders. *Landscape and Urban Planning* 69, 385–401.

Digiovinazzo P., Ficetola G. F., Bottoni L., Andreis C. and Padoa-Schioppa E. 2010. Ecological thresholds in herb communities for the management of suburban fragmented forests. *Forest Ecology and Management* 259: 343–349.

Emlen, J. T. 1974. An urban bird community in Tucson, Arizona: Derivation, structure, regulation. *Condor* 76, 184–197.

Ficetola, G. F. and De Bernardi, F. 2004. Amphibians in a human-dominated landscape: The community structure is related to habitat features and isolation. *Biological Conservation* 119(2), 219–230.

Ficetola, G. F., Thuiller, W. and Padoa Schioppa, E. 2009. From introduction to the establishment of alien species: Bioclimatic differences between presence and reproduction localities in the slider turtle. *Diversity and Distributions* 15(1), 108–116.

Ficetola, G. F., Sacchi, R., Scali, S., Gentilli, A., De Bernardi, F. and Galeotti, P. 2007. Vertebrates respond differently to human disturbance: Implications for the use of a focal species approach. *Acta Oecologica* 31(1), 109–118.

Fischer J., Lindenmayer D. and Cowling A. 2003. Habitat models for the four-fingered skink (Carlia tetradactyla) at the microhabitat and landscape scale. *Wildl Res* 30, 495–504.

Florgård, C. 2010. Integration of natural vegetation in urban design: Information, personal determination and commitment. In Müller, N., Werner P., and Kelcey, J. G. (eds), *Urban Biodiversity and Design* (pp. 477–496). Wiley-Blackwell, Oxford. doi: 10.1002/9781444318654.ch26

Friesen, L. E., Eagles, P. F. and Mackay, R. J. 1995. Effects of residential development on forest-dwelling neotropical migrant songbirds. *Conservation Biology* 9(6), 1408–1414.

Fujita, A., Maeto, K., Kagawa, Y. and Ito, N. 2008. Effects of forest fragmentation on species richness and composition of ground beetles (Coleoptera: Carabidae and Brachinidae) in urban landscapes. *Entomological Science* 11(1), 39–48.

Garden, J., McAlpine, C., Peterson, A., Jones, D. and Possingham, H. 2006. Review of the ecology of Australian urban fauna: A focus on spatially explicit processes. *Austral Ecol* 31, 126–148.

Garden, J. G., McAlpine, C. A., Possingham, H. P. and Jones, D. N. 2007. Habitat structure is more important than vegetation composition for local level management of native terrestrial reptile and small mammal species living in urban remnants: A case study from Brisbane, Australia. *Austral Ecology* 32(6), 669–685.

Gavareski, C. A. 1976. Relation of park size and vegetation to urban bird populations in Seattle, Washington. *Condor* 375–382.

Gehrt, S. D. and Chelsvig, J. E. 2003. Bat activity in an urban landscape: Patterns at the landscape and microhabitat scale. *Ecological Applications* 13(4), 939–950.

Getzin, S., Wiegand, T., Wiegand, K. and He, F. 2008. Heterogeneity influences spatial patterns and demographics in forest stands. *Journal of Ecology* 96(4), 807–820.

Godefroid, S. and Koedam, N. 2003. How important are large vs. small forest remnants for the conservation of the woodland flora in an urban context? *Global Ecology and Biogeography* 12, 287–298.

Gómez-Baggethun, E., Gren, Å., Barton, D. N., Langemeyer, J., McPhearson, T., O'Farrell, P., ... and Kremer, P. 2013. Urban ecosystem services. In Elmqvist, T., Fragkias, M., Goodness, J., Güneralp, B., Marcotullio, P. J., McDonald, R. I., Parnell, S., Schewenius, M., Sendstad, M., Seto, K. C. and Wilkinson, C. (eds), *Urbanization, Biodiversity and Ecosystem Services: Challenges and Opportunities* (pp. 175–251). Springer Netherlands, Dordrecht.

Jaggard, A. K., Smith, N., Torpy, F. R. and Munro, U. 2015. Rules of the roost: Characteristics of nocturnal communal roosts of rainbow lorikeets (*Trichoglossus haematodus*, Psittacidae) in an urban environment. *Urban Ecosystems* 18(2), 489–502.

Jim, C. Y. and Liu, H. T. 2001. Species diversity of three major urban forest types in Guangzhou City, China. *Forest Ecology and Management* 146, 99–114.

Keller, R. P., Frang, K. and Lodge, D. M. 2008. Preventing the spread of invasive species: Economic benefits of intervention guided by ecological predictions. *Conservation Biology* 22, 80–88.

Kellert, S. R. and Wilson, E. O. 1993. *The Biophilia Hypothesis*. Island Press, Washington, DC.

Koskimies, P. 1989. Birds as a tool in environmental monitoring. *Annales Zoologici Fennici* 1989, 153–166.

Kraxner, F., Aoki, K., Kindermann, G., Leduc, S., Albrecht, F., Liu, J. and Yamagata, Y. 2016. Bioenergy and the city – What can urban forests contribute? *Applied Energy* 165, 990–1003.

Larsen, S., Sorace, A. and Mancini, L. 2010. Riparian bird communities as indicators of human impacts along mediterranean streams. *Environmental Management* 45, 261–273.

Lindenmayer, D. B., Franklin, J. F. and Fischer, J. 2006. General management principles and a checklist of strategies to guide forest biodiversity conservation. *Biological Conservation* 131(3), 433–445.

Lugo, A. E. 2002. Can we manage tropical landscapes? An answer from the Caribbean perspective. *Landscape Ecology* 17, 601–615

Lugo, A. E. and Helmer, E. 2004. Emerging forests on abandoned land: Puerto Rico's new forests. *Forest Ecology Management* 190:145–161.

MacArthur, R. H. and Wilson, E. O. 1967. *The Theory of Island Biogeography*. Princeton University Press, Princeton, NJ.

Mace, G. M., Norris, K. and Fitter, A. H. 2012. Biodiversity and ecosystem services: A multilayered relationship. *Trends in Ecology and Evolution* 27(1), 19–26.

Marzluff, J. M., Bowman, R. and Donnelly, R. 2001. *Avian Ecology and Conservation in an Urbanizing World*. Kluwer Academic, Boston, MA.

McDonald, R. I., Kareiva, P. and Forman, R. T. 2008. The implications of current and future urbanization for global protected areas and biodiversity conservation. *Biological Conservation* 141(6), 1695–1703.

McKinney, M. L. 2008. Effects of urbanization on species richness: A review of plants and animals. *Urban Ecosystems* 11(2), 161–176.

Mills, G. S., Dunning Jr., J. B. and Bates, J. M. 1989. Effects of urbanization on breeding bird community structure in southwestern desert habitats. *Condor* 91, 416–428.

Niemelä, J. 1999. Ecology and urban planning. *Biodiversity and Conservation* 8(1), 119–131.

Niemelä, J. 2001. Carabid beetles (Coleoptera: Carabidae) and habitat fragmentation: A review. *European Journal of Entomology* 98(2), 127–132.

Niemelä, J. and Halme, E. 1998. Effects of forest fragmentation on carabid assemblages in the urban setting: Implications for planning and management. In Breuste, J., Feldmann, H. and Uhlmann, O. (eds), *Urban Ecology* (pp. 692–695). Springer, Berlin.

Padoa-Schioppa, E., Baietto, M., Massa, R. and Bottoni, L. 2006. Bird communities as bioindicators: The focal species concept in agricultural landscapes. *Ecological Indicators* 6, 83–93.

Pearman, P. B. 1997. Correlates of amphibian diversity in an altered landscape of Amazonian Equador. *Conservation Biologty* 11, 1211–1225.

Pignatti, S. 2005. Valori di bioindicazione delle piante vascolari della flora d' Italia. *Braun-Blanquetia* 39, 3–97.

Primack, R. B. 2000. *A Primer of Conservation Biology*. Sinauer Associates, Sunderland, MA.

Rodrigues, A. S. L., Andelman, S. J., Bakarr, M. I., Boitani, L., Brooks, T. M. …Yan, X. 2004. Effectiveness of the global protected area network in representing species diversity. *Nature* 428, 640–643.

Sanesi, G., Padoa-Schioppa, E., Lorusso, L., Bottoni, L. and Lafortezza, R. 2009. Avian ecological diversity as an indicator of urban forest functionality. Results from two case studies in northern and southern Italy. *Journal of Arboriculture* 35(2), 80.

Saunders, D. A., Hobbs, R. J. and Margules, C. R. 1991. Biological consequences of ecosystem fragmentation: A review. *Conservation Biology* 5(1), 18–32.

Wilson, E. O. 2002. *The Future of Life*, 1st edn. Alfred A. Knopf, New York.

10

URBAN FOREST BENEFITS IN DEVELOPING AND INDUSTRIALIZING COUNTRIES

Fabio Salbitano, Simone Borelli,
Michela Conigliaro, Noor Azlin Yahya,
Giovanni Sanesi, Yujuan Chen and Germán Tovar Corzo

Introduction

When reflecting on the significance and potential of urban and peri-urban forests in developing and industrializing countries, some questions immediately come to mind. Does it make sense to speak of urban and peri-urban forestry (UPF) by marking differences among countries? To what extent does the stage of economic development determine/influence the way the potential benefits from urban and peri-urban forests are perceived and valued?

On the one hand, national and local policies, decentralization and technical expertise and capacity at the different levels of government, influence the way in which each country regulates the institutional relationships between the central authorities and municipal government, particularly on the management of open spaces. On the other hand, the historical and cultural heritage of a country as well as lifestyles, attitude to outdoor life, and spiritual and religious views, deeply influence the way people perceive the benefits of urban trees and forests, especially when comparing developed and developing/industrializing countries.

The range of needs and quality of life expectations in developing countries is much wider than in developed ones, with some basic benefits (often overlooked in developed countries) playing a key role for the livelihood of poor urban dwellers. Issues such as income, poverty, unemployment, migration, access to sanitation or health services, access to and use of food and energy resources, or access to education and capacity building opportunities deeply influence the way in which local communities value and utilize urban and peri-urban forests.

This chapter highlights the special needs and requirements for the development of UPF in developing countries. It identifies specific governance approaches to optimize the contribution of urban and peri-urban forests to improved livelihoods, human health and

136

well-being. However, the reflections and indications that follow are not only relevant to developing countries. Indeed, in many cities around the world, independently from their stage of development, the divide between rich and poor shapes cities and heavily affects livelihoods, public health, and quality of life.

In Manila, like in Naples or Baltimore and Bogota (e.g. Escobedo et al., 2015), the gap in character of places, living conditions, and access to basic services between rich and poor neighbourhoods can be as wide as the differences between developed and developing countries, and different approaches may be needed to respond to the different needs.

Urbanization in developing and industrializing countries

Most projections indicate that by 2050, the world's urban population will increase by 2.5 billion people, with nearly 90 per cent of the increase in Asia and Africa. Three countries alone – India, China and Nigeria – will account for 37 per cent of this projected growth (United Nations, 2014). The urbanization rate in high-income countries is strongly decreasing while it is expected to remain higher in middle-income countries and much higher in low-income ones (Figure 10.1).

Although close to half of the world's urban dwellers reside in cities of less than 750,000 inhabitants (Figure 10.2), the last decades have witnessed the emergence of a number of megacities (i.e. cities with more than 10 million inhabitants). About one in eight people live in the latter today. According to the United Nations, there are 28 megacities at the time (United Nations, 2014), with China alone hosting six megacities and ten other cities with populations above five million in 2014. India follows closely with four cities projected to become megacities in the coming years (Ahmadabad, Bangalore, Chennai and Hyderabad).

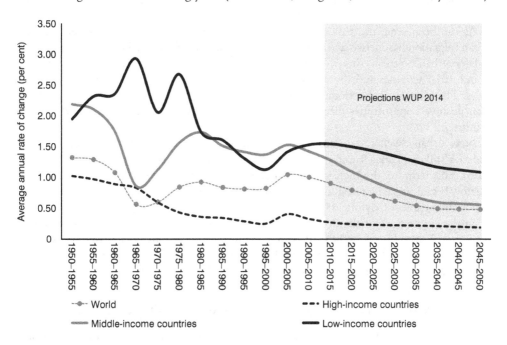

Figure 10.1 Urbanization rate in development regions of the world

Sources: United Nations (2014, 2015)

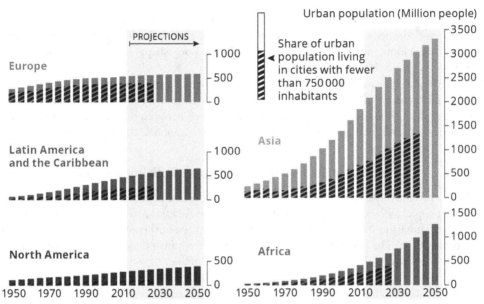

Figure 10.2 Share of population in urban areas and relative frequency of cities with fewer than 750,000 inhabitants

Source: modified from EEA (2015)

Outside of China and India, Asia has seven other megacities and Africa has three (Cairo, Kinshasa and Lagos). There are currently four in the Latin America region: Buenos Aires, Mexico City, Rio de Janeiro and São Paulo.

As can been seen, most megacities are located in developing countries where rural to urban migration has increased dramatically over the last few decades. The massive migration to large urban centres is driven by a dramatic shift in economic structure from agricultural to industrial activities and the widening gap in living conditions between urban and rural areas. Therefore, it is deteriorating rural conditions that have driven urban growth rather than positive opportunities offered by urban areas, as was the case in developed western countries.

New urban dwellers coming from rural areas are quite often displaced or refugees, and they still maintain a 'rural' attitude and feel 'diverse' in an urban context (Salbitano et al., 2015). Furthermore, the increasingly complex connections between urban and rural areas are now beginning to be recognized, but still have a relatively limited impact on development policy and practices (Mylott, 2009).

In the city regions of the developing world, the assumption that there is a clear divide between urban and rural can be misleading with respect to the actual reality of what is urban, rural or peri-urban. The latter, where both urban and rural features converge – is becoming increasingly important in these countries (Mylott, 2009).

Main consequences of urbanization in developing countries

The livelihood and well-being of urban dwellers strongly depends on the ecological services provided by healthy natural ecosystems in and around cities. This applies to all cities, both in less developed countries and in the richest cities of high-income countries.

For the most part, the rapid expansion of cities takes place without any real land use planning strategy, and the resulting human pressure causes highly damaging effects on forests and whole landscapes in and around cities. Without sound urban planning and governance, natural and agricultural land areas succumb to pressure for conversion to urban land uses or are depleted in meeting demands for food, fuel, and building materials in adjacent cities. In less developed countries in particular, the rapid urbanization process has outpaced the capacity of cities to provide dwellers with essential goods and services to sustain their livelihoods as well as their health and well-being.

The environmental impacts of urbanization, often intensified by climate change, can threaten whole city sustainability resulting in a lack of basic infrastructure, housing, employment, sanitation, and health services (see Box 10.1). In unplanned urban areas, residents are more likely to be exposed to pollution from cars, trucks, and industries in close proximity, resulting in higher risk of respiratory infections, asthma, and lead poisoning that contribute to learning disabilities and overall reductions in life expectancy and community resilience.

The main consequences of unplanned urbanization in the developing world follow three main streams:

- *Environmental*: surrounding landscape fragmentation and habitat loss; higher frequency and vulnerability to extreme weather events including floods, droughts, landslides and extreme winds; vulnerability of soil erosion and watershed degradation; heat island effect; air pollution.
- *Economic*: rising urban poverty; resource and food insecurity; loss in supply and increasing cost of forest and wood fuel products for industrial and domestic purposes.
- *Social and cultural*: public disconnection from nature; loss of identity; ethnic discrimination and exclusion of vulnerable groups; decreased physical activities and consequent increase of non-communicable diseases (e.g. cancer, cardiovascular diseases, allergies, obesity).

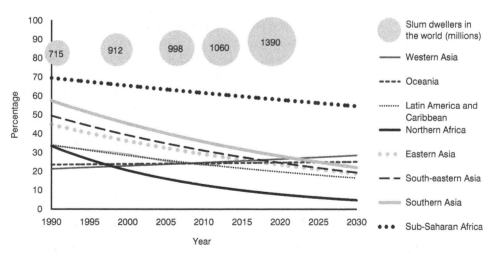

Figure 10.3 Slums between growth and decrease. The lines show the trend of the proportion between people slum dwellers and total urban population in eight regions and subregions of the world. The grey circles report the number (in million) of slum dwellers along time

Sources: United Nations Statistics Division, UN-Habitat database (http://urbandata.unhabitat.org); WHO Global Health Observatory dataset (www.who.int/gho/en)

Box 10.1 Of slums and trees

The proportion of slum dwellers is decreasing as compared to the world urban population; however, the net number of people living in slums is increasing (Figure 10.3). According to UN-Habitat, 55 million new slum dwellers have been added to the global population since 2000, especially in developing countries. In 2012, one third of the developing world's urban population was living in slums.

Typing the word 'slum' in an image search engine, one gets an astonishing array of desolating and distressing pictures; images that show the harsh reality of parts of cities where an estimated one billion people live, the vast majority in the developing world. Slums are characterized by poor infrastructure and housing conditions, widespread poverty, and public health concerns. However, another characteristic is the almost total absence of trees and green spaces. Green is rare and smothered by muddy brown, the dominant colour of the slums.

Flying over Nairobi, one gets the impression of a very green city, but huge parts are completely devoid of trees. The slum of Kibera (meaning forest or jungle in Nubian language) is the largest in Kenya, housing 170,000 to 250,000 residents in an area of 2.38 km^2 (Desgroppes and Taupin, 2011) – a population density (ranging from 71,400 to 105,000 residents per km^2) that is twice that of Manila, the city with the highest population density in the world.

Checking the number of trees in the entire area of Kibera (personal estimate), the count totals 230 to 250 trees. This means roughly 1 tree per hectare, or less than 0.0015 tree per person.

The presence of trees is certainly not the first concern for slum dwellers. Most of the people who live in Kibera have no access to basic services such as clean water, sanitation, or electricity. Nevertheless, the absence of trees and green spaces is felt as a misfortune in both cultural and physical terms. In May and June 2015, the charity 'Trees for Cities' planted 6500 trees across multiple sites in the Kibera aiming to provide numerous benefits to the slum community, including improved air quality, increased canopy cover, reduced impacts of flooding in the rainy season, and a source of sustainable firewood, and environmental education.

Gopal and Nagendra (2014) highlighted the fact that trees and greenery can improve social capital, livelihoods, health, and nutrition in the slums of Bangalore, India. They pointed out that the density of trees in Bangalore slums was much lower as compared to other residential areas (11 versus 28 trees per hectare; Gopal et al., 2015). However, the slums contain a greater proportion of native trees compared to other parts of the city, where introduced species tend to be dominant over native vegetation. Among these species, *Moringa oleifera* and *Cocos nucifera* are the most common and both are important sources of food. The authors also discovered that the sacred *Ficus religiosa* trees are very popular, playing an important role in social community structure, providing shade for social congregations and recreation activities.

The concept of urban and peri-urban forestry in developing countries

The rate and extent of urbanization vary considerably from country to country, and so do the nature and character of urban settlements, including the structure, diversity, extent, and perceived contribution of the urban and peri-urban forest, which is strongly influenced by the history and the socioeconomic, demographic, cultural, and political evolution of individual cities.

For example, while in some countries UPF is still a new concept, other countries have a long history of urban tree cultivation. Beyond the long-standing relationships between trees and human settlements and cities (see Chapter 2), the development of UPF in developing countries is intimately related to the urbanization types. In many developing countries, colonialism deeply influenced the urbanization process, including the social, economic, and spatial separation of the different societal classes. The incorporation of parks and trees was part of the colonial policy of residential segregation, reinforced by avenue trees in the European neighbourhoods. According to Onganga (1992), many local Kenyans negatively associate urban amenity trees with the colonial past. In a different view, Johannesburg in South Africa has the distinction of being one of the largest manmade urban forests and this evidence constitutes an absolute pride for the citizens (see also Chapter 2).

Traditionally, people in West Africa planted trees around their homes for fruits, nuts, seeds, leaves, fuelwood, fodder, and raw materials for handicrafts and building, and for shade and windbreaks. This rural practice was transferred to cities. However, this custom is difficult to maintain in cities, where urbanization does not leave space for planting trees. In addition, trees have traditionally been planted along rivers and stream banks, major city roads and around cultural and religious centres (Fuwape and Onyekwelu, 2011).

UPF is progressively becoming part of the development agenda of developing cities, and the necessary financial resources are being made available to municipal governments. In China, for example, the government has embarked on a very ambitious programme to develop National Forest Cities that meet criteria of urban sustainability. Launched in 2004 under the slogan 'forests in city and city in forest', 96 cities were honoured with the title by 2015. The tree cover of urban communities has increased to 40 per cent and more, from less than 10 per cent in 1981. Urban forests in China are becoming part of a fundamental urban landscaping norm (Li and Yang, 2016).

In this context, a broader *green infrastructure* approach including urban forestry, urban agriculture, urban agroforestry, urban ecology to designing and managing sustainable urban and peri-urban forests through inclusive planning can be considered as a new paradigm for developing cities (Borelli et al., 2015; see also Chapter 14 of this volume).

Environmental, economic, and socio-cultural benefits of urban and peri-urban forests in developing cities

By providing ecosystem services, goods, and public benefits in and around urban settlements, well-managed urban and peri-urban forests can help respond to the needs and challenges posed by an increasing urban population. They can increase food provisioning, mitigate local climate, regulate ecosystem processes, support natural ecosystem conservation, and generate income to cities. To help illustrate these benefits the Ecosystem Services Framework is used here. This framework divides services into provisioning, supporting, regulating and cultural (TEEB, 2011; see also Chapter 4 of this volume).

Provisioning services

Increasing urban populations implies an increasing demand for food and basic services, posing major infrastructural, social, environmental, and economic challenges for local administrations. When properly planned and managed, urban and peri-urban forests can play an important role in providing food, water, wood, medicines, raw materials, and energy to cities.

In several cities, urban forestry practices such as the collection of wild edible plants, the planting of fruit-bearing street trees, and the establishment of multifunctional or medicinal public parks contribute to an improvement in the availability of food within cities (Fuwape and Onyekwelu, 2011). For example, jamun trees (*Syzygium cumini*) alongside roads of Delhi yield about 500 tons of fruit each year, harvested and sold to passing pedestrians and motorists during monsoon season, when fruits ripen (Singh et al., 2010). The availability of fresh and nutritious food within cities also contributes to diversify the diets and increase the vitamin intake for urban dwellers. In Africa, children living in areas with more tree cover have more diverse and nutritious diets than those living in poorly treed areas (Ickowitz et al., 2014). In addition to increasing food accessibility, localized food production makes food more affordable, reduces environmental costs of distribution systems, and increases community resilience to food shortages.

Urban and peri-urban forests also provide wood, both as timber and wood fuel, thus helping generate income and ensure energy supply to cities. In 2013, an estimated 38 per cent of the world's population (and 50 per cent in developing countries) still relied on the traditional use of solid biomass for cooking (International Energy Agency, 2015). In Africa alone, wood fuel accounts for over 80 per cent of all domestic fuel and it is projected to remain the main source of domestic energy in the region over the next decades (FAO, 2010). Demand for wood fuel is mostly satisfied at the expense of natural forests, often unsustainably exploited by local communities. Apart from increasing local livelihood, the creation and sustainable management of peri-urban forest systems for wood fuel production could also help protect natural forests from overexploitation.

Many cities rely on healthy peri-urban forest systems for their drinking water supply. In Nairobi, the overexploitation of the tree resources around the city has led to serious depletion of the peri-urban forest that is crucial in the water catchment and purification system of the city. Having experienced the consequences of such a loss, the local government of the Kiambu County is now implementing restoration activities to rehabilitate three peri-urban forests (Kenya Forest Service, 2010) critical for protecting the local water catchments and in ensuring an adequate water supply to the city. Quito (Ecuador) is an example of a Latin American city that has implemented a water consumption fee to protect its watershed forests (Echavarria, 2001).

Supporting services

Urban and peri-urban forests provide habitat for plant and animal species, playing a critical role in supporting conservation of natural landscapes within and beyond city boundaries. Natural and semi-natural areas, in particular, can help preserve local biodiversity (both native and endemic species) and increase ecological connectivity, thus reducing forest fragmentation and increasing the resilience of natural ecosystems to human pressure. If properly planned and managed, urban green spaces can host surprising levels of biodiversity. The BioCity programme implemented by the City of Curitiba, Brazil, is a leading example of urban planning that takes biodiversity-related issues into consideration through the re-introduction of indigenous plant species, the establishment of biodiversity conservation units, the preservation of water resources, and the creation of major transportation corridors with alternative transportation options (Cuquel et al., 2009).

Many cities, especially in drylands, have adopted UPF as a valuable practice to prevent land and soil degradation, and restore their function and productivity. Trees can contribute to soil formation, increase soil productivity, and improve infiltration and nutrient content. By blocking winds and stabilizing soils, trees can also prevent erosion and reduce soil compaction.

In light of this, 400 hectares of green belt have been implemented around the City of Ouarzazate, Morocco. The project plans to establish 1350 hectares of forest plantation in peri-urban contexts, 125 hectares of urban green spaces, 105 hectares of street trees belts, and 300 hectares of green belt to lower the evaporation and erosion of the Ouarzazate dammed reservoir (Nakhli, 2014). In drylands, trees are also valued for their role as physical barriers against wind and sand storms, contributing to fight desertification and erosion (Sène, 1993). For example, the Korea–Mongolia Greenbelt Plantation Project is a 10-year project (2007–2016) aimed at planting 3000 ha of trees in the Gobi Desert to prevent desertification and mitigate effects of dust and sand storms in urban and rural areas (Kang et al., 2010).

Regulating services

Urban and peri-urban forests have a decisive role in regulating ecosystem processes and in increasing the resilience of the urban environment. By shading and cooling the air, trees can help mitigate microclimate (mitigating or reversing the urban heat island effect) and support local communities' adaptation to climate-change effects. When strategically placed around buildings, trees can save the energy needed for cooling and heating. In Sahelian countries like Burkina Faso, trees are planted to mitigate the localized high temperatures around houses and public institutions (Fuwape and Onyekwelu, 2011).

Urban forests, trees, and soils can also potentially increase carbon sequestration in and around urban areas, depending, however, on a number of variables including (primarily) the species and size of the tree. In fact, while large urban trees have been proven to play a key role as carbon sinks, for other urban vegetation types (dominated by palms, small trees, and/or turf grass), the net balance between carbon absorption and emission can be in favour of the latter. Several studies have illustrated that if not properly planned and managed, green spaces can indeed act as a net source of carbon dioxide (Zetter and Watson, 2006).

Water flow and stormwater regulation are also closely dependent on healthy urban and peri-urban forests. Trees protect and improve the quality of watersheds and water reservoirs by combating erosion, limiting evapotranspiration, and filtering sediments. In addition, by capturing water and increasing soil infiltration and stability, urban trees reduce and mitigate the impact of severe flooding events.

Declining air quality in rapidly urbanizing regions places a growing burden on people's health. Outdoor PM2.5 levels are four times the recommended limit in South Asia and are also high in East Asia and the Pacific region. In both regions, air quality has worsened since 1990. Air pollution is rapidly increasing also in Africa and Latin America (Brauer et al., 2016). Trees are excellent air filters, as they intercept gaseous pollutants and particulates from urban activities and vehicular traffic thus contributing to improved air quality. In 2002, the 2.4 million trees in the centre of highly polluted Beijing removed more than 1260 tons of pollutants from the air. In 2012, the city initiated a US$4.7 billion programme to plant 67,000 hectares of trees around Beijing to improve the urban air quality (Yang et al., 2005).

Cultural services

Urban and peri-urban forests also contribute to social equity, can promote a sense of community among dwellers, and ensure the preservation of local spiritual and cultural values essential to place making.

UPFs are ideal settings for implementing programmes that enhance local environmental education and raise urban dwellers' awareness of the importance to conserve trees. In

Malaysia, the Forest Research Institute Malaysia hosts a government-based forest reserve responsible for biological, botanical, and other scientific research and environmental education; the centre also attracts visitors who want to experience the Malaysian tropical rain forest without traveling too far from the city (Ramlan et al., 2012).

By beautifying central and suburban areas, urban forests and trees also help reduce social, environmental, and housing inequity between urban dwellers. In Lima, the residents of an impoverished neighbourhood on the city outskirts were asked what they felt was needed to improve their area. The reply was that they wanted a green space (Hodson, 2012). A survey conducted in Lomé highlighted that its dwellers perceived the role and the importance of trees in urban and suburban areas for enhancement (61%) and aesthetics (33%) but also for improvement of living conditions (4%) (Polorigni et al., 2014).

By providing green open spaces for recreational and sport activities, urban forests and trees also contribute to reducing stress and improving urban dwellers' health and well-being (both physical and mental; see Chapter 6, this volume), and can support convalescence programmes in local hospitals. In light of this, several countries and cities are starting to set minimum green-cover standards for hospitals and convalescence homes (Jim and Liu, 2000).

Urban and peri-urban parks also provide urban communities with open-air settings for implementing local activities and events, thus increasing social cohesion. In West Africa, most schools in urban areas are adorned with trees for students to sit under and relax during break period (Fuwape and Onyekwelu, 2011).

Furthermore, urban and peri-urban forests and trees are often associated with strong cultural, social, and religious values. In 2002, a survey in Bangkok found that 261 heritage trees were conserved over the decades thanks to a religious tradition prohibiting the cutting of certain tree species considered to be holy species (i.e. *Ficus religiosa*; Thaiutsaa et al., 2008). In addition, many traditional practices (e.g. t'ai chi, qigong) are performed in urban parks and other open green spaces in Asia since ancient times.

Additional socio-economic benefits

In addition to ecosystem services, urban and peri-urban forests also provide direct and indirect socio-economic benefits to communities, and make significant contributions to the creation of a local model of green economy (see also Chapter 4).

Urban forests generate jobs related to the establishment, management, and maintenance of green spaces and their products, thus contributing to boosting a green economy model for the city. In the Democratic Republic of the Congo, the wood-fuel sector employs over 300,000 people for the supply of Kinshasa alone (Schure et al., 2014). Urban green can also increase property and land values and rental prices, with a direct revenue for the government in terms of taxes.

Urban green contributes to the branding of the city, attracting investment, businesses, and tourism. For example, in Sabah, Malaysia, urban forests play an important role in the local eco-tourism industry by providing nature at the doorstep for visitors to enjoy (Lee et al., 2004). Finally, the wood and non-wood forest products (e.g. timber, fruits, nuts, berries, mushrooms, medicinal plants, and wood fuel) provided by urban forests also help to increase local income and markets, and improve community resilience.

In terms of indirect economic benefits, urban and peri-urban forests also reduce public costs overall. By protecting buildings, they lower heating and cooling costs. By improving physical and mental health, cooling the environment, reducing pollution, and producing oxygen, they contribute to reducing public health costs.

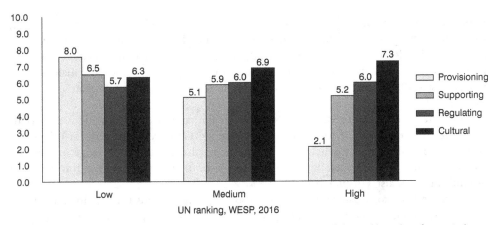

Figure 10.4 Perceived importance (score 1 to 10) of urban ES groups delivered by urban forests in low-, middle- and high-income countries

Source: Expert workshop by the authors

Box 10.2 The perception of importance of urban forest benefits in developing countries

The relative importance of the benefits – both real and perceived – provided by urban and peri-urban forests deeply differ between rich and poor cities. The authors discovered this through an evaluation of the relative importance of the urban and peri-urban forests benefits perceived in low-, middle- and high-income countries. Respondents provided a score between 1 and 10 to quantify the perceived importance of ecosystem services, grouped in Provisioning, Supporting, Regulating, and Cultural. While all the groups of services were considered of interest, the Provisioning ES was found to be perceived as the most important for low-income countries (Figure 10.4), with Supporting ES still of high interest. Regulating and Cultural ES were of higher importance for middle- and high-income countries. Sometimes this difference is visible also in different districts of the same city (Escobedo et al., 2015; Scopelliti et al., 2016). High-income residential areas in cities of developing countries perceive the availability of green spaces and trees similarly to those living in cities of developed countries.

Addressing main challenges of governance of urban and peri-urban forests in developing cities

Governance is commonly defined as a shift in strategic decision-making in which a range of government actors to an increasing extent share (or even transfer) decision making and rule setting with civic society and businesses (see Chapter 16).

While the general principles of governance should apply to all situations, the governments of rapidly growing cities often have little time to adjust to the changing circumstances and to the increasing pressure on land and resources resulting from uncontrolled urbanization. This in turn may lead to inadequate planning and ineffective governance resulting in significant economic, social and environmental costs, threatening the sustainability of urban development.

The failure of governments to provide appropriate infrastructure and public services is at the core of many urban challenges in developing countries, which range from the incapacity to provide adequate housing, public safety and welfare, green public spaces, or clean water, to the larger challenges of building climate-resilient cities. Furthermore, in many developing cities, slum dwellers can number more than 50 per cent of the population. Streets, squares, and parks, especially in the informal city, are often chaotic, and poorly planned and maintained – if they exist at all. Lack of space and lack of planning are endemic to many urban settlements in the developing world.

Adequate governance of green public spaces and of urban forests can provide an important model for the resolution of some of these challenges. Effective governance of urban and peri-urban forests requires policies and laws aimed at harmonizing the range of interests and developing and strengthening a common vision for urban and peri-urban forests in and around cities (see also Chapter 16).

Policies related to urban forests are often *sectoral*, aimed exclusively at either urban or rural areas, leading to conflicts with other sectors over the use of open spaces. Indeed, public policies, and particularly in the developing world, are still very often aimed exclusively at either urban or rural areas.

To address the urban–rural imbalance and achieve faster development in under-developed areas of developed countries, governments at the national and local level must also recognize the growing importance of *urban–rural linkages*. UN-Habitat is currently working to promote the urban–rural linkage development approach and has adopted a number of related resolutions (Okpala, 2003). For example, although they traditionally only target rural areas, development projects based on non-timber forest products in the northern Bolivian Amazon have also benefited peri-urban populations in the region. A survey of 120 households at the periphery of Riberalta revealed 'that peri-urban livelihoods depend significantly on both the extraction of Brazil nut and palm heart and their urban-based processing' (Stoian, 2005).

The governance of urban green spaces requires that planning departments acquire the necessary technical *skills and knowledge* to be able to include urban forestry aspects in the overall planning processes. The UPF strategy developed with the support of FAO for the city of Bangui, Central African Republic, for example, included a strong component of capacity building in its framework. The activities included the development of agreements with the University of Bangui and other international universities to offer training, environmental education programmes for schools and for rural communities, and an *awareness* raising campaign directed at municipal authorities, technicians and farmers' organizations (FAO, 2009b).

Building public awareness of the goods and services provided by urban forests is necessary to help address the lack of specific *public funds* allocated to management. Municipal budgets are often insufficient to fund important urban greening programmes. Therefore, innovative management must rely heavily on local volunteers, not only to raise funds, but also to provide programme leadership and physical labour. Strategies for funding include public investment; cost avoidance, reduction, and recovery; and use of trusts and private funds (Kuchelmeister, 1997). Income-generating activities linking recreation and/or goods derived from the urban forest, could also contribute to funding tree and park maintenance. Of course, whenever fees or other forms of service payments are introduced, it is always important to give due consideration to public access and social equity aspects.

A failure of governance often results from a *lack of coordination* among public authorities (i.e. between state and local governments or between agencies of a single government) and *inadequate involvement of various stakeholders* such as village leadership, community organizations, and urban residents in different phases of planning and establishment. Urban forestry involvement programmes should start with an assessment of management needs and opportunities. People support what they believe to be valuable, especially if they receive direct benefits from projects. Kuchelmeister (1998) argues that urban trees (in developing countries) survive best when people have begged for them. Neighbourhood associations and local NGOs are key stakeholders to work with. Public involvement requires looking beyond the normal city park boundaries to the communities that surround urban open spaces. In the Philippines, since the early 1990s, foundations and other NGOs manage vast areas of green space (e.g. Rizal Park, Quezon Circle, Ninoy Aquino Park; Palijon, 2001). In Delhi (India) efforts of citizens and volunteer groups have achieved nature reserve status for the Delhi Ridge Forest, a tropical thorn forest type (Asgish and Rao, 1997).

Furthermore, *private/public partnerships* represented, in some cases, a key solution to the involvement of stakeholders in urban forestry governance. Business owners in Brazil care for street trees or sponsor tree planting in front of their stores in exchange for advertising on the tree protectors. São Paolo's public–private partnerships for tree planting is an example. Companies are contracted – through public bids – to plant trees along roadways. In exchange for planting the trees, the company receives a permit to sell small advertising spots placed on seedling-protection rails (Coleman et al., 2013).

In many developing countries, urban plans are *ineffective, obsolete, or simply non-existent*. As a result, there is difficulty in identifying specific objectives and priorities, particularly for urban forest and green spaces. In a bid to address these problems, in 2007, the government of Ghana initiated a planning reform under the Land Use Management and Planning Project (LUMP). The aim of LUMP was to develop a decentralized and sustainable land use management and planning system based on consultative and participatory approaches to manage human settlement. Approaches included developing and testing land use planning models and controls in consultation with communities and land owners; establishing a coherent and modernized legal framework for town and country planning, including model guidelines and regulations; and developing an information/public awareness campaign strategy and materials to support implementation of the reforms.

Fragmentation of responsibilities and technical and administrative services in policy and planning documents and across levels of government is another factor limiting the governance of urban and peri-urban forests. 'Integration' can solve many problems. For example, the City of Johannesburg reorganized its park services – previously fragmented across Greater Johannesburg's five councils – into a single agency called Johannesburg City Parks. The goal of the new agency, with a single managing director and board of advisors, is to build and maintain more parks within the existing budget. In this single action, the city established clear responsibility, improved efficiency, and enabled application of coherent standards across the Johannesburg region (FAO, 2005).

Conclusions and perspective

The global community has recently adopted new goals to fight climate change and poverty through sustainable development, but urban growth threatens to undermine these aims, as cities become the largest sources of carbon emissions and income inequality. In recognition of this, urbanization was the main theme of the State of the World's Forest (SOFO; FAO, 2009a). The regional outlook studies of SOFO 2009 pointed to the need for taking action through the establishment of UPF programmes.

Sustainability and the prosperity of cities was also the theme of the State of the World's Cities (SWC) for 2013 (UN-Habitat, 2013). The introduction to SWC 2013 states that 'Urban areas consume huge amounts of environmental goods and services like food, water, energy, forests, building materials, and "green" or open spaces often beyond their boundaries.' Indeed, prosperity and environmental sustainability of cities go hand in hand; we cannot have one without the other.

In 2015, the United Nations launched the 2030 Agenda for Sustainable Development (United Nations, 2015). This remarkable document established 17 Sustainable Development Goals (SDGs) and 169 targets, many of which can be achieved through UPF. For example, as a key element of the strategies on air pollution removal and climate change mitigation, UPF is a concrete tool for the targets of SDG 3 'Good health and wellbeing' and SDG 13 'Climate action'. Most importantly, UPF can be a key contributor to achieving SDG 11 'Sustainable cities and communities' focused on making cities safe, resilient, and sustainable. Among the targets of SDG 11, UPF is potentially a leading tool for *increasing resource efficiency, mitigation, and adaptation to climate change, resilience to disasters in cities*, and *providing universal access to safe and accessible green and public spaces* by 2030.

The attention to nature and green spaces in cities of developing countries continues to increase in conceptual frameworks not limited to urban forestry, but also urban agriculture, urban planning. One example of this is in the new urban agenda in development for the Quito, Ecuador meeting of Habitat III, where nature and green spaces in cities will be an emerging component.

Therefore, to be truly successful in the developing world, UPF needs to be part of a shared vision that permeates every level of the planning and design process of cities. The fact that there might be a higher demand for goods rather than services in cities of developing countries (Figure 10.5) does not reduce the contributing value of ecosystem services provided by urban and peri-urban forests to urban quality of life.

Cities are ecological systems that consume, transform, and release materials and energy, in particular, they are very dynamic, interactive, and human-oriented. Therefore, cities should be managed and protected like any other ecosystems regardless of a nation's level of development. Nature-based solutions like urban and peri-urban forests are cost-effective,

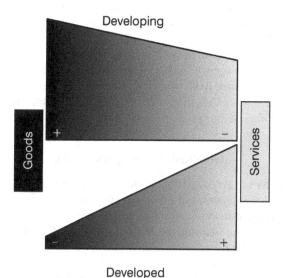

Figure 10.5 Perceived relative importance of goods and services as delivered by urban forestry in developing and developed countries

Source: Expert workshop by the authors

Figure 10.6 Children and trees in Bangui

In 2008, on a cloudy morning in the equatorial June of Bangui, the capital city of the Central African Republic, we planted mango trees at the edge of a city street with school students. It was not an anniversary or a celebration, rather it was an opportunity to sing a song that the students had written: 'Come, friend, come, let us go together to sing nature, to plant our future.' The children sang holding hands and making a circle around the trees they had just planted.

simultaneously providing environmental, social and economic benefits and helping to build community resilience to rapidly changing environments like the cities in developing and industrializing regions are. Urban forests and trees can definitively serve the communities of developing cities to plant a better and more sustainable future (Figure 10.6).

References

Asgish, K., Rao, S. (1997) 'How are we managing? Saving Delhi's natural ecosystems. A model of citizen participation', *Ecosystem Health* vol. 3, no. 2:124–126.

Borelli, S., Chen, Y., Conigliaro, M., Salbitano, F. (2015) 'Green infrastructure: A new paradigm for developing cities', technical paper presented at the XIV World Forestry Congress, Durban, South Africa.

Brauer, M., Freedman, G., Frostad, J., van Donkelaar, A., Martin, R. V., et al. (2016) 'Ambient air pollution exposure estimation for the global burden of disease 2013', *Environmental Science and Technology* 50, 1, pp. 79–88.

Coleman, G., Kontesi, A., Li, X., Masliah, A., Renwick, D., Torà, L., Vargas, L. (2013) *Building Successful Public Private Partnerships in São Paulo's Transportation Sector*, Columbia University, School of International and Public Affairs, New York, retrieved from https://sipa.columbia.edu/sites/default/files/C40%20Cities%20Final%20Report.pdf (accessed 7 August 2016).

Cuquel, F. L., Mielke, E. C., do Valle, F. J. R., Poliquesi, C. B. (2009) 'Biocity project', *Acta Horticulturae* vol. 813, pp. 95–100.

Desgroppes, A., Taupin, S. (2011). 'Kibera: The biggest slum in Africa?' *Les Cahiers de l'Afrique de l'Est*, vol. 44, pp. 23–34.

Echavarria M. (2001) *FONAG: The Water-Based Finance Mechanism of the Condor Bioreserve in Ecuador*, case study in online Conservation Finance Guide, Conservation Finance Alliance, Washington, DC, retrieved from http://guide.conservationfinance.org (accessed 15 July 2016).

EEA (2015) The European environment: State and outlook 2015, European Environment Agency, Copenhagen, retrieved from www.eea.europa.eu/soer (accessed 21 July 2016).

Escobedo, F. J., Clerici, N., Staudhammer, C. L., Corzo, G. T. (2015) 'Socio-ecological dynamics and inequality in Bogotá, Colombia's public urban forests and their ecosystem services', *Urban Forestry and Urban Greening*, vol. 14, no. 4, pp. 1040–1053.

FAO (2005) Legal and institutional aspects of urban, peri-urban forestry and greening: A working paper for discussion, authored by Knuth, L., FAO Legal Papers Online, http://www.fao.org/3/a-bb081e.pdf

FAO (2009a) *State of the World's Forests 2009*, (SOFO) FAO, Rome.

FAO (2009b) *Stratégie de développement et Plan d'action pour la promotion de la foresterie urbaine et périurbaine de la ville de Bangui*, Foresterie urbaine et périurbaine, Document de travail no 3, FAO, Rome.

FAO (2010) *Foresterie urbaine et périurbaine en Afrique: Quelles perspectives pour le bois-énergie?* Foresterie urbaine et périurbaine, Document de travail no 4, FAO, Rome.

Fuwape, J. A., Onyekwelu, J. C. (2011) 'Urban forest development in West Africa: Benefits and challenges', *Journal of Biodiversity and Ecological Sciences*, vol 1, no 11, pp. 77–94.

Gopal, D., Nagendra, H. (2014) 'Vegetation in Bangalore's slums: Boosting livelihoods, well-being and social capital', *Sustainability*, vol. 6, no. 5, pp. 2459–2473.

Gopal, D., Nagendra, H., Manthey, M. (2015) 'Vegetation in Bangalore's slums: Composition, species distribution, density, diversity, and history', *Environmental Management*, vol. 55, no. 6, pp. 1390–1401.

Hodson, J. (2012) 'A Peruvian slum gets a massive green makeover', retrieved from www.washington. edu/news/2012/02/13/a-peruvian-slum-gets-a-massive-green-makeover-with-slide-show (accessed 10 June 2016).

Ickowitz, A., Powell, B., Salim, M. A., Sunderland, T. C. H. (2014). 'Dietary quality and tree cover in Africa', *Global Environmental Change*, vol. 24, pp. 287–294.

IDB (1997) *Good Practices for Urban Greening*, Social Programme and Sustainable Development Department, Environment division, Env-109, Washington DC.

International Energy Agency (2015) 'World energy outlook', retrieved from www.worldenergyoutlook. org/resources/energydevelopment/energyaccessdatabase (accessed 12 June 2016).

Jim, C. Y., Liu, H. H. T. (2000) 'Statutory measures for the protection and enhancement of the urban forest in Guangzhou City', *China Forestry*, vol. 73, no. 4, pp. 311–329.

Kang, M.-K., Park, D., Chun, Y.-W. (2010) 'The performance analysis of Korean NGOs' tree plantation projects in Mongolia', *Journal of Korean Forest Society*, vol. 99 no. 5, pp. 655–662.

Kenya Forest Service (2010) 'Aberdare Forest Reserve management plan 2010–2019', retrieved from www.kenyaforestservice.org/documents/Aberdare.pdf (accessed 8 July 2016).

Konijnendijk, C. C., Sadio, S., Randrup, T. B., Schipperijn, J. (2004) 'Urban and peri-urban forestry in a development context – strategy and implementation', *Journal of Arboriculture*, vol. 30, pp. 269–275.

Kothari, A., Sunita, R. (1997) 'How we managing? Saving Delhi's natural ecosystems. A model of citizen participation', *Ecosystem Health*, vol 3, no. 2, pp. 124–126.

Kuchelmeister, G. (1997). 'Urban trees in arid landscapes: Multipurpose urban forestry for local needs in developing countries', *Arid Lands Newsletter*, no 42 ALN, retrieved from https://ag.arizona.edu/OALS/ALN/aln42/kuchelmeister.html (accessed 23 May 2016).

Kuchelmeister, G. (1998). *Urban Forestry in the Asia-Pacific Region: Status and Prospects*, Asia-Pacific Forestry Sector Outlook Study Working Paper Series no. 44, Rome, FAO.

Lee, Y. F., Ligunjang, J., Yong, S. C. (2004) Urban forestry and its relevance to tourism development, paper presented at the Asia Europe Meeting (ASEM) Symposium on Urban Forestry, Suzhou and Beijing, China.

Li, J., Yang, T. (eds) (2016) *China's Eco-city Construction*, Springer, Berlin.

Mylott, E. (2009) 'Urban–rural connections: A review of literature', scholars-archive @OSU, Papers (rural–urban connections), retrieved from http://ir.library.oregonstate.edu/xmlui/handle/1957/10574 (accessed 20 May 2016).

Nakhli, F. (2014) 'Une ceinture verte pour protéger Ouarzazate', *L'Economiste*, 5 May, p. 21.

Okpala, D. (2003) *Promoting the Positive Rural-Urban Linkages Approach to Sustainable Development and Employment Creation: The Role of UN-Habitat*, retrieved from www.fi g.net/pub/morocco/proceedings/PS1/PS1_1_ (accessed 20 May 2016).

Onganga, O. (1992) 'Urban forestry development in Kenya', in Rodbell, P. (ed), *Proceedings of the Fifth National Urban Forest Conference*, Los Angeles, CA.

Palijon, A. M. (2001) 'An analysis of green space management strategies in Metro Manila', in Sievänen, T., C.C. Konijnendijk, L., Langler and K. Nilsson, K. (eds), *Forest and Social Services: The Role of Research*, Proceedings of IUFRO Research Groups 6.01, 6.11.04 and 6.14 sessions at the XXI IUFRO World Congress 2000, Kuala Lumpur, FFRI, Vantaa.

Polorigni, B., Radji, R., Kokou, K. (2014) 'Perceptions, tendances et préférences en foresterie urbaine: cas de la ville de Lomé au Togo', *European Scientific Journal*, vol. 10, no. 5, pp. 261–277.

Ramlan, M. A., Aziz, A., Yahya N. A., Kadir, A. A., Yacob, M. R. (2012) 'Potential of canopy tourism in Malaysia', *The Malaysian Forester*, vol. 75, no. 1, pp. 87–96.

Salbitano F., Borelli S., Sanesi G. (2015) 'Urban forestry and agro-forestry', in De Zeeuw, H. (ed), *Cities, Food and Agriculture: Towards Resilient Urban Food Systems*, Routledge Earthscan, Oxford, pp. 285–311.

Schure, J., Levang, P., Wiersum, K. F. (2014) 'Producing woodfuel for urban centers in the Democratic Republic of Congo: A path out of poverty for rural households?' *World Development*, vol. 64, supp. 1, pp. 80–90.

Scopelliti, M., Carrus, G., Adinolfi, C., Suarez, G., Colangelo, G., Lafortezza, R., Panno, A., Sanesi, G. (2016) 'Staying in touch with nature and well-being in different income groups: The experience of urban parks in Bogotá', *Landscape and Urban Planning*, vol. 148, pp. 139–148.

Sène, E. H. (1993) 'Urban and peri-urban forests in sub-Saharan Africa: The Sahel', *Unasylva*, vol. 173, no. 44, pp. 45–51.

Singh, V. S., Pandey, D. N., Chaudhry, P. (2010) *Urban Forests and Open Green Spaces: Lessons for Jaipur*, Occasional Paper, RSPCB, Rajasthan, India.

Stoian, D. (2005) 'Making the best of two worlds: Rural and peri-urban livelihood options sustained by non timber forest products from the Bolivian Amazon', *World Development*, vol. 33, no. 9, pp. 1473–1490.

TEEB (2011) *TEEB Manual for Cities: Ecosystem Services in Urban Management*, TEEB, Geneva, retrieved from www.teebweb.org/publication/teeb-manual-for-cities-ecosystem-services-in-urban-management/ (last accessed 12 July 2016).

Thaiutsaa, B., Puangchita, L., Kjelgren, R., Arunpraparut, W. (2008) 'Urban green space, street tree and heritage large tree assessment in Bangkok, Thailand', *Urban Forestry and Urban Greening*, vol. 7, no. 3, pp. 219–229.

UN-Habitat (2013) *State of the World's Cities: Prosperity of Cities*, Earthscan, New York.

United Nations (2014) *World Urbanization Prospects: The 2014 Revision*, ST/ESA/SER.A/352, United Nations, Department of Economic and Social Affairs, Population Division, New York.

United Nations (2015) Transforming our world: the 2030 Agenda for Sustainable Development, A/RES/70/1, retrieved from https://sustainabledevelopment.un.org/post2015/transformingourworld (accessed 15 July 2016).

Yang, J., McBride, J., Zhou, J., Sun, Z. (2005) 'The urban forest in Beijing and its role in air pollution reduction', *Urban Forestry and Urban Greening*, vol. 3, no. 2, pp. 65–78.

Zetter, R., Watson, G. B. (eds) (2006) *Designing Sustainable Cities in the Developing World*, Routledge, New York.

11

ASSESSING THE BENEFITS AND ECONOMIC VALUES OF TREES

David J. Nowak

Introduction

Understanding the environmental, economic, and social/community benefits of nature, in particular trees and forests, can lead to better vegetation management and designs to optimize environmental quality and human health for current and future generations. Computer models have been developed to assess forest composition and its associated effects on environmental quality and human health. While research is still needed regarding many of the environmental services that trees provide, resource managers can utilize existing models to better understand the role of vegetation in improving human health and environmental quality, lower costs of maintenance, and increase resource stewardship as an effective means to provide substantial economic savings to society.

Understanding the myriad of potential services and costs associated with forests are critical to estimating net benefits of vegetation and for guiding appropriate vegetation management plans. However, while many of the ecosystem services and costs of vegetation cannot be adequately quantified or valued at this time, it is important to understand within decision -making processes that these services or costs do exist. Discounting nature or vegetation as having no value leads to uninformed decisions regarding nature (e.g. Costanza et al., 1998). Quantifying or understanding monetary and non-monetary values of nature in a given context, though difficult, will lead to more informed environmental and economic decisions.

Trees provide numerous economic and ecosystem services that produce benefits to a community, but also incur various economic or environmental costs. Through proper planning, design, and management, trees can improve human health and well-being in urban areas by moderating climate, reducing building energy use and atmospheric carbon dioxide (CO_2), improving air quality, providing an aesthetic environment and recreational opportunities, mitigating rainfall runoff and flooding, lowering noise levels and producing other social/environmental services. However, inappropriate landscape designs, tree selection, and tree maintenance can increase environmental costs such as pollen production, chemical emissions from trees and maintenance activities that contribute to air pollution, and can also increase building energy use, waste disposal, infrastructure repair, and water consumption. These potential costs must be weighed against the environmental benefits in developing natural resource management programmes.

Specific attributes of the vegetation resource structure such as abundance, size, species, health and location affect the amount of services and costs provided by vegetation. Many of the services and costs provided by vegetation and their management affect human health. Thus, designing nature and management to maximize these benefits and minimize the costs can help improve human health.

There are four main steps needed to quantify ecosystem services and values from forests:

1 Quantify the forest structural attributes that provide the service for the area of interest (e.g. number of trees, tree cover). These structural data are essential as they quantify the resource attributes that provide the services.
2 Quantify how the structure influences the ecosystem service (e.g. tree density, tree sizes, and forest species composition are significant drivers for estimating carbon storage).
3 Quantify the impact of the ecosystem service. In many cases, it is not the service itself that is important, rather the impact that the service has on human health or other attributes of the environment that provide value to society.
4 Quantify the economic value of the impact provided by the ecosystem service.

There is an interdependence between forest structure and ecosystem services and values as valuation is dependent upon good estimates of the magnitude of the service provided, and the service estimates are dependent upon good estimates of forest structure and how structure affects services. The key starting point to valuing services provided by forests is quality data on forest structure. Services and values cannot be adequately valued without good forest data. Combining accurate forest data with sound procedures to quantifying ecosystem services will lead to reliable estimates of the magnitude of ecosystem services provided by the forest. Finally, with sound estimates of forest ecosystem services, values of the services can be estimated using valid economic estimates and procedures. Thus three critical elements in sequence are needed to value forest ecosystem services: structure → services → values. Errors with precursor elements will lead to errors in subsequent estimates (e.g. errors in forest structure will lead to errors in estimating services and valuation). All current estimates and means of estimation can be improved to varying degrees.

Assessing forest structure

Structure is a key variable as that is what managers manipulate to influence forest services and values. Managers often choose what species to plant, where and when to plant it, and what trees are removed from the landscape. These actions directly influence forest structure and consequently the values derived from the forest.

While managers make these structural decisions, it is important to understand that nature often has a more substantial influence on urban forest structure and that management decisions that affect these natural interactions will influence structure. Natural influences include natural regeneration, climate, insects and diseases, invasive tendency of various species, etc. Management decisions to allow regeneration (e.g. limiting mowing) and regarding species selections, which have varying susceptibilities to insects and climate, and varying invasive characteristics, will influence forest structure. In the United States, it is estimated that two-thirds of the existing urban forest is from natural regeneration (Nowak, 2012a). However, the influence of tree planting tends to increase in cities in grassland and desert areas, more densely populated cities, and on land uses that are highly managed in relation to trees (e.g. residential lands).

There are two basic means to quantify urban forest structure:

1 top-down aerial-based approaches; or
2 bottom-up ground-based assessments.

Top-down assessments provide basic metrics on tree and other cover types (e.g. percentage tree cover) and can include specific locations of these cover elements when maps are produced. Tree cover can often be estimated by interpreting aerial photographs or by developing tree cover maps using moderate to high resolution imagery (e.g. Nowak, 2012b). If just the amount or percentage tree cover is needed, photo-interpretation provides the most cost-effective and accurate means of assessing tree and other cover attributes, but lacks specific cover location information. If tree and other cover locations are needed, tree cover maps can provide both tree cover estimates and specific locations of cover elements (e.g. to be integrated within Geographic Information Systems). Tree cover and distribution are important elements of urban forest structure as they provide a simple means to convey the magnitude and distribution of the forest resource. However, more detailed data are often needed on forest structure to assess ecosystem services and values, and to help guide forest management (e.g. species composition, number of trees, tree sizes, tree condition, leaf area, leaf biomass, tree biomass). To obtain these more detailed forest structural data, field measurements are often required.

In assessing the structure of the urban forest, there are various steps that should be followed:

1 *Delimit the study area.* The first step is to bound the region of analysis – in what area do you want to assess the forest services and values? This area could be any size (e.g. backyard, park, city boundary), but the area or tree population must be delimited.
2 *Inventory or sample?* After the area or tree population is delimited, the next question on assessing structure is: are all trees in the area to be measured (inventory) or only a small proportion of the total population sampled. Inventories have the advantage of increased precision as all trees are measured and are useful for individual tree management (e.g. street tree population), but inventories have higher costs and are often inefficient for large tree populations. Sampling of large tree populations provide population information at a more reasonable cost, but lack the comprehensive individual tree information that is useful for individual tree management. Sampling is more useful for population management that is not focused on every individual tree.
3 *What services or values do you want to quantify?* There are several variables that can be measured on individual trees or forests, with each variable used to quantify specific services and values. As each variable adds additional cost to the assessment of structure, variable selection is critical to meet the specific needs for the inventory or sample and reduce assessment costs. Often most of the cost of measurement is getting to the trees, but extraneous measured tree variables will add cost with no benefit (e.g. if effects on building energy use are not desired, measuring variables specific to energy conservation estimates are not needed).
4 *Select tree and/or plot variables to be measured.* There are numerous variables that can be measured on trees. The core variables are typically tree species, diameter at breast height (dbh – diameter at 1.37 m), crown variables and tree condition. Other variables can be measured depending upon needs, including tree damage, maintenance needs, risk, etc.
5 *Collect field data and assess services and values.* Once field (e.g. species, number of trees, tree size) or aerial data are collected, various ecosystem services and values can be quantified using various tools. This process will be described in the section on modelling urban

forest ecosystem services and values. Field data collection procedures are detailed in another chapter.

While various aerial-based approaches are being researched and developed to derive specific tree information, the current best methods to derive many of the tree variables are field measurements. While future technologies may allow complete assessments using remote sensing technologies, many variables are currently measured more accurately from ground-based assessments. Some measurements of tree attributes (e.g. tree cover, leaf area) may be more accurately assessed using aerial images, but some key tree variables (e.g. species, condition) are often more accurately collected using field measurements. More research is needed to adequately and cost-effectively estimate these variables using aerial platforms. At the individual tree scale, field measurements are the best means to assess tree variables. For large tree populations, field data in conjunction with aerial based assessments will likely provide the best and most cost-effective means to assess forest structure. In urban forestry, field measurements are particularly important, as not only do they provide data, but they put the manager in the field where citizens reside. Interactions between urban foresters and citizens are critical for proper urban forest management.

Focusing on tree variables that are measured in the field, the most important variables are tree species, diameter, crown dimensions, and condition. This information is helpful to managers regarding population management, but is also essential for estimating ecosystem services. With regards to most services, the most important tree attribute is leaf area. While not directly measured in the field, this variable can be directly estimated from species, crown and condition information. In addition, several aerial-based methods can be used to estimate overall leaf area, but are typically not done to the species level. Diameter measures are also essential for estimating carbon storage. Leaf and tree biomass are other important variables that can be estimated from the core tree variables. Other information that is important for estimating ecosystem services is crown competition (important for tree growth estimation and carbon sequestration) and location around buildings (important for energy conservation). Numerous ecosystem effects and values can currently be estimated from these tree variables, along with information on local cover types and other local information (e.g. weather, pollution concentrations, population data). In addition to variables needed to assess ecosystem services from trees, other variables can also be collected to aid in urban forest management (e.g. soils data, risk assessments, maintenance needs) or assessing ecosystem services from other vegetation types (e.g. grass and shrub cover).

As urban forests can provide numerous services and disservices, it is important to understand how urban forest structure affects these various services and values.

Quantifying urban forest benefits and disbenefits

As detailed in separate chapters of this handbook (see especially Chapter 13), there are numerous benefits and disbenefits/costs (or disservices) associated with urban vegetation. In general, the costs are easier to quantify as many of these are direct costs borne by the land manager. These costs include planting, pruning, maintenance, tree removal, property repair (e.g. lifted sidewalks, damage from falling branches, clogged drains), injuries from vegetation, damage from forest fires, and leaf raking. Often these costs are known and can be quantified by the land manager. However, there are other costs incurred by the land manager or society that are often not directly paid for in managing vegetation. These disbenefits include costs associated with allergies from tree pollen, emission of plant volatile organic compounds

(VOCs) that can lead to the formation of ozone, carbon monoxide and particulate air pollutants, trees limiting pollutant dispersion near polluted roadways, increases in winter building energy use due to tree shade, invasive plant altering local biodiversity, increased tax rates due to increased property values, water use in arid regions and increased fear of crime (e.g. Brasseur and Chatfield, 1991; Gromke and Ruck, 2009; Tallamy and Shropshire, 2009; Carinanos et al., 2014). Although the list of these disbenefits is smaller than the list of potential benefits, these disbenefits need to be understood and quantified to help minimize these negative attributes of vegetation. Many of these disbenefits have been or are currently being input into modelling/management systems to quantify these disservices based on measured field data (e.g. pollen, VOC emissions, winter energy effects, invasive species).

In addition to the costs and disservices of vegetation, there are a myriad of services and values derived from vegetation. Unlike most costs, many of these beneficial services do not directly affect the cash balance of land managers (i.e. they do not directly pay the land manager like the manager directly pays for costs). Rather, these services indirectly save the land owner money or provide services that affect the health and well-being of the land owner and surrounding residents, and have a monetary value to the land owner and to society at large.

The Millennium Ecosystem Assessment (Hassan et al., 2005) describes four categories of ecosystem services:

1 supporting (e.g. nutrient cycling, primary production);
2 provisioning (e.g. food, fuel);
3 regulating (e.g. climate regulation, water purification); and
4 cultural (e.g. aesthetic, spiritual).

Many of these services are described in other chapters, with many services not able to be easily quantified from data on urban forest structure (e.g. tourism, recreation, aesthetics, social and physiological benefits, noise reduction). While science continues to advance in understanding and quantifying the relationships between forest structure and many of these services, several of these services can be currently quantified based on local urban forest, environmental and human population data. These services include tree effects on:

- *Air temperature.* By transpiring water through leaves and shading surfaces, trees reduce air temperatures. Reduced air temperatures can have a direct impact on human health (e.g. Martens, 1998) and can also reduce pollutant emissions from various sources (e.g. power plants) (Nowak et al., in review), which will consequently affect human health. Models have been developed to estimate urban tree effects on air temperature (e.g. Yang et al., 2013; Heisler et al., 2016) and illustrate that doubling tree cover in Baltimore, MD, would reduce daily temperature by 0.35°C (±0.02) (Ellis, 2009).
- *Pollution removal.* Vegetation directly removes air pollution through leaf stomata (Nowak et al., 2014) or intercepting particles on plant surfaces (Nowak et al., 2013b). Woody plants also sequester and store carbon in their biomass, reducing levels of the greenhouse gas carbon dioxide (Nowak et al., 2013a). They also emit oxygen (Nowak et al., 2007). Models have been developed to estimate these effects and show that in the United States, urban forests remove 651,000 metric tons (tonnes) of air pollution in 2010, store 643 million tonnes of carbon, annually sequester 25.6 million tonnes of carbon and annually produce 61 million tonnes of oxygen.
- *Building energy use.* Trees near buildings alter building energy use by cooling air temperatures, blocking winds and shading building surfaces (Heisler, 1986). Energy use

is decreased during the summer season, but depending on location and species, can be increased or decreased during the winter season due to altering wind speeds and solar access around buildings. Changes in energy use will consequently alter pollutant and greenhouse gas emissions from power plants and thus air quality and human health. Procedures have been developed to estimate urban forest effects on building energy use (e.g. McPherson and Simpson, 1999) and reveal energy savings of 38.8 million Mwh and 246 million MMBtus in urban/community areas of the United States, which equates to a 7.2 per cent reduction in building energy use (Nowak et al., in review).

- *Water cycles and quality.* Through intercepting rainfall, absorbing soil moisture and chemicals, transpiring water, and increasing soil infiltration, tree and forests can reduce storm water runoff and improve water quality. These hydrologic effects can reduce risk to flooding and improve human health related to sediments, chemicals and pathogens found with waterways. Various hydrologic models exist to estimate the effect of trees or land cover on stream flow (e.g. Bicknell et al., 1997; Tague and Band, 2004; Wang et al., 2008).
- *Ultraviolet radiation.* Tree leaves absorb 90–95 per cent of ultraviolet (UV) radiation and thereby affect the amount of UV radiation received by people under or near tree canopies (Na et al., 2014). This reduction in UV exposure affects incidence of skin cancer, cataracts and other ailments related to UV radiation exposure (Heisler and Grant, 2000).
- *Wildlife populations.* Tree species composition and structure directly affect wildlife habitat and local biodiversity. Various procedures estimate the relationship between local forest structure and wildlife species habitat suitability and insect biodiversity (Tallamy and Shropshire, 2009; Lerman et al., 2014).

Trees affect many other attributes of the physical and social environment in cities, many of which remain to be quantified at the local site scale through the regional scale. Understanding the effects of trees at both of these scales is important in terms of developing vegetation policies and specific landscape designs. For example, while trees reduce overall pollution concentration in cities through pollution removal, at the local scale pollution concentrations may be increased or decreased more than city wide estimates depending upon local vegetation designs (e.g. Gromke and Ruck, 2009; Nowak et al., 2014). Once the magnitude of a service is quantified, the next step, if so desired, is to convert the service to a monetary value.

Estimating the economic values of ecosystem services

Once the ecosystem services are quantified, various methods of market as well as non-market valuation can be applied to characterize their monetary value. Some valuing procedures use direct market costs. For example, for altered building energy use, the local cost of electricity (USD/kWh) and heating fuels (USD/MBTU) can be applied to changes in energy use due to local vegetation. For other ecosystem services, proxy values often need to be used as many of the services derived from trees are not accounted for in the cost of a market transaction. That is, many forest benefits produce externalities. An externality arises whenever the actions of one party either positively or negatively affect another party, but the first party neither bears the costs nor receives the benefits. Externalities are not reflected in the market price of goods and services. A classic example of a negative externality is air pollution, where the associated health costs are paid by society and not the producer of the pollutant. Trees often produce positive externalities (e.g. cleaner air). There are various ways to estimate externality costs including general systems analysis, the social fabric matrix, direct cost, contingent valuation, travel cost, and the property approach (Hayden, 1989).

In urban forestry, various methods of valuation have been used depending on the service or value being estimated. For air pollution removal, common methods include health care costs and externality values (Nowak et al., 2014), for carbon storage and sequestration – the social cost of carbon (Nowak et al., 2013a), for energy effects – utility costs (Nowak et al., in review), for water values – storm water control or treatment costs (e.g. Peper et al., 2009), and for structural values – tree appraisal methods (e.g. Nowak et al., 2002). Many services remain to be quantified and valued.

These values can vary locally depending on local forest structure and its impact. Nationally, these values can be in the billions of dollars annually. In the United States, it is estimated that the annual values of urban forests is 4.7 billion USD from energy conservation and 2.3 billion USD from avoided pollutant emissions (Nowak et al., in review), 4.7 billion USD from air pollution removal (Nowak et al., 2014), 2 billion USD from carbon sequestration (Nowak et al., 2013a) and is negligible for oxygen production (Nowak et al., 2007). While annual service values are on the order of billions of dollars per year in the United States, the structural asset value of US urban forests is in the trillions of dollars (Nowak et al., 2002).

By understanding how vegetation affects numerous services, values and costs, better decisions can be made relating to landscape management to improve environmental quality and human health. To this end, tools can be used that incorporate local data to estimate ecosystem services and its economic value to help guide management and sustain optimal vegetation structure through time.

Modelling urban forest ecosystem services and values

There are various models that quantify ecosystem services. Some free models include InVEST (Natural Capital Project, 2016), Biome-BGC (Numerical Terradynamic Simulation Group, 2016) and numerous tools to assess forest carbon (e.g. US Forest Service, 2016). To date, the most comprehensive model developed to quantify urban forest structure, ecosystem services and values is i-Tree (www.itreetools.org). This free suite of tools was developed by the U.S. Forest Service through a public-private partnership. The model has and can be used globally with over 125,000 users in over 120 countries, but tools vary in their ease-of-use and functionality globally. i-Tree was designed to accurately assess local vegetation composition and structure and its impacts on numerous ecosystem services and values (Table 11.1). The model focuses on estimating forest structure and the magnitude of services received (e.g. tonnes removed). It then relies on economic valuation (e.g. $/tonne removed) to estimate a value of the service. These values can vary depending upon how the receivers of the benefits (e.g. humans) are distributed across the landscape relative to the trees. i-Tree tools can also be used to target tree species and locations to sustain or enhance ecosystem services and human health. Threats to trees and associated services can also be assessed.

The core program is i-Tree Eco – this model uses sample or inventory data to assess forest structure, ecosystem services and values for any tree population (including number of trees, diameter distribution, species diversity, potential pest risk, invasive species, air pollution removal and health effects, carbon storage and sequestration, runoff reduction, VOC emissions, building energy effects) (Nowak et al., 2008). It runs on local field data and hourly meteorological and pollution data. The program includes plot selection programs, data entry programs or mobile application data entry, table and graphic reporting and exporting, and automatic report generation. Not all ecosystem services are, or can be, evaluated due to scientific limitations. However, this model was created as a continuously developing platform to incorporate newly available science on vegetation services. This tool can be used globally if users provide local geographic data, but is currently set to easily run in the United States, Canada, Australia and the United Kingdom (i.e. required city and environmental data are preloaded).

Other tools in i-Tree include:

- *i-Tree Forecast* uses tree data from i-Tree Eco to simulate future tree population totals, canopy cover, tree diversity, dbh distribution and ecosystem services and values by species based on user-defined planting rates and default or user-defined mortality rates (e.g. user can simulate effect of emerald ash borer by specifically killing off ash trees).
- *i-Tree Streets* is similar to Eco, but focuses on street tree populations. The program is only designed to work in the United States.
- *i-Tree Species* is a free-standing utility designed to help users select the most appropriate tree species based on desired environmental functions and geographic area. Tree species are mainly temperate tree species.
- *i-Tree Storm* helps assess widespread street tree damage in a simple and efficient manner immediately after a severe storm. It is adaptable to various community types and sizes and provides information on the time and funds needed to mitigate storm damage.
- *i-Tree Hydro* is designed to simulate the effects of changes in tree and impervious cover within a watershed on hourly stream flow and water quality. It contains auto-calibration routines to help match model estimates with measured hourly stream flow and produces tables and graphs of changes in flow and water quality due to changes in tree and impervious cover within the watershed. This tool can be used globally where gauged stream data exist.
- *i-Tree Canopy* allows users to easily photo-interpret Google aerial images of their area to produce statistical estimates of tree and other cover types along with calculations of the uncertainty of their estimates. This tool provides a simple, quick and inexpensive means for cities and forest managers to accurately estimate their tree and other cover types. i-Tree Canopy can be used anywhere in the world where high-resolution, cloud-free Google images exist (most urban areas). Use of historical imagery can also be used to aid in change analyses. From US cover data, users can also estimate pollution removal and carbon storage/sequestration amounts and values.
- *i-Tree Design* links to Google maps and allows users to outline their home and see how the trees around their home affect energy use and savings, and provide other environmental services. Users can use this tool to assess which locations and tree species will provide the highest level of benefits. This is a simple tool geared toward homeowners, school children or anyone interested in tree benefits. This program allows users to add multiple trees, illustrate future and past benefits, and display priority planting zones around buildings to conserve energy use. i-Tree Design is currently set to work in the United States and Canada.
- *i-Tree Landscape* is a web-based tool that allows users to explore tree canopy, land cover, and basic demographic information anywhere in the conterminous US. With the information provided by i-Tree Landscape, users can learn about their location (e.g. tree and impervious cover, population statistics), the benefits of trees (carbon storage, air pollution removal, reduced runoff) in their area, and map areas in which to prioritize their tree planting and protection efforts.

While some of these tools are currently limited in the global context due the requirement for tree cover maps (i.e. Landscape) or national processing of environmental data (i.e. Design), these tools are continuously updated with new features and can be developed to work globally. Many new ecosystem services are currently being added to the model, including tree effects on air temperature (Yang et al., 2013), ultraviolet radiation (Na et al., 2014), wildlife habitat (Lerman et al., 2014), pollen and human comfort (Table 11.1). New features are also in development including mobile apps, a web-based global importer

Table 11.1 Ecosystem effects of trees currently quantified and in development in i-Tree. Many of the listed ecosystem effects are both positive and negative depending on specific conditions or perspective. For example, trees can increase or decrease energy use depending upon location; pollen can be positive in terms of food production or negative in terms of allergies depending upon species

Ecosystem effect	Attribute	Quantified	Valued
Atmosphere	Air temperature	○	○
	Avoided emissions	●	●
	Building energy use	●	●
	Carbon sequestration	●	●
	Carbon storage	●	●
	Human comfort	○	
	Pollen	○	
	Pollution removal	●	●
	Transpiration	●	
	UV radiation	●	○
	VOC emissions	●	
Community/Social	Aesthetics / property value	○	○
	Food / medicine	○	
	Health Index[1]	○	
	Forest products[2]	○	○
	Underserved areas	●	
Terrestrial	Biodiversity	○	
	Invasive plants	●	
	Nutrient cycling	○	
	Wildlife habitat	●	
Water	Avoided runoff	●	●
	Flooding	○	○
	Rainfall interception	●	
	Water quality	●	○

● Attribute currently quantified or valued in i-Tree.

○ Attribute in development in i-Tree.

[1] Developing a health index based on mapping of green viewing ('forest bathing').

[2] Estimating product potential based on forest structure (e.g. timber, wood pellets, ethanol).

of species and city data to aid international users in operating i-Tree Eco, climate change projections, green infrastructure impacts (e.g. rain gardens, retention ponds) on water flows and water quality in i-Tree Hydro, automatic assessment of change between remeasured field data sets, new species in i-Tree Species to aid in selecting appropriate species, and new map layers and querying abilities in i-Tree Landscape.

i-Tree is built based on a collaborative effort among numerous partners to better understand and quantify how changes in forest structure will affect ecosystem services and values. Some of the goals of the program are to (a) provide global standards for data collection and analyses, (b) aid managers in optimizing ecosystem services from their forest for current and future generations, (c) better understand risks to forest and human health, and (d) integrate multiple ecosystem services and values to allow users to see trade-offs among various services based on proposed or actual changes to forest structure (e.g. designs to improve water quality may reduce wildlife habitat or also enhance air quality). Through a better understanding of how forest structure affects numerous services, better management strategies can be designed to maximize benefits and minimize costs associated with forests and their impacts on environmental quality and human health and well-being.

Conclusion

By understanding and accounting for the ecosystem services provided by trees, better planning, design and economic decisions can be made toward utilizing trees as a means to improve environmental quality and human health and well-being. A key to this improvement is data on urban forest structure and how structure (i.e. species composition, tree locations) affects services and values. i-Tree tools offer a means to assess and value the impact of trees and forests from the local parcel level to a regional landscape scale for several key ecosystem services. While more research is needed regarding several ecosystem services and costs, and associated impacts on human health and well-being, landscape management plans and designs that better incorporate the impacts of vegetation could lower costs and improve human health and environmental quality, and thereby provide substantial economic savings to society.

Acknowledgements

The use of trade names in this chapter is for the information and convenience of the reader. Such does not constitute an official endorsement or approval by the United States Department of Agriculture or Forest Service of any product or service to the exclusion of others that may be suitable.

References

Bicknell, B. R., Imhoff, J. C., Kittle, J. L., Jr., Donigian, A. S., Jr., Johanson, R. C. (1997) *Hydrological Simulation Program – Fortran*, user's manual for version 11, EPA/600/R-97/080, National Exposure Research Laboratory, US Environmental Protection Agency, Athens, GA.

Brasseur, G. P., Chatfield, R. B. (1991) 'The fate of biogenic trace gases in the atmosphere', in T. D. Sharkey, E. A. Holland, H. A. Mooney (eds), *Trace Gas Emissions by Plants*. Academic Press, New York.

Carinanos, P., Casares-Porcel. M., Quesada-Rubio, J. (2014) 'Estimating the allergenic potential of urban green spaces: A case-study in Granada, Spain', *Landscape and Urban Planning*, vol 123, pp. 134–144.

Costanza, R., d'Arge, R., de-Groot, Farber, S., Grasso, M., Hannon, B., Limburg, K., Naeem, S., O'Neill, R. V., Paruelo, J., Raskin, R. G., Sutton, P., van den Belt, M. (1998) 'The value of ecosystem services: Putting the issues in perspective', *Ecological Economics*, vol 25, pp. 67–72.

Ellis, A. (2009) Analyzing canopy cover effects on urban temperatures. MS thesis, SUNY College of Environmental Science and Forestry, Syracuse, NY, retrieved from http://gradworks.umi.com/14/82/1482101.html (accessed 15 January 2016).

Gromke C., Ruck B. (2009) 'On the impact of trees on dispersion processes of traffic emissions in street canyons', *Boundary-Layer Meteorology* vol 131, no 1, pp. 19–34.

Hassan, R., Scholes, R., Ash, A. (2005) *Ecosystems and Human Well-being: Current State and Trends, Volume 1*, Island Press, Washington, DC.

Hayden, F. G. (1989) *Survey of Methodologies for Valuing Externalities and Public Goods*, EPA-68-01-7363, Environmental Protection Agency, Washington, DC.

Heisler, G. M. (1986) 'Energy savings with trees', *Journal of Arboriculture*, vol 12, pp. 113–125.

Heisler, G. M., Grant, R. H. (2000) 'Ultraviolet radiation in in urban ecosystems with consideration of effects on human health', *Urban Ecosystems*, vol 4, pp. 193–229.

Heisler, G., Ellis, A., Nowak, D., Yesilonis, I. (2016) 'Modeling and imaging land-cover influences on air-temperature in and near Baltimore, MD', *Theoretical and Applied Climatology*, vol 124, no 1, pp. 497–515.

Lerman, S. B, Nislow, K. H., Nowak, D. J., DeStefano, S., King, D. I., Jones-Farrand, D. T. (2014) 'Using urban forest assessment tools to model bird habitat potential', *Landscape and Urban Planning*, vol 122, pp. 29–40.

Martens, W. J. (1998) 'Climate change, thermal stress and mortality changes', *Social Science and Medicine*, vol 46, pp. 331–344.

McPherson, E. G., Simpson, J. R. (1999) *Carbon Dioxide Reduction through Urban Forestry: Guidelines for Professional and Volunteer Tree Planters*, PSW-GTR-171, Forest Service, Pacific Southwest Research Station, US Department of Agriculture, Berkeley, CA.

Na, H. R., Heisler, G. M., Nowak, D. J., Grant, R. H. (2014) 'Modeling of urban trees' effects on reducing human exposure to UV radiation in Seoul, Korea', *Urban Forestry and Urban Greening*, vol 13, pp. 785–792.

Natural Capital Project (2016) InVEST: Integrated valuation of ecosystem services and tradeoffs, retrieved from www.naturalcapitalproject.org/invest (accessed 21 January 2016).

Nowak, D. J. (2012a) 'Contrasting natural regeneration and tree planting in 14 North American cities', *Urban Forestry and Urban Greening*, vol 11, pp. 374–382.

Nowak, D. J. (2012b). *A Guide to Assessing Urban Forests*, NRS-INF-24-13, Forest Service, Northern Research Station, US Department of Agriculture, Newtown Square, PA.

Nowak, D. J., Crane, D. E., Dwyer, J. F. (2002) 'Compensatory value of urban trees in the United States', *Journal of Arboriculture,* vol 28, no 4, pp. 194–199.

Nowak, D. J., Hoehn, R. H., Crane, D. E. (2007) 'Oxygen production by urban trees in the United States', *Arboriculture and Urban Forestry*, vol 33, no 3, pp. 220–226.

Nowak, D. J., Appleton, N., Ellis, A., Greenfield, E. (2017) 'Residential building energy conservation and avoided power plant emissions by urban and community trees in the United States', *Urban Forestry and Urban Greening*, vol 21, pp. 158–165.

Nowak, D. J., Greenfield, E. J., Hoehn, R., LaPoint, E. (2013a) 'Carbon storage and sequestration by trees in urban and community areas of the United States', *Environmental Pollution*, vol 178, pp. 229–236.

Nowak, D. J., Hirabayashi, S., Bodine, A., Hoehn, R. (2013b) 'Modeled PM2.5 removal by trees in ten U.S. cities and associated health effects', *Environmental Pollution*, vol 178, pp. 395–402.

Nowak, D. J., Hirabayashi, S., Ellis, A., Greenfield, E. (2014) 'Tree and forest effects on air quality and human health in the United States', *Environmental Pollution*, vol 193, pp. 119–129.

Nowak, D. J., Hoehn, R. E., Crane, D. E., Hoehn, R. E., Walton, J.T., Bond, J. (2008) 'A ground-based method of assessing urban forest structure and ecosystem services', *Arboriculture and Urban Forestry*, vol 34, pp. 347–358.

Numerical Terradynamic Simulation Group (2016) Biome-BGC, retrieved from www.ntsg.umt.edu/project/biome-bgc (accessed 15 January 2016).

Peper, P. J., McPherson, E. G., Simpson, J. R., Vargas, K. E., Xiao, Q. (2009) *Lower Midwest Community Tree Guide: Benefits, Costs and Strategic Planting*, PSW-GTR-219, Forest Service, Pacific Southwest Research Station, US Department of Agriculture, Albany, CA.

Tague, C. L., Band, L. E. (2004) RHESSys: Regional Hydro-Ecologic Simulation System – an object-oriented approach to spatially distributed modeling of carbon, water, and nutrient cycling, *Earth Interactions*, vol 8, no 19, pp. 1–42.

Tallamy, D. W., Shropshire, K. J. (2009) 'Ranking lepidopteran use of native versus introduced plants', *Conservation Biology*, vol 23, pp. 941–947.

US Forest Service (2016) Carbon: Tools for carbon inventory, management, and reporting, retrieved from www.nrs.fs.fed.us/carbon/tools (accessed 15 January 2016).

Wang, J., Endreny, T. A., Nowak, D. J. (2008) 'Mechanistic simulation of urban tree effects in an urban water balance model', *Journal of American Water Resource Association*, vol 44, no 1, pp. 75–85.

Yang Y., Endreny, T. A., Nowak, D. J. (2013) 'A physically-based local air temperature model', *Journal of Geophysics Research-Atmospheres*, vol 118, pp. 1–15.

12

DISSERVICES OF URBAN TREES

Jari Lyytimäki

Introduction

Ecosystem services provided by urban green areas have been recognised to an increasing degree following the turn of the millennium (MEA, 2003; Gómez-Baggethun and Barton, 2013). Urban trees in particular provide urban dwellers with a variety of ecosystem services (see Chapter 4 of this volume). However, urban trees are also the source of various types of harm, nuisance and costs. These 'bad' aspects may be labelled as ecosystem disservices. The concept of ecosystem disservice is a recent one and there is no widely agreed definition for it. On a general level, ecosystem disservices can be defined as the functions, processes and attributes generated by the ecosystem that result in perceived or actual negative impacts on human wellbeing (Shackleton et al., 2016). Both ecosystem services and disservices are inherently anthropogenic concepts, putting emphasis on the human valuation of ecosystem properties and functions. What is perceived as beautiful and beneficial by one person may be considered ugly, useless, unpleasant or unsafe by another. For example, biodiversity-rich, semi-natural areas inside city limits are often experienced as suffering from a lack of maintenance, as opposed to intensively maintained but biodiversity-poor urban parks.

Relatively few studies have focused on ecosystem disservices in urban areas. However, this paucity of research does not mean there is an absolute absence of knowledge. On the contrary, various disciplines have long traditions of describing different types of harm caused by natural forces – without mentioning the term ecosystem disservice. Other labels have been used to highlight the fact that nature can be scary, disgusting, or uncomfortable (Bixler and Floyd, 1997). Scholarly contributions include management studies focusing on the effects of natural disasters, botanical research focusing on pests and parasites, medical research analysing ecosystem-based health risks or socio-psychological studies scrutinising different fears and risk perceptions related to natural elements. Many of these studies describe single and isolated disservices, such as the inconvenience caused by a certain nuisance animal or an illness caused by a vector-borne pathogen.

This chapter argues for the recognition of both the ecosystem services and disservices of urban trees. A balanced assessment resulting in the net ecosystem services is required in order to guide urban green planning and management. It is also required as a basis for attempts to resolve unavoidable social controversies and conflicts related to urban green management. Urban areas are characterised by different lifestyles, values and attitudes with different levels of tolerance towards nuisances related to urban green areas. In some cases, concerns related to urban ecosystem disservices may be exaggerated by the news media and social media debates (Lyytimäki, 2014). This social amplification of risks is especially likely if overly optimistic public expectations are created about the benefits of urban greening, without paying proper attention to the possible nuisances. In such cases, the disservice may come as a surprise to the public. This, in turn, is likely to create frustrated or angry public responses, and demands for swift and effective countermeasures. Such critique may be avoided altogether if potential ecosystem disservices are taken into account during the early phases of the urban green management. Identification of disservices is also important as in some cases benefits to human well-being may be cost-effectively achieved through the reduction or mitigation of ecosystem disservices, rather than promoting ecosystem services.

As discussed in this chapter, various kinds of disservice can be produced by urban trees. The chapter starts with concrete cases of disservices and proceeds towards more abstract methodological issues. First, a typology of disservices based on available examples is presented. Second, different methods and data sources used to identify and analyse disservices are outlined. Third, criteria for frameworks aimed at aiding the assessment of the disservices are discussed. It is argued that in the long term, successful urban planning and management should be based on integrated knowledge of services and disservices as well as continuous communication and interaction aimed at increasing public acceptance and policy awareness of different aspects of urban trees and urban biodiversity.

The many faces of disservices

Ecosystem disservices represent a relatively novel research area. Only a few studies focusing on the relationship between urban trees and ecosystem disservices have been published (e.g. Camacho-Cervantes et al., 2014; Delshammar et al., 2015), while some reviews have focused on ecosystem disservices more generally (von Döhren and Haase, 2015; Shackleton et al., 2016). Ecosystem disservices have also been discussed as a side topic in review papers focusing on urban trees or urban ecosystem services (e.g. Dobbs et al., 2011; Roy et al., 2012). However, the most comprehensive pieces of literature comprise case studies focusing on urban ecosystem services and addressing the actual or potential negative effects or costs of urban trees as a side topic.

Together, these studies show that a wide variety of disservices may be produced by urban trees depending on different factors, such as species composition, location of the tree in relation to other trees and built structures, the growth patterns and life phase of the tree, stress caused by external conditions and the intensity of maintenance activities. Many of the disservices are dependent on the particular qualities of the built infrastructure and specific characteristics of the urban ecosystem, as well as socio-cultural aspects influencing how people value trees. Table 12.1 presents examples of different types of disservice. The table does not aim to provide a complete and comprehensive overall picture, instead, it intends to illustrate the different but intertwined categories of ecosystem disservices.

Ecosystem disservices operate on various spatial, temporal and functional scales. The frequency at which they occur may be highly irregular or they may be permanently present, at low or high background levels. They can be direct impacts of ecosystem properties and processes on human well-being, such as pollen allergens. They can also be present as the diminished flow of an ecosystem service, such as pests decreasing the recreational value of a tree and leading to the loss of a cultural ecosystem service.

Ecosystem disservices originate, by definition, in or from an ecosystem and are manifest in social-ecological systems. Abiotic phenomena such as earthquakes or volcanic eruptions

Table 12.1 Examples of ecosystem disservices related to urban trees

Type of disservice	Examples
Aesthetic issues	• Trees perceived as ugly (e.g. unmanaged trees with dead branches, trees suffering pest invasions). • Trees growing in unsuitable places (e.g. trees blocking views from windows or trees distorting architectonic ensembles). • Indirect effects of tree growth decreasing the aesthetic value of built structures (e.g. moisture damaging painted walls, debris, leaves or pollen littering the environment). • Trees hosting species producing aesthetic discomfort (e.g. bird excrement and unwanted birdsong or other aural behaviour).
Safety and security issues	• Direct physical risks related to trees and tree growth (e.g. roots causing tripping, leaves making surfaces slippery or blocking storm water drainage, trees falling, branches dropping, vegetation blocking visibility). • Safety and security issues related to other natural or semi-natural species (e.g. fears related to bats, rodents or urban carnivores). • Urban parks as places of fear related to human misconduct (e.g. perceived risk of night-time crime, uncontrolled pet dogs).
Health issues	• Trees causing direct health effects (e.g. pollen causing allergic reactions). • Trees producing air pollutants or precursors of air pollutants affecting health (e.g. volatile organic compounds). • Trees providing habitats for other species causing health effects (vectors of diseases).
Economic issues	• Direct costs caused by planting, maintaining and removing plant coverage. • Direct costs caused by attempts to remove unwanted species (e.g. weeds, birds nesting in inappropriate places, invasive species). • Indirect costs caused by land use restrictions (especially if a green area or certain species is protected).
Mobility and infrastructure issues	• Urban trees and parks forestalling fast and comfortable transportation and movement, especially the use of motorised transportation or the movement of people with disabilities or elderly people. • Roots causing blockages of sewer pipes, branches causing electric and other wires to short circuit.
Environmental and energy issues	• Biogenic volatile organic compounds and secondary aerosol emissions from trees, carbon and methane emissions from decomposition affecting air quality and climate change. • Increased pollution levels due to reduced air exchange (blocking wind). • Displacement of native species and introduction of invasive species. • Decreased possibilities for utilisation of sunlight because of shade. • Energy consumption, resource use and pollution from maintenance activities.

Sources: Lyytimäki et al. (2008); Escobedo et al. (2011); Gómez-Baggethun and Barton (2013); Delshammar et al. (2015); Säumel et al. (2016); von Döhren and Haase (2015)

cannot be counted as ecosystem disservices. However, the boundaries between abiotic and biotic – as well as ecological and social – systems are blurred. As noted by Shackleton et al. (2016) the dropping of litter in urban parks should not be considered an ecosystem disservice since this act falls clearly under the social domain, with no direct origin in the ecosystem. However, organic litter such as paper tissue decomposing as a result of the functioning of microbiota can be considered an ecosystem disservice if it causes aesthetic discomfort for park users. This example illustrates the importance of the temporal scale of the assessment. Other problems associated with urban forests that should not necessarily be counted as ecosystem disservices include parks as places for people to loiter or sources of neighbourhood conflict (Baur et al., 2014). For example, litigation costs related to conflicts over vegetation damage to buildings or urban infrastructure may sometimes exceed the direct costs caused by vegetation.

Many of the disservices can be classified as *social issues* or *health risks*. Some of them may be considered minor and temporary nuisances while others are long-lasting health hazards (Dunn, 2010). These issues, ranging from psychological to physical problems, overlap in many cases. For example, certain animal species such as bats or stray dogs occupying urban green areas can be vectors of diseases such as rabies. Fear related to such areas, species and diseases may create well-being losses exceeding the direct health effects of the disease itself. Avoiding such unnecessary well-being losses requires successful implementation of carefully tailored communication and interaction strategies (Decker et al., 2012). Active communication is also important in cases where urban trees pose health risks that remain unnoticed by susceptible groups of people.

Ecosystem disservices have occasionally been referred to as *missed opportunities* to enjoy ecosystem services. Examples of missed opportunities include water retention, urban air cooling and resilience against pests. A lack of urban parks can increase the intensity of urban flooding, a lack of trees or green roofs can make urban heath island effects worse, and monocultures of decorative plants can increase the risk of pest attacks. Some species that increase urban biodiversity may cause damage to those species that are cared for (e.g. herbivores using ornamental plants as food). Factors preventing the production or use of certain ecosystem service may include both natural variability in ecosystems and anthropogenic environmental deterioration (Power, 2010). Access to green areas providing the services may be also restricted or denied. However, some authors argue that such cases should be understood in terms of constrained supply of ecosystem services, rather than as disservices (Shackleton et al., 2016). In other words, some ecosystem services may be missing because of trade-offs between different services provided by urban trees, rather than disservices related to trees.

Another line of reasoning focuses on disservices as *increased costs*. Besides the costs of lost opportunities, disservices can be understood as management costs, such as the costs related to monitoring and restoration of damaged ecosystems. A well-known example of this kind of management cost is the resources used for management of invasive species, such as the fungus causing the Dutch elm disease (*Ophiostoma ulmi* and *O. novo-ulmi*) (Delshammar et al., 2015). Considerable costs can be caused by maintenance aimed at forestalling or removing urban ecosystem disservices such as bird excrement accelerating corrosion, tree roots damaging pavements, or animals digging nesting holes. Costs can result from recurring management actions, such as the removal of fallen leaves and debris or repairs to one-off damages such as the decomposition of construction wood due to microbial activity.

It should be noted that such direct costs are relatively easy to assess, whereas economic benefits originating from urban biodiversity are more difficult to assess since they are more often externalities not captured by current market mechanisms. Costs are also generated indirectly as maintaining urban biodiversity and green areas often restricts or prevents other land uses. Despite being highly relevant for urban planning, it is not clear whether these indirect costs should be accounted as ecosystem disservices.

The activities aimed at curbing ecosystem disservices may themselves be an additional source of disturbance or pollution. Therefore, they can be categorised as part of a larger class of *environmental effects* of urban tree management. Stressful urban environments typically require intensive maintenance measures, such as irrigation, use of fertilisers, pest-disease control, pruning and removal and replacement of damaged or old trees. These activities may lead to increased use of natural resources, air and soil pollution, nutrient runoff or increased traffic and noise, but again, careful consideration is needed in order to judge whether these should be considered ecosystem disservices. A clearer case of disservice can be seen in the trees themselves as sources of air emissions such as volatile organic compounds or precursors of particles and tropospheric ozone.

Methods and data sources for assessing ecosystem disservices

Comparing the importance of various ecosystem disservices with each other and with other issues is often complicated. In order to identify and assess different types of disservices, various research methods and data sources are needed. The temporal and spatial focus of the assessment strongly influences the selection of a suitable research method. Different methods are likely to be needed in order to study the generation of disservices, human exposure and effects of disservices, as well as possible management options. Overreliance on any single method or data source should be avoided in order to maintain the capability to provide a rich picture with all the relevant nuances.

Assessments based on natural sciences approaches are essential in order to produce reliable information on the ecosystem properties and functions that result in ecosystem services or disservices (see also Chapters 7 and 8 of this volume). A lack of resources and readily available data often limits the assessment of disservices. Much of the existing information is based on case studies covering relatively small areas and giving anecdotal evidence focusing on disservices related to single species (von Döhren and Haase, 2015; Shackleton et al., 2016).

Many of the assessments have focused on present disservices and few have aimed to describe the long-term past trends or future scenarios. Long-term data on disservices is typically scarce, therefore methods based on the use of proxy indicators indirectly illustrating the historical development of certain ecosystem disservices can be highly useful. For example, complaints found in municipal records may be used to illustrate what kind of disservices are produced by urban parks (Delshammar et al., 2015), or newspaper archives may be used to identify what kind of issues have been publicly raised as disservices related to urban nature (Lyytimäki, 2014). Using such data sources involves various caveats: archived information may be incomplete, the collection and storage of data may be inconsistent, and recorded cases may reflect the level of civic activity or public interest rather than actual level of disservice.

Identifying future disservices presents considerable methodological challenges. Some disservices are characterised by ecological thresholds, hysteresis, and points of no return. The concept of an ecological threshold refers to the level of a stressor that triggers an abrupt change in ecosystem quality, property, or phenomenon. It highlights that even small changes in stressors can produce large responses in ecosystems. The occurrence of many ecosystem disservices (such as outbreaks of pests) may be dependent on particular threshold conditions that need to be met (Escobedo et al., 2011). Hysteresis refers to processes with

Table 12.2 Examples of methods and materials related to ecosystem disservices

Type of disservice	Examples of research methods	Examples of data sources
Aesthetic issues	Interviews, surveys, media and document analysis	Lay people and experts as informants, documents and records, artistic works, recorded complaints to municipalities
Safety and security issues	Interviews, public surveys, media analysis	Social statistics and surveys, crime records, media and social media representations
Health issues	Epidemiological studies, laboratory tests, field studies	Health statistics, test data
Economic issues	Economic modelling, cost-benefit analysis, direct (revealed preferences) and indirect (stated preferences) valuation methods	Economic statistics, rent levels, property prices, consumer behaviour
Mobility and infrastructure issues	Traffic and transport analysis, GIS-based research, land-use studies	Traffic and transport statistics, geographical information systems
Environmental and energy issues	Life-cycle analysis, laboratory tests, field studies	Environmental monitoring data, energy consumption

significant time lags between a driving force and a corresponding change in an ecosystem. Points of no return refer to permanent regime shifts between different alternative stable states, characterised by modified feedbacks in the system. The unpredictability in the timing or magnitude of such nonlinear changes presents substantial monitoring, modelling and management challenges.

In addition to natural sciences-based assessments, social sciences and humanities methods are required in order to understand what ecosystem functions have been or may be considered disservices. Depending on the context of the valuation, different values, norms and attitudes can be involved and different ecosystem services and disservices can be highlighted as the relevant ones. Therefore, it is of utmost importance to be aware of the wide scope of different methods and materials complementing the natural scientific ones (Table 12.2).

What is considered a service or disservice varies over time and space. Therefore, ecosystem services and disservices should be studied by taking the qualities of different local contexts, cultures and population groups into consideration (Lyytimäki and Sipilä, 2009). The inclusion of local knowledge and interaction with people is essential, as the question is fundamentally one of residents' personal values, beliefs and knowledge bases. Local knowledge of disservices should be systematically collected and processed, using public participation methods adjusted for charting the disservices. However, caution is needed also here. For example, shifting baseline syndrome may influence what people view as the normal or preferred state of the ecosystem. Shifting baseline syndrome refers to changing human perceptions of biological systems, due to a loss of experience of past conditions. It can involve generational amnesia, where knowledge extinction occurs because younger generations are unaware of past conditions, or personal amnesia, where knowledge extinction occurs as individuals forget their own experiences (Papworth et al., 2009). As a result, perception of disservice may be as a result of lack of knowledge of normal functioning of the ecosystem.

Citizen science provides promising opportunities for reducing the cost of labour and data-intensive monitoring and research (Dickinson et al., 2012). Citizen science is also a particularly promising approach because it is expected to provide legitimate and more socially robust knowledge, increase awareness of environmental problems, empower citizens to participate and increase their scientific literacy. Volunteer engagement is also associated with improved science-society-policy interaction and more democratic research and governance. Citizen science approaches have been widely used to monitor environmental changes and to chart ecosystem services. New digital tools offer novel platforms for collaboration and present new features for interaction that may be utilised in the assessment of disservices as well.

Comprehensive assessment of ecosystem disservices is obviously an interdisciplinary task. Interdisciplinary expertise is needed not only to cross the boundaries between natural and social sciences but also to cross the boundaries between different sub-disciplines. In many cases, transdisciplinary expertise capable of integrating academic and lay knowledge is required in order to fully utilise all available data and to avoid unnecessary gaps in knowledge generation, leading to better management decisions (Lyytimäki and Petersen, 2014). A transdisciplinary approach may also prove highly useful for the appropriate utilisation of the research results. Early-phase participation by stakeholders decreases the risk of misunderstandings, increases the possibilities for efficient uptake of research results and gives important possibilities for incorporating local tacit knowledge into the assessment. Inclusion of lay knowledge can also increase trust and social cohesion and lessen the likelihood of legal challenges.

Criteria for frameworks for assessing ecosystem disservices

Conceptual frameworks are needed to guide the selection of data and methods aimed at assessing ecosystem disservices. Good conceptual frameworks improve the organisation and analysis of information and minimise the risks of gaps in analyses and assessments. Such frameworks range from theoretically informed and detailed ones to practically oriented heuristics and rules-of-thumb aimed at providing useful general-level guidelines. So far the research on ecosystem disservices has been characterised by the lack of robust conceptual frameworks.

Various criteria exist for a good conceptual framework. On a general level, they should include identification of the socio-ecological system, anchor the assessment in theory, provide an organisational structure, help to identify relevant information and data gaps, ensure comparability and facilitate communication with the public and decision makers. Importantly, they assist in judging what issues should be categorised as ecosystem disservices (Shackleton et al., 2016). In any case, a good framework should provide a basis for structured and consistent practice aimed at operationalising the data collection, analysis and knowledge utilisation.

A good conceptual framework helps to identify the relevant scale for the assessment. Choosing the relevant spatial and temporal focus requires case-specific tailoring, taking into account both the ecological and governance contexts. For example, urban trees can create habitats for species that cause harm far outside urban parks. Birds or rodents taking shelter in urban parks might search for food in rubbish bins and litter the environment outside the park area, a squirrel falling into a water tower can induce health epidemics affecting a whole city, and migratory birds nesting or resting in an urban park may cause problems in distant countries. Such instances of harm may remain unnoticed if the assessment is confined to solely the park area.

Because of the dynamics of socio-ecological systems, the services and disservices are also temporally variable. Diurnal, lunar and annual cycles influence what kind of services and disservices are produced by ecosystems, and social cycles influence how they are encountered and experienced. One of the distinctive characteristics of the modern urban environment is the complete or partial absence of natural cycles of diurnal and nocturnal time. Some ecosystem disservices are a result of disturbance caused by night-time lighting. For example, garden lights with certain spectral compositions strongly attract insects and potentially increase the risk of vector-borne diseases such as malaria (Longcore et al., 2015). However, the urban population in affluent societies is accustomed to continuous night-time outdoor illumination and may perceive natural darkness as unnatural and scary. This partly explains why urban night-time forests and parks are often perceived as unpleasant and unsafe, especially by women.

Conceptual frameworks describing ecosystem disservices should include both physical and social aspects in order to produce a realistic and policy-relevant overall picture. Perceptions of dangers lurking in a dark park can be social constructs with little or no correspondence with actual security risks. Furthermore, the origin of such risks can typically be found within human behaviour rather than ecosystem properties or functions. However, this does not mean that such risks are any less real for the people suffering from them. Even in cases where such risks fall outside of the concept of ecosystem disservices, they may be highly relevant for urban green management, as public opinion of urban parks and trees can be strongly influenced by culturally shaped and emotionally charged perceptions.

A conceptual framework clearly defining the key concepts and their relationships is a prerequisite for successful communication and interaction. As shown by the examples presented above, different things may be inferred with the terms such as ecosystem services or disservices. A lack of a common vocabulary between different actors is a key factor making interdisciplinary knowledge production and green urban planning and management complicated and prone to misunderstandings.

Only a few conceptual frameworks specifically aimed at the integrated analysis of both ecosystem services and disservices exist, but various conceptual frameworks have been developed in order to organise the assessment of ecosystem services, spanning from general level check-lists to more complicated and nuanced frameworks. Some frameworks are aimed at harmonising global level assessments while others are adapted to certain unique contexts. Conceptual frameworks focusing specifically on the management of ecosystem services have also been developed (Primmer et al., 2016). The need to study trade-offs between different ecosystem services has been increasingly acknowledged (Hauck et al., 2013). In addition to this, there is a need to study the trade-offs between services and disservices and between different ecosystem disservices. Importantly, in order to anticipate and avoid unwanted surprises, the conceptual framework should also help to identify potential synergies between different ecosystem disservices.

Figure 12.1 presents an example of a simple conceptual framework, the aim of which is to comprehensively capture the different dimensions that should be taken into account when assessing ecosystem services and disservices. This integrative and holistic framework highlights the role of human individuals both as biological creatures with evolutionary developed physical capabilities and as social creatures with technological and cultural assets emergent from ecosystems (Tapio and Willamo, 2008). The framework differentiates between the major categories of intrapersonal, interpersonal and non-human factors. First, personal psychological and physiological factors determine how urban trees are valued and what effects they may have on individuals. Second, interpersonal social factors include the relationships between urban green areas and the social lifestyles of urban residents. Cultural factors concern urban green areas as part of urban history and place-based identities. The

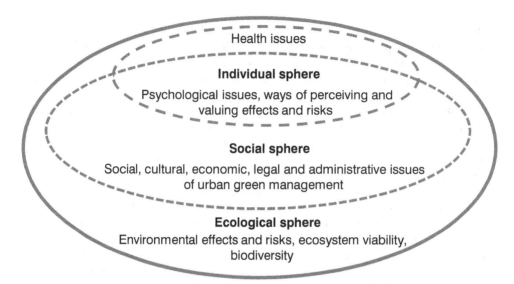

Figure 12.1 A simple conceptual framework for holistic assessment of urban tree disservices, differentiating between three main spheres

Source: modified from Tapio and Willamo (2008); Bezák and Lyytimäki (2011)

economic factors include both direct and indirect monetary benefits and costs of urban green areas. The legal and administrative factors involve the status of urban green areas in management practices. Third, the ecological sphere, including fundamental bio-geo-chemical processes, is the basis for other spheres and the origin of ecosystem services and disservices.

Simplified conceptual frameworks may be useful for scoping the overall situation or for overall assessments performed with limited resources. They may also help decision-making by providing easy-to-use heuristics and check-lists (Huutoniemi and Willamo, 2014). However, more detailed frameworks are needed for in-depth assessments. Figure 12.2 presents an example of a framework aimed at assessment of both ecosystem services and disservices in urban green settings (Escobedo et al., 2011). The framework assumes that an ecosystem service or disservice is a result of a certain ecosystem function based on a particular ecosystem structure and composition of biodiversity. Context, scale and heterogeneity determine whether a particular end product of an ecological system is a service or disservice. The context refers to different uses of urban forests and trees by different people living in different surroundings. Scale refers to the relationship between the size of urban green area and the value of a particular ecosystem service or disservice. It also refers to the importance of taking into account a broad range of economic, social and temporal scales. Management intensity refers to different requirements for management posed by different urban ecosystems, ranging from artificial green walls or roofs to naturally grown urban trees.

Wide applicability is a key criterion for a good conceptual framework. However, aiming for a universally applicable framework for ecosystem services and disservices is a task plagued with difficulties because of the multi-faceted and dynamic nature of the socio-ecological systems. This problem is even more pressing regarding procedural frameworks aimed at guiding the utilisation of information. Procedural frameworks can be built based on experiences gained from elsewhere, but they must be adapted to the specific context in order to be able to take into account the different decision-making situations and knowledge needs. As noted

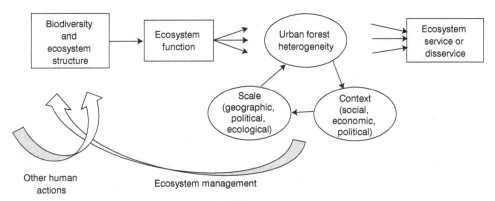

Figure 12.2 An example of conceptual framework focusing on identification of urban ecosystem services and disservices and their management

Source: modified from Escobedo et al. (2011)

by Rinne and Primmer (2015), rather than aiming for a universally applicable analysis, operationalisation of the ecosystem services approach requires case-specific customisation and deliberative co-operation between parties involved.

Factors influencing the management of disservices

All classifications of ecosystem disservices are inherently value-based. What is counted as harmful may differ widely depending on the context of the valuation and the person making the valuation. Age, gender, personal experiences, knowledge level, social settings and cultural background influence people's preferences. For example, density of the vegetation is one of the key features influencing how parks and other green areas are valued, however people of different ages, genders and levels of physical fitness may prefer different densities of vegetation. The tall and leafy trees that are appreciated by many pedestrians may cause annoyance to some pedestrians or to the residents of nearby houses if their view is blocked by the trees. Shading provided by a tree may be highly valued in an urban community located in a tropical climate but not in an urban community located in a cold climate zone.

Successful urban green management must be sensitive not only to different but also changing perceptions from different social groups. Both culturally shared expectations and personal beliefs change over time. Ignorance can swiftly turn into disgust or acceptance when new knowledge of ecosystems is obtained, and new ecosystem disservices may be encountered as a result of changes in urban biodiversity, or due to changes in human perceptions and knowledge alone.

Expectations regarding urban trees are often contradictory. In some cases people may have opposing views as to whether a certain ecosystem function is providing a service or disservice. For example, information about new bird species invading an urban park may be welcomed by a bird watcher but not by those who are concerned about bird excrements potentially forestalling the use of the park. The visiting history can also make a difference: frequent visitors to urban green areas may be more tolerant of disservices associated with urban trees than infrequent visitors would be (Baur et al., 2014).

To make management even more complicated, urban trees typically produce both services and disservices, and their timing, intensity and extent may vary substantially. Some of the disservices may be considered irrelevant or of negligible importance, when compared with ecosystem services that make it possible to fulfil basic human needs such as breathing, eating or resting. However, even minor disservices may become important when considered in relation to everyday practices of urban people in affluent societies. For example, aesthetic discomfort that is not a direct threat to the survival of urban residents can still raise demands for management options that are detrimental to urban biodiversity and ecosystem services, such as demands for intensively managed, artificially illuminated, largely paved and barren green open spaces.

In some cases the best management strategy for minor ecosystem disservices is to learn to tolerate them. In other cases, disservices may cause more serious harm, but they should be tolerated nevertheless because of bigger, related benefits. For example, less intensive urban green management may provide savings in terms of public resources and lead to higher urban biodiversity. However, uncontrolled growth of natural vegetation in urban green spaces may be perceived as an undesirable lack of control (Skandrani and Prévot, 2015). In such cases the key challenge for urban green management is to create adequate public and policy awareness about the relationship between ecosystem services and disservices.

Careful assessment of ecosystem services and disservices helps with their management by giving clear advice as to whether it is the ecosystem or the human behaviour that should be altered. Interactive and participatory approaches are important for understanding and influencing public reactions. However, it should be noted that both the exclusion of public opinion and the uncritical inclusion of resident voices may pave the road for urban green mismanagement.

Even though ecological conditions and the political and societal background to planning and management activities differ considerably across the world, there are some grounds for building commonly shared approaches. Some key trends, such as urbanisation, are advancing throughout the world, and global environmental changes such as climate change require globally coordinated countermeasures. Globally adopted information and communication technologies create opportunities for internationally shared cultural influences that are likely to create more uniform expectations of urban nature.

Media representations, such as high-quality nature documentaries focusing on rare and exotic species in wild natural surroundings, may increase appreciation of nature, but there is also a risk that people may devalue ordinary everyday landscapes, or even perceive them as a disservice. High-quality digital representations of nature can encourage people to stay inside in order to avoid the less desirable aspects of real-world nature experiences, including possible exposure to stinging and biting insects or to unpleasant odours, noise, and sights (Lyytimäki, 2012; Stanley et al., 2015). On the other hand, wireless communication and technologies in the realm of augmented reality provide unforeseen opportunities to make the benefits of urban trees more visible to the people visiting urban parks. The challenges and opportunities for urban green management are manifold.

Conclusion

The current debate around ecosystem disservices can be criticised as ahistorical, as it takes the framework of ecosystem services as an explicit reference point. Both concepts are relatively recent ones, but their roots extend deep into history. It can be maintained that the concept of ecosystem services emerged partly as a response to a traditional view of nature as wild, dangerous and unpredictable – something to be tamed, cultivated and even eradicated (Cronon, 1996). Against this background, the concept of ecosystem services can be seen as highly useful in highlighting our profound dependence on ecosystem processes and goods.

However, research and management focusing solely on ecosystem services may also produce unwanted results by creating overly optimistic expectations of the capabilities of urban trees and ecosystems to improve human well-being and generate other benefits.

The aim of the concept of ecosystem disservices is not to critique but to complement the concept of ecosystem services. It is intended to bring balance to the assessment and management of urban trees, urban green areas and other ecosystems. Therefore, assessments of ecosystem disservices should always be conducted in an integrated manner, also taking into account the services produced by the ecosystem and the social settings of the people involved. This brings forth many challenges as services and disservices may unfold during different temporal and spatial scales and they may affect urban dwellers different ways.

Further research is needed in order to comprehensively map the ecosystem services in relation to the disservices produced by urban trees. Urban ecosystem disservices can be best managed by focusing both on the bio-physical aspects of ecosystems and on the ways in which people perceive and value ecosystem functions. Instead of maximising the gross amount of urban ecosystem services, the aim should be to find the optimal balance between services and disservices, taking into account the whole life-cycle of urban trees.

References

Baur, J. W. R., Tynon, J. F., Ries, P., and Rosenberger, R. S. (2014) 'Urban parks and attitudes about ecosystem services: Does park use matter?', *Journal of Park and Recreation Administration,* vol. 32, pp. 19–34.

Bezák, P. and Lyytimäki, J. (2011) 'Complexity of urban ecosystem services in the context of global change', *Ekológia,* vol. 30, pp. 22–35.

Bixler, R. D. and Floyd, M. F. (1997) 'Nature is scary, disgusting, and uncomfortable', *Environment and Behavior,* vol. 29, pp. 443–467.

Camacho-Cervantes, M., Schondube, J. E. and Castillo, A. (2014) 'How do people perceive urban trees? Assessing likes and dislikes in relation to the trees of a city', *Urban Ecosystems,* vol. 17, pp. 761–773.

Cronon, W. (1996) 'The trouble with wilderness, or, getting back to the wrong nature', in W. Cronon (ed), *Uncommon ground: Rethinking the human place in nature,* Norton, New York, pp. 69–90.

Decker, D. J., Siemer, W. F., Evensen, D. T. N., Stedman, R. C., McComas, K. A., Wild, M. A., Castle, K. T. and Leong, K. M. (2012) 'Public perceptions of wildlife-associated disease: Risk communication matters', *Human–Wildlife Interactions,* vol. 6, pp. 112–122.

Delshammar, T., Östberg, J. and Öxell, C. (2015) 'Urban trees and ecosystem disservices: A pilot study using complaints records from three Swedish cities', *Arboriculture and Urban Forestry,* vol. 41, 187–193.

Dickinson, J. L., Shirk, J., Bonter, D., Bonney, R., Crain, R. L., Martin, J., Phillips, T. and Purcell, K. (2012) 'The current state of citizen science as a tool for ecological research and public engagement', *Frontiers in Ecology and the Environment,* vol. 10, pp. 291–297.

Dobbs, C., Escobedo, F. J. and Zipperer, W. C. (2011) 'A framework for developing urban forest ecosystem services and goods indicators', *Landscape and Urban Planning,* vol. 99, pp. 196–206.

Dunn, R. R. (2010) 'Global mapping of ecosystem disservices: The unspoken reality that nature sometimes kills us', *Biotropica,* vol. 42, pp. 555–557.

Escobedo, F. J., Kroeger, T. and Wagner, J. E. (2011) 'Urban forests and pollution mitigation: Analyzing ecosystem services and disservices', *Environmental Pollution,* vol. 159, pp. 2078–2087.

Gómez-Baggethun, E. and Barton, D. N. (2013) 'Classifying and valuing ecosystem services for urban planning', *Ecological Economics,* vol. 86, pp. 235–245.

Hauck, J., Görg, C., Varjopuro, R., Ratamäki, O. and Jax, K. (2013) 'Benefits and limitations of the ecosystem services concept in environmental policy and decision making: Some stakeholder perspectives', *Environmental Science and Policy,* vol. 25, pp. 13–21.

Huutoniemi, K. and Willamo, R. (2014) 'Thinking outward: Heuristics for systemic understanding of environmental problems', in K. Huutoniemi and P. Tapio (eds), *Transdisciplinary sustainability studies: A heuristic approach,* Routledge, Abingdon, pp. 23–49.

Longcore, T., Aldern, H. L., Eggers, J. F, Flores, S., Franco, L., Hirshfield-Yamanishi, E., Petrinec, L. N., Yan, W. A. and Barroso, A. M. (2015) 'Tuning the white light spectrum of light emitting diode lamps to reduce attraction of nocturnal arthropods', *Philosophical Transactions of the Royal Society B*, vol. 370, article 20140125.

Lyytimäki, J. (2012) 'Indoor ecosystem services: Bringing ecology and people together', *Human Ecology Review*, vol. 19, pp. 70–76.

Lyytimäki, J. (2014) 'Bad nature: Newspaper representations of ecosystem disservices', *Urban Forestry and Urban Greening*, vol. 13, pp. 418–424.

Lyytimäki J. and Petersen, L. K. (2014) 'Ecosystem services in integrated sustainability assessment: A heuristic view', in K. Huutoniemi and P. Tapio (eds), *Transdisciplinary sustainability studies: A heuristic approach*, Routledge, Abingdon, pp. 50–67.

Lyytimäki, J. and Sipilä, M. (2009) 'Hopping on one leg: The challenge of ecosystem disservices for urban green management', *Urban Forestry and Urban Greening*, vol. 8, pp. 309–315.

Lyytimäki, J., Petersen, L. K., Normander, B. and Bezák, P. (2008) 'Nature as a nuisance? Ecosystem services and disservices to urban lifestyle', *Environmental Sciences*, vol. 5, pp. 161–172.

MEA (2003), *Ecosystems and human well-being: A framework for assessment*, Island Press, Washington, DC.

Papworth, S. K., Rist, J., Coad, L., and Milner-Gulland, E. J. (2009) 'Evidence for shifting baseline syndrome in conservation', *Conservation Letters*, vol. 2, pp. 93–100.

Power, A. G. (2010) 'Ecosystem services and agriculture: Tradeoffs and synergies', *Philosophical Transactions of the Royal Society B*, vol. 365, pp. 2959–2971.

Primmer, E., Jokinen, P., Blicharska, M., Barton, D. N., Bugter, R., and Potschin, M. (2016) 'A framework for empirical analysis of ecosystem services governance', *Ecosystem Services*, vol. 16, pp. 158–166.

Rinne, J. and Primmer, E. (2015) 'A case study of ecosystem services in urban planning in Finland: Benefits, rights and responsibilities', *Journal of Environmental Policy and Planning*, vol. 18, no. 3, 286–305.

Roy, S., Byrne, J. and Pickering, C. (2012) 'A systematic quantitative review of urban tree benefits, costs, and assessment methods across cities in different climatic zones', *Urban Forestry and Urban Greening*, vol. 11, pp. 351–363.

Säumel, I., Weber, F. and Kowarik, I. (2016) 'Toward livable and healthy urban streets: Roadside vegetation provides ecosystem services where people live and move', *Environmental Science and Policy*, vol. 62, pp. 24–33.

Shackleton, C. M., Ruwanza, S., Sinasson Sanni, G.K., Bennett, S., De Lacy, P., Modipa, R., Mtati, N., Sachikonye, M. and Thondhlana G. (2016) 'Unpacking Pandora's Box: Understanding and categorising ecosystem disservices for environmental management and human wellbeing', *Ecosystems*, vol. 19, no. 4, pp. 587–600.

Skandrani, Z. and Prévot, A.-C. (2015) 'Beyond green-planning political orientations: Contrasted public policies and their relevance to nature perceptions in two European capitals', *Environmental Science and Policy*, vol. 52, pp. 140–149.

Stanley, M. C., Beggs, J. R., Bassett, I. E, Burns, B. R, Dirks, K. N., Jones, D. N., Linklater, W. L., Macinnis-Ng, C., Simcock, R., Souter-Brown, G., Trowsdale, S. A. and Gaston, K. J. (2015) 'Emerging threats in urban ecosystems: A horizon scanning exercise', *Frontiers in Ecology and the Environment*, vol. 13, pp. 553–560.

Tapio, P. and Willamo, R. (2008) 'Developing interdisciplinary environmental frameworks', *Ambio*, vol. 32, pp. 125–133.

von Döhren, P. and Haase, D. (2015) 'Ecosystem disservices research: A review of the state of the art with a focus on cities', *Ecological Indicators*, vol. 52, pp. 490–497.

PART III

Urban forest landscapes
A STRATEGIC PERSPECTIVE

13

STRATEGIC GREEN INFRASTRUCTURE PLANNING AND URBAN FORESTRY

Raffaele Lafortezza, Stephan Pauleit,
Rieke Hansen, Giovanni Sanesi and Clive Davies

Introduction

The green infrastructure approach, which takes an integrative perspective on a city's or city region's green (and blue) spaces, has rapidly become accepted as a policy and planning tool. This chapter looks into the history and concept of green infrastructure, tracing some of its origins, for instance, in landscape and urban ecological thinking. The focus is on the relevance of the urban forest as a fundamental element of green infrastructure from the neighbourhood to city-regional level.

What is green infrastructure?

The integration and strategic planning and delivery of networks of connected greenspace, described as *green infrastructure* (GI), has become a major discourse in urban greening and is increasingly accepted as a policy and planning approach (European Environment Agency, 2011; Lennon, 2014; Davies et al., 2015). Indeed, in parts of Europe, North America and Asia, GI has been receiving attention as essential urban infrastructure. Among the reasons to substantiate this trend is the increasing acceptance that functional GI is supportive of ecological processes while it simultaneously contributes to improved human health and well-being (Lafortezza et al., 2013). There is also growing interest in the role that GI plays as a planning and delivery framework for nature-based solutions (NBS) that address the impacts of detrimental global change, such as extreme flooding events, storm surges and urban heat waves (Kabisch et al., 2016).

'Green infrastructure' as a concept has gained notable prominence since the 1990s, but its roots go back decades earlier (Davies et al., 2006). It first emerged in North America as a distinct terminology and arrived in Europe via the United Kingdom. GI is based on two important concepts: (i) linking green spaces for the benefit of society, and (ii) preserving and linking natural areas to benefit biodiversity and counter habitat fragmentation (Benedict and McMahon, 2002). An early proponent of this line of reasoning was the landscape architect Frederick Law Olmsted, who in the late nineteenth and early twentieth centuries in his work on parks affirmed that a connected system of parks is more useful than a series of isolated parks. Olmsted's work was fundamental to the development of connectivity thinking (Davies et al., 2006). The idea of linking the ecological capacity and social opportunities of an area is also present in landscape planning. Ebenezer Howard's contributions were essential to shaping modern GI thinking and his work promoted values similar to Olmsted's, but was directed at investigating the physical and psychological benefits of green spaces in proximity to residential zones (Mell, 2008). Howard advanced the notion of creating and maintaining spaces that provide green and service infrastructures to support the communities residing therein, thus reducing urban expansion and the conversion of green belt lands into housing or industry (Mell, 2008).

There is no single accepted definition for 'green infrastructure'; hence, it should be considered as a contested term. An example of this is its use in infrastructure engineering, notably in connection with renewable energy generation from wind and tide. However, the concept can be broadly defined as a strategically planned network of high quality natural and semi-natural areas with other environmental features, which is designed and managed to deliver a wide range of ecosystem services (ES) and protect biodiversity in both rural and urban settings (Hansen et al., in preparation). A further instance of a broad definition comes from the UK Government, who state that:

> Green infrastructure is a network of multifunctional green space, urban and rural, which is capable of delivering a wide range of environmental and quality of life benefits for local communities and that GI is not simply an alternative description for conventional open space. As a network, GI can include parks, open spaces, playing fields, woodlands, but also street trees, allotments and private gardens. It can also include streams, canals and other water bodies and features such as green roofs and walls.
>
> *UK Government, 2016, para. 2*

In principle, every green space in urban areas either on the ground or at canopy level can contribute to GI regardless of type, ownership, management, current use and functionality. However, in doing so, it needs to be considered in planning and comply with the principles, such as connectivity and multifunctionality, set out later in this chapter. Moreover, it should be noted that GI is interpreted in different ways in practice. While there are examples of GI where the concept is applied broadly, such as New York and Barcelona (Rall et al., 2012; Ajuntament de Barcelona, 2013), in other instances certain functions and scales of urban GI are emphasised. In the US, in particular, GI is often used synonymously with decentralised approaches to stormwater management whereby runoff from hard surfaces is retained, cleaned and eventually infiltrated in, for example, swales or rain gardens within local green spaces (e.g. United States Environmental Protection Agency, 2010; Carlet, 2015).

This breadth of the concept and its different interpretations are based on the fact that GI has evolved from other concepts, notably those used in landscape and green space planning. This has led to GI being described as 'old wine in new bottles' (Davies et al., 2006). Equally though, it has been highlighted that GI has the potential to advance the planning and management of urban green space as a melting pot for innovative ideas (Hansen and Pauleit, 2014).

A strength of the GI approach lies in the fact that it accounts for ecological and social values and combines these with other land use developments (Faehnle et al., 2015; Lafortezza and Konijnendijk, in press). Indeed, GI is being increasingly applied through planning processes (Davies et al., 2015) to provide goods and services to people and to reverse negative trends (e.g. habitat fragmentation) (Lafortezza et al., 2013). The GI approach introduces strategies that overcome land fragmentation and enhance functionality; for example, the use of green spaces to buffer sensitive areas close to urban settlements.

As with other umbrella concepts the breadth of the GI approach can be simultaneously a strength and a weakness. For example, GI can be applied in a variety of settings (e.g. urban gardens to forests), at different levels (e.g. neighbourhood to landscape scale) and for numerous purposes (e.g. enhancing biodiversity and social cohesion, reversing the impacts of climate change). However, this flexibility can lead to arguments about scale, functionality, and even the very purposes of strategic green areas. Nevertheless, GI can claim to be grounded in sound science, judging by the breadth and depth of peer-reviewed papers, and also fits in with land use planning theories and practices (Benedict and McMahon, 2002). Multiple scientific and land planning professions including forestry, agronomy, conservation biology, landscape architecture, landscape ecology, urban and regional planning, geography and civil engineering all have a contributory role in the successful design and planning of GI systems, depending on the specific purposes/issues in focus.

There is also growing interest in placing a value on GI. In the UK, Natural England has conducted a review of the Micro-Economic Evidence for the Benefits of Investment in the Environment, which focused around 'green infrastructure' interventions and was structured using the Ecosystem Approach (Rolls and Sunderland, 2014). There is also strong consideration being given to the role of GI in the health care sector; for example, access to green space has been viewed as a tool to lower the cost of health care to taxpayers in the UK (Mell, 2007). Also related to this benefit is the mental and physical well-being that access to GI offers individuals (for more in-depth studies on this topic, see Tzoulas et al., 2007; Davies et al., 2015; Lafortezza and Konijnendijk, in press).

It is well established that GI acts as a relevant planning instrument to sustain and enhance ES (e.g. air filtering, carbon sequestration) to benefit the urban population. This notion is already implicit in the definition of GI given by the European Environment Agency (2011, p. 6): 'a concept addressing the connectivity of ecosystems, their protection and the provision of ecosystem services, while also addressing mitigation of and adaptation to climate change'. The GI approach takes an integrative perspective on a city's or city region's green and blue spaces. This is key to promoting efficient and sustainable use of land, especially in dense and rapidly expanding cities where pressures on land are particularly extreme (Lafortezza et al., 2013).

Green infrastructure planning

According to the European Union's DG Environment Biodiversity Unit (European Union, 2010), GI promotes integrated spatial planning by identifying multifunctional zones and by incorporating habitat restoration measures and other connectivity elements into various land use plans and policies (e.g. linking urban and peri-urban areas; marine spatial planning policy; Lafortezza et al., 2013). Beyond this nature conservation-centred view, GI can also foster social and territorial cohesion among communities, promote polycentric and balanced territorial development (European Environment Agency, 2011) and encourage integrated development in urban, rural and specific areas (Lafortezza et al., 2013). Developing and delivering GI involves the adoption of an integrated territorial planning approach supporting not only ecological coherence between protected and unprotected areas but also a wide range of functions and benefits to society (Lafortezza et al., 2013). It is also the case that GI planning, when fine-tuned, can be integrated with public health planning (Lafortezza and Konijnendijk, in press).

Urban green infrastructure (UGI) planning, as the preceding section implies, addresses the usually public sector-led process of planning and implementing green space-related policy goals, such as achieving more sustainable management of natural resources and biodiversity conservation, adapting to and mitigating climate change, as well as increasing social coherence and supporting the transformation towards a green economy. Planning cannot be separated from governance, as the process of planning and its implementation involves a wide range of stakeholders and individual citizens. The processes of co-design, co-creation and co-management of urban green space are rapidly evolving as, for instance, the results from a European study have shown (Buizer et al., 2015; Buijs et al., 2016). For governance issues related to urban forestry the reader is referred to Chapter 15.

UGI planning and management is differentiated from other green space planning approaches by being based on a specific set of principles that relate to the content as well as the process of planning. While authors differ in the number of such principles (Benedict and McMahon, 2006; Kambites and Owen, 2006; Ahern, 2007; Mell, 2008; Pauleit et al., 2011; European Commission, 2013; Lafortezza et al., 2013), the European Union collaborative project entitled 'Green Surge' in a major study identified four principles related to the content of UGI planning, and three in relation to planning processes within urbanised areas (see Table 13.1).

Table 13.1 Core principles of urban green infrastructure (UGI) planning

Principles of the planning content	*Principles of the planning process*
• *Network/connectivity*: UGI planning aims for added values derived from interlinking green spaces functionally and physically.	• *Strategic*: UGI planning is based on long-term spatial visions supplemented by actions and means for implementation, but remains flexible over time. The process is usually led by the public sector, but that does not mean that non-state actors are excluded (see 'socially inclusive' below).
• *Multifunctionality:* the ability of UGI to provide several ecological, socio-cultural, and economic benefits concurrently. It means that multiple ecological, social and also economic functions, goods and services shall be explicitly considered instead of being a product of chance. UGI planning aims at intertwining or combining different functions to enhance the capacity of urban green space to deliver valuable goods and services. The ecosystem services concept is suggested for operationalising multifunctionality.	• *Inter- and transdisciplinary:* UGI planning aims at linkages between disciplines as well as between science, policy and practice. It integrates knowledge and demands from different disciplines such as landscape ecology, urban and regional planning, and landscape architecture and is developed in partnership with different local authorities and stakeholders.
• *Grey–green integration:* UGI planning considers urban green as a kind of infrastructure and seeks the integration and coordination of urban green with other urban infrastructures in terms of physical and functional relations (e.g., built-up structure, transport infrastructure, water management system).	
• *Multi-scale*: UGI planning can be considered for different spatial levels ranging from city-regions to local projects. UGI planning aims at linking different spatial scales within and above city-regions.	• *Socially inclusive*: UGI planning aims for collaborative, socially inclusive processes.

Source: adapted from Davies et al. (2015), based on Benedict and McMahon (2006), Kambites and Owen (2006), Ahern (2007), Pauleit et al. (2011) and European Commission (2013)

Green infrastructure planning in practice

Globally, some cities have adopted advanced strategic plans for GI, most prominently in the US (Rouse and Bunster-Ossa, 2013) and spearheaded by cities such as Philadelphia and Seattle. While in the US the focus of GI planning is often on stormwater management, in the UK and Ireland the GI approach has been adopted to overcome limitations of green belt planning (Thomas and Littlewood, 2010; Lennon, 2014).

Across Europe, the GI concept is promoted by a European Green Infrastructure Strategy (European Commission, 2013) and more and more taken up by cities, like Barcelona with its 'Green Infrastructure and Biodiversity Plan 2020' as a prominent example (Ajuntament de Barcelona, 2013; Davies et al., 2015). Barcelona's strategy has the ambitious aim to combine the conservation of biodiversity in this densely built city via green networks and ecological restoration with the promotion of multiple ecological and social benefits provided by urban nature. The comprehensive assessment of a range of ES provided by the city's 13 green space types (including both public and private green spaces) is a foundation of the strategy. It emphasises the importance of nature for people living in the city. Therefore, communication and environmental education are salient elements of the strategy. Importantly, the strategy aims to adapt Barcelona's green spaces to the adverse impacts of climate change.

Compared to front-runner urban areas that frequently appear in academic literature, for other cities the GI concept partly holds potential for innovation. In the European Union research project Green Surge, the current planning practices in 20 cities from 14 European countries have been compared to planning principles related to the GI approach (see Table 13.1) based mainly on planning document analyses and interviews with municipal green space planners.

The study reveals that most European cities preserve and develop green networks, such as historic green ring or belt structures, habitats, or other ecological networks and/or systems of recreational corridors and paths. Thus, the idea of the structural and functional connectivity of GI is not new for most cities. While multiple functions or services delivered by urban green space are considered in most planning documents, enhancing multifunctionality (e.g. by increasing synergies among different functions) is rarely considered as a planning objective. As exemplified by Barcelona's GI and biodiversity strategy, an assessment of ES appears to be a suitable approach to underpin multifunctionality planning (Hansen and Pauleit, 2014).

The possibility of integrating green with grey infrastructures is noted by several cities mainly in relation to stormwater management, but also in relation to sustainable mobility through a network of green corridors. However, the integration of green with different kinds of infrastructures, such as energy supply and demand systems, could be strongly increased in most cities. Likewise, the coordination of green space plans at multiple administrative scales could be strengthened.

In relation to the UGI planning process, most cities follow a long-term strategic planning approach. However, regarding inter- and transdisciplinarity and social inclusion ample scope has been detected for cooperation among departments and with other organisations as well as earlier and wider inclusion of stakeholder groups.

From this Green Surge study it is clear that strategic green space planning is already widely practised in cities, albeit under different names. However, greater potential for enhancement can be detected when comparing the above-mentioned approaches with the principles of GI planning. Most of all, there is an apparent need to strengthen social inclusion, in particular in the Mediterranean countries where citizen participation in UGI planning ranked the lowest in the study (Davies et al., 2015). Some instructive examples exist, such as neighbourhood green planning in the Netherlands and the UK. When connected to city-wide strategic planning and supported by funding for implementation, such approaches can effectively take into account local needs and link them to the strategic development of the urban green resource.

Green infrastructure planning and urban forests

As described in Chapter 1, 'urban forests' is a collective term used to describe all the trees located in urban areas. Frequently, this term is extended to include trees located in the periphery of urban areas (the urban fringe) and in this case they are sometimes referred to as peri-urban forests. The typology of urban forests ranges from individual street trees through to city forests and community forests. In urban forestry, trees 'set the scene' for a number of uses ranging from wood resources through to socio-economic benefits, cultural uses, and destinations for recreation. The role of urban forests in developed societies tends to be different from that in developing societies, for example, in connection with food security; this can be illustrated by the role of urban and peri-urban forests as a source of charcoal used for cooking and heating as well as a source for edible fruits and nuts. In industrialized countries, urban forests are considered for their social and environmental services, such as recreation and air/water regulation, and less for food provision or other livelihood necessities. On the contrary, in developing countries urban forests contribute more to the subsistence of individuals (e.g. food, fodder, wood for fuel and construction) (Konijnendijk et al., 2005). While urban forests may have commercial potential this is not the principle reason why they are managed and maintained.

The Food and Agriculture Organization of the United Nations (FAO, 2016) reports that urban and peri-urban forests and trees, if properly managed, can make important contributions to the planning, design and management of sustainable and resilient cities; furthermore, the FAO has suggested that urban forests help make cities:

- Safer – by reducing stormwater runoff and the impacts of wind and sand storms, mitigating the heat island effect and contributing to the mitigation of and adaptation to climate change;
- More pleasant – by providing space for recreation and venues for social and religious events and by ameliorating weather extremes;
- Healthier – by improving air quality, providing space for physical exercise and fostering psychological well-being;
- Wealthier – by providing opportunities for the production of food, medicines and wood and by generating economically valuable environmental services; and
- More diverse and attractive – by providing natural experiences for urban and peri-urban dwellers, increasing biodiversity, creating diverse landscapes and maintaining cultural traditions.

It is clear from this list that there is considerable synergy between urban forestry and GI at all levels. In many cities the urban forest is also the largest and most extensive GI typology. There are historical reasons for this as the concepts embedded in urban forestry predate GI by several decades. As pointed out by Davies et al. (2006), urban forestry is a predecessor concept which GI has in some cases adopted. There are also spatial synergies that can be created between urban forestry and GI, and these are set out in Table 13.2.

In some cities urban forestry planning dominates over GI, although the opposite is equally true in others. In practice, the central concepts adopted for UGI planning set out in Table 13.1 can be adapted for use in urban forestry planning.

Table 13.2 The spatial equivalents of the green infrastructure approach and urban forestry

Urban forest descriptor	Green infrastructure descriptor	Geographical descriptor
Arboriculture approach/ individual tree management	Street scene (e.g., trees, swales, rain gardens, green roofs, sustainable drainage systems)	Street, house, property
Discernible groups of urban trees, tree-lined boulevards, spinneys and copses	Parks, areas of natural vegetation, managed amenity green space	Whole neighbourhood
Individual urban woodland	Nature reserves, green 'nodes'	City district
Urban or peri-urban forest or conjoined urban woodlands	Landscape features managed for ecosystem services	Whole city or metropolitan region

Source: Clive Davies

In their seminal work 'Planning the Urban Forest: Ecology, Economy, and Community Development', Kollin and Schwab (2009) argue that GI surrounding an urban core is a good technique and that connecting trees, parks, and other UGI at site and neighbourhood scales is a frontier in planning and government services. They go on to propose that a means to secure GI is by establishing a tree canopy target and that the first step in reincorporating GI into a community's planning framework is to measure the urban forest canopy and set these goals. This approach is finding its way into practice as seen in the Wyre Green Infrastructure Strategy in England (MD2 Consulting, 2014). The proposal included in this study is that management of the municipality's trees should be framed as canopy management rather than the tree-by-tree management practised hitherto.

To exemplify the synergy existing between urban forestry and GI, the following two case studies are presented: Regional urban forest parks in Milan and the Melbourne Urban Forest Strategy.

Regional urban forest parks in Milan, Italy

Long before the arrival of the concept of 'green infrastructure' as a planning term, the creation of a strategic green network based on urban forestry had already begun in Milan. The post-World War II period resulted in rapid urban expansion and the fragmentation of the regional suburban landscape (Hansen et al., 2015). This included the fragmentation of forest areas (Carovigno et al., 2011). In response, and starting in the early 1980s, urban forestry interventions were gradually introduced to offset this trend. A number of projects were carried out to preserve remaining forest patches in the rural landscape and create new urban and peri-urban forests (UPFs). From an early stage, planners sought outcomes for what are now considered as ES. These range from 'provisioning' services (e.g. edible products, timber) through 'regulatory' (e.g. prevention of soil erosion, carbon sequestration), to 'cultural' services (e.g. well-being, recreation) (Barbante et al., 2014; Lafortezza and Chen, 2016; Mariani et al., 2016; Pesola et al., 2017).

The creation of the Milan UPFs follows a number of steps, from general masterplans to applicative projects (Carovigno et al., 2011) (see Figure 13.1). The province of Milan has led most of the steps through some important and iconic projects, such as 'Parco Nord Milano', 'Boscoincittà' and 'Bosco delle Querce'. Besides these main projects, a majority of local municipalities had started to implement smaller plans with the support

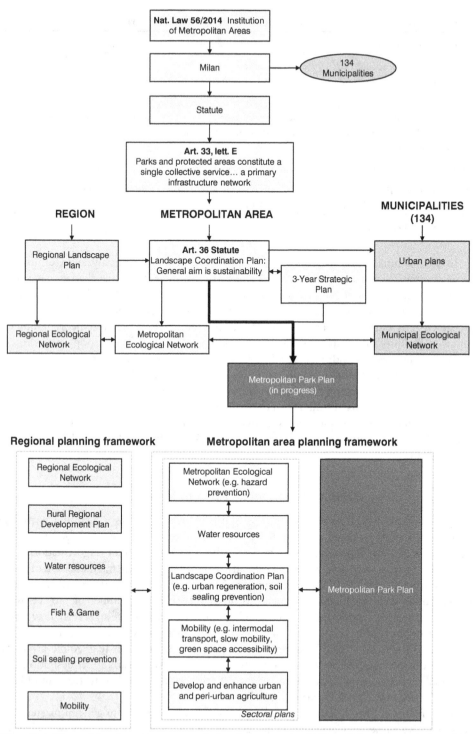

Figure 13.1 The multi-scale and inter-sectoral planning approach of Milan. Urban forestry is the golden thread that links the Regional Planning Framework, Metropolitan Area Planning Framework and the Metropolitan Park Plan

Source: Giovanni Sanesi

of environmental groups and NGOs, such as 'Italia Nostra' ('Our Italy') and the WWF (Carovigno et al., 2011).

In support of the developmental process of UPFs, a number of legislative policies were instituted to fine-tune strategic green space planning and the creation of the urban forests. The Province of Milan and the Lombardy Region, using regional and state laws, supported the initial projects of urban afforestation. Later in 2000, the Lombardy Region promoted the 'Ten large forests for the plains' project at regional level, and some of these afforestation projects were implemented within the provincial boundaries of Milan (Carovigno et al., 2011). In early 2010, with the support of the European Union project Emonfur LIFE+ 10/ ENV/IT/000399 (www.emonfur.eu), it was agreed that urban forests should be strategically connected within a forest network. The proposal was made that this urban forest network should be identified as critical GI and a fundamental component of the Province's GI system (Barbante et al., 2014). As a consequence of complex legislative changes, a system of protected green spaces was established in the Metropolitan Area of Milan. The system consists of six regional parks (hubs) aimed at maintaining the character of the landscape and its habitats.

While in the Metropolitan Area of Milan the landscape-scale pattern of UGI remains fragmented, a radial pattern of GI can be discerned (Giannico et al., 2016; Sanesi et al., 2016). Nevertheless, GI is now much more accessible to residents than hitherto and this is consequent upon many years of investment in urban and peri-urban forestry. There has been an increase in the number and extent of bike paths (25% run through parks and green spaces; www.comune.milano.it), walkways and intermodal passenger public transport. An example of the latter is the recent introduction of metro line 5 on the occasion of the Milan Expo exhibit, which enables greater accessibility to the peri-urban metropolitan park Parco Nord Milano. Important UPFs such as Parco Nord Milano (see Figure 13.2) are, however, still under development and the benefits they provide will continue to accrue in the years to come. Although UPF resources do not constitute the majority of Milan's GI, they do play a pivotal role. In the Metropolitan Area of Milan, GI comprises 14 parks of local interest covering 7,000 ha (Sanesi et al., 2016). Woodlands are the most visible product of policies aimed at conserving, protecting and guiding the management of privately owned land; they are also the main focus for publicly funded projects (Sanesi et al., 2016). The process of afforestation of agricultural and brownfield land has occurred within a timeframe of less than 40 years and this includes the creation of UPFs.

The Lombardy region has been investing in new forms of governance (e.g. community-led, public-private partnerships) in relation to UGI planning in the Metropolitan Area of Milan. This interest is based on the notion that a complex GI system cannot be supported solely by public funding, given its inherent high costs. Indeed, this is the biggest challenge that decision makers and planners face in building the future metropolitan park concept in Milan. To meet the demand of high costs and to lighten the burden of GI maintenance, planners and policy makers have been involving citizens in green space planning and management activities (Sanesi et al., 2016). This bottom-up governance approach includes the participation of associations, NGOs, and farmers besides citizens.

In Milan there is a role for private land within the urban forest and UGI network. At the larger scale, private GI consists of privately owned and managed farmland with some woodland. In suburban neighbourhoods, conjoined private gardens act as green corridors. One attraction of privately managed GI is that for the most part it is cost neutral to public authorities since the owners pay directly for the management of these areas. However, public access is limited so in most cases private GI is principally of ecological and landscape interest.

Figure 13.2 Parco Nord Milano

Source: courtesy of ERSAF – Regional Agency for Agriculture and Forestry Services, Milan

Melbourne Urban Forest Strategy, Melbourne, Australia

Although urban forestry in Australia is an established discipline, targeted research and its practical application are of a more recent date. Between November 2011 and April 2012, Melbourne's Urban Forest Strategy was developed in consultation with the local community (City of Melbourne, 2014). More specifically, the plan was conceived based on the climate change predictions of hotter summers and lower precipitation for the city of Melbourne made in 2007 by the Intergovernmental Panel on Climate Change (IPCC) (City of Melbourne, 2014). Elaboration of the strategy, which involved analyses on, for instance, vegetation health, diversity and distribution, ended in 2011 and was subsequently presented and discussed among citizens and local associations. The Urban Forest Strategy sits alongside an Open Space Strategy. It sets out to address three significant challenges, namely, climate change, population growth and urban heating. The urban forest is seen as a way to ensure a healthy and liveable environment for future generations. The Melbourne urban forest typology consists of parks, gardens and green spaces of different origin and nature (e.g. street trees, riverbanks, allotments).

A driver for the Urban Forest Strategy is the City of Melbourne's awareness that climatic conditions are changing and becoming progressively warmer, dryer, and liable not only to more frequent heat extremes but also to flooding events caused by varying rainfall regimes (Dai, 2011; City of Melbourne, 2014). Local administrators are also forecasting that the urban heat island (UHI) effect will intensify. The strategic plan seeks to build a healthy, resilient and diverse urban forest that can contribute to the health and well-being of communities.

The strategy also foresees proactive and adaptive management approaches. Specifically, the Urban Forest Strategy plan pursues the following approaches:

- mitigation of and adaptation to climate change;
- reduction of the UHI;
- creation of a 'water sensitive' city, meaning that citizens are aware of the importance of water and its rational use;
- creation of healthier ecosystems;
- designing urban green to promote health and well-being;
- designing for liveability and cultural integrity; and
- positioning Melbourne as a leader in urban forestry.

The plan identifies six strategies and targets for building a healthy, resilient and diverse urban forest, as shown in Table 13.3.

While the Open Space Strategy is not necessarily synonymous with GI strategy, the Council is adopting the GI approach through an integrated framework as shown in Figure 13.3. The Council has put in place a wide variety of supportive measures ranging from a Growing Green Guide – Australia's first guide to green roofs, walls and façades – through to a Greening Laneways project, a Rooftop Project, and a Canopy Green Roof Forum. All of these can be considered as pertinent to the relationship between GI and urban design, a key issue in growing cities. Furthermore, these measures have been added through an economic valuation of the urban greening framework, which provides a method to help local governments in Australia understand and develop the business case for urban greening.

Conclusions

Green infrastructure planning has come to the fore over the last two decades and now dominates the planning discourse on strategic urban greening. There are already good

Table 13.3 The strategies and targets for Melbourne's Urban Forest Strategy

Strategy	Target
Increase canopy cover	Increase public realm canopy cover from the current 22% to 40% by 2040
Increase urban forest diversity	The urban forest will be composed of ≤5% of any tree species, ≤10% of any genus and ≤20% of any one family
Improve vegetation health	90% of the tree stock in the City of Melbourne will be healthy by 2040
Improve soil moisture and water quality	Soil moisture will be kept at levels that provide healthy growth of vegetation
Improve urban ecology	Protect and enhance a biodiversity level that contributes to a healthy ecosystem
Inform and consult the community	Enhance the community's awareness of the importance of the urban forest, increase the connectivity between the community and the urban forest, and engage with the forest's process of evolution

Source: adapted from City of Melbourne (2014)

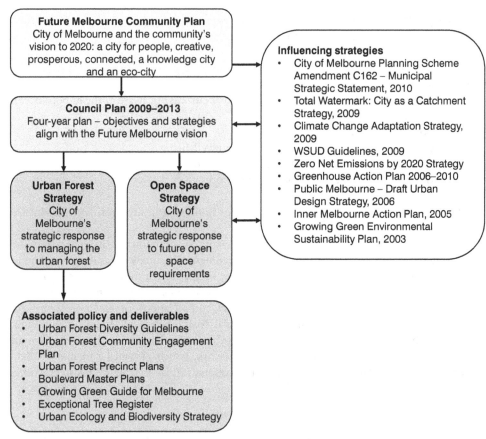

Figure 13.3 Flowchart showing the relationships between the Urban Forest Strategy, the City of Melbourne policy documents and other strategies that underpin and inform it as well as documents that support its implementation

Source: adapted from City of Melbourne (2014)

examples of where urban forestry has been successfully incorporated into GI planning and this is demonstrated in the two case studies in this chapter. There is a historical precedent for this, as urban forestry is a predecessor concept from which GI planning has borrowed much. In some cities the urban forest is the dominant typology of its GI, and in these situations it is desirable for GI to be seen as the major driver for it. Even in cities where urban forest cover is presently low, GI planning can be the vehicle to drive efforts to conserve the remaining tree resource left as a historical legacy and the spur to create new urban forests, which in time increases the percentage of woodland cover in and around the urban settlement.

GI planning can also encourage a strategic approach within key agencies to the management of trees in urban areas. This can facilitate a better understanding of the wider benefits of trees in other professions, for example, in relation to their role in climate change adaptation. GI planning also presents an opportunity to argue for a canopy management approach to urban forestry as opposed to ad-hoc reactive methods, which are little more than responses to tree crises but common in practice. The principles upon which GI planning is grounded (see Table 13.1) can also be applied in urban forestry planning. For example, linking the urban forest to other green systems, such as water bodies or greenways, can contribute greatly to GI and will also create synergistic benefits to the urban forest (e.g. improved habitat connectivity and greater recreational use).

Preparation of GI strategies is also an opportunity for forestry professionals including tree officers to meet with urban professionals (planners, landscape architects, urban designers, civil engineers, etc.) who they might not otherwise influence. This is generally referred to as 'having a seat at the table' and is invaluable in promoting urban forestry practice. Urban foresters have much to offer those engaged in GI planning, for example: (i) their experience in involving citizens and communities; (ii) insights into long-term land management; (iii) skills in creating new habitats and joining habitats together; and (iv) ability to tackle problem sites with innovative solutions such as phytoremediation.

Urban forestry and GI can be seen as two sides of the same coin; so while there are differences in emphasis and approach, commonalities easily exceed the differences. Each approach has much to gain from the other and many of the main advocates are the same. For this reason key international meetings, such as the European Forum on Urban Forestry (EFUF), now give space and encouragement to GI contributions.

Acknowledgements

This work was carried out under the research project 'Green Infrastructure and Urban Biodiversity for Sustainable Urban Development and the Green Economy' (Green Surge)', funded by the European Union FP7 programme, collaborative project, FP7-ENV.2013.6.2-5-60356 (grant agreement number 603567). The work was also supported by grants from the research project 'Development of innovative models for multiscale monitoring of ecosystem services indicators in Mediterranean forests (MiMoSe)', funded by the FIRB 2012 programme of the Italian Ministry of Education, Universities and Research (grant: RBFR121TWX).

References

Ahern, J. (2007) 'Green infrastructure for cities: The spatial dimension', in V. Novotny and P. Brown (eds), *Cities for the Future: Towards Integrated Sustainable Water and Landscape Management* (pp. 265–283), IWA Publishing, London.

Ajuntament de Barcelona (2013) *Medi Ambient i Serveis Urbans: Hàbitat Urbà, Barcelona* [*Barcelona Green Infrastructure and Biodiversity Plan*], retrieved from https://w110.bcn.cat/MediAmbient/Continguts/Documents/Documentacio/BCN2020_GreenInfraestructureBiodiversityPlan.pdf (accessed 8 May 2014).

Barbante, E., Calvo, E., Sanesi, G., Selleri, B., Verlič, A., Vilhar, U. (eds) (2014) 'Urban and peri-urban forests: Management, monitoring and ecosystem services', in D. Gatti and A. Omodei (eds), *Emonfur Life+ Project Experiences,* 1st edn, retrieved from www.emonfur.eu (accessed 2 March 2016).

Benedict, M. A., McMahon, E. T. (2002) 'Green infrastructure: Smart conservation for the 21st century', *Renewable Resources Journal*, vol 20, pp. 12–17.

Benedict, M. A., McMahon, E. T. (2006) *Green Infrastructure: Linking Landscapes and Communities*, Island Press, Washington, DC.

Buijs, A., Elands, B., Havik, G., Ambrose-Oji, B., Gerőházi, É., van der Jagt, A., Mattijssen, T., Møller, M. S., Vierikko, K. (2016) 'Innovative governance of urban green spaces – learning from 18 innovative examples across Europe', European Union FP7 project Green Surge (ENV.2013.6.2-5-603567), Deliverable 6.2, retrieved from http://greensurge.eu/working-packages/wp6/files/Innovative_Governance_of_Urban_Green_Spaces_-_Deliverable_6.2.pdf (accessed 6 March 2016).

Buizer, M., Elands, B., Mattijssen, T., van der Jagt, A., Ambrose-Oji, B., Gerőházi, É., Santos, A., Moller, M. (2015) 'The governance of urban green spaces in selected European Union-cities', European Union FP7 project Green Surge (ENV.2013.6.2-5-603567), Deliverable 6.1, retrieved from http://greensurge.eu/working-packages/wp6/files/Buizer_et_al_2015_D6.1_GREEN_SURGE_The_ governance_of_urban_green_spaces_in_selected_European Union_cities.pdf (accessed 6 March 2016).

Carlet, F. (2015) 'Understanding attitudes toward adoption of green infrastructure: A case study of US municipal officials', *Environmental Science and Policy*, vol 51, pp. 65–76.

Carovigno, R., Calvo, E., Colangelo, G., Dentamaro, I., Lafortezza, R., Sanesi, G. (2011) 'The afforestation of rural landscape in Northern Italy: New benefits and services to society', retrieved from http://ec.europa.eu/agriculture/fore/events/28-01-2011/carovigno_en.pdf (accessed 3 February 2016).

City of Melbourne. (2014) *Urban Forest Strategy: Making a Great City Greener 2012–2032*, retrieved from http://participate.melbourne.vic.gov.au/application/files/5714/1273/7437/COM_SERVICE_ PROD-_8591078-v1-Urban_Forest_Strategy_Document_2014_Online.PDF (accessed 3 April 2016).

Dai, A. (2011) 'Drought under global warming: A review', *WIREs Climate Change*, vol 2, no 1, pp. 45–65, doi: 10.1002/wcc.81

Davies, C., MacFarlane, R., McGloin, C., Roe, M. (2006) *Green Infrastructure Planning Guide. 2 Volumes: Final Report and GI Planning Guide*, University of Northumbria, North East Community Forests, Newcastle University, Countryside Agency, English Nature, Forestry Commission, Newcastle, pp. 1–45.

Davies, C., Hansen, R., Rall, R., Pauleit, S., Lafortezza, R., De Bellis, Y., Santos, A., Tosics, I. (2015) 'Green infrastructure planning and implementation. The status of European green space planning and implementation based on an analysis of selected European city-regions', Green Surge Deliverable 5.1, retrieved from http://greensurge.eu/working-packages/wp5/files/D_5.1_Davies_et_ al_2015_Green_Infrastructure_Planning_and_Implementation_v2.pdf (accessed 8 March 2016).

European Commission. (2013) 'Green infrastructure (GI) – Enhancing Europe's natural capital. Communication from the Commission to the European Parliament, the Council, the European Economic and Social Committee and the Committee of the Regions', COM 249 final.

European Environment Agency. (2011) 'Green infrastructure and territorial cohesion. The concept of green infrastructure and its integration into policies using monitoring systems', Technical Report No. 18/2011, Copenhagen, Denmark.

European Union. (2010) 'LIFE building up Europe's green infrastructure: Addressing connectivity and enhancing ecosystem functions', Technical Report European Commission, Environment Directorate-General, Brussels, Belgium, pp. 60.

Faehnle, M., Söderman, T., Schuylman, H., Lehvävirta, S. (2015) 'Scale-sensitive integration of ecosystem services in urban planning', *GeoJournal*, vol. 80, no. 3, pp. 411–425.

FAO. (2016) *Guidelines on Urban and Peri-urban Forestry*, Food and Agriculture Organization of the United Nations, Rome.

Giannico, V., Lafortezza, R., John, R., Sanesi, G, Pesola, L., Chen, J. (2016) 'Estimating stand volume and above-ground biomass of urban forests using LiDAR', *Remote Sensing*, 8(4), 339: 1–14.

Hansen, R., Pauleit, S. (2014) 'From multifunctionality to multiple ecosystem services? A conceptual framework for multifunctionality in green infrastructure planning for urban areas', *AMBIO*, vol 43, no 4, pp. 516–529.

Hansen, R., Buizer, M., Rall, E., DeBellis, Y., Davies, C., Elands, B., Wiersum, F., Pauleit, S. (2015) *Green Surge: Report of Case Study City Portraits*, retrieved from http://greensurge.eu/filer/GREEN_ SURGE_Report_of_City_Portraits.pdf (accessed 1 March 2016).

Hansen, R., Rall, E., Pauleit, P., Fohlmeister, S., Erlwein, S., Davies, C., DeBellis, Y., Lafortezza, R., Gerőházi, E., Száraz, L, Tosics, I., Santos, A., Luz, A., Santos-Reis, M., Branquinho, C., Vierikko, K., Delshammer, T., van der Jagt, A., Cvejić, R., Železnikar, Š., Nastran, M., Pintar, M., Andersson, E., Kronenberg, J., Caspersen, O. H., Stahl Olafsson, A., Steen Møller, M., Gentin, S. (in preparation) 'Urban green infrastructure planning: A guide for practitioners', Green Surge Deliverable 5.3.

Kabisch, N. et al. (2016) 'Nature-based solutions to climate change mitigation and adaptation in urban areas: Perspectives on indicators, knowledge gaps, opportunities and barriers for action', *Ecology and Society*, vol 21, no 2, p. 39, doi: http://dx.doi.org/10.5751/ES-08373-210239.

Kambites, C., Owen, S. (2006) 'Renewed prospects for green infrastructure planning in the UK', *Planning Practice and Research*, vol 21, no 4, pp. 483–496.

Kollin, C., Schwab, J. (2009) 'Planning the urban forest: Ecology, economy, and community development', in J. C. Schwab (ed.), *Report Number 555*, American Planning Association, Planning Advisory Service, retrieved from http://na.fs.fed.us/urban/planning_uf_apa.pdf (accessed 16 March 2016).

Konijnendijk, C.C., et al. (2005) 'Research on urban forests and trees in Europe', in C. C. Konijnendijk, K. Nilsson, T. B. Randrup, J. Schipperijn (eds), *Urban Forests and Trees: A Reference Book*, Chapter 16, p. 46, Springer, Berlin.

Lafortezza, R., Chen, J. (2016). 'The provision of ecosystem services in response to global change: Evidences and applications', *Environmental Research*, vol 147, pp. 576–579.

Lafortezza, R., Konijnendijk, C. C. (in press) 'Green infrastructure: Approach and link to public health benefits', in W. Bird and M. van den Bosch (eds), *Nature and Public Health: The Role of Nature in Improving the Health of a Population*, Oxford, Oxford University Press.

Lafortezza, R., Davies, C., Sanesi, G., Konijnendijk, C. C. (2013) 'Green infrastructure as a tool to support spatial planning in European urban regions', *iForest – Biogeosciences and Forestry*, vol 6, pp. 102–108.

Lennon, M. (2014) 'Green infrastructure and planning policy: A critical assessment', *Local Environment: The International Journal of Justice and Sustainability*, vol 20, no 8, pp. 957–980, doi: 10.1080/13549839.2014.880411.

Mariani, L., Parisi, S. G., Cola, G., Lafortezza, R., Colangelo, G., Sanesi, G. (2016) 'Climatological analysis of the mitigating effect of vegetation on the urban heat island of Milan, Italy', *Science of the Total Environment*, 569, 762–773.

MD2 Consulting. (2014) 'Wyre green infrastructure strategy', retrieved from www.md2.org.uk/news/page/2 (accessed 17 March 2016).

Mell, I. C. (2007) 'Green infrastructure planning: What are the costs for health and well-being?', *Journal of Environment, Culture, Economic and Social Sustainability*, vol 3, no 5, pp. 117–124.

Mell, I. C. (2008) 'Green infrastructure: Concepts and planning', *FORUM: International Journal for Postgraduate Studies in Architecture, Planning and Landscape*, vol 8, no 1, pp. 69–80.

Pauleit, S., Liu, L., Ahern, J., Kazmierczak, A. (2011) 'Multifunctional green infrastructure planning to promote ecological services in the city', in J. Niemelä, G. Breuste, N. Guntenspergen, T. McIntyre, T. Elmqvist, P. James (eds), *Urban Ecology. Patterns, Processes, and Applications* (pp. 272–286), Oxford, Oxford University Press.

Pesola, L., Cheng, X., Sanesi, G., Colangelo, G., Elia, M., Lafortezza, R. (2017) Linking aboveground biomass and biodiversity to stand development in urban forest areas: A case study in northern Italy. *Landscape and Urban Planning*, vol 157, pp. 90–97.

Rall, E. L., Hansen, R., Pauleit, S. (2012) 'The current landscape of green infrastructure planning and ecosystem services: The cases of Berlin and New York', in *Proceedings of the Symposium Designing Nature as Infrastructure, 28–29 November* (pp. 160–180), Technical University of München, Munich.

Rolls, S., Sunderland, T. (2014) *Microeconomic Evidence for the Benefits of Investment in the Environment 2 (MEBIE2)*, Research Reports 057, Natural England, York.

Rouse, D. C., Bunster-Ossa, I. F. (2013) *Green Infrastructure: A Landscape Approach*, Planning Advisory Service Report 571, American Planning Association, Chicago, IL.

Sanesi, G., Colangelo, G., Lafortezza, R., Calvo, E., Davies, C. (2016). 'Urban green infrastructure and urban forests: A case study of the metropolitan area of Milan', *Landscape Research* (Special Issue), doi: 10.1080/01426397.2016.1173658

Thomas, K., Littlewood, S. (2010) 'From green belts to green infrastructure? The evolution of a new concept in the emerging soft governance of spatial strategies', *Planning Practice and Research*, vol 25, pp. 203–222.

Tzoulas, K., Korpela, K., Venn, S., Yli-Pelkonenm, V., Kazmierczak, A., Niemela, J., James, P. (2007) 'Promoting ecosystem and human health in urban areas using green infrastructure: A literature review', *Landscape and Urban Planning*, vol 81, pp. 167–178.

UK Government. (2016) 'What is green infrastructure?', retrieved from http://planningguidance.communities.gov.uk/blog/guidance/natural-environment/green-infrastructure (accessed 7 March 2016).

United States Environmental Protection Agency. (2010) *Green Infrastructure Case Studies: Municipal Policies for Managing Stormwater with Green Infrastructure*, United States Environmental Protection Agency, Washington, DC, retrieved from nepis.epa.gov/Exe/ZyPURL.cgi?Dockey=P100FTEM.txt (accessed 6 March 2016).

14

A LANDSCAPE AND URBANISM PERSPECTIVE ON URBAN FORESTRY

Alan Simson

Introduction

The history of human culture suggests that 'the landscape' is one of the earliest concepts for perceiving and describing our changing environment, and trees and woodlands have played a central role in this. Our early landscapes were 'natural' of course but human beings, being the creative and inquisitive creatures that they are, quickly engaged in changing their natural landscapes into 'designed' landscapes, in order to improve the quality and efficiency of their existence and accommodate the evolving demands expected of their landscapes. Thus change is, and always has been, an inherent aspect of human civilisation, and it is rarely comfortable.

Once people began to congregate together to create towns and villages, the resultant demands upon the new townscapes refined still further. Trees were still a valuable aspect of such townscapes, but would be selected according to their specific attributes, be they for food, fuel or shelter. Ever since, towns, cities and their urban landscapes throughout the world have been, and continue to be, subject to constant change, and no urban area is likely to be immune from the forces that bring this about. Indeed, as the twenty-first century progresses, the urban landscapes of our towns and cities are going to have to change far more quickly and radically than they are at present, if we want them to be resilient and provide and maintain a healthy setting for people, places, biodiversity and investment. It is known that the pace of urbanisation will accelerate as towns and cities respond to this change – changes in population, in the economy, in ethnic composition, climate change, and in people's expectations and demands of the places they inhabit (Brotchie et al., 1995).

In spite of being aware of these expectations and demands, it could be argued that the planning, design and management of the urban landscapes of the late twentieth and early twenty-first centuries have all too often not created the liveable places of quality that their designers claimed they would do. Too much contemporary architecture and urban design has been criticised as being inhuman and repressive, despite the high social and political ideals shared by so many of the influential planners and designers of the time. The advent of the concept of 'green infrastructures' towards the end of the twentieth century, initially as a counter to the loss of 'nature' and 'natural landscapes' and more recently as an alternative approach to the more negative aspects of mediocre, uniform urban development, has had some success in Europe and the USA in improving the quality of the urban offer (see also Chapter 13, this volume).

Urban design, landscape design and green infrastructures are at a crossroads however in many countries, as the concept of globalisation increases apace, post-industrialism takes hold and the quality of city centre urban life continues to decline as a result. Thus, many cities continue to expand, often with standardised, low-quality developments, which encourages those citizens who have the available resources and opportunities to seek safer, greener, more pleasant, edge-of-town surroundings to inhabit, with an appreciably higher canopy cover of trees. The concept of urban sustainability is increasingly compromised as a result. This chapter will consider some of the issues posed by contemporary urban landscapes and urban design, and highlight the role played by trees and urban forestry in engaging with these issues.

Urban forestry has long moved on from being seen as an 'urban green cosmetic', and this chapter will consider the role that urban forestry has played in influencing some of the beneficial changes that have taken place in landscape and urban planning and design, with urban embellishments such as model villages, garden cities, garden suburbs and new towns and settlements. Other urban concepts that incorporate an urban forestry approach to urban greening have emerged more recently, such as landscape urbanism, biophilic cities, biomimicry and green urbanism, and these concepts will also be considered.

The present chapter will conclude by suggesting that, in spite of the increasing political and scientific recognition and acceptance of the need to plan and design resilient, liveable urban communities that incorporate the concept of a viable urban forest, there cannot be a 'one size fits all' solution to achieving this. While local community involvement is a crucial element of urban forestry, decision making on planning and design cannot be made solely at the local level – strategic thinking is required. It might be an opportune moment therefore to refine the concept of urban forestry still further by bringing together aspects of the built environment, grey infrastructure, blue infrastructure and the broad concept of green infrastructures into an all-embracing, trans-disciplinary approach to responsive urban forestry, thereby creating resilient communities and viable urban futures worldwide.

Trees, people, and designed landscapes

Human beings have had a long, deep cultural relationship with their landscapes. This relationship transcends national cultures, and happily sits as an equal alongside our scientific, economic, ecological and our spiritual relationships. This relationship is reflected in the text of the European Landscape Convention, where landscape is defined as 'an area, as perceived by people, whose character is the result of the action and interaction of natural and / or human factors' (Council of Europe, 2000, p. 4).

Such landscapes comprise a wide variety of elements, such as 'natural' areas, roads, tracks, fields, hedges, settlements, private gardens and, of course, trees and woodlands. The trees and woodlands are therefore of great significance, as they tend to be the key three-dimensional components of the landscape scenery in their own right. Most of the landscapes of Europe, and indeed the World, have derived much of their personality from the presence of trees and woodlands, and within any landscape, trees can provide vital clues about other facets of the place. Trees in an urban park may, for example, be derived from the hedgerows of an older, working countryside; or rows of pollards might mark the former limits of a woodland now removed, while ancient free-standing trees may well have been boundary markers for centuries. We should always look for the 'ghosts in the landscape', and trees have been a vital component of these ghosts and our cultural relationships with them since time began. Indeed, as W. H. Auden reminded us, 'A culture is no better than its woods' (Auden, 1955).

Although our relationship with the landscape, trees and woodland was initially more to do with survival, shelter and spirituality, it could be argued that the relationship quickly became an economic one, where trees and the landscape provided many of the essentials of life – food, building materials, household materials, fuel, chemicals and other raw materials, etc. This happy relationship between people, trees and their landscapes began to get more refined, as scientific and economic cultures tended to gain the upper hand; an upper hand that resulted in revolution – the industrial revolution and the agrarian revolution.

The landscape was more than just scenery therefore: it was the place where people lived, loved, worked, recreated and had their being. Thus it was just as much an urban concept as it was rural, and this fact is even more relevant to our urban lives today. Indeed, the designed landscape could be deemed to be the bedrock upon which all our societies are built.

The industrial and agrarian revolutions imposed significant changes upon the landscapes of Europe, with the expansion of cities and the de-population of the countryside. The poor living conditions of the working classes, the new class conflicts that were arising and the pollution of the environment all contributed to a rising reaction against urbanism and industrialisation, and brought about a new emphasis on the beauty and value of nature, trees and the landscape (see also Chapter 2, this volume). These ideas were not lost upon the key industrial entrepreneurs of the time. Robert Owen, for example, the entrepreneur behind the development of New Lanark in Scotland stated that 'the presence of trees is pleasant to the eye, refreshes the workers and improves the health of the district' (Owen, 1816, p. 97).

Such ideas, propounded by Owen and other inspired industrialists, began to promote other roles for trees, woodlands and urban green space at this time to counter the lack of the quality of life experienced by the majority of workers in the Industrial Revolution cities. Urban green was seen as providing potential 'green lungs' for the city and its inhabitants, and the fact that healthier workers meant healthier profits was not lost on these industrial entrepreneurs.

Different countries became urbanised at different times. For the UK, the date 1851 was significant, as that is when the country officially became an urban nation, as the census of that year proved, for the first time, that urban people outnumbered rural dwellers. As a result, attention turned to landscape design – the design and quality of the landscapes that comprised the public realm in these expanding towns and cities. The popular conception of landscape design was that it was an art form confined to private gardens and parks. Up to this point in history, this was understandable, as many of the industrial nouveau riche employed famous and not so famous landscapists to improve the design of their country estates. Such landscapes were primarily aesthetic however, and it was really only in the early nineteenth century that the 'collective landscape' emerged as a social necessity, because such landscapes were not just aesthetic, but were a crucial part of ordinary people's everyday life.

Urbanism and urban forestry

One of the reactions to the rapid social, economic and physical change in the nineteenth century, and the resulting social unrest, was a concerted attempt to try to make things better by 'greening the collective urban landscape'. Trees were often at the epi-centre of these initiatives, and thus street trees, public parks and green space all started to become features of many towns and cities at this time. All across Western Europe, wooded parks were becoming stages for the theatre of social order. Different social classes had their own way and their own time for using the wooded areas, thus reaffirming their place in the social pecking order. When the railway reached Fontainebleau for example, some 60 km from Paris, the woodland was turned into a 'promenade parisienne', and mass recreation commenced (Kalaora, 1981). In the UK, the annual Dunlop Fair in Epping Forest attracted crowds of over 100,000 Londoners during the mid-1850s (Green, 1996), although there were still undertones of class distinction however. Dyos (1982), for example, commenting on the location and choice of tree species in the streets of Camberwell, London, observed that 'the choice of trees had its social overtones – planes and horse chestnuts for the wide avenues and the lofty mansions of the well-to-do; limes, laburnums and acacias for the middle incomes; unadorned macadam for the wage-earners'.

The concept of urbanism was an early twentieth century invention, although only about 14 per cent of the world's human inhabitants lived in cities at the turn of the twentieth century (United Nations, 2016). The concept was initially devised to consider the implications of the increasing migration of people from rural to urban areas, and the resulting social interaction that took place within such expanding urban areas. The positive relationships between trees, people, their quality of life and the resulting social benefits were increasingly being appreciated by urban theorists, one of whom was Sir Ebenezer Howard. Howard strongly advocated the value of incorporating trees and green space within urban development, and he subsequently developed the concept of the Garden City in his book 'Garden Cities of Tomorrow', published in 1902. This book stimulated the Garden City Movement and, as well as creating several such towns in the UK, influenced many model suburbs in the United States, including Radburn, Forest Hills and the Suburban Resettlement Programme. It also influenced the 'humane design principles' adopted by some urban developments in the Weimar Republic in Germany, as well as countless other developments throughout the world, and is currently experiencing something of a renaissance in political thinking in the UK (Moss-Eccardt, 1973).

The pace of urbanisation greatly increased as the twentieth century progressed, and this has accelerated in the twenty-first century. Towns and cities have responded – and continue to respond – to changes in population levels, ironically both increases and decreases, the ethnic composition of such populations, climate change, changes in industry and employment and changes in people's expectations and demands. Thus, we truly live in times of great change, and this is having a profound effect upon population densities, lifestyles and the design, sustainability and governance of our towns and cities, including the design of our urban landscapes (Brotchie et al., 1995). Such changes are already occurring across Europe, where there have been noticeable realignments over recent years in the physical growth and subsequent sprawl of some of our urban areas.

European urban sprawl is something of a paradox however. In most countries – other than in the UK – national populations are either stable or declining (the recent influx of refugees from wars in the Middle East apart), which would suggest that, at best, cities should either be static in area or even shrinking. They are not, however. Many cities are appreciably expanding, and the prime reason for this is a desire by many sectors of the community to seek a better, quieter, healthier and greener quality of life, which is hard to find in so many contemporary city centres (European Environment Agency, 2006).

In many of our towns and cities, in spite of our so-called 'urban design' skills, our increasingly hard urban perspective has tended to insulate and isolate us from our surroundings. So many of us see the world as passive observers, safe in our homes, cars, offices or universities, through the lens of our smart phones, or through the intermediary of our TV or cinema screens. A better quality of life can be found however in the greener landscapes of the suburbs, and the fact that the registration of private motor vehicles in Western Europe exceeds the registration of births by a ratio of four to one (European Environment Agency, 2006, p. 40) suggests that, even with the high quality of most European public transport systems, people are opting for up-to-date personal transportation to support them in their new suburban lifestyles, rather than having more children to support the population or themselves in their old age.

It cannot be denied that many people find a powerful attraction in 'natural' green, treed landscapes, even in the city. They can give us deep aesthetic experiences of sublimity and beauty, or even perhaps that elusive return to a pre-Eden innocence which enables us to transcend, for a short moment in time at least, the pressures of our daily lives. That said, urban sprawl continues to eat away at our more rural and natural landscapes, partially through direct building and construction, but also through the extension of communication infrastructures to serve such developments, such as new roads, road widening, new or the re-instatement of rail tracks and cell phone masts – the classic 'grey infrastructure' – and this has taken its toll on the quality of our peri-urban landscapes, including our peri-urban woodland areas and trees, as well as compromising the sustainability of towns and cities as a result.

The publication in 1990 of the European Union's Green Paper on the Urban Environment (European Union, 1990) attempted to confront the issues of urban sprawl by extolling the virtues of urban densification as a way of creating more mixed land uses and related multifunctional, liveable open spaces in our urban landscapes. Subsequent policies were promoted by the Green Paper throughout the Member States to achieve this aim, in spite of the fact that contemporary urban design policies at both international and national levels advocated versions of the 'Compact City' as a strategy to address urban sprawl and achieve a more sustainable approach to urban development. It has become apparent however that, far from halting the spread of urbanism, it has been increasing with renewed vigour.

Neuman (2005) has called into question the whole concept of the compact city. He has suggested that the evolution of the planning profession at the end of the nineteenth century was largely due to the need to improve the health and well-being of urban inhabitants by relieving the overcrowding of the industrial towns and cities by letting in more light, air and 'green', the results of which have been largely successful, until now. Faced with the current issues of increasing urban sprawl, the contemporary urban design response now seems to want to reverse these ideas and 're-compact' the city.

The use of the term 'urban design' emanates from the mid-twentieth century, but it has in fact been practised throughout history. There are very many examples of carefully planned and designed cities throughout the world, many going back to classical times. There is however a growing amount of research that suggests that contemporary urban design is not creating the 'liveable urbanism' that it claimed it would, in spite of the high social and design ideals of many of its promoters and practitioners. McGlynn (1993) has claimed that its recent iteration coincided with two interconnected developments. First, it had become apparent in the 1960s that the existing environmental disciplines – especially architecture and town planning – had become professions with increasingly specialised and jealously protected areas of activity, and a gap had opened up between them. Architecture's clear concern was with the design and construction of a building or buildings on a specifically designed site. Planning on the other hand took responsibility for the general disposition of land use through policy formation and plan making. As these professional boundaries hardened and were institutionalised, it became clear that the gap between them was the public realm itself – the voids between the buildings, the streets and the spaces which constitute our everyday experience of urban places. This public realm – the urban public landscape – is made up of the very places that potentially accommodate the urban forest, and are therefore of critical interest to promoters of an urban forestry approach to urban greening.

The European Union is well aware that the future of Europe depends on having successful, sustainable, resilient cities and city regions (Potočnik, 2013). Further, the report 'Building a Green Infrastructure in Europe' (European Commission, 2013) very much promoted the idea of a European Urban Green Infrastructure, including urban green space, street trees, other urban forestry plantings and sustainable urban drainage. The advent of the concept of green infrastructures, arguably as an initial counter to the loss of 'nature' and 'natural landscapes' and, more recently, in conjunction with the concept of urban forestry as a counter to the more negative aspects of urban development, has had some success (see Chapter 13, this volume). There is also a growing canon of research that proves that the inclusion of trees in and around the city has a broad spectrum of benefits. They can greatly improve our health and well-being for example (see Chapter 6, this volume), improve learning, increase property values, provide focal points to improve social cohesion, improve air quality, offset carbon emissions, promote biodiversity, limit the risk of flooding, cool our towns and cities, promote inward investment and job creation and even make us drive more safely (Konijnendijk et al., 2005).

Urban forestry and urban areas – some of the strategic issues

Europe was the first continent to become truly urbanised. In the second decade of the twenty-first century, some two thirds of the European population can be classified as being urban or urban orientated. Many large cities, including defiant urban centres such as Paris, are still losing people however, and thus the urban footprint of many cities is steadily expanding. Parts of some countries such as the Netherlands, Flanders, the UK and Germany for example are rapidly approaching 90 per cent urbanisation.

Urbanism has been back on the political agenda for some while. As most of us in the West now live and work in urban areas, we could be forgiven for thinking that it has never really been away – but it has. For many of the latter years of the twentieth century, urban areas were all too often perceived as being a problem and a drain upon our national economies, because they were in economic and demographic decline. This view has now generally been reversed, and it needed to be, for without a doubt, one of the biggest problems that faces us in Europe as a whole is the problem of urban change. Places that once were prosperous may slip into decay and physically or commercially decline as a result, while other areas that are currently deemed to be poor or run-down may benefit from regeneration or a revival. The reasons for this state of affairs are many and varied of course, and paradoxically can have as much to do with the urban image of an area – either real, imagined or invented – as with physical re- or degeneration. Urban trees and urban forestry, or the lack of it, can play a significant part in this (Shaw and Robinson, 1998).

Ignoring this urban change is not a viable medium or long-term option for successful politicians, policymakers, investors or environmental professionals. It could be argued that a failure to engage convincingly with this changing urban agenda has, in some parts of the world, led to a worsening of urban problems. Some sectors of our communities are developing a growing disenchantment with the fact that the new urban rhetoric of the politician, the media and the professional journals fails to live up to the reality of the new urbanism. In extreme cases, this has led to a rise in the fortunes of political extremism as a desperate, if rather simplistic, way of trying to remedy the situation.

The pace of urban change across the world has been accelerating over recent decades, and shows no sign of slowing down. Europe is in the process of undergoing profound change, change that was originally concerned with the unification of the continent – and thus was essentially economically driven – but now which considers matters as diverse as the suspension of national boarders, easier pan-continental travel and the accommodation of hundreds of thousands of migrants from the Middle East. From a continent of competing countries, we are becoming a continent of competing regions and cities. The rules of the game are changing, and it could be argued that the conventional approaches to strategic planning, designing and regulating urbanisation are failing, as they are all too often seen as formulaic and unsustainable, and are thus unable to attune to the increasingly unstable urban conditions and the globalisation of our world.

The current phase of globalisation that is sweeping our planet is a multi-faceted process of economic and social restructuring that is having a significant effect upon the design and quality of urban life, particularly on the established urban areas that can be found in our older cities, such as those that we can find in Europe, the Far East, parts of Asia, etc. Contemporary developments are all too often seen as a 'standard product', or 'the McDonaldisation of society' as Ritzer has so succinctly called it (Ritzer, 1993). Arguably, this has contributed to what some have deemed to be a crisis in European urban areas of 'local distinctiveness', the increasing uniformity of our towns and cities and the rather negative effect this has on local populations, communities and local economies. A definition of this often elusive 'quality of place' has been produced by Clifford (2003, p. 1), who suggests that:

> local distinctiveness is about differentiating our everyday surroundings and the significance that places have to us. It is an expression of the accumulation of story upon history upon natural history that results as we work with (or against) nature to shape the land, making a living and live our lives. It embraces not only the differences between places, their buildings and landscapes, city street and farming patterns, but recipes, legends, language, customs and festivals as well.
>
> *Common Ground, 2016*

Urban design, urban landscapes and the 'quality of place' of the urban public realm are at a crossroads, however, as the quality of urban life continues to decline, cities continue to expand as a result, and sustainability is at best static. A range of new ideas, concepts and sub-strands of urbanism are emerging as a result, concepts such as landscape urbanism, biophilic cities, biomimicry and green urbanism to name but a few. All these concepts incorporate a central role for trees and urban forestry. What should be borne in mind however is that as new thinking takes place and new scenarios are established, they do not wholly replace the previous ideas but simply add another layer to the depth of our knowledge. This applies as much to the concept of urban forestry as it does to anything else.

It has been argued by Simson (2008) that urban forestry was 'rediscovered' at the end of the twentieth century as a philosophical approach to having a strategic overview of the planning, design, establishment and prospective management of the whole mosaic of a city's urban and peri-urban green space, and all the implications that had to engaging with the associated social, economic and environmental issues and strategies that were to be found there. Urban forestry is a 'big idea', and now that the concept is recognised and established globally, can it adapt and respond to the myriad of changes that are occurring across the world-wide urban stage? The late Hubert Humphrey hit the nail on the head when he said 'We are in danger of making our cities places where business goes on but where real life, in the real sense, is lost'.

But what makes life real in the city? It is considered 'officially' that a liveable city needs appropriate amounts of:

- energy;
- food;
- water;
- transport;
- jobs/development;
- retail;
- telecommunications;
- public services;
- emergency services;
- health; and
- finance.

Even if adequate supplies of these services are available to us, we cannot solve the problems of our urban landscapes however by deploying slightly different versions of the same thinking that created them in the first place. We need to move on, to change our approach, and one of the things we need to change is our approach to nature, to things natural. Bringing nature or the opportunity for people to be influenced by natural processes into our towns and cities is vital of course, as is making the links and connections between green spaces in and around the places where we live, work and recreate. So what makes a really liveable city? This is not an exhaustive list, but it might include:

- ready access to the urban forest, parks, green spaces, etc.;
- ready access to music, art, culture;
- proximity to friends;
- access to daylight and sunlight;
- the opportunity to walk unimpeded;
- a sense of personal safety; and
- the availability of fresh food and clean water.

Just as urbanism is changing, so is the natural world. We need to accept and embrace the fact that, in the twenty-first century, there is an increasing number of hybrid ecosystems developing across our landscapes and that they will influence sustainable, resilient development, and urban forestry. They are the result of the interacting forces of urbanisation, globalisation and climate change, and are made up of organisms that have been brought together by the removal or modification of barriers that have kept them apart for thousands of years. Such hybrid eco-systems are not only to be found in and around our towns and cities, but also in many of the landscapes that have been disturbed by intensive agriculture, industrial processes and mining, and it is unrealistic to assume that we can turn the ecological clock back to bring 'old nature' back into our towns and cities, any more than we can turn back the economic clock that created these disturbed landscapes in the first place.

A need is emerging therefore for a responsive concept for planning, designing, constructing and managing sustainable, resilient development to emerge that is trans-disciplinary, cost effective, and easy to both understand and communicate to all levels of the community. If the conventional approaches to regulating urbanism are no longer reliable, an opportunity exists for an alternative approach to develop. It might be an opportune time therefore for urban forestry to step forward and start getting involved in proposing cogently argued strategic, regional plans to seduce and convince the policy-makers, just as urban planners had to do for city developments in the late twentieth century. Urban foresters have the credentials to deal with these unstable conditions because urban forestry continually adapts and transforms itself and can accommodate a myriad of forces and initiatives. Urban forestry is getting more sophisticated and has moved beyond being seen as a 'green cosmetic' that all too often was used to legitimate poor planning, to becoming an integral part of a new, more resilient urbanism.

What we have already learned is that cities are not only economic assets, not merely market places. They have a great capacity to promote community development, social cohesion and civic and cultural identity. The somewhat narrow pursuit of market-led success has not led to the elimination of social problems in our different countries, even though we all have different institutional, economic and social arrangements and policies. Achieving economic success with social justice in sustainable, resilient cities remains a challenge to all governments and organisations – local, regional, national and international, and the concept of urban forestry can assist in this quest.

If urban foresters are to deliver under this new resilient urbanism agenda, we will from time to time have to leave the security of the trees and engage with concepts that are new and perhaps alien to us. For example, central to this new resilient urbanism is the concept of 'physical capital'. This can be defined as 'the potential value – financial, social and cultural – of the built environment', of which of course the urban forest is a vital part.

There is undoubtedly a trend across Europe for dominant flows of capital to shift from governments and municipalities to the market and the private sector. The concept of 'capital' is well understood in such circumstances – it's about economics, and thus capital is an asset which creates value or has the potential to create value. Even a cursory examination of such assets however tells us that there is no such thing as intrinsic value – nothing has value unless someone values it. Value is socially created. In the 'market', value can only be judged by the prices that people are willing to pay.

Urban forestry is about creating and / or maintaining a good physical environment, and there is no doubt that it is socially, environmentally and economically desirable and valuable. But what is the nature of that value? How does the value accruing to the owners of property relate to the values of the wider public? Can any of these kinds of value be measured? What is their relationship to other types of wealth, income or capital?

Fortunately there has been considerable research and resulting progress over recent years in being able to put a provable value upon such a quality physical environment, although many urban foresters still wonder why they should engage with such issues. We must engage however because, in order for us to promote urban forestry as a metaphysic of urban regeneration – which it is – and thus play in the difficult decision-making process and the dialogue involved with this, we need to better understand the nature of the value of the built environment, as well as the natural environment. Whether we like it or not, urban forestry is part of the development process, and this constantly has to:

- judge the best levels of spending and investment;
- judge between alternative projects;
- manage investments, with the right depreciation, portfolios of risk and reward; and
- determine the balance of risk and reward between public and private players.

This is familiar territory for the economist; less so for the urban forester, particularly as our work can span a considerable number of years. Trees can last a long time; many other built environments on the other hand can come and go, or be re-furbished or re-built within a relatively short space of time. This can make our life complicated, but if we genuinely believe that trees make great places and great places can make trees, and that urban foresters too make great places, and that such great places boost value for the public realm, we will have to learn the tricks of the urbanism trade, understand and subscribe to the subtleties of concepts such as physical capital, and participate in research to analyse and test these concepts so that we can construct a convincing case for urban forestry in the emerging polycentric cities of the world.

The responsive urban forest

One of these tricks that we might borrow from the concept of Responsive Environments, an approach to successful 'place-making' proposed by Bentley et al. (1985), is the idea of a 'responsive urban forest'. Bentley et al. suggested that the good design of a place affects the choices that people can make there. An urban forest is also a place, and thus the design of a responsive urban forest would affect the choices people can make, at many levels:

- It would affect where people can go, and where they cannot – a quality we could call *permeability*.
- It would affect the range of uses available to people – a quality we could call *variety*.
- It would affect how easily people could understand what opportunities it offers – a quality we could call *legibility*.
- It would affect the degree to which people could use a given place for different purposes – a quality we could call *robustness*.
- It would affect whether the detailed appearance of the place makes people aware of the choices available – a quality we could call *visual appropriateness*.
- It would affect people's choice of sensory experiences – a quality we could call *richness*.
- And it would affect the extent to which people could put their own stamp on a place – a quality we could call *personalisation*.

This is not an exhaustive list of choices, but they are perhaps the key issues in making our urban forests 'responsive'.

At the heart of responsive urban forestry lies a belief in the quadripartite approach of research, experimentation, practice and reflection. This is a 'big idea', and it may be one of the 'big ideas' that will be part of the driving force behind the twenty-first century landscape and urbanism perspectives in all our cities, whether they are in Europe or elsewhere. As Daniel Burnham, the visionary planner of Chicago said, 'make no little plans; they have no magic to stir men's blood ... make big plans ... aim high in hope and work' (quoted in Moore, 1921).

In our responsive urban forestry mission to shape a better world, we must shape better cities. Helping to create places where people can aim high in hope and work is the potential role that urban forestry can take on in the emerging, world-wide new urbanism, based very much on the premise that the true wealth of our towns and cities can only really be measured in terms of the well-being and the culture of our people, and the sustainability and resilience of their environment. Some would say that these responsive urban forestry promises are unrealistic, maverick views that cannot be realised in the contemporary urban debate. It could be equally argued however that today's maverick views are tomorrow's orthodoxy, and Evans (2002) is correct in his belief that 'the challenge for all of us who understand that trees matter is to find ways of reaching that promise'.

References

Auden, W. H. (1955) 'Woods', from the Bucolics sequence, in his *The Shield of Achilles*, Faber & Faber, London.

Bentley, I., Alcock, A., Murrain, P., McGlynn, S., Smith, G. (1985) *Responsive Environments: A Manual for Designers*, Architectural Press, London.

Brotchie, J., Batty, M., Blakeley, E., Hall, P., Newton, P. (1995) *Cities in Competition*, Longman, Melbourne.

Clifford, S (2003) 'Celebrating local distinctiveness, in social history in museums', *Journal of the Social History Curators Group*. Edited by Rebecca Fardel, Vol 28, p. 1.

Council of Europe (2000) 'European landscape convention', retrieved from www.coe.int/en/web/landscape/home (accessed 25 April 2016).

Dyos, H. K. (1982) 'The objects of street improvement in regency and early Victorian London', in D. Cannadine, D. Reeder (eds), *Exploring the Urban Past; Essays in Urban History* (pp. 81–86), Cambridge University Press, Cambridge.

European Commission (2013) *Building a Green Infrastructure in Europe*. European Commission, Brussels.

European Environment Agency (2006) *Urban Sprawl in Europe*, EEA Report No. 10/2006, EEA, Copenhagen.

European Union (1990) *Green Paper on the Urban Environment*, COM [90] 218 final, European Commission, Brussels.

Evans, P. (2002) 'A night of dark trees', *Arboricultural Journal – the International Journal of Urban Forestry*, vol 26., no. 3, pp. 249–256.

Green, G. (1996) *Epping Forest through the Ages*, 5th edn, Woodford Bridge, Ilford.

Howard, E. (1898) *Tomorrow, a Peaceful Path to Real Reform*, Swan Sonnenschein, London.

Kalaora, B. (1981) 'Les Salons Verts: parcours de la ville à la forêt', in M. Anselme, J.-L. Parisis, M. Péraldi, Y. Ronchi, B. Kalaora (eds), *Tant qu'il y aura des arbres: pratiques et politiques de nature 1870–1960*, Recherches No. 45, Paris.

Konijnendijk, C., Nilsson, K., Randrup, T., Schipperijn, J. (eds) (2005) *Urban Forests and Trees: A Reference Book*, Springer, Berlin.

McGlynn, S. (1993) 'Reviewing the rhetoric', in R. Haywood, S. McGlynn (eds), *Making Better Places Urban Design Now*, Butterworth Architecture, Oxford.

Moore, C. (1921) *Daniel H Burnham: Architect, Planner of Cities*, vol. 2, Houghton Mifflin, Boston, MA, retrieved from http://archive.org/stream/danielhburnhamar02moor#page/n7/mode/2up (accessed 17 July 2016).

Moss-Eccardt, J. (1973) *An Illustrated Life of Sir Ebenezer Howard 1850–1928*, Shire Publications, Aylesbury.

Neuman, M. (2005) 'The compact city fallacy', *Journal of Planning Education Research*, vol. 25, pp. 11–26.

Owen, R. (1816) *New View of Society or Essays on the Principle of Formation of the Human Character, preparatory to the Development of a Plan for gradually Ameliorating the Condition of Mankind*, 2nd edn, Longman, London.

Potočnik, J. (2013) 'Cities will define our future', speech given to the 7th European Conference of Sustainable Cities and Towns, Geneva, 17 April, retrieved from www.europa.eu/rapid/pressrelease_SPEECH.13-326_en.htm (accessed 25 April 2016).

Ritzer G (1993) *The McDonaldization of Society*, Pine Forge Press, Thousand Oaks, CA.

Shaw, K., Robinson, F. (1998) 'Learning from experience?', *Town Planning Review*, vol. 69, no 1, pp. 49–63.

Simson, A. J. (2008) 'The place of trees in the city of the future', *Arboricultural Journal – the International Journal of Urban Forestry*, vol. 31, pp. 97–108.

United Nations (2016) 'Human population: urbanization' www.prb.org/Publications/Lesson-Plans/HumanPopulation/Urbanization.aspx (accessed 25 April 2016).

15

URBAN FOREST GOVERNANCE AND COMMUNITY ENGAGEMENT

Stephen R. J. Sheppard, Cecil C. Konijnendijk van den Bosch, Owen Croy, Ana Macias and Sara Barron

Introduction

This chapter focuses on concepts, issues and practices of governance and public participation and how these apply to urban forestry. It highlights trends in governance and community engagement to date, and explores how emerging challenges and opportunities in a changing world may influence governance and engagement for urban forestry in the future.

Governance has been defined as efforts 'to direct human action towards common goals, and more formally as the setting, application and enforcement of generally agreed to rules' (Konijnendijk van den Bosch, 2014, p. 36). Community engagement on forestry has been defined as processes where 'individuals, communities, and stakeholder groups can exchange information, articulate interests, and have the potential to influence decisions or the outcome of forest management issues' (Beckley et al., 2006, p. 3). These two domains intersect amid a complex web of actors and sectors of society, in a rapidly changing world.

Modern urban forestry governance is often associated with local government programmes for planning and managing trees in parks, streets, and private land. However, based in part on historical precedents, Konijnendijk (2003) has framed urban forestry in general as being:

1 integrative;
2 socially inclusive;
3 strategic; and
4 embracing its urban mandate.

These key dimensions also relate to urban forest governance. Integration ideally is both 'horizontal', crossing traditional sectoral and natural resource boundaries, as well as 'vertical' in terms of collaboration between public and other actors at different levels of governance and management. Urban forestry involves government, business and civil society actors (Lawrence et al., 2013). Urban forestry's social-inclusive nature relates to equity issues and the wider involvement of stakeholders and urban residents. Crucial aspects of urban forest governance include who has access (and who has not) to decision-making (Konijnendijk van den Bosch, 2014).

These aspects of urban forestry governance play out in a context that is changing rapidly due to several major trends:

- Rapid urbanization due to rural flight, immigration, and intensification of urban land uses in both developed and developing nations: Statistics Canada, for example, reports that 81 per cent of Canadians now live in cities.
- Global climate change and its increasing local impacts, with a growing need for climate-proofing against rising risks and costs, and for capacity building across all sectors of society.
- Increasing attention to the importance and costs of maintaining and improving public health and wellbeing (see Chapter 6, this volume).
- Budget constraints and the increasing costs of maintaining and renewing urban infrastructure (e.g. Hauer and Peterson, 2016).
- Growing public expectations for involvement in decision-making, with movements in some sectors from governance by government to governance with or without government (e.g. Van Tatenhove et al., 2000; Konijnendijk van den Bosch, 2014).
- Advancing technology in terms of more sophisticated and powerful 'smart tools' for monitoring, analysis, planning and visualization of urban landscapes (e.g. 3D LiDAR datasets, drones, modelling of climate projections), and for interaction through social media.

Together, these trends represent shifting ground for the governance of urban forestry, adding to the pressure on those who would govern and manage urban forests, and calling for better decision-making for an uncertain future. In other words, the stakes for urban forestry are rising as solutions to societal challenges become increasingly complex.

The evolving nature of urban forestry governance amid these contextual trends is explored below. The next section introduces a framework for understanding and analysing urban forest governance. This is followed by a description of the range of regulatory, policy and guidance instruments typically used in urban forest governance. Next, community engagement and participatory processes applicable to urban forestry are examined. Selected case studies illustrate how urban forestry governance issues are being addressed in various countries. The final section summarizes the current state of urban forest governance, and identifies needed and promising new approaches.

A framework for understanding urban forest governance

Recent years have seen an increasing emergence of the concept of 'governance' in the literature, where 'government' and 'policy-making' were more in focus earlier. Kjær (2004, p. 10) defines governance as 'the setting of rules, the application of rules, and the enforcement of rules'. Definitions of governance vary widely, but all recognize a shift in strategic decision-making in which (a range of) government actors to an increasing extent share (or even transfer) decision-making and rule-setting with civic society and businesses (Lawrence et al., 2013; Konijnendijk van den Bosch, 2014).

Several scholars have applied the so-called *policy arrangement approach* originally developed by Van Tatenhove et al. (2000) specifically to the context of forest governance (Liefferink,

2006) and even urban forest governance (Krajter Ostoic, 2013; Konijnendijk van den Bosch, 2014). Governance arrangements have been defined as temporary stabilization of the substance and organization of the governance and policy domain. These arrangements can change according to four interlinked dimensions (see Figure 15.1):

1 actors and their coalitions involved in the policy domain;
2 division of power and other resources between the actors;
3 rules of the game (covered below in more depth); and
4 current policy discourses.

'Rules of the game' refer to institutions and the regulations, legislation and procedures relevant to a certain policy domain. Discourses have been defined as 'an ensemble of ideas, concepts and categories through which meaning is given to social and physical phenomena, and which is produced and reproduced through an identifiable set of practices' (Hajer and Versteeg, 2005, p. 176).

Analysis of urban forest governance can focus on one of the four dimensions of the tetrahedron (shown in Figure 15.1), and then use this as a 'lens' to analyse the other components, in terms of how they change in response to changes in the chosen dimension. For example, a discourse that promotes greater involvement of local residents in decisions about a nearby park changes the actor constellation as well as the 'rules of the game' (how decisions are taken).

Urban forest governance today

Recently, the number of studies on urban forest governance has increased, at both the local scale (e.g. Krajter Ostoic, 2013) and the national (Molin, 2014) or even international scale (e.g.

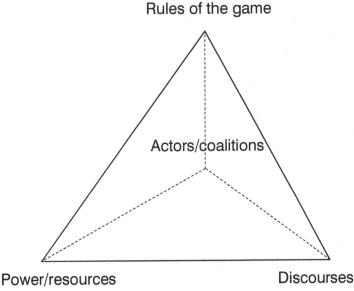

Figure 15.1 Visualization of the four dimensions of the policy arrangement approach

Source: Liefferink (2006)

Lawrence et al., 2013). All studies show an increasing complexity in urban forest governance, driven by various discourses, with changing actor constellations, shifting rules of the game, and changing roles and (access to) resources such as knowledge, funding and power. Discourses driving urban forestry are many, ranging from sustainability to a neoliberal discourse that places a monetary value on urban forest benefits (Gulsrud, 2015). Debates over density versus urban sprawl, as well as over density versus less dense, greener cities, represent an important discourse in urban planning (Haaland and Konijnendijk van den Bosch, 2015).

The principal actors and alliances in urban forest governance, and their typical roles, fall into the following broad categories (Buizer et al., 2015; Lawrence et al., 2013):

1 Government ('top-down' or state actors) at various levels and scales, with a crucial role for local government. However, in countries such as the US, national governments have also been involved in urban forestry, through policy, funding and research programmes (e.g. the USDA Forest Service provision of i-Tree tools; see Chapter 11).
2 Civil society, including 'bottom-up' actors such as citizen and stakeholder groups, volunteers, etc. (see next section).
3 Businesses, such as water and utility companies, developers, companies investing in tree planting and managing assets for social corporate responsibility, private contractors, landscapers and other parts of the 'green industry'.
4 Other third party intervenors such as non-governmental organizations (NGOs), professional organizations, and researchers, often playing pivotal roles where governments are not strongly involved (Sheppard et al., 2015). University staff for example are often involved in urban forest resource assessments as well as advising on policy-making and management.

There are of course actors who cross over these boundaries, and it should not be assumed that urban forestry expertise resides exclusively with the government or private practitioners. There are many different kinds of urban foresters. While the terms 'Urban Forester' or 'City Forester' have often been used to describe an employee of local or regional government responsible for municipal forestry operations, the term has broadened to include various types of organizations and individuals involved in community greening and operations. For instance, Casey Trees, a non-profit organization in Washington DC, has a mandate that includes 'restoring, enhancing and protecting the tree canopy of the nation's capital for the long term' (executive director M. Buscaino, personal communication, 2016). Key employees of this organization are bona fide urban foresters, whose roles are not limited to tree planting, but also include tree advocacy and education, research and mapping, and conservation efforts.

Governance and rules of the game for urban forestry vary across the world. Molin (2014), referring to the work of Arnouts et al. (2012), uses a set of four 'ideal-typical' modes of governance (see Table 15.1) in her work on public involvement in green space maintenance in Denmark. Although green space management here is still seen generally as a public sector activity, models of open co-governance and self-governance are also now frequently applied; governance without government is becoming more common.

Scale is crucial in environmental and urban forest governance, as highlighted by Lawrence et al. (2013). Molin (2014) argues that governance is not limited to the higher geographical scales, but can be applied to individual green spaces and sites, as well as their hands-on maintenance. She argues for a place-based perspective on governance, where the interactions and relations between local communities and local green spaces are emphasized (discussed further below).

Table 15.1 Ideal-typical governance arrangements according to Arnouts et al. (2012)

	Hierarchical	*Closed co-governance*	*Open co-governance*	*Self-governance*
Actors	Mainly governmental actors	Select mixed group of actors	Large mixed group of actors	Mainly non-governmental actors
Power	With government	Pooled	Diffused	With non-government
Rules	Governmental coercion	Restricted cooperation	Flexible collaboration	Non-governmental forerunning

---

Box 15.1 Changing green space governance in Denmark

Denmark has seen a call for more involvement of local residents in the governance and management of local green spaces. City governments have started to engage local dwellers not only in planning, but also in the maintenance of green spaces. Thus place-based governance includes strategic decision-making as well as day-to-day maintenance decisions. In interviews, green space managers in 10 Danish municipalities all acknowledged the discourse of greater public involvement, as well as its potential benefits. However, they felt ill-prepared for engaging with the public because of lack of training and strategy. The rules of the game were gradually changing, with more cases of public involvement emerging. Good practice was often built on the personal ambitions and contacts of the green space manager. Managers preferred collaboration with residents through already existing organizations (e.g. local neighbourhood groups or NGOs).

Source: Molin (2014)

Regulations, policies, guidelines and programmes

This section reviews the mechanisms and instruments available in regulating or guiding urban forestry planning and practice. Urban forest governance on public lands has typically been by government, though sometimes the public expects government to supply urban forestry programmes and services on private land too. In urban areas of North America, governance of urban forests is provided mostly at the municipal level, with some governance occurring at the regional/county/district level. Here, the regulatory basis for most municipal and regional governance of private urban forests arises from state or provincial legislation, but in other countries it may come from national legislation. In Europe however, the regulatory base for large parts of the urban forest often lies at the municipal level (e.g. Konijnendijk, 2003).

Overview of local/regional government policy and regulatory tools

Municipalities and regional governments may be legislatively empowered to create *by-laws* (ordinances, codes) that regulate activities impacting urban forests on both private and public lands. Governments may play roles in *direct delivery* of urban and community forestry programmes (e.g., through physical interventions, plans, and regulatory permits and tree preservation orders); or *indirect delivery* (influencing the actions of others) through:

Table 15.2 Summary of where urban forestry governance actions commonly occur or potentially could occur

Scale of governance	Governance activities by government						Community engagement		Direct physical interventions
	Policies/ guidelines	Regulations, rights, permits	Plans	Incentives/ grants	Fees/fines	Programmes	Outreach/ interaction	Volunteer training	
National	✗	✓		✓		✓			
State/provincial	✗	✓	✓	✓		✓			
Regional	✓	✓	✗						
City/county/municipal	✓	✓	✓	✓	✓	✓	✓	✓	✓
Neighbourhood/local area/major corridor	✗		✗			✗	✗	✗	✓
Block	✗	✗	✗	✗		✗	✗	✗	✓
Park	✓	✓	✓			✓	✓	✓	✓
School/public facility	✓	✓					✓	✓	✓
Business	✓	✓			✓		✗	✗	
Household/lot	✓	✓		✓	✓		✓		✓

✓ = Commonly occurring governance actions

✗ = Rarely occurring governance actions that potentially could advance urban forestry and community engagement

- grants with conditions (as commonly practised in the USA through state community forestry programmes);
- requiring private land-owners to manage government-owned lands (e.g. requiring residents adjacent to road rights-of-way to prune and manage street trees at their own cost, where these services are not provided by the municipal authority, as in Portland, Oregon); or
- education and engagement (see below).

Table 15.2 provides a list of direct and indirect governance actions by government occurring at various scales.

Municipal or regional governance is typically divided into either legislative authority (a body of elected officials) or administrative authority (appointed civil servants) to implement policies, regulations and enforcement. The legislative authority approves by-laws and may also set policy based on recommendations from staff.

The range of instruments typically used include:

1 *Regulations* that are contained within by-laws, which carry the force of municipal law and generally contain: definitions of terms; the administrative position(s) empowered to provide approvals; explicit rules that permit or prohibit actions; and penalties for violations. Enforcement is typically carried out by police officers or by-law/code enforcement officials. Many municipal and regional governments have approved by-laws that regulate tree cutting on both private and public lands. Regulations may include the requirement to obtain a permit prior to undertaking any tree-related work.

2 To support the issuance of building and tree-cutting permits, *tree surveys* are frequently required. The survey may include information such as a plan drawing showing tree locations relative to infrastructure, descriptions of the size, species and condition of the trees, and proximity to planned infrastructure. The permit may also have requirements for tree protection measures and post development re-planting efforts. Tree cutting by-laws may require fees to be paid, based on the number, quality and size of trees to be removed under permit. By-laws may also prohibit the cutting, pruning or removal of trees based on species rarity, heritage value and habitat value. In some municipal by-laws, trees of significance in the community may be specifically identified for protection whenever future operations are being considered.

3 Urban forest *policies* are statements adopted by elected or appointed officials that guide future decision-making, preferably before there are critical site-specific urban forestry decisions to be made. Policies are not rules, and are subject to interpretation and modification based on specific circumstances. Urban forest policies should have: (a) a statement of intent that describes what is to be achieved by application of the policy; (b) the rationale for development of the policy; and (c) a decision-making framework for policy application (City of Surrey, 2000). For example, many local governments have adopted tree removal policies regarding requests for removal of public trees. Criteria for considering removal could include the degree of tree risk, heritage values, and the extent of aesthetic, economic or environmental damage. Some policies set specific quantitative thresholds for attainment, such as canopy percent cover targets to be reached by a specific date; however, unless these are tied to real consequences (such as loss of funding sources if targets are not met), they remain subject to weak political will among elected officials.

4 *Management plans or strategies* have been produced by many but not all cities, to guide prioritization and overall delivery of programmes in different areas of the city (see Chapter 13, this volume). European cities often have no urban forest policies as such, but do have urban forest management plans that can have a more strategic, 'policy like' component.

5 Urban forest *guidelines* are action-oriented outlines of methods used to deliver urban forest programmes or services (e.g. Diamond Head Consulting, 2016). Guidelines can address issues ranging from determining species composition in new neighbourhood plantings, to engaging volunteers in urban forest activities. For instance, a community might have a guideline that no species make up more than 10 per cent of the inventory, no genus more than 20 per cent and no family more than 30 per cent, in order to ensure diversity and resilience of plantings. However, within a specific neighbourhood, a particular desired aesthetic effect that somewhat exceeds the guidelines might be accepted if the guidelines could still be achieved for the community as a whole.

Governmental urban forestry actors and interactions with other governance bodies

How local government controls urban green spaces may be strongly influenced by its internal structure and relationships with other bodies. Within municipal or regional governments, the urban forest programme may be located within a parks department, especially if the majority of the public urban forest is located within the park system. However, where street

---

Box 15.2 Evolving guidelines to resolve conflicts between sustainable density and urban forestry in Surrey, British Columbia

In North America, a growing trend is to create dense, liveable, suburban neighbourhoods as an alternative to sprawling development. The City of Surrey in British Columbia planned and built a model sustainable suburban neighbourhood for about 13,000 residents in East Clayton. The goal was to provide enough residential density to support local walkable businesses and transit, while mimicking the site's natural hydrology and targeting a 40 per cent tree canopy on private lots, parks, and school sites. Density was achieved mostly through small-lot single-family homes, with additional units in basement suites, laneway houses, and townhouses. The neighbourhood is successful in terms of generally satisfied citizens, high profitability for developers, and reasonable affordability for homeowners. However, the neighbourhood falls far short of the intended urban forest cover, with few mature trees saved during development, little room to grow larger trees, and harsh conditions for establishing new plantings.

Discussions with local residents revealed a desire for more and larger trees within the neighbourhood. Neighbourhoods offer little shade or views of nature, and children will grow up without trees to climb. Accordingly, the City has recently developed modified guidelines to create wide landscaped buffers and boulevard plantings along major corridors, to give space to grow and protect larger trees, and has applied these guidelines to new developments in a similar neighbourhood but with higher canopy coverage (see Figure 15.2).

Sources: Barron (2016); City of Surrey (2015)

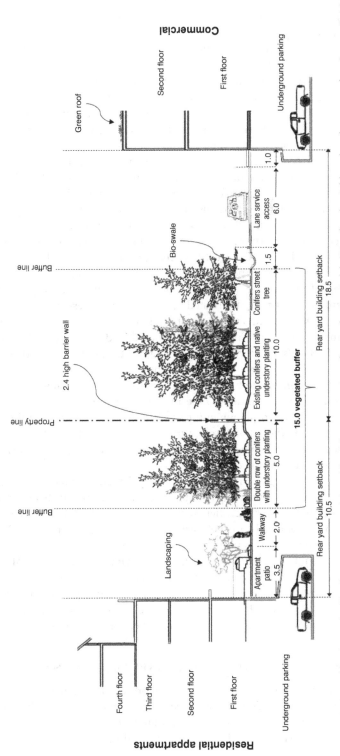

Figure 15.2 Typical landscape buffer along a commercial–residential apartment interface in the Anniedale–Tynehead neighbourhood concept plan, Surrey, British Columbia

trees make up the majority of a community's urban forest, the urban forest function may be located in an engineering or public works department. Some communities have their own stand-alone forestry departments that report either directly to elected officials or to a chief city administrator/manager. It is common in local governments for urban forestry professionals with different responsibilities to be based in different departments (e.g. public versus private realms, parks versus street trees), with the danger that 'silo-thinking' can constrain effective system-wide planning and management (Krajter Ostoic, 2013).

Within government, there are many other departments and disciplines that may play a strong role in urban forestry governance. While some analysts have placed urban foresters as the first tier of the planning process (Schwab, 2009), a common discourse among urban forestry practitioners is that other disciplines (such as engineers, architects, planners, etc.) often have more authority or influence over decision-making, and therefore key values and opportunities associated with urban forestry may be lost (Macias, 2016). Trained urban foresters should have a close working relationship with land use planners and engineers that plan infrastructure, developers who create new neighbourhoods, officials responsible for protective services, etc. The modern urban forester must understand the technical aspects of managing the urban forest, but equally must have mastery of skills to work with residents, volunteers, non-profit organizations, news media and the business community.

In some cases, an elected council (legislative authority) may establish commissions, boards and other authorities specifically related to urban trees. These authorities may be advisory in nature or, less frequently, may be empowered to set policies, approve budgets, and authorize work programmes. In either case, most boards and commissions have a mix of elected and appointed officials who receive delegations from the public about urban forestry matters, and have a primary advocacy role for community forestry programmes (see below). Other types of boards or commissions may impact urban forests and should be involved in developing urban forest programmes. These may include: planning commissions that recommend direction for future development; parks boards and heritage commissions that act to conserve heritage and other public trees; utility commissions that install, operate and maintain utilities such as power-line rights of way or reservoirs; and airport authorities with jurisdiction over tree height near runways.

Urban forestry teams in local government have responsibility (and liability) for many direct interventions on the urban forest, including maintenance, tree risk assessment and mitigation, and replanting. These activities typically adhere to practices, standards and norms such as Tree Risk Assessment protocols adopted by many North American cities (Dunster et al., 2013) and training and arborist certification programmes such as those administered by the International Society of Arboriculture and often required of contractors. Standards such as the ANSI A300 in the United States, are an important component of urban forest governance. Standards often describe best management practices, including specifications needed to achieve sustainable, efficient and cost-effective services. For instance, in tree planting operations, standards may specify planting hole dimensions, staking methods, and other post planting actions.

Community engagement and participatory processes

Beyond the governance measures just described, urban foresters have other ways to advance urban forestry goals. These include outreach to and involvement of the public, through means such as literacy campaigns, organization of volunteers, etc. However, community engagement in urban forestry can also be initiated by third party organizations or by the community themselves.

Participatory processes and engagement tools

True community engagement in urban forestry has been characterized as 'a two-way process between experts/managers and the public, and should not be confused with the one-way flow of information in public relations exercises. There are many diverse "publics," and most "stakeholders" hold multiple stakes in any planning process' (Beckley et al., 2006, p. 3). Public participation is important because these processes can lead to better decision-making, by 'providing local or independent sources of information and by examining alternative management strategies; they also build trust, educate and inform all involved, and can reduce long-term delays and uncertainty' (Beckley et al., 2006, p. 3). Meaningful community engagement can confer legitimacy to final outcomes, but it is important to integrate a wide range of voices, interested or affected parties, and public values into the process (Beierle and Cayford, 2002).

Public participation processes are implemented through certain techniques or tools for communication and interaction. These range from limited representation (e.g. small workshops) to broad-scale representation (e.g. random sample surveys engaging thousands of citizens). Techniques can also be classified as direct versus indirect. *Indirect techniques* involve little face-to-face contact, employing tools such as comment sheets, toll-free lines, referenda, and surveys. These allow for larger samples which are representative of the broad public, for statistical analysis. They often do not allow in-depth analysis of the reasons for certain responses, or much two-way flow of information between experts and lay-people.

Direct techniques allow for interaction between the audience and those organizing the engagement, through tools such as advisory committees, focus groups, and citizen juries. These smaller group, face-to-face processes provide stronger opportunities for learning and are more flexible; they can sometimes establish dialogue between individuals with diverse interests, and help reach workable solutions under conditions of complexity and uncertainty. It can, though, be harder to extrapolate to the wider public, unless careful selection of stakeholder groups is undertaken (Sheppard and Meitner, 2005).

Engagement processes vary in their depth, degree of influence on decision-making, and effectiveness. Depending on the intended outcome of the exercise and level of engagement committed to, one of four conceptual types of process can be selected (adapted from Beckley et al., 2006):

1 *Information exchange or directive participation,* where information is communicated mainly in one direction, with little dialogue. Information exchange tools are often indirect, including discussion papers and informational websites. These processes are often seen as 'educating the public'.

2 *Consultation,* where public opinions are sought by decision-makers in their deliberation. Information flows in two directions, via techniques such as public hearings, town hall meetings, surveys, and toll free numbers for callers. These can be a mixture of direct and indirect techniques.

3 *Collaboration,* where representatives of the public are actively involved in directly influencing decisions and developing solutions. This is usually an iterative process, centred on dialogue among people with joint responsibilities. Collaboration tools are mostly direct techniques, including task forces, round tables, public advisory committees, and workshops.

4 *Co-management/control,* where decision-making authority and sometimes responsibility for public engagement lies partly or wholly with the public or their representatives.

Co-management/control approaches are mostly direct, such as community forest boards, co-management teams, and steering committees.

While there are relationships between these levels of engagement and the governance modes described in Table 15.1, with hierarchical governance arrangement often pursuing more indirect techniques for information exchange or consultation, in fact any of these techniques can be employed with any governance model. A suite of such tools is usually necessary to implement a public involvement programme successfully.

Where the intent of community engagement is to motivate behaviour change or engage citizens in action, research shows that it is vital to get beyond 'education' and simple consultation methods, because information alone is often ineffective (Kollmuss and Agyeman, 2002). More engaging and interactive methods are necessary that allow for experiential learning and relate to peoples' values and emotions (Sheppard, 2012). Emerging techniques such as resident-employed photography of valued forest resources (Baldwin and Chandler, 2010), community-based GIS (Kenney and Puric-Mladenovic, undated), and landscape visualization (Sheppard, 2012) can be very effective in eliciting meaningful public comment (see Figure 15.3 below).

The role of volunteers in urban forestry

Much of the general literature on governance assumes a somewhat hierarchical model of government policy, decision-making, and community consultation. Urban forestry, however, features another major channel of public involvement through *volunteerism*, where those who are engaged carry out management activities in the public realm, work that would otherwise be done by government staff or contractors (Molin, 2014; Miller et al., 2015). This is rare in other fields such as urban design, building management or highway maintenance. It is common for urban foresters to organize volunteer teams and for NGOs to take responsibility for tasks such as control of invasive species, planting, or watering street trees. These arrangements might be considered a special kind of collaborative engagement within a system of hierarchical governance.

Benefits from volunteerism in urban forestry include better tree health and structure from cultural practices carried out by volunteers and increased biodiversity from planting native trees and shrubs. Volunteer activities can result in cost savings in programme delivery, especially in labour-intensive operations such as weeding and mulching, though offsetting costs of training and coordination need to be considered. Hauer and Peterson (2016) estimated that Americans annually volunteered almost 1.5 million hours on municipal tree activities, worth approximately $35 million and amounting to 4.8 per cent of the total time required for tree care in a community. There are also less tangible but significant benefits from working with volunteers, such as improved relationships with local people and increased advocacy for urban forestry from partners, which can lead to increased programme recognition and funding. Volunteers who serve on tree boards or committees sometimes use their volunteer experience as a platform for launching political careers, where they are in a position to strongly influence urban forestry programmes in their communities. Neighbourhood-based volunteerism in urban forestry can also lead to increased community pride, which then positively affects the quality of life for residents (Wolf and Rozance, 2013). The volunteers themselves may benefit from enhancing their knowledge and career skills, and making new social connections.

In urban forestry volunteer programmes organized by local government, roles (in terms of power relationships and capabilities) need to be carefully planned. It is important to match the skills, knowledge and motivation of the volunteers to available activities. Major operations requiring large equipment need skilled operators, but activities such as planting bare-root trees, establishing tree wells, and watering trees can be carried out by volunteers with minimal training. Volunteer activities requiring training and auditing by trained arborists can include outreach to local residents, tree inventories, and inspecting trees for pests. Considerable effort also needs to be put into recognition, regular feedback, and sharing of information, in order to recruit and retain volunteers.

It should be noted that volunteers can be a key part of other governance models, including co-governance, co-management, and even self-governance. For example, UK parish councils are almost completely staffed by volunteers who play a major decision-making role at local scales on issues not covered by higher levels of government, e.g. managing churchyards, maintaining recreation areas, and tree planting. Volunteers may have high levels of skill, strong local knowledge and traditional cultural methods of managing vegetation, as with First Nation communities in peri-urban forest areas of Canada. Increasingly, there are moves to involve much larger numbers of citizens in urban forest monitoring and management, through crowd-sourcing with mobile phones. Precedents for such grass-roots activities at neighbourhood or block scales (which historically have often been neglected in planning; see Table 15.2), include: systematic mapping of urban trees ('Neighbourwoods') in Ontario (Kenney and Puric-Mladenovic, undated); and citizen toolkits on urban forestry and climate change, using fun exercises, community mapping, and do-it-yourself visualization of future conditions (Sheppard, 2015).

Future perspectives on urban forest governance

This chapter has reviewed both traditional and emerging approaches to governance and participation in urban forestry, across public and private realms. In the context of emerging and projected trends, challenges and opportunities, this final section draws together threads from the above discussions and text-box case studies, to suggest *new directions for urban forestry governance and engagement*.

Many local governments, especially in developed counties, are dealing with urban forests in a fairly systematic manner, often along traditional lines of hierarchical 'top-down' and expert

---

Box 15.3 New community engagement techniques for London tree officers

Urban foresters often feel that traditional communication with the public is time consuming and inefficient, and that the public has a poor understanding of critical issues such as the health benefits of trees or the need for climate change adaptation. Tree Officers in some boroughs have set up 'citizen science' initiatives, using volunteers to develop an inventory with i-Tree London or the Treezilla online platform. Success with these techniques varies, but established programmes report multiple benefits including reduced vandalism and improved establishment rates for newly planted trees (e.g. the Borough of Hackney has brought down its tree failure rate to 1%). Many interviewed Tree Officers expressed interest in the use of credible visualization to engage communities prior to new planting schemes or management actions, though were sceptical of inaccurate visualizations used by developers to promote new development. Very few of the practitioners surveyed produce or use visualizations themselves, but those who had used visualizations in the past reported a high level of satisfaction (see Figure 15.3).

Figure 15.3 Visualizing tree growth. Sequence of visualizations depicting the projected change in street-tree replacement and growth following the removal of a tree in poor health in Buckingham Palace Road (London, UK): (i) existing street trees; (ii) critical tree is felled; (iii) felled tree is replaced; (iv) younger tree still evident after 10 years; (v) all trees in the tree line reach similar height at maturity after 25 years.

Source: Macias (2016)

driven management and decision-making. However, more inclusive modes of governance are becoming more common: 'public participation has emerged as a key component of forest management and policy decision-making … Forest managers are now faced with enacting a transition … to a more inclusive and socially responsive model of decision-making' (Beckley et al., 2006, p. 3).

At present, the state of urban forestry governance appears to be highly variable within and between nations, often with inconsistent policies and practices among municipalities within a given region or state, e.g. large versus small cities. The scale and effectiveness of volunteerism and citizen inclusion in urban forestry planning and management also appears to vary substantially. One common governance trend in growing cities is the failure of existing democratic governance mechanisms to limit development pressures which are steadily reducing canopy cover on private land. This loss of critical urban forest values, and the rarity of large scale interventions (outside of China) to restore green infrastructure networks and urban ecosystems, means that many cities are increasingly vulnerable to global and local change.

However, we can identify various approaches for making urban forestry governance and participatory processes more effective in an uncertain future, by making them more inclusive, developing multiple partnerships beyond a single 'silo', applying powerful new engagement methods, and improving the capacity of urban foresters. Accordingly, the following governance and engagement strategies should be considered:

1 To *address the impacts on society* of urbanization/density, urban heat islands, climate change, and loss of restorative green space:

   a Raise the overall priority given by governments and business to green spaces and trees on public and private land, in order to safeguard and where possible enhance citizens' happiness, health, property values, and social cohesion.

   b Develop and enforce stronger policies to limit further development in larger denser cities with low canopy cover, and divert further urban growth to smaller, less densely developed centres.

   c Conduct visioning studies with long-term projections of various future scenarios affecting urban forests, with identification of vulnerabilities, adaptation response options and mitigation solutions.

d Implement key adaptation measures such as increasing canopy levels to offset warming trends, enhancing stormwater and green infrastructure networks to reduce flooding impacts and store groundwater, and diversifying tree species and greenspace types.

e Implement mitigation measures such as establishing active transportation 'green' corridors, promoting cooling of buildings by trees, increasing carbon sequestration, use of local bioenergy resources from vegetation management, etc.

f Strengthen coordination of urban forestry efforts at the regional level across municipal boundaries.

Without such measures, decision-makers could be accused of failing to exercise due diligence in the face of evident threats, and pay a heavy political price. If such measures were more systematically adopted, urban forestry could help deliver sustainable urban development and promote cities' image and competitiveness.

2 To *address risks to the health and viability of the urban forest itself*, particularly when management/maintenance budgets are shrinking or inadequate:

a Develop, adopt and disseminate regional guidelines for adaptation/diversification of urban forests and urban forestry practices (e.g. Diamond Head Consulting, 2016).

b Build public literacy about urban forestry and promote the responsibility of all citizens to protect local trees, both in gardens and the public realm.

c Scale-up social mobilization and shared responsibilities of citizen volunteers through community-led action (Sheppard et al., 2015) at the neighbourhood and city block level, to improve care for the urban forest and offset management costs.

d Explore cross-jurisdictional methods to compensate urban forestry operations from healthcare cost savings derived from the urban forest (Hotte et al., 2015).

3 To meet demand from the public and stakeholders for *more inclusion in decision-making and design*, and to avoid embarrassing public conflicts:

a Improve practitioners' skills in engaging the community, through social media, training programmes, and revamped degree programme curricula.

b Promote engagement through citizen science and online interactive tools to encourage learning and collaboration on monitoring the urban forest.

4 To *support the professionals who protect, manage, and design urban forests* to meet governmental and public expectations and deliver needed ecosystem services:

a Develop leadership and coordination roles for urban foresters, to help them advocate for urban forestry land allocation, influence decision-making, and access resources. This calls for university curriculum development, broader training, and leadership certification programmes.

2 Promote a higher profile for urban foresters and an equal footing with other disciplines involved in urban planning and management. Establish requirements for early involvement of urban foresters in strategic planning and design of infrastructure.

The above strategies represent some new priorities for governance of urban forests. There is an increased urgency for proactive canopy establishment and other resilience measures for both people and trees. Those who govern need the political will and ability to be out in front of the public sometimes on these complex and serious issues, while working hard to bring the public along with them. The public and elected officials seem less aware of accelerating challenges for urban forests than is the academic and practitioner community. Governments and third parties need to build capacity among citizens, staff, and politicians, for dealing with the uncertain future of the urban forest. There is a growing need for the promotion and dissemination of successful examples of volunteer and co-management schemes. Such moves are likely to affect how best practices are spread, and how urban forestry career paths develop in the future.

References

Arnouts, R., van der Zouwen, M., Arts, B. (2012) 'Analysing governance modes and shifts: Governance arrangements in Dutch nature policy', *Forest Policy and Economics*, vol. 16, pp. 43–50.

Baldwin, C., Chandler, L. (2010) 'At the water's edge: Community voices on climate change', *Local Environment*, vol. 15, no. 7, pp. 637–649.

Barron, S. (2016) 'Happy people, healthy forests: A framework for developing future (sub)urban forest scenarios', presentation, Urban Tree Diversity 2, Melbourne, Australia, 22 February.

Beckley, T., Parkin, J. R., Sheppard, S. R. J. (2006) *Public Participation in Sustainable Forest Management: A Reference Guide*, Sustainable Forest Management Network, Edmonton, AL.

Beierle, T. C., Cayford, J. (2002) *Democracy in Practice: Public Participation in Environmental Decisions*, RFF Press, Washington, DC.

Buizer, M., Elands, B., Mattijssen, T., van der Jagt, A., Ambrose, B., Gerőházi, E., Møller, M. S. (2015) 'The governance of urban green spaces in selected EU-cities', Deliverable 6.1, EU 7th Framework project Green Surge, retrieved from http://greensurge.eu/products/planning-governance (accessed 21 July 2016).

City of Surrey (2000) *Guide to Policy Development*, City of Surrey, Surrey, BC.

City of Surrey (2015) *Anniedale–Tynehead Neighbourhood Concept Plan, 2015*, City of Surrey, Surrey, BC.

Diamond Head Consulting. (2016) *Urban Forest Climate Adaptation Framework for Metro Vancouver Tree Species Selection, Planting and Management*, Prepared for Metro Vancouver, Vancouver, BC.

Dunster, J. S., Smiley, E. T., Matheny, N., Lilly, S. (2013) *Tree Risk Assessment Manual*, International Society of Arboriculture, Champaign, IL.

Gulsrud, N. (2015) 'The role of green space in city branding: An urban governance perspective', PhD dissertation, Department of Geosciences and Natural Resource Management, University of Copenhagen, Frederiksberg.

Haaland, C., Konijnendijk van den Bosch, C. (2015) 'Challenges and strategies for urban green space planning in cities undergoing densification: A review', *Urban Forestry and Urban Greening*, vol. 14, no. 4, pp. 760–771.

Hajer, M., Versteeg, W. (2005) 'A decade of discourse analysis of environmental politics: Achievements, challenges, perspectives', *Journal of Environmental Policy and Planning*, vol. 7, no. 3, pp. 175–184.

Hauer, R. J., Peterson W. (2016) *Municipal Tree Care and Management in the United States: A 2014 Urban and Community Forestry Census of Tree Activities*, Special Publication 16-1, College of Natural Resources, University of Wisconsin–Stevens Point.

Hotte, N., Nesbitt, L., Barron, S., Cowan, J., Cheng, Z. C., Sheppard, S. R. J., Neuvonnen, J. (2015). *The Social and Economic Values of Canada's Urban Forests: A National Synthesis*, prepared for Canadian Forest Service, Ottawa, Canada.

Kenney, W. A., Puric-Mladenovic, D. (undated) *Neighbourwoods: Community-Based Urban Forest Stewardship*, University of Toronto, Toronto, retrieved from http://www.forestry.utoronto.ca/neighbourwoods/web (accessed 21 July 2016).

Kjær, A. M. (2004) *Governance*, Polity Press, Cambridge.

Kollmuss, A., Agyeman, J. (2002) 'Mind the gap: Why do people act environmentally and what are the barriers to pro-environmental behavior', *Environmental Education Research*, vol. 8, no. 3, pp. 239–260.

Konijnendijk, C. C. (2003) 'A decade of urban forestry in Europe', *Forest Policy and Economics*, vol. 5, no. 3, pp. 173–186.

Konijnendijk van den Bosch, C. (2014) 'From government to governance: Contribution to the political ecology of urban forestry', in L. A. Sandberg, A. Bardekjian, A. and S. Butt (eds), *Urban Forests, Trees and Greenspace: A Political Ecology Perspective*, Routledge, New York, pp. 35–46.

Krajter Ostoic, S. (2013) 'Analysis of current urban forest governance in the city of Zagreb', PhD dissertation, Faculty of Forestry, University of Zagreb.

Lawrence, A., De Vreese, R., Johnston, M., Sanesi, G., Konijnendijk van den Bosch, C. C. (2013) 'Urban forest governance: Towards a framework for comparing approaches', *Urban Forestry and Urban Greening*, vol. 12, no. 4, pp. 464–473.

Liefferink, D. (2006) 'The dynamics of policy arrangements: Turning round the tetrahedron', in B. Arts, P. Leroy (eds), *Institutional Dynamics in Environmental Governance*, Springer, Dordrecht, pp. 45–68.

Macias, A. (2016) 'The role of visualizations in urban forestry: Conclusions from managers' perspectives', PhD dissertation, Universidad Politecnica de Madrid.

Miller, R. W., Hauer R. J., Werner L. P. (2015) *Urban Forestry Planning and Managing Urban Greenspaces*, 3rd edn, Waveland Press, Long Grove, IL.

Molin, J. F. (2014) 'Parks, people and places: Place-based governance in urban green space maintenance', PhD thesis, Department of Geosciences and Natural Resource Management, University of Copenhagen, Copenhagen.

Schwab, J. A. (2009) (ed.) *Planning the Urban Forest: Ecology, Economy, and Community Development*, American Planning Association, Chicago, IL.

Sheppard, S. R. J. (2012) *Visualizing Climate Change: A Guide to Visual Communication of Climate Change and Developing Local Solutions*, Earthscan/Routledge, Abingdon.

Sheppard, S. R. J. (2015) 'Making climate change visible: A critical role for landscape professionals', *Landscape and Urban Planning*, Special Issue: Critical Approaches to Landscape Visualization, vol. 142, pp. 95–105.

Sheppard, S. R. J., Meitner, M. J. (2005) 'Using multi-criteria analysis and visualization for sustainable forest management planning with stakeholder groups', *Forest Ecology and Management*, vol. 207, no. 1–2, pp. 171–187.

Sheppard, S. R. J., Mathew Iype, D., Cote, S., Salter, J. (2015) *Special Report – A Synthesis of PICS-Funded Social Mobilization Research*, prepared for Pacific Institute for Climate Solutions (PICS). Climate Change: Impacts and Responses Conference, April 2015, Victoria, BC, retrieved from http://pics.uvic.ca/synthesis-pics-funded-social-mobilization-research (accessed 23 July 2016).

Van Tatenhove, J. P. M., Arts, B., Leroy, P. (2000) *Political Modernization and the Environment: The Renewal of Environmental Policy Arrangements*, Kluwer Academic Publishers, Dordrecht.

Wolf, K. L., Rozance, M. A. (2013) 'Social strengths: A literature review', College of the Environment, University of Washington, retrieved from http://depts.washington.edu/hhwb/Thm_Community.html.

PART IV

Trees in the urban environment

16

URBAN TREE PHYSIOLOGY

Methods and tools

Carlo Calfapietra, Gabriele Guidolotti,
Galina Churkina and Rüdiger Grote

The interactions between urban environment and trees

Although the physiological processes in urban trees are the same as in trees growing in forests, the ambient conditions under which these processes occur in cities are more variable and extreme. Environmental conditions in urban areas are tremendously diverse because of differences in climate, building density and structure, building material, and fraction of green and blue space. Generally, urban areas are warmer than their surroundings by up to 1–3°C because the dominating stone structures absorb radiation more than vegetated surfaces and lack the cooling effect from evapotranspiration (Arnfield, 2003). This phenomenon is known as the urban heat island (UHI) effect, which is usually larger at night than during the day and in winter than in summer (Gaffin et al., 2008). In addition, UHI can alter regional precipitation and increase the frequency of severe weather events (Changon, 2001). Water and nutrient availability might be decreased compared to forests or pastures due to increased runoff (on impervious surfaces such as roofs, streets, sidewalks, parking lots, etc.) and soil compaction (during construction, by vehicles and people). More specifically, soil compaction can lead to aeration problems, as well as increased mechanical damage. In addition, urban forests are often grown on former rubble or garbage dumps and thus might be exposed to soil contaminants (Craul, 1991). Furthermore, leaves are often removed especially in parks and residential yards altering the natural cycling of litter and nutrients between plants and soils (Kjelgren and Clark, 1992), whereas to compensate for low soil fertility, the vegetation is sometimes fertilized.

Urban trees are generally subjected to lower light, due to the frequent occurrence of urban haze, which is formed from the primary and secondary aerosols and particles emitted from traffic, households and industry (Huang et al., 2014). This also increases diffuse radiation compared to direct radiation, which increases the carbon uptake efficiency of plants (Mercado et al., 2009). In addition, vegetation close to high building structures such as is often the case for street trees and gardens, are directly shaded and thus may suffer from a diminished light availability (Kjelgren and Clark, 1992). The light environment affects developmental, morphological and physiological attributes of individual leaves such as leaf thickness, nitrogen per unit area, chlorophyll per unit area, and net photosynthesis (Herrick and Thomas, 1999).

Soil moisture, air temperature, leaf temperature, relative humidity, and vapor pressure deficit are often less favorable for street trees than for their rural counterparts, resulting in slower growth, lower density root systems, significant terminal growth differences, and earlier tree senescence (Close et al., 1996). In contrast, trees in gardens and parks are often irrigated and fertilized which benefits physiology and growth at those places. Irrigation plays a major role in maintaining vegetation in cities located in semi-arid and desert climate zones, and the intensive irrigation with subsequent change of land use types and evapotranspiration can even lead to a 'cool island effect' (Frey et al., 2007). Another environmental feature for urban areas is a specific air chemical composition due to emissions from traffic, households and industry, which results in higher CO_2 concentration and more air pollution. Apart from this, wind speed is also mostly different compared to natural environments but depends very much on regional specifications. For example wind speed is higher due to channeling in urban street canyons but lower in wind shielded areas (Arnfield, 2003).

Not only the specific urban climatic environment is significantly different, trees that are growing along streets are subjected to a microenvironment characterized by higher pollution levels due to traffic emission, additional soil contamination by input of heavy metals or salts (Ugolini et al., 2013), as well as a restricted area for root extension which in turn decreases water availability (Craul, 1991).

The study of the specific impacts of these environmental conditions on urban trees, which is also thoroughly discussed in Chapter 4 of this book and summarized in Figure 16.1 contributes to understanding the mitigation capacity of plants regarding air pollution or carbon uptake, heat or noise reduction, because trees are not only affected by the urban environment, but in turn the environment is modified by the abundance of trees (Georgi and Zafiriadis, 2006). A particular target of this research is thus to evaluate the importance of trees for crucial ecosystem services and the risk that this performance might be compromised if urban conditions get too stressful (Calfapietra et al., 2015).

One of the main services of urban vegetation is the mitigation of the UHI effect. In particular cooling by evapotranspiration is much larger from vegetation than from sealed surfaces (Gaffin et al., 2012). In most cases canopies will also reduce gaseous and particle pollution by increased deposition due to their large surfaces and comb-like structures (Pugh

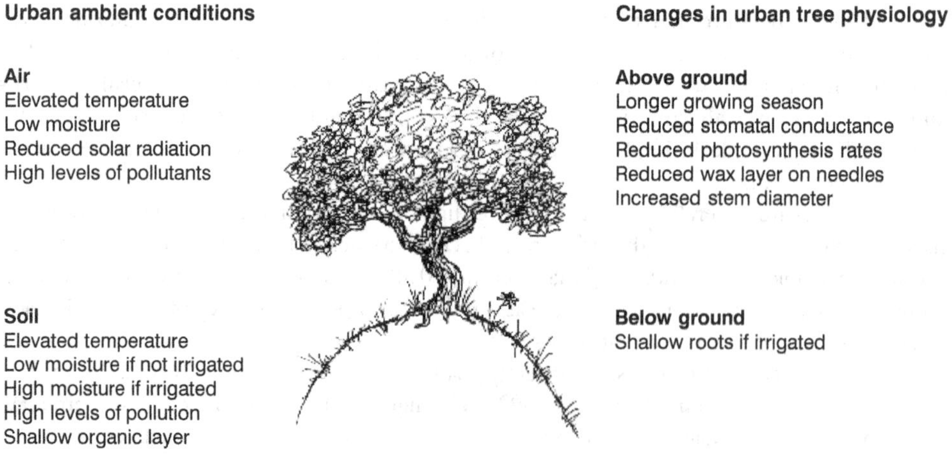

Urban ambient conditions

Changes in urban tree physiology

Air
Elevated temperature
Low moisture
Reduced solar radiation
High levels of pollutants

Above ground
Longer growing season
Reduced stomatal conductance
Reduced photosynthesis rates
Reduced wax layer on needles
Increased stem diameter

Soil
Elevated temperature
Low moisture if not irrigated
High moisture if irrigated
High levels of pollution
Shallow organic layer

Below ground
Shallow roots if irrigated

Figure 16.1 Urban ambient conditions and possible changes in tree physiology

et al., 2012). However, the same trees also reduce the exchange between the (sometimes heavily polluted) air close to the ground and higher atmospheric layers, thus increasing the risk of air pollution damages (Gromke and Blocken, 2015). In addition, some tree species are strong emitters of volatile organic compounds (VOCs), which are highly reactive and contribute to photochemistry, eventually increasing ozone concentrations. Understanding to which degree urban trees contribute to people's wellbeing, energy saving and air pollution mitigation thus recalls the consideration of a dynamic interaction that accounts for the plants being not only passively exposed to the stressful urban conditions but are also able to influence them (Niinemets and Peñuelas, 2008).

Given that the physiological processes of trees in urban forests, parks, and along the streets are the same as of trees growing in natural forests, forest or land ecosystem models may be applicable to urban trees and forests too. These models are driven by climatic variables such air temperature, air humidity, precipitation, and solar radiation and can provide estimates of urban tree-related processes (carbon- nitrogen- and water-cycling) and their responses to changing environmental conditions. Several studies suggest that the specific urban settings impact tree growth, allocation, and phenology, and indicate that allometric relationships developed for trees in traditional forest settings may misrepresent urban tree biomass (McHale et al., 2009). Since influential environmental variables can fluctuate from one city to another, actual allometric relationships may also change among cities. General models that do not account for these specific settings (e.g. fraction of impervious surface area, building shadow) may not be applicable for simulation of urban tree patches at fine spatial scale. Therefore, models have been developed specifically for urban forests or urban conditions (including buildings and different vegetation types, e.g. Envi-MET). One of the most widely used models of this kind is the Urban FORest Effects model (UFORE) nowadays known as i-Tree Eco (Nowak et al., 2006). This model estimates various forest variables ranging from tree species diversity and stem diameter to filtering effects of urban forests on air pollution. When used for a new city, this model requires extensive field data collection for initialization. However, not all interactions of urban forests with urban pollutants can yet be considered in such models. For example, recent studies point to complex interactions between urban trees and airborne pollutants that require the coupling of atmospheric chemistry models and urban ecosystem models (Guidolotti et al., 2016). Only if this is achieved, a better understanding and forecast of feedbacks between urban ambient conditions, air pollution, and urban forests will be possible.

Setting experimental boundary conditions for eco-physiological investigations

During recent decades, several manipulative experiments have investigated the effects of abiotic factors on urban trees either under laboratory conditions or in the field (e.g. Wittig et al., 2007). In contrast to experiments with crops and herbaceous plants, trees generally require more expensive facilities to vary and control their environments. Therefore, many experiments are limited to few hours, days or weeks that can provide important insights regarding some processes and mechanisms but do not allow investigating the acclimation processes that often occur in several months or years. It should also be noted that experiments can be biased by light conditions and nutrient supply that may be of a different quality compared to field conditions, or by a constrained root system, especially when plants are grown in pots.

While in manipulative experiments in the laboratory all environmental factors can be controlled, field experiments only allow to modify one or very few factors. This is a particular weakness for the investigation of developments under urban conditions since important

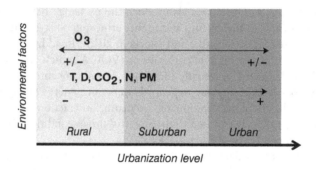

Figure 16.2 The urban–rural gradient represents a unique 'open lab' for concurrently studying the adaptation of plants to changing conditions and the effect of this adaptation on the mitigative capacity of the environmental conditions. '+' indicates that the levels or concentrations of respective environmental factor increase in the direction of the arrow. '-' refers to decreasing levels or concentrations of the corresponding factor in the direction of the arrow. '+/-' indicate that either increase or decrease is possible under suitable conditions (O_3, ozone; T, temperature; CO_2, carbon dioxide; N, nitrogen; D, drought; PM, particulate matter)

factors are manifold (including anthropogenic influences) and are often interrelated (e.g. microclimate and air chemistry). On the other hand, field experiments are needed because studying each factor separately does not assess the combined impact of simultaneous interactions. Therefore, a better understanding of the effects of urban environment on urban trees might be obtained by large-scale experiments in which it is possible to control many factors simultaneously. Experiments at smaller scales might be particularly useful to identify specific responses and for parameterizing mathematical descriptions of specific mechanisms.

Investigations under urban environmental conditions may represent an opportunity to study long-term plant responses to environmental factors and thus to focus on adaptation mechanisms of plants growing permanently in a manipulated plot. In other words, urban conditions that change with anthropogenic impacts represent an 'open laboratory' where environmental manipulations are carried out at no costs (Calfapietra et al., 2015). This mainly applies, but not only, to stressful conditions such as heat waves, droughts, or high pollutants concentrations (either air, water or soil pollutants). Since urban atmospheric and climatic conditions vary considerably from the borders to the centers of cities, environmental gradients, for example regarding to temperature, CO_2 or air pollutants such as O_3, can be evaluated (Figure 16.2) (Calfapietra et al., 2015). The drawbacks in these investigations are that firstly, the levels of the target factor in each study area are not constant over time and secondly, a large number of secondary factors may vary, which complicates the evaluation of dose-response relationships. Nevertheless, urban environmental studies may represent exceptional opportunities to investigate tree responses under predicted climate change conditions.

Tools and methods to investigate responses of urban trees

A large amount of literature is available about plant physiological responses to environmental stresses (mainly CO_2, temperature, water shortage, soil salinity, air and soil pollutants) (Wittig et al., 2007; Bussotti, 2008). While investigations about individual impacts mostly gain straightforward results, the combination of various factors may result in surprising physiological responses. Another issue for experimental design is that responses may depend on the temporal and spatial scale of investigation, reaching from short-term biochemical and physiological investigations over seasonal leaf scale measurements to decadal observations at the canopy/forest scale. Consequently, laboratory techniques, field observations with continuous data recording, as well as discontinuous observations that are repeated over long periods need to be applied. In some cases, these have to be specifically adapted for urban conditions that are characterized by a close vicinity to various kinds of other activities.

Most of the techniques presented in the next paragraphs are applied at a leaf, branch or plant scale and provide only limited temporal and spatial information about the urban trees functionality in terms of photosynthesis, evapotranspiration, pollutants uptake and deposition, and more general in terms of tree growth, development and stress evaluation. To understand and analyze the functional responses of the entire urban forest community in relation to the urban environment it is necessary to scale up to ecosystem level. This can be based on the results from physiological measurements using appropriate models (not accounted for in this overview) or combining techniques from different scales such as the micrometeorology and/or remote sensing (Figure 16.3).

Photosynthesis and evapotranspiration

Small-scale CO_2 assimilation and H_2O transpiration can be assessed by means of chamber measurement techniques which provide net exchange rates in real time. The technique is based on the enclosure of a (small) plant or patch of soil, branch, leaf or a portion of a leaf into a chamber that can be either transparent or opaque (provided of artificial illumination) as in

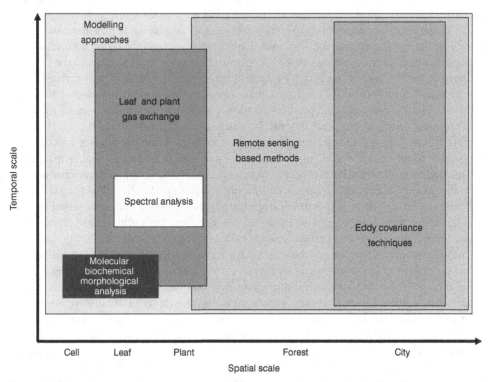

Figure 16.3 Combination of temporal and spatial scales to determine the level of investigation and experimental approaches to be carried out in order to study urban tree physiology

most of the current commercial systems (e.g. LI-6400, Li-Cor, Lincoln, NE, USA). Inside the chamber, air circulation is controlled by an air supply unit with a precision flow-meter, and the CO_2 and H_2O concentrations are measured by Infrared Gas Analysers (IRGA). Depending on the recirculation of the air inside the chamber, these can be differentiated in 'closed systems' and 'open systems'. In 'closed systems', air is recirculated inside the chamber in a closed loop and the change of CO_2 and H_2O concentrations are measured within a short time interval. The more common system is the 'open system', where an air flow passes through the chamber and the CO_2 and H_2O concentrations are measured in the air entering and leaving the chamber, with the difference indicating the net gas exchange. Most commercial systems are equipped with devices that allow the control of light, temperature, humidity, and CO_2 which avoids artefacts caused by the chamber setup itself (e.g. increasing temperature due to a radiation trapped in the chamber). These systems provide CO_2 assimilation (A), transpiration (E), leaf conductance (gs) and intercellular CO_2 concentration (Ci).

In addition, gas exchange measurements can provide important information about Rubisco activity (V_{cmax}) and maximum rate of electron transport for regenerating of Ribulose-bisphosphate (J_{max}) which are both important to determine the limiting conditions for photosynthesis. As described by Long and Bernacchi (2003), the response curve can be divided into three sections: a first phase with Ci below the ambient conditions and a high dA/dCi where V_{cmax} is the limiting factor, a second phase where Ci increases while dA/dCi is close to zero which indicates a limitation by J_{max}, and a final phase that occurs at further increase of Ci while dA/dCi drops below zero during which the limiting factor is the rate of triose-phosphate utilization (V_{TPU}). A mathematical fit of the three phases allows the determination of V_{cmax}, J_{max} and V_{TPU}. The gas exchange technique is the simplest method to estimate changes of photorespiration, the process that releases CO_2 due to the oxygenation of O_2 by Rubisco. To obtain photorespiration rates, CO_2 exchange is measured switching photorespiratory (supplying the leaf with a mixture of air consisting of 20% of O_2) to non-photorespiratory conditions (the O_2 concentration is lowered to 2% and the air flow is compensated by increasing N_2). The specificity of Rubisco for CO_2 and carboxylation can be reduced in favor of O_2 and oxygenation with a consequent increase of the CO_2 released. The measurement of photorespiration can be particularly important in urban areas because extreme temperatures can be easily reached considering the UHI effect. Differences in canopy temperature are furthermore modified by species-specific leaf properties, water use strategies, and location. For example considerably higher maximum foliage temperatures have been found at typical street trees such as plane trees and maples than at oaks and beeches that are more commonly grown in parks and urban forests (Meier and Scherer, 2012). Also, temperatures are getting higher with the degree of sealed ground area, leading street trees to grow in a warmer environment than other urban trees where higher evaporation leads to a certain amount of cooling (Leuzinger et al., 2010).

Another small-scale technique that can be applied for more detailed process-knowledge about the activity of the photosynthesis apparatus, in particular the light harvesting system, is the measurement of chlorophyll fluorescence based on the re-emission of light energy at longer wave lengths. The light energy is absorbed by the chlorophyll and can have three fates: used in the photochemistry of photosynthesis, dissipated as heat at longer wave lengths, or as light-chlorophyll fluorescence. Because the three processes are in competition, measuring of chlorophyll fluorescence provides information about the efficiency of photochemistry (Maxwell and Johnson 2000). The measurement of chlorophyll fluorescence allows the estimation of several photochemical (e.g. quantum yield and maximum quantum yield of PSII) and non-photochemical quenching parameters. Maxwell and Johnson (2000) describe the procedure to obtain these parameters: a fluorometer is positioned above the leaf with minimum light, providing the minimum level of fluorescence (F_o). Afterwards a flash of saturating light is applied, obtaining the measurement of F_m in the dark-adapted state. After this, actinic light is provided to the leaf, and at different intervals flashes of saturating light are applied obtaining the fluorescence maximum in the light (F'_m). F_t and F_o represent the values of fluorescence immediately before and after the flash, removing actinic light, respectively. In addition to these classical fluorometers, there are modulated measuring systems (switching on and off at high frequencies) that allow continuous field measurements in presence of full sunlight (Maxwell and Johnson, 2000).

In addition, the maximum quantum yield can be obtained by means of light response curves that describe the net CO_2 assimilation by a plant leaf as function of photosynthetic photon flux density, generally increasing from no light to saturating light. In a light response curve several phases can be distinguished: in absence of light dark respiration is obtained, and the compensation point represents the light level at which photosynthesis equals respiration. At low light there is a linear increase of assimilation with irradiance, the slope of this linear increase representing maximum quantum yield. At higher irradiance levels, the assimilation reaches a light-saturated plateau (Ögren and Evans, 1993).

This approach can also be applied to understand the effect of abiotic stressors and is thus particularly useful in the urban environment. For example, it has been shown that assimilation is generally increased by urban CO_2 concentrations, which in turn depend on the intensity of anthropogenic emissions, turbulent transport efficiency as well as CO_2 removal activity (Gratani and Varone, 2014).

Plant water use on the individual level can be quantified by means of sap-flow measurements. The sap flows are typically obtained using heat as tracer of the sap movement. The two methods commonly used are the 'stem heat balance' and the 'trunk sector heat balance'. The first method performs continuous heating on the entire circumference of the stem, and the heat balance from which the mass flow is calculated is solved for the amount of heat taken up by the sap stream. This method is usually used for small trees only. With the heat-pulse method, short pulses of heating are applied into the sapwood, and the mass flow of sap is determined measuring the velocity of the heat pulse along the stems (Smith and Allen, 1996).

Larger spatial scales are targeted using the eddy covariance technique (EC), which allows direct measurement of various trace gas fluxes between ecosystems and the atmosphere (Foken et al., 2012). The EC is basically a micro-meteorological technique that accounts for the trace gases masses transported by turbulent upward and downward movements of the air (eddies) without altering the surrounding environment. The basic technical requirements of EC are: a sonic anemometer to measure vertical wind speed and temperature variations and a scalar sensor for density variations (gas or mass analyzer). The high frequencies (>10 Hz) of those measurements ensure registration of all type of eddies (Foken et al., 2012). Although the heterogenic surface of urban landscapes is not ideal for the application of EC, some attempts with promising results have been done worldwide in urban green contexts, quantifying the CO_2 exchange rates of the urban ecosystems also in response to the environmental perturbations (e.g. Ward et al., 2015).

Air pollution deposition and emission of biogenic volatile organic compounds

Small-scale estimates of the deposition of particulate matter (PM) are normally determined by leaf sampling and spectroscopy of the leaf ashes. Uptake and deposition on the surface can be distinguished by washing the foliage prior to spectroscopy (Ugolini et al., 2013). With regard to gaseous deposition, chamber measurements can be applied to estimate small scale fluxes of O_3, SO_2, NO_X or biogenic volatile organic compounds (BVOCs) into or out of the stomata (Calfapietra et al., 2016). In order to evaluate not only the quantity but also the influences that are responsible for the observed fluxes, gas exchange techniques are often used in combination with other measurements. For instance the estimation of stomatal conductance can be integrated with measurements of leaf water potential (Ψ): the sum of turgor pressure, osmotic and matric potentials. This can easily be measured with a pressure chamber such as the Scholander-Hammel pressure bomb, which is probably the most widely used method: a leaf is excised, inserted inside a pressure chamber leaving the petiole out of it, the pressure inside the chamber is gradually increased until the water appears in the cut area of the petiole. This pressure represents the force which with water is bound to leaf cells and varies from 0 to negative values (Tyree, 1997).

Chamber measurements are also used for the determination of BVOC emissions in combination with sampling air probes in order to define the specific compounds later on in the laboratory by gas chromatography (see Kim et al., (2013) for a closer discussion on various techniques). If only the overall impact on air chemistry is of interest, these probes can also be analyzed for their reactivity, which might be a relatively simple way to judge vegetation impacts on air quality in the future (Matsumoto, 2014). In fact, it has been found that quantity as well as composition of BVOC emission varies widely with species and environmental conditions. In particular, temperature, radiation, CO_2 concentration and air pollutants seem to be important determinants in urban surroundings that are not yet sufficiently characterized.

On the ecosystem scale, EC methods also have a significant role in determining deposition and emission. Regarding deposition, condensational particle counter are used for particles greater than 10 nm in diameter. If particle size distributions are known, deposition models can then be applied to estimate size-dependent deposition velocities (Deventer et al., 2015). An alternate method is the 'relaxed eddy accumulation' (REA) that is based on accumulated concentrations of compounds, which are sampled separately for upwards and downwards movements. Here, sensors with slower response time are sufficient. REA technique has been used successfully for measurements of the atmospheric particle fluxes (Gaman et al., 2004). Due to the development of fast detection methods such as proton-transfer-reaction (PTR) time-of-flight (TOF) mass spectrometry (MS), continuous measurements of biogenic volatile organic compounds are now also possible (Park et al., 2013). The removal of air pollutants represents an important ecosystem service of urban trees but at the same time it directly or indirectly affects the primary metabolism of the plant and thus eventually decreases the efficiency of this ecosystem service. Therefore, it is an important research issue in the urban context to determine tree species that are on the one hand efficient in improving air quality and on the other hand resistant against pollution. While the emission of BVOCs is mainly species-specific and thus suitable tree species can be selected relatively easily, deposition depends very much on tree architecture, leaf area and longevity, as well as on leaf morphology (e.g. presence of wax or hairs) (Grote et al., 2016). These issues have to be weighed against air pollution resistance, the provision of other ecosystem services and the general suitability for the specific site conditions.

Growth, development and stress indication

Similarly to other ecosystems, the methodology to determine growth and development in urban trees are:

1   destructive sampling;
2   repeated optical or mechanical measurement; and
3   continuous observation methods.

The difference to the investigation of forest trees, however, is that destructive sampling is restricted by the number of individuals, the value of the individual, legal and safety issues. Also continuous observation methods are difficult to establish since they often require large space for installation or they are susceptible to interferences that are difficult to prevent. Therefore, repeated observations by individuals or remote sensing are the most common choice to investigate long-term developments in urban trees.

Due to the value of urban trees, many of them are regularly monitored by visual observation to judge their vitality and health in order to ensure the safety of pedestrians and vehicles. Besides the traditional field survey, various tools are now available: apart from some remote sensing applications mostly used for defining the location and type of the trees only, it gets increasingly easier to carry out quick assessments of the physiological status of trees using the unmanned aerial vehicles (UAVs). A UAV can be easily equipped with a multi-spectral camera for determination of various indicators such as the chlorophyll content, biomass, or the degree of damages (Kulhavy et al., 2016). UAV observations in combination with new remote sensing options from satellites may represent a powerful biomonitoring tool for the future of our cities.

Since urban areas are generally warmer, temperature related developments are of particular concern for urban foresters. For example local observations as well as remotely sensed data provide extensive evidence for an earlier onset of growing season in vegetation of cities. This has been demonstrated for urbanized areas located in the temperate climate zone (Zhang et al., 2004). It has been shown that on most occasions leaf bud burst was earlier in cities than in rural areas; moreover, higher temperatures in spring and autumn also decrease the risk of frost damage in urban trees and prolong their vegetation period compared with their rural counterparts (Jochner and Menzel, 2015, and references therein).

Monitoring dedicated to drought stress might become an important issue in the future because warmer conditions sometimes imply that water availability is crucial for tree growth. Therefore urban street trees often show a drought-induced reduction of growth – particularly in regions with little precipitation such as the Mediterranean areas (Sanders and Grabosky, 2014). Apart from long-term growth observation and selected investigations of evaporation (see above), also the use of electrical admittance/impedance can help to detect the vitality of an urban tree based on the principle that more vital trees have a higher moisture content, higher concentration of mobile cations and lower concentration of ions (Johnstone et al., 2013). This technique is based on a couple of probes that, once inside the cambium, measure the electrical resistance between them. Other important indicators to monitor stresses at foliar level are to look at leaf morphological traits. For example leaf area, leaf mass per area, leaf thickness and nitrogen content per leaf area unit increased with oxidative stress (Bussotti, 2008). Also, leaf size and stomata density have been reported to decrease under polluted environmental conditions (Bussotti, 2008, and references therein). This particularly applies to soil contaminations such as increased concentrations of salts and heavy metals at roadsides (Ugolini et al., 2013). It should be noted that soil pollution does not only act directly on tree health but also indirectly by damaging the soil microbiota and preventing the formation of mycorrhiza, thus leading to less nutrient provision and uptake capacity (Alzetta et al., 2012).

Implications of urban tree physiology for urban tree selection

In Europe, about 600 years ago, tree planting in cities was already considered as an added value for the citizens, providing beauty and allowing recreation. Since then, the aesthetic function was traditionally the most used criterion to plant trees in the urban context. This aesthetic function had to be established considering limiting conditions such as light, water and nutrient availability, stress from cold or heat, air or soil pollution, diseases or parasites. Over the last decades the lifetime of urban trees dropped due to changes of environmental conditions. Apart from the improvement of site conditions by decreasing pollution and applying appropriate management practices, using stress tolerance as a preferred selection criterion seems to be crucial in the future.

The impact on urban climate and air quality are among the most important ecosystem services that urban trees provide. This has increased the demand for more green areas in cities, particularly in order to mitigate the expected impacts of climate change. To optimize these services by avoiding disservices is the main goal of city managers (Escobedo et al., 2011, and references therein). Therefore, it is of utmost importance to correlate tree-species traits with particular ecosystem services and disservices, as well as with particular environmental requirements in order to optimize the success of new urban green areas (Roloff et al., 2009). Measuring and studying tree physiological performances, morphological and anatomical traits, as well as growth development thus is useful for at least two reasons: first, it presents a unique opportunity to observe and understand the potential impacts of climate change on vegetation since urban conditions today simulate the expected climate conditions of the future; second, it supports strategies to optimize the management of urban green areas in order to maximize ecosystem services provision. Specific information on urban tree selection is reported in Chapter 25 of this volume.

References

Alzetta, C., Scattolin, L., Scopel, C. and Mutto Accordi, S. (2012) The ectomycorrhizal community in urban linden trees and its relationship with soil properties. *Trees-Structure and Function*, vol 26, no 3, pp. 751–767.

Arnfield, A. J. (2003) Two decades of urban climate research: A review of turbulence, exchanges of energy and water, and the urban heat island. *International Journal of Climatology*, vol 23, no 1, pp. 1–26.

Bussotti, F. (2008). Functional leaf traits, plant communities and acclimation processes in relation to oxidative stress in trees: A critical overview. *Global Change Biology*, vol 14, no 11, pp. 2727–2739.

Calfapietra, C., Peñuelas, J. and Niinemets, Ü. (2015) Urban plant physiology: Adaptation-mitigation strategies under permanent stress. *Trends in Plant Science*, vol 20, no 2, pp. 72–75.

Calfapietra, C., Morani, A., Sgrigna, G., Di Giovanni, S., Muzzini, V., Pallozzi, E., Guidolotti, G., Nowak, D. and Fares, S. (2016). Removal of ozone by urban and peri-urban forests: Evidence from laboratory, field, and modeling approaches. *Journal of Environmental Quality*, vol 45, no 1, pp. 224–233.

Changon, S. A. J. (2001) Assessment of historical thunderstorm data for urban effects: the Chicago case. *Climate Change*, vol 49, pp. 161–169.

Close, R. E., Nguyen, P. V. and Kielbaso, J. J. (1996) Urban vs. natural sugar maple growth: I. Stress symptoms and phenology in relation to site characteristics. *Journal of Arboriculture*, vol 22, no 3, pp. 144–150.

Craul, P. J. (1991) Urban soil: Problems and promise. *Arnoldia*, vol 51, no 1, pp. 23–32.

Deventer, M. J., El-Madany, T., Griessbaum, F. and Klemm, O. (2015) One-year measurement of size-resolved particle fluxes in an urban area. *Tellus B*, 67, retrieved from www.tellusb.net/index.php/tellusb/article/view/25531.

Escobedo, F. J., Kroeger, T. and Wagner, J. E. (2011) Urban forests and pollution mitigation: Analyzing ecosystem services and disservices. *Environmental Pollution*, vol 159, no 8–9, pp. 2078–2087.

Foken, T., Aubinet, M. and Leuning, R. (2012) The Eddy covariance method. In M. Aubinet, T. Vesala and D. Papale (eds) *Eddy Covariance: A Practical Guide to Measurement and Data*. Springer Dordrecht, The Netherlands, pp. 1–19.

Frey, C. M., Rigo, G. and Parlow, E. (2007) Urban radiation balance of two coastal cities in a hot and dry environment. *International Journal of Remote Sensing*, vol 28, no 12, pp. 2695–2712.

Gaffin, S. R., Rosenzweig, C. and Kong, A. Y. Y. (2012) Adapting to climate change through urban green infrastructure. *Nature Climate Change*, vol 2, no 10, p 704.

Gaffin, S. R., Rosenzweig, C., Khanbilvardi, R., Parshall, L., Mahani, S., Glickman, H., Goldberg, R., Blake, R., Slosberg, R. and Hillel, D. (2008) Variations in New York city's urban heat island strength over time and space. *Theoretical and Applied Climatology*, vol 94, no 1, pp. 1–11.

Gaman, A., Rannik, P., Aalto, P., Pohja, T., Siivola, E., Kumala, M. and Vesala, T. (2004) Relaxed eddy accumulation system for size-resolved aerosol particle flux measurements. *Journal of Atmospheric and Oceanic Technology*, vol 21, no 6, pp. 933–943.

Georgi, N. J. and Zafiriadis, K. (2006) The impact of park trees on microclimate in urban areas. *Urban Ecosystems*, vol 9, no 3, pp. 195–209.

Gratani, L. and Varone, L. (2014) Atmospheric carbon dioxide concentration variations in Rome: Relationship with traffic level and urban park size. *Urban Ecosystems*, vol 17, no 2, pp. 501–511.

Gromke, C. and Blocken, B. (2015) Influence of avenue-trees on air quality at the urban neighborhood scale. Part II: Traffic pollutant concentrations at pedestrian level. *Environmental Pollution*, vol 196, pp. 176–184.

Grote, R., Samson, R., Alonso, R., Amorim, J. H., Cariñanos P., Churkina G., Fares S., Thiec D. L., Niinemets Ü., Mikkelsen T. N., Paoletti E., Tiwary A. and Calfapietra C. 2016. Functional traits of urban trees in relation to their air pollution mitigation potential: A holistic discussion. *Frontiers in Ecology and the Environment*, vol 14, pp. 543–550.

Guidolotti, G., Salviato, M. and Calfapietra, C. (2016) Comparing estimates of EMEP MSC-W and UFORE models in air pollutant reduction by urban trees. *Environmental Science and Pollution Research*, vol 23, no 19, pp. 19541–19550.

Herrick, J. D. and Thomas, R. B. (1999) Effects of CO_2 enrichment on the photosynthetic light response of sun and shade leaves of canopy sweetgum trees (*Liquidambar styraciflua*) in a forest ecosystem. *Tree Physiology*, vol 19, no 12, pp. 779–786.

Huang, R.-J., Zhang, Y., Bozzetti, C., Ho, K., Cao, J., Han, Y., Daellenbach, K., Slowik J., Platt, S., Canonaco, F., Zotter, P., Wolf, R., Pieber, S., Brunus, E., Crippa, M., Ciarelli, G., Piazzalunga, A., Schwikowski, M., Abbaszade, G., Schnelle, J., Zimmermann, R., An, Z., Szidat, S., Baltenspergere, U., El Haddad, I. and Prevot, A. (2014) High secondary aerosol contribution to particulate pollution during haze events in China. *Nature*, vol 514, pp. 218–222.

Jochner, S. and Menzel, A. (2015) Urban phenological studies – past, present, future. *Environmental Pollution*, vol 203, pp. 250–260.

Johnstone, D., Moore, G., Tausz, M. and Nicolas, M. (2013) The measurement of plant vitality in landscape trees. *Arboricultural Journal*, vol 35. no 1, pp. 18–27.

Kim, S., Guenther, A. and Apel, E. (2013) Quantitative and qualitative sensing techniques for biogenic volatile organic compounds and their oxidation products. *Environmental Science: Processes and Impacts*, vol 15, pp. 1301–1314.

Kjelgren, R. and Clark, J. (1992) Microclimates and tree growth in three urban spaces. *Journal of Environmental Horticulture*, vol 10, no 3, pp. 139–145.

Kulhavy, D. L., Unger, D. R., Hung, I. and Zhang, Y. (2016) Comparison of AR. Drone quadricopter video and the visual CTLA method for urban tree hazard rating. *Journal of Forestry*, 114: doi: 10.5849/jof.15-005.

Leuzinger, S., Vogt, R. and Körner, C. (2010) Tree surface temperature in an urban environment. *Agricultural and Forest Meteorology*, vol 150, no 1, pp. 56–62.

Long, S. P. and Bernacchi, C. J. (2003) Gas exchange measurements, what can they tell us about the underlying limitations to photosynthesis? Procedures and sources of error. *Journal of Experimental Botany*, vol 54, no 392, pp. 2393–2401.

Matsumoto, J. (2014) Measuring biogenic volatile organic compounds (BVOCs) from vegetation in terms of ozone reactivity. *Aerosol and Air Quality Research*, vol 14, no 1, pp. 197–206.

Maxwell, K. and Johnson, G. N. (2000) Chlorophyll fluorescence – a practical guide. *Journal of Experimental Botany*, vol 51, no 345, pp. 659–668.

McHale, M., Burke, I., Lefsky, M., Peper, P. and McPherson, E., 2009. Urban forest biomass estimates: Is it important to use allometric relationships developed specifically for urban trees? *Urban Ecosystems*, vol 12, no 1, pp. 95–113.

Meier, F. and Scherer, D. (2012) Spatial and temporal variability of urban tree canopy temperature during summer 2010 in Berlin, Germany. *Theoretical and Applied Climatology*, vol 110, no 3, pp. 373–384.

Mercado, L. M., Bellouin, N., Sitch, S., Boucher, O., Huntingford, C., Wild, M. and Cox, P. (2009) Impact of changes in diffuse radiation on the global land carbon sink. *Nature*, vol 458, no 7241, pp. 1014–1017.

Niinemets, Ü. and Peñuelas, J. (2008) Gardening and urban landscaping: Significant players in global change. *Trends in Plant Science*, vol 13, no 2, pp. 60–65.

Nowak, D., Crane, D. and Stevens, J. (2006) Air pollution removal by urban trees and shrubs in the United States. *Urban Forest and Urban Greening*, vol 4, pp. 115–123.

Ögren, E. and Evans, J. R. (1993) Photosynthetic light response curves. I. The influence of CO_2 partial pressure and leaf inversion. *Planta*, vol 189, p. 182.

Park, J., Goldstein, A., Timkovsky, J., Fares, S., Weber, R., Karlik, J. and Holzinger, R. (2013) Active atmosphere-ecosystem exchange of the vast majority of detected volatile organic compounds. *Science*, vol 341, pp. 643–647.

Pugh, T. A. M., Mackenzie, R., Whyatt, J. D. and Hewitt, C. N. (2012) Effectiveness of green infrastructure for improvement of air quality in urban street canyons. *Environmental Science and Technology*, vol 46, no 14, pp. 7692–7699.

Roloff, A., Korn, S. and Gillner, S. (2009) The Climate-Species-Matrix to select tree species for urban habitats considering climate change. *Urban Forestry and Urban Greening*, vol 8, no 4, pp. 295–308.

Sanders, J. and Grabosky, J. (2014) 20 years later: Does reduced soil area change overall tree growth? *Urban Forestry and Urban Greening*, vol 13, no 2, pp. 295–303.

Smith, D. M. and Allen, S. J. (1996) Measurement of sap flow in plant stems. *Journal of Experimental Botany*, vol 47, no 12, pp. 1833–1844.

Tyree, M. T. (1997). The Cohesion-Tension theory of sap ascent: Current controversies. *Journal of Experimental Botany*, vol 48, no 10, pp. 1753–1765.

Ugolini, F., Tognetti, R., Raschi, A. and Bacci, L. (2013) *Quercus ilex* L. as bioaccumulator for heavy metals in urban areas: Effectiveness of leaf washing with distilled water and considerations on the trees distance from traffic. *Urban Forestry and Urban Greening*, vol 12, no 4, pp. 576–584.

Ward, H. C., Kotthaus, S., Grimmond, C., Bjorkengren, A., Wilkinson, M., Morrison, W., Evans, J., Morison, J. and Iamarino, M. (2015) Effects of urban density on carbon dioxide exchanges: Observations of dense urban, suburban and woodland areas of southern England. *Environmental Pollution*, vol 198, pp. 186–200.

Wittig, V. E., Ainsworth, E. A. and Long, S. P. (2007) To what extent do current and projected increases in surface ozone affect photosynthesis and stomatal conductance of trees? A meta-analytic review of the last 3 decades of experiments. *Plant, Cell and Environment*, vol 30, no 9, pp. 1150–1162.

Zhang, X., Friedl, M. A., Schaaf, C. B., Strahler, A. H. and Schneider, A. (2004) The footprint of urban climates on vegetation phenology. *Geophysical Research Letters*, vol 31, p. L12209.

17

ABIOTIC STRESS

Glynn C. Percival

Introduction

Globally, 54 per cent of the world's population reside in urban areas. The potential of trees to mitigate the urban landscape in ways that are beneficial to the physiological and psychological health of the human population are now widely recognized (Konijnendijk et al., 2005; Nowak et al., 2014). Consequently, great expense is spent annually on the creation of urban treescapes worldwide. Trees planted and growing in urban landscapes are, however, routinely exposed to a range of environmental stresses that pose a serious threat to their biology. Prolonged drought, root deoxygenation (waterlogging, soil compaction), salt contamination of soil, atmospheric pollutants and high temperature episodes, singly and in combination can negatively impact tree growth, development and distribution (Hirons and Percival, 2012). Of concern is the fact that climate studies have shown significant summer warming on an annual basis and a trend towards an increased frequency of summer drought events throughout Europe (Schaer et al., 2004; Seneviratne et al., 2010). Moreover, climate models predict a stronger inter- and intra-annual weather variability, which will cause an increased risk of extreme heatwave, drought and precipitation events. Consequently an understanding of how urban trees tolerate and adapt to harsh environmental conditions is necessary in order to sustain ecosystem services (Pincetl et al., 2013).

Because trees are sessile and continue to develop over many growing seasons they possess a tremendous capacity to adjust to the environment in which they have been planted. Trees can adjust continuously to a changing light environment within minutes (i.e. photosynthetic induction), within days (i.e. photosynthetic adjustments), within weeks or months (i.e. morphological changes; specific leaf area, leaf nutrient content, chlorophyll; nitrogen ratio) or years (i.e. architectural changes; crown plasticity, root architecture) (Rozendaal et al., 2006). Studies show that these acclimation responses enhance growth, survival and, ultimately, plant fitness (Rozendaal et al., 2006). In addition, significant progress in anatomical, physiological, biochemical and molecular studies coupled with advances in analytical and DNA sequencing technologies has facilitated discoveries of abiotic stress-associated proteins, enzymes and genes involved in tree stress responses (Harfouche et al., 2014; Niinemets and Valladares, 2006). Likewise while the inter-related nature of multiple stressors within urban landscapes factors is fully appreciated, in many instances specific stresses tend to induce an explicit physiological and biochemical response (Xiong et al., 2002). For example, herbicides result in mechanical constraints, changes in activities of macromolecules and enhanced cellular leakage, while salinity involves damage to both cellular ionic and osmotic cellular components. Similarly, specific stresses are associated with specific adaptive responses. Waterlogging tolerance, for example, is linked to up-regulation of key enzymes involved with lactic acid fermentation to maintain glycolysis, while heat tolerance is associated with induction and synthesis of heat shock proteins and scavenging of toxic by-products of uncontrolled electron transfer reactions (Kreuzwieser et al., 2009; Königshofer et al., 2008). For the purposes of this review the major environmental stressors involved with urban landscapes and subsequent biochemical and physiological responses within trees are discussed separately. For detailed reviews of stress interactions in plants see Mittler (2006), Niinemets (2010), Niinemets and Valladares (2006).

Sensing of abiotic stress and downstream signalling pathways

A better understanding of the mechanisms by which trees perceive environmental cues and transmit signals to activate adaptive responses is of fundamental importance in stress-biology research. Knowledge about stress-signal transduction is also crucial for breeding and engineering tree tolerance to abiotic stress. Given the multiplicity of stress signals, many different sensors can be expected (Xiong et al., 2002). While identification of proteins involved in abscisic acid (ABA) signalling in crop plants is ongoing, recent work aimed at elucidating equivalents in tree species is still in its infancy and data available is primarily through studies with forest trees (see Harfouche et al., 2014 for detailed review).

Abiotic stress

Stress responses and tolerance mechanisms involve the prevention or alleviation of cellular damage, the re-establishment of homeostasis and growth resumption. In its simplest form, stress resistance mechanisms within trees fall into three categories:

1   avoidance mechanisms: prevents exposure to stress;
2   tolerance mechanisms: permits the plant to withstand stress; or
3   escape: plants alter their phenology/physiology in response to stress (Bussotti et al., 2014).

Figure 17.1 diagrammatically shows how trees respond to urban landscape stresses. However, a recurrent difficulty in studying tree performance in urban landscapes is that performance is the result of a large number of interacting factors. For the purposes of this review the most significant environmental factors are photo-oxidative damage (atmospheric pollution (ozone), drought, soil deoxygenation (waterlogging/soil compaction), salinity and heat).

Photo-oxidative stress: air pollution, ozone

Studies indicate that anthropogenic activities will result in elevated concentrations of air pollutants in urban areas world-wide; especially ozone (O_3) regarded as one of the most widespread air pollutants in the world (Nowak et al., 2014). The majority of tropospheric O_3 comes from photochemical reactions of O_3 precursors, namely methane (CH4), carbon

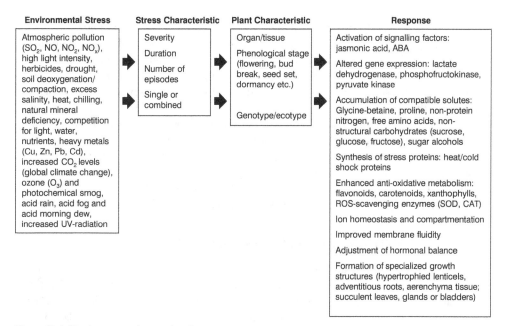

Figure 17.1 Environmental stress in plants

monoxide (CO), volatile organic compounds (VOCs), and nitrogen oxides (NO$_x$). The majority of researchers conclude that tree transpiration and canopies beneficially reduce concentrations of O$_3$ precursors that contribute to O$_3$ formation (Xu et al., 2015). However, it has been demonstrated that trees themselves can produce VOCs that may potentially contribute to the formation of O$_3$ as well as carbon monoxide. Species selection is important here as VOC emission rates vary by species (Nowak et al., 2014). Elevated ozone can lead to sudden increase in intracellular levels of reactive oxygen species (ROS) within leaf tissue (Mittler et al., 2011). Reactive oxygen species (ROS) are unavoidable by-products of plant cell metabolism, generated by chloroplast, mitochondria, and plasma membrane electron transport activities (Blokhina and Fagerstedt, 2010). Under normal growth condition, ROS production in plant cell compartments are low. At high concentrations, however, ROS cause oxidative damage to lipid, protein, and nuclear acid leading to cellular membrane damage, loss of enzyme activity, inhibition of protein synthesis and cell death (Mittler et al., 2004). In order to reduce oxidative damage, tree leaves possess a complex protective system comprising non-enzymatic and enzymatic components such as carotenoids, polyphenols (flavonoids), the ROS-scavenging enzymes superoxide dismutase (SOD) and calatase (CAT), and secondary metabolites such as proline and glycine betaine (Kangasjärvi and Kangasjärvi, 2014; Jaleel et al., 2007; Fini et al., 2012). Higher activities of ROS-scavenging enzymes are associated with stress-tolerant tree genotypes compared to susceptible ones; suggesting inherently high antioxidant enzyme content plays an important role in tolerance against photo-oxidative damage (Apel and Hirt, 2004; Jaleel et al., 2006). For example naturally inherent SOD content has been used as an indirect selection criterion for evaluating drought-resistance in plants (Zaefyzadeh et al., 2009) while stress analysis studies revealed increased susceptibility of CAT-deficient plants to herbicide and ozone (Guan et al., 2009). Carotenoids and xanthophylls function as non-enzymatic protective photo-oxidative pigments responsible for the quenching of chlorophyll excited states, singlet oxygen and interception of deleterious free oxygen and organic radicals (Fini et al., 2012). Gomathi and Rakkiyapan (2011) observed that high carotenoids content favours better adaptation of plants to photo-oxidative stress while carotenoids and xanthophylls were identified as important contributors to stress tolerance (Jaleel et al., 2006). Phenylpropanoids, particularly flavonoids, have been suggested as playing primary antioxidant functions in the responses of trees to abiotic stresses. Furthermore, flavonoids are effective endogenous regulators of auxin movement, thus behaving as developmental regulators. Flavonoids are capable of controlling the development of individual organs and the whole plant; and, hence, to contribute to stress-induced morphogenic responses of plants (Brunetti et al., 2013).

Quantification of naturally inherent stress-protective compounds and enzymes within urban trees and rates of accumulation in response to environmental stress stimuli has received little investigation despite the fact that these compounds appear to play a pivotal role in determining the stress tolerance of trees. Further research in this area is warranted.

Soil deoxygenation (waterlogging/soil compaction)

Soil deoxygenation as a result of excess water application, impeded drainage, mechanical compaction or impermeable surface coverings is a ubiquitous factor in the urban landscape (Percival and Keary, 2007). In waterlogged soils, air spaces become filled with water delaying the diffusion of gases between the atmosphere, rhizophere and roots (Calvo-Polanco et al., 2012). Dissolved oxygen is further depleted from the soil solution by respiration of roots and soil microorganisms. Depletion of dissolved oxygen in waterlogged soils can occur in several hours leading to hypoxia (deficiency in oxygen reaching root tissue) or anoxia (total depletion of soil oxygen), depending on soil temperature, root respiration activity and the frequency and duration of soil saturation. Once oxygen concentrations fall below 15 per cent within soil pores-root functioning is impaired affecting numerous physiological and metabolic processes (Kreuzwieser and Rennenberg, 2014). These include disintegration of leaf ultra-structure, leading to wilting, chlorosis, abscission and reduced photosynthetic capacity, as well as stem hypertrophy, blackening of roots and death (Percival and Keary, 2007). Cytological investigations conducted in tree roots indicated water stress damages the cellular membranes of root tips. At different development stages the occurrence of damage such as lack of protoplasm, formation of membrane vesicles and coagulation of cytoplasmic content has been demonstrated (Chiatante et al., 2000). In addition, reduced soil oxygen causes beneficial aerobic microbes to be outcompeted by anaerobic microbes. Anaerobic microbes produce phytotoxic substances such as reduced iron and manganese ions. These reduced ions are highly soluble in water so increased availability to plants rapidly reaches phytotoxic levels. Anaerobic microbes also denitrify soils resulting in a reduced pH which, in turn, further reduces nutrient availability.

Trees possess several strategies to combat prolonged root deoxygenation, with different species utilizing different adaptations or combinations of them (Glenz et al., 2006). Adventitious roots typically form near the soil surface when the primary root system of a tree is impaired by soil oxygen deficiency. Adventitious roots are highly porous structures that possess intercellular spaces facilitating longitudinal oxygen transport (i.e. uptake of oxygen along the root growing at the water surface while being in contact with the atmosphere; Haase and Rätsch, 2010). Such oxygen uptake allows for the maintenance of mitochondrial respiration, radial oxygen loss which contributes to the oxidation of the rhizosphere, and absorption of water and nutrients to replace the submerged primary root system (Calvo-Polanco et al., 2012). Hypertrophied lenticels formed along the tree trunk just above the water level or compacted surface allow oxygen uptake into the tree, and partially contribute to the release of gaseous compounds (carbon dioxide, acetaldehyde, ethanol) from the stem into the atmosphere. Hypertrophied lenticels also form on submerged stems and roots with their main role appearing to be maintenance of water uptake in place of roots (see Glenz et al., 2006 for compilations of tree species that show this adaptation). Aerenchyma are intercellular spaces formed within root and trunk tissue that circulate oxygen throughout the tree and potentially contribute to the export of phytotoxic volatile metabolites (acetaldehyde, ethanol) from the plant (Parent et al., 2008). While tree roots have a limited capacity to respire anaerobically, this process produces harmful by-products such as acetaldehyde and cyanogenic compounds which accumulate in the soil causing acidification of the root cell cytoplasm and vacuole.

If stress avoidance strategies such as hypertrophied lenticels, adventitious roots or aerenchyma do not exist or are poorly formed, molecular and physiological tolerance mechanisms become essential for survival when roots become oxygen deficient (Christianson et al., 2010). Molecular and physiological studies using flood tolerant and sensitive tree species have demonstrated that fermentative pathways are stimulated under oxygen deficient conditions. Because oxygen is the final electron acceptor of mitochondrial respiration, this crucial process of plant energy metabolism is slowed down under hypoxia or completely inhibited under anoxia. Thus, ATP cannot be produced by aerobic respiration and cells affected suffer from an energy crisis (Bailey-Serres and Voesenek, 2008). Consequently trees tolerant of root deoxygenation have a greater ability to switch from mitochondrial respiration to lactic acid fermentation in response to hypoxia. This switch has been documented at both the metabolite and molecular level by monitoring lactate dehydrogenase (LDH) gene expression (Kreuzwieser et al., 2009). In addition to the up-regulation of LDH, glycolytic fluxes (the catabolism of carbohydrates, as glucose and glycogen, by enzymes, with the release of energy and the production of lactic or pyruvic acid) are enhanced, in order to maintain ATP production under conditions of inhibited mitochondrial respiration. Enhanced glycolytic flux is suggested from an up-regulation of key enzymes of this pathway (i.e. phosphofructokinase and pyruvate kinase; Kreuzwieser et al., 2009; Christianson et al., 2010). Maintaining enhanced glycolysis for a steady and sufficient supply of carbohydrates seems to be crucial for tree survival under hypoxia as clear experimental evidence exists demonstrating that flood tolerant and sensitive species differ in their ability to maintain adequate carbohydrate supply over prolonged periods of waterlogging (Ferner et al., 2012; LeProvost et al., 2012).

Other factors involved with root deoxygenation adaptations include changes in transcript levels genes within root tissue leading to down-regulation of energy-intensive processes such as nutrient uptake, biosynthesis of cell wall components (cellulose, hemicellulose), proteins, and lignin (Kreuzwieser et al., 2009; Christianson et al., 2010; LeProvost et al., 2012). Such gene down-regulation of energy-consuming processes results in reduced shoot and root growth. However, the consensus of research indicates the initiation of fermentative pathways together with enhanced glycolytic flux is of greatest importance for tree survival under periods of oxygen deprivation. Significant progress on molecular technologies such as transcriptomics, proteomics and metabolomics now offers the potential to rapidly identify tree species and ecotypes for urban landscapes that possess inherent capacity to tolerate prolonged root deoxygenation stress (Harfouche et al., 2014; Kreuzwieser and Rennenberg, 2014). See Pallardy (2008), Kozlowski and Pallardy (2002), and Niinemets and Valladares (2006) for tree species associated with tolerance to prolonged root de-oxygenation that offer potential for urban landscape selection.

Water deficit

Poor growth and death of newly transplanted and established trees can in many instances be attributed to internal water deficits within a tree resulting in loss of leaf turgor, stomatal closure, decreased photosynthesis and metabolic function (Sjöman et al., 2015). In areas where newly planted trees are not irrigated, initial establishment relies heavily on precipitation. If the transplant does not receive sufficient precipitation during the period of new root regeneration, its internal water deficits increase considerably due to excessive water transpiration and non-absorption of water from the soil (Kozlowski and Pallardy, 2002). In addition, physical loss of root material is another cause of internal water deficits within a tree. At harvest, as little as 5 per cent of a tree's root system may be moved with a tree, even when accepted nursery practices are adopted (Fraser and Percival, 2003). Following leafing out, the capacity of the roots to supply the leaves with water can be severely restricted. Water deficits are therefore regarded as the major causes of failure of newly planted trees planted within urban landscapes (Percival et al., 2006). Prolonged water deficits can also induce large-scale tree decline episodes of established trees and enhance susceptibility to pests and disease attack. Water stress disrupts water fluxes in the xylem, leading to cavitation with the resulting embolisms limiting the tree's ability to conduct water, and thus limit growth (Rice et al., 2004). In addition, reduced soil water availability limits microbial activity in soils and, depending on the intensity and duration of the drought event, may lead to total inhibition of microbial metabolism (Borken and Matzner, 2009).

Acclimation to drought are based on a tree's capacity to maintain water status (water potential, Ψ and/or relative water content, RWC). Trees are classified as drought tolerant because they can either withstand dehydration of protoplasm (low RWC) or avoid low Ψ and/or RWC (Kozlowski and Pallardy, 2002; Rennenberg et al., 2006). Drought acclimation of trees includes rapid leaf shedding, smaller or few leaves, sunken stomata, rapid stomatal closure, waxy or hairy leaf cuticle, highly developed palisade mesophyll (altered leaf morphology) and tissue elasticity (Pallardy, 2008; Munné-Bosch and Alegre, 2004). In stems, acclimations include twig and stem photosynthesis; low resistance to water flow in vascular tissues (Nguyen and Lamant, 1989); while in roots, acclimations include extensive root growth, high root : shoot ratios and a strong root-regenerating potential after transplanting (Chaves et al., 2003). Other responses include increases in soluble non-protein nitrogen, increases in the levels of free amino acids (Fotelli et al., 2002), non-structural carbohydrates (sucrose, glucose, fructose) and sugar alcohols. The presence of raffinose and/or stachyose in buds or stem tissues have also been suggested as good indicators of drought stress tolerance in woody plants (Labeke et al., 2004) while a primary role for mannitol as a osmoprotectant in several plant families (*Oleaceae, Fraxinus, Rubiaceae*) has been documented (Conde et al., 2011). Downstream of water-deficit sensing, ABA is undoubtedly the plant hormone most intimately involved in stress-signal transduction in trees. It is thought to play important roles in root-to-shoot signalling (Davies and Zhang, 1991) in the regulation of stomatal movement in response to drought (reviewed by Belin et al., 2010; Popko et al., 2010), and in the maintenance of root growth under mild-to-moderate drought stress, while restricting leaf growth (Hsiao and Xu, 2000). In response to drought, growth allocation patterns in trees are altered, and the variable growth rates observed in roots and shoots are shown to be correlated with ABA levels, although this regulation of growth may be modulated by ethylene (Sharp et al., 2004). All these adaptations promote water homeostasis either by restricting water loss from the plant body or by increasing water absorption to replace losses by transpiration. See Pallardy (2008) for tree species demonstrating these adaptations and potential for urban landscape plantings.

The different mechanisms employed between drought-avoiding and drought-tolerant species during two consecutive summers in four evergreen Mediterranean tree species was studied by Martinez-Ferri et al (2000). In response to midday water stress, the drought-avoiding species *Pinus halepensis* exhibited marked stomatal closure (*g*s) but no changes in stem water potential (s), whereas the drought-tolerant species *Quercus coccifera, Q. ilex* ssp. *ballota* and *Juniperus phoenicea* L. displayed declines in midday *g*s and Ψs. The higher resistance to CO_2 influx in needles of *P. halepensis* compared with the other species did not result in either a proportional increase in non-radiative dissipation of excess energy or photo-inactivation of photosystem II (PSII). No significant differences were found among species either in the de-epoxidation state of the xanthophyll cycle or in the pool of its components on a total chlorophyll basis. Despite contrasting midday assimilation rates, the three drought-tolerant species all exhibited a pronounced drop in photochemical efficiency at midday that was characterized by a decrease in the excitation capture efficiency of the open PSII centers. Although photo-inhibition was not fully reversed before dawn, it apparently did not result in cumulative photo-damage. Thus, the drought-avoiding and drought-tolerant species employed different mechanisms for coping with excess light during the midday depression in photosynthesis that involved contrasting midday photochemical efficiencies of PSII and different degrees of dynamic photo-inhibition as a photo-protective mechanism.

Recent studies indicate that drought tolerance within and between tree genera can be identified under laboratory conditions, potentially negating the need for expensive field trials (Sjöman et al., 2015). The water potential at turgor loss (Ψ_{p0}) provides a robust measure of plant drought tolerance since a more negative Ψ_{p0} allows a leaf to continue functioning over an increased range of leaf water potential (Sack et al., 2003; Lenz et al., 2006). Plant genotypes that have a low, i.e. more negative, Ψ_{p0} tend to maintain leaf gas exchange, hydraulic conductance and growth at lower soil matric potentials (Ψ_{soil}) so are at a competitive advantage where soil water deficits occur (Mitchell et al., 2008; Blackman et al., 2010). The Ψ_{p0} also provides a measure for the Ψ_{soil} below which a plant cannot recover from wilting (Bartlett et al., 2012). Consequently, Ψ_{p0} is a highly instructive plant trait to measure when actual tolerance of water deficits is required rather than information on the persistence of a tree species under an amalgamation of climatic conditions (Sjöman et al., 2015). Such research is particularly pertinent given the fact that selection of tree species with avoidance strategies such as deep rooting are limited, as urban landscape planting sites generally possess limited soil volumes. Consequently the value in assessing leaf turgor loss point is that it provides a measure of drought tolerance that is particularly relevant to the urban environment.

Salinity stress

De-icing salts in the form of sodium chloride (NaCl) are used widely throughout Europe and the US in order to maintain roads free from ice and snow to ensure public safety (Zimmerman and Jull, 2006). However, NaCl can be a major chemical pollutant in urban landscapes having a detrimental effect on roadside vegetation (Li et al., 2014). Salt damage occurs via direct toxicity of the chloride ion, resulting in a reduction in leaf chlorophyll concentration and concomitant photosynthetic performance, breakdown of leaf structure at the cellular level, necrosis, bud failure and twig and branch die-back (Li et al., 2014). Sodium (Na^+) tends to damage soil structure by competition with other cation exchange sites, causing nutrient deficiency symptoms and increased soil pH (Zimmerman and Jull, 2006). In addition NaCl imposes ion imbalance and hyper-osmotic stress, leading to cell membrane disorganization, ion toxicity and oxidative damage (Li et al., 2014). Indeed, de-icing salts are estimated to be directly responsible for the deaths of more than 700,000 urban trees annually in Western Europe alone (Percival et al., 2003) and are particularly devastating to young spring growth that is unable to acclimate and therefore extremely susceptible. With increases in traffic volume and the expansion of road networks throughout Europe and the US, the quantity of salt used for de-icing operations has increased correspondingly.

Trees have evolved several adaptation mechanisms to cope with salt stress based on tolerance (salt-includer) or avoidance (salt-excluder) (Tattini et al., 2006). Salt includers withstand high tissue salt concentrations largely through osmotic adjustment resulting from synthesis in the cytoplasm of compatible organic solutes such as proline, glycine, betaine, and sugars to re-establish osmotic balance. Importantly these low-molecular-mass compounds do not interfere with cellular biochemical and enzymatic reactions (Gucci et al., 1997). Other species sequester sodium and other ions within cellular vacuoles compartmentalized from metabolic sites such as cytoplasm and chloroplasts (Flowers, 2004). Striking differences in salinity tolerance exists within species of poplar (Fung et al., 1998; Brosché et al., 2005). Key salt-tolerance mechanisms identified in this species include compartmentalization of chloride in root cortex cell vacuoles, diminished xylem loading of sodium and chloride ions, sodium extrusion into the soil solution, and simultaneous avoidance of excessive potassium ion loss by depolarization–activation of cation channels. Salt excluding mechanisms involve passive salt exclusion, active salt extrusion, or dilution of salt as it enters a plant (Kozlowski and Pallardy, 2002). Species such as *Populus euphratica* or *Rhizophora mucronata* develop leaf succulence to prevent an excessively high concentration of salt in the cell sap resulting in salt dilution (Chen and Polle, 2010), while trees belonging to the genus *Tamarix, Aegiceras* and *Laguncularia* that thrive in salt-laden mangrove swamps possess specialized salt glands or salt bladders that exude salt ions such as sodium and chloride from the plant via apoplastic pathways. Intra-species comparative studies may provide additional information regarding salt-tolerance mechanisms in trees. For example, white poplar (*Populus alba* (L.)) has considerable genetic variation for salt tolerance resulting from natural selection in contrasting ecological habitats (Beritognolo et al., 2011). Proline and glycine betaine function as osmolytes to reduce leaf osmotic potential so that the water potential gradient favours water uptake and plants can maintain turgor under low moisture or high salinity conditions (Ehmedov et al., 2002). An association between increased proline concentration and tolerance to osmotic stress has been shown (Ehmedov et al., 2002; Jiang et al., 2012; Yamada et al., 2005). Evidence also shows a role for proline in reducing free radical levels caused by osmotic stress and significantly improving the ability of seedlings to grow in medium containing NaCl (Hong et al., 2000).

Studies evaluating the salt tolerance of a range of *Acer* and *Crataegus* genotypes commonly planted into urban landscapes identified wide variation in leaf necrosis/chlorosis responses varying from none to severe leaf burn (Percival et al., 2003). None of the tree genotypes evaluated possessed the physiological and anatomical characteristics commonly associated with salt tolerance. In this instance tolerance was based on operational electron flow through photosystem II of the leaf photosynthetic system during salt stress. Maintenance of leaf photosynthetic integrity has been shown to be an important factor in determining a plant's ability to survive prolonged harsh environmental conditions (Levitt, 1980). In tree species such as holly (*Ilex aquifoilum*), evergreen oak (*Quercus ilex*) and Sitka spruce (*Picea sitchensis*), moderate to severe stress tolerance was associated with increased rates of photosynthesis with greater rates occurring in hardier varieties than in less hardy ones (Levitt, 1980). Maintenance of leaf photosynthetic integrity allows for continual metabolic activity within the plant (synthesis of sugars, amino acids, proteins, nucleic acids lipids), essential for the growth and repair of damaged tissue. Developments in technology to quantify leaf photosynthetic integrity in response to salinity as a means of identifying salt-tolerant trees for urban landscapes have been discussed (Percival et al., 2006; Johnstone et al., 2013). See Pallardy (2008) and Kozlowski and Pallardy (2002) for tree species adapted to tolerate salinity episodes.

Heat stress

Higher temperatures in urban environments caused by the lack of evapotranspirational cooling, heat convection and long-wave radiation from non-vegetative surfaces (i.e. buildings and roads) can be detrimental to tree biology (Ladjal et al., 2000). Under prolonged heat stress, high temperature injury is manifest by leaf and wood desiccation inducing water stress throughout the canopy. With prolonged desiccation, tree limbs and trunks can break and shed prematurely (Harris et al., 2004). Such a response is undesirable in highly populated urban areas. At the leaf level, photosynthesis is one of the most heat-sensitive processes in plant cells, leading to numerous changes of the structure and function of the photosynthetic apparatus (Georgieva et al., 2000). Within the photosynthetic system, it has been recognized that photosystem II (PSII) is the most thermally labile component of the electron transport chain (Schrader et al., 2004). Among partial reactions of PSII, the oxygen-evolving complex is particularly heat sensitive (Georgieva et al., 2000). In addition, heat stress reduces the capacity of chloroplasts to carry out electron transport efficiently and quickly down-regulates the content of ribulose 1,5-bisphosphate (RuBP), an organic substance essential for the functioning of the Calvin Cycle (Bita and Gerats, 2013). As a consequence carbohydrate metabolism is impaired, leading to a reduced ability of the tree to defend against attack by biotic agents such as insects or fungi. At the whole tree level, scorching of leaves and stems, leaf abscission and senescence, shoot and root growth inhibition and fruit damage, leads to reduced tree biomass and loss of aesthetics.

In response to heat stress, leaves of some tree species have developed mechanisms to minimize heat absorption and maximize dissipation of latent heat through stomatal aperture adjustment. In addition, membranes are photo-protected by mechanisms that reduce photo-oxidation and membrane denaturation, such as the xanthophyll cycle, synthesis of phenolics (flavonoids, anthocyanins), and plant steroids (Bita and Gerats, 2013) while a potentially under-estimated role for volatile isoprenoid content as heat ptotectant compounds has recently been suggested (Loreto et al., 2014). Thus, successful compensation of heat damage requires osmotic adjustment, down-regulation of electron flux, and scavenging of toxic by-products of uncontrolled electron transfer reactions such as ROS (see photo-oxidative stress). Interestingly, ROS production contributes to the transduction of the heat signal and expression of specific heat shock genes that induce synthesis of heat shock proteins (HSPs) (Königshofer et al., 2008). Heat stress results in the mis-folding of newly synthesized proteins and the denaturation of existing ones. Under standard growth conditions HSPs control cellular signaling, protein folding, translocation, and degradation, but under heat stress they prevent protein mis-folding and aggregation as well as protect cellular membranes. An increased production of HSPs occurs when plants experience either abrupt or gradual increases in temperature (Nover et al., 2001). Research with crops has shown that considerable variations in the pattern of HSP gene expression occurs in different species and even among genotypes within species. With predicted increases in summer warming and urban heat island effects, trees that are genetically capable of tolerating high temperatures should be selected for hot sites. While species tolerant of heat stress have been identified (Appleton et al., 2015), the basis of resistance has yet to be elucidated. Developments in molecular DNA technologies offer a means of identifying heat-tolerant tree genotypes based on HSP synthesis, while recent work has related heat tolerance of trees to ecological adaptations and geographical distribution (Wertin et al., 2001; Ghannoum and Way, 2011). Ecological studies such as these potentially offer a means of identifying an abundance of untapped genetic resource to select for heat tolerance and are of practical importance for ecotype and species selection within urban landscapes.

Conclusions

The significance of abiotic stressors on tree performance in urban landscapes are broadly appreciated by professionals involved in tree use and management; when, however, it is necessary to make decisions on tree selection for a given site species suitability it is still, in many instances, poorly appreciated. In these circumstances many trees are selected for use on the basis of what can be seen to grow well locally. This approach has negative implications in that it excludes from consideration other species that are equally or better suited. The net effect is to reduce or restrict the diversity of taxa that are planted; an undesirable characteristic when this occurs on a metropolitan or regional scale. With recent technological advances in the areas of genomics, eco-physiological studies and molecular DNA technologies, we are now well poised to use systems-biology approaches to elucidate the molecular mechanisms of abiotic stress in trees at the whole-tree, organ and molecular level. Integrating data collected at different levels into networks, in the context of the whole tree, may uncover complex molecular mechanisms and pathways that underpin the abiotic stress response. Ultimately, this holistic approach will deepen our understanding of stress responses and enhance our ability to develop and select desired tree phenotypes.

Acknowledgements

The author is grateful for helpful comments and suggestions by Dr Andrew Hirons (Myescough College/University of Central Lancashire), Mr Jon Banks (Bartlett Tree Research Laboratory) and two anonymous reviewers.

References

Apel, K., Hirt. H., 2004. Reactive oxygen species: metabolism, oxidative stress, and signal transduction. *Annual Review of Plant Biology*, 55: 373–399.

Appleton, B., Rudiger, E. L. T., Harris, R., Sevebeck, K., Alleman, D., Swanson, L., 2015. *Trees for Hot Sites*. Technical Note Publication 430-024, Virginia Cooperate Extension, Virginia Tech, Virginia State University.

Bailey-Serres, J., Voesenek, L. A. C. J., 2008 Flooding stress: acclimations and genetic 597 diversity. *Annual Reviews of Plant Biology*, 59: 313–339.

Bartlett, M. K., Scoffoni, C., Sack, L., 2012. The determinants of leaf turgor losspoint and prediction of drought tolerance of species and biomes: a globalmeta-analysis. *Ecology Letters*, 15: 393–405.

Belin, C., Thomine, S., Schroeder, J. I., 2010. Water balance and the regulation of stomatal movements. In Pareek, A., Sopory, S. K., Bohnert, H. J., Govindjee (eds), *Abiotic Stress Adaptation in Plants*. Springer, Netherlands, pp 283–305.

Beritognolo, I., Harfouche, A., Brilli, F., Prosperini, G., Gaudet, M., Brosché, M., Salani, F., Kuzminsky, E., Auvinen, P., Paulin, L., Kangasjärvi, J., Loreto, F., Valentini, R., Mugnozza, G. S., Sabatti, M., 2011. Comparative study of transcriptional and physiological responses to salinity stress in two contrasting *Populus alba* L. genotypes. *Tree Physiology*, 31: 1335–1355.

Bita, C. E., Gerats, T., 2013. Plant tolerance to high temperature in a changing environment: scientific fundamentals and production of heat stress-tolerant crops. *Frontiers in Plant Science*, 4: 273–283.

Blackman, C. J., Brodribb, T. J., Jordan, G. J., 2010. Leaf hydraulic vulnerability is related to conduit dimensions and drought resistance across a diverse range of woody angiosperms. *New Phytologist*, 188: 1113–1123.

Blokhina, O., Fagerstedt, K. V., 2010. Reactive oxygen species and nitric oxide in plant mitochondria: origin and redundant regulatory systems. *Physiologia Plantarum*, 138: 447–462.

Borken, W., Matzner, E., 2009. Introduction: impact of extreme meteorological events on soils and plants. *Global Change Biology*, 15: 781–782.

Bray, E. A., 2001. Plant response to water-deficit stress. *Els*. doi: 10.1038/npg.els.0001298

Brosché, M., Vinocur, B., Alatalo, E. R., 2005. Gene expression and metabolite profiling of *Populus euphratica* growing in the Negev desert. *Genome Biology*, 6: R101.

Brunetti, C., Di Ferdinando, M., Fini, A., Pollastri, S., Tattini, M., 2013. Flavonoids as antioxidants and developmental regulators: relative significance in plants to humans. *International Journal of Molecular Science*, 14: 3540–3555.

Bussotti, F., Ferrini, F., Pollastrina, M., Fini, A., 2014. The challenge of Mediterranean sclerophyllous vegetation under climate change: from acclimation to adaptation. *Environmental Experimental Botany*, 103: 80–98.

Calvo-Polanco, M., Señorans, J., Zwiazek J. J., 2012. Role of adventitious roots in water relations of tamarack (*Larix laricina*) seedlings exposed to flooding. *BMC Plant Biology*, 12: 99–100.

Chaves, M. M., Maroco, J. P., Pereira, J. S., 2003. Understanding plant responses to drought: from genes to the whole plant. *Functional Plant Biology*, 30: 239–264.

Chen, S., Polle, A., 2010. Salinity tolerance of *Populus*. *Plant Biology*, 12: 317–333.

Chiatante, D., Di Iorio, A., Maiuro, L., Scippa S. G., 2000. Effect of water stress on root meristems in woody and herbaceous plants during the first stage of development. In Stokes, A. (ed.), *The supporting Roots of Trees and Woody Plants: Form Function and Physiology*. Kluwer Academic Publishers, Netherlands, pp 245–258.

Christianson, J. A., Llewellyn, D. J., Dennis, E. S., Wilson, I. W., 2010. Comparisons of 629 early transcriptome responses to low-oxygen environments in three dicotyledonous 630 plant species. *Plant Signalling and Behaviour*, 30: 1006–1009.

Conde, A., Manuela Chaves, M., Gerós, H., 2011. Membrane transport, sensing and signaling in plant adaptation to environmental stress. *Plant and Cell Physiology*, 52: 1583–1602.

Davies, W. J., Zhang, J., 1991. Root signals and the regulation of growth and development of plants in drying soil. *Annual Review of Plant Physiology and Plant Molecular Biology*, 42: 55–76.

Ehmedov, V., Akbulut M., Sadiqov, S. T., 2002. Role of Ca^{2+} in drought stress signaling in wheat seedlings. *Biochemistry*, 67: 491–497.

Ferner, E., Rennenberg, H., Kreuzwieser, J., 2012. Effect of flooding on C metabolism of flood-tolerant (*Quercus robur*) and non-tolerant (*Fagus sylvatica*) tree species. *Tree Physiology*, 32: 135–145.

Fini, A., Guid, L., Ferrini, F., Brunetti, C., Di Ferdinando, M., Biricolti, S., Pollastri, S., Calamai, L., Tattini, M., 2012. Drought stress has contrasting effects on antioxidant enzymes activity and phenylpropanoid biosynthesis in *Fraxinus ornus* leaves: an excess light stress affair? *Journal of Plant Physiology*, 169: 929– 939.

Flowers, T. J., 2004. Improving crop salt tolerance. *Journal of Experimental Botany*, 55: 307–319.

Fotelli, M. N., Rennenberg, H., Gessler, A., 2002. Effects of drought on the competitive interference of an early successional species (*Rubus fruticosus*) on *Fagus sylvatica* L. seedlings: N-15 uptake and partitioning, responses of amino acids and other N compounds. *Plant Biology*, 4: 311–320.

Fraser, G. A., Percival, G. C., 2003. The influence of biostimulants on growth and vitality of three urban tree species following transplanting. *Arboricultural Journal*, 27: 43–57.

Fung, L. E., Wang, S., Altman, A., Hüttermann, A., 1998. Effect of NaCl on growth, photosynthesis, ion and water relations of four poplar genotypes. *Forest Ecology Management*, 107: 135–146.

Georgieva, K., Tsonev, T., Velikova, V., Yordanov, I., 2000. Photosynthetic activity during high temperature of pea plants. *Journal Plant Physiology*, 157: 169–176.

Ghannoum, O., Way, D. A., 2011. On the role of ecological adaptation and geographic distribution in the response of trees to climate change. *Tree Physiology*, 31: 1273–1276.

Glenz, C., Schlaepfer, R., Iorgulescu, I., Kienast, F., 2006. Flooding tolerance of Central European tree and shrub species. *Forest Ecology and Management*, 235: 1–13.

Gomathi, R., Rakkiyapan, P., 2011. Comparative lipid peroxidation, leaf membrane thermostability, and antioxidant system in four sugarcane genotypes differing in salt tolerance. *International Journal of Plant Physiology and Biochemistry*, 3: 67–74.

Guan, Z. Q., Chai, T. Y., Zhang, Y. X., Xu, J., Wei, W., 2009. Enhancement of Cd tolerance in transgenic tobacco plants overexpressing a Cd-induced catalase cDNA. *Chemosphere*, 76: 623–630.

Gucci, R., Lombardini, L., Tattini, M., 1997. Analysis of leaf water relations in leaves of two olive (*Olea europaea*) cultivars differing in tolerance to salinity. *Tree Physiology*, 17: 13–21.

Haase, K., Rätsch, G., 2010. The morphology and anatomy of tree roots and their aeration strategies. In Junk, W. J., Piedade, M. T. F., Wittmann, F., Schöngart, J., Parolin, P. (eds), *Central Amazonian Flood Plain Forests: Ecophysiology, Biodiversity and Sustainable Management*. Springer, New York, pp. 141–162.

Harfouche, A., Meilan, R., Altman, A., 2014. Molecular and physiological responses to abiotic stress in forest trees and their relevance to tree improvement. *Tree Physiology*, 34: 1181–1198.

Harris, R. W., Clark, J. R., Matheny, N. P., 2004. *Arboriculture: Integrated Management Of Landscape Trees, Shrubs and Vines*, 4th edn. Prentice-Hall, New York.

Hirons, A., Percival, G., 2012. Fundamentals of tree establishment: a review. In Johnston, M., Percival, G. (eds), *Trees, People and the Built Environment*. Forestry Commission, London, pp 51–62.

Hong, Z., Lakkineni, K., Zhang, Z., Verma, D. P. S., 2000. Removal of feedback inhibition of Δ^1-pyrroline–5-carboxylate synthetase results in increased proline accumulation and protection of plants from osmotic stress. *Plant Physiology*, 122: 1129–1136.

Hsiao, T. C., Xu, L. K., 2000. Sensitivity of growth of roots versus leaves to water stress: biophysical analysis and relation to water transport. *Journal of Experimental Biology*, 51: 1595–1616.

Jaleel, C. A., Gopi, R., Alagu Lakshmanan, G. M., Panneerselvam, R., 2006. Triadimefon induced changes in the antioxidant metabolism and ajmalicine production in paclobutrazol enhances photosynthesis and ajmalicine production in *Catharanthus roseus* (L.). *Plant Science*, 171: 271–276.

Jaleel, C. A., Gopi, R., Manivannan, P., Kishorekumar, A., Sridharan, R., Panneerselvam, R., 2007. Studies on germination, seedling vigour, lipid peroxidation and proline metabolism in *Catharanthus roseus* seedings under salt stress. *South African Journal of Botany*, 73: 190–195.

Jiang, H., Peng, S., Li, X., Korpelainen, H., Li, C., 2012. Transcriptional profiling analysis in *Populus yunnanensis* provides insights into molecular mechanisms of sexual differences in salinity tolerance. *Journal of Experimental Biology*, 63: 3709–3726.

Johnstone, D., Moore, G., Tausz, M., Nicolas, M., 2013. The measurement of plant vitality in landscape trees. *Arboricultural Journal*, 35: 18–27.

Kangasjärvi, S., Kangasjärvi, J., 2014. Towards understanding extracellular ROS sensory and singalling systems in plants. *Advances in Botany*, 2014: 1–10.

Kludze, H. K., Pezeshki, S. R., DeLaune, R. D., 1994, Evaluation of root oxygenation and growth in bald cypress in response to short-term soil hypoxia. *Canadian Journal of Forest Research*, 24: 804–809.

Königshofer, H., Tromballa, H. W., Löppert, H. G., 2008, Early events in signaling high-temperature stress in tobacco BY2 cells involve alterations in membrane fluidity and enhanced hydrogen peroxide production. *Plant Cell Environment*, 31: 1771–1780.

Konijnendijk, C. C., Nilsson, K., Randrupp, T. B., Schipperijn, J. 2005. *Urban Forests and Trees*. Springer, Berlin.

Kozlowski, T. T., Pallardy, S. G., 2002. Acclimation and adaptive responses of woody plants to environmental stresses. *The Botanical Review*, 68: 270–334.

Kreuzwieser, J., Rennenberg, H., 2014. Molecular and physiological responses of trees to waterlogging stress. *Plant Cell Environment*, 37: 2245–2259.

Kreuzwieser, J., Hauberg, J., Howell, K. A., Carroll, A., Rennenberg, H., Millar, A. H., Whelan, J., 2009. Differential response of grey poplar leaves and roots underpins stress adaptation during hypoxia. *Plant Physiology*, 149: 461–473.

Labeke, M. C., van, Degeyter, L., Fernandez, T., Davidson, C. G., 2004. Non-structural carbohydrate content as an aid for interpreting quality testing of nursery stock plants. *Acta Horticulturae*, 630: 191–198.

Ladjal, M., Epron, D., Ducrey, M., 2000. Effects of drought preconditioning on thermotolerance of photosystem II and susceptibility of photosynthesis to heat stress in cedar seedlings. *Tree Physiology*, 20: 1235–1241.

Lenz, T. I., Wright, I. J., Westoby, M., 2006. Interrelations among pressure–volume curve traits across species and water availability gradients. *Physiologia Plantarum*, 127: 423–433.

LeProvost, G., Sulmon, C., Frigerio, J. M., Bodénès, C., Kremer, A., Plomion, C., 2012. Role of waterlogging-responsive genes in shaping interspecific differentiation between two sympatric oak species. *Tree Physiology*, 32: 119–134.

Levitt, J., 1980. *Responses of Plants to Environmental Stress*, vol. 1. Academic Press, New York.

Li, Z., Liang, Y., Zhou, J., Sun, X., 2014. Impacts of de-icing salt pollution on urban road greenspace: a case study of Beijing. *Frontiers of Environmental Science and Engineering*, 8: 747–756.

Loreto, F., Pinelli, P., Manes, F., Kollist, H. 2004. Impact of ozone on monoterpene emissions and evidences for an isoprene-like antioxidant action of monoterpenes emitted by Quercus ilex (L.) leaves. *Tree Physiology,* 24, 361–367.

Martinez-Ferri, E., Balaguer, L., Valladares, F., Chico, J. M., Manrique, E., 2000. Energy dissipation in drought-avoiding and drought-tolerant tree species at midday during the Mediterranean summer. *Tree Physiology*, 20, 131–138.

Mitchell, P. J., Veneklaas, E. J., Lambers, H., Burgess, S. S. O., 2008. Leaf water relations during summer water deficit: differential responses in turgor maintenance and variations in leaf structure among different plant communities in southwestern Australia. *Plant Cell Environment*, 31: 1791–1802.

Mittler, R., 2006. Abiotic stress, the field environment and stress combination. *Trends in Plant Science*, 11: 15–19.

Mittler, R., Vanderauwera, S., Gollery, M., Van Breusegem, F., 2004. Reactive oxygen gene network of plants. *Trends in Plant Science*, 9: 490–498.

Mittler R., Vanderauwera, S., Suzuki, N., Miller, G., Tognetti, V. B., Vandepoele, K., 2011. ROS signaling: the new wave? *Trends in Plant Science*, 16: 300–309.

Munné-Bosch, S., Alegre, L., 2004. Die and let live: leaf senescence contributes to plant survival under drought stress. *Functional Plant Biology*, 31: 203–216.

Nguyen, A., Lamant, A., 1989. Variation in growth and osmotic regulation of roots of water-stressed maritime pine (*Pinus pinaster* Ait.) provenances. *Tree Physiology*, 5: 123–133.

Niinemets, Ü., 2010. Responses of forest trees to single and multiple environmental stresses from seedlings to mature plants: past stress history, stress interactions, tolerance and acclimation. *Forest Ecology and Management*, 260: 1623–1639.

Niinemets, Ü., Valladares, F., 2006. Tolerance to shade, drought, and waterlogging of temperate northern hemisphere trees and shrubs. *Ecological Monographs*, 76: 521–547.

Nover, L., Bharti, K., Döring, P., Mishra, S. K., Ganguli, A., Scharf, K. D., 2001. *Arabidopsis* and the heat stress transcription factor world: how many heat stress transcription factors do we need? *Cell Stress Chaperones*, 6: 177–189.

Nowak, D. J., Hirabayashi, S., Bodine, A., Greenfield, E., 2014. Tree and forest effects on air quality and human health in the United States. *Environmental Pollution*, 193: 119–129.

Pallardy, S. G., 2008. *Physiology of Woody Plants*, 3rd edn. Academic Press, London.

Parent, C., Capelli, N., Berger, A., Crevecoeur, M., Dat, J. F., 2008. An overview of plant responses to waterlogging. *Plant Stress*, 2: 20–27.

Percival, G. C., Keary, I. P., 2007. The influence of nitrogen fertilization on waterlogging stresses in *Fagus sylvatica* L. and *Quercus robur* L. *Arboriculture and Urban Forestry*, 34: 29–41.

Percival, G. C., Fraser, G. A., Oxenham, G., 2003. Foliar salt tolerance of *Acer* genotypes using chlorophyll fluorescence. *Journal of Arboriculture*, 29: 61–66.

Percival, G. C., Keary, I. P., Habsi Sulaiman, A. L. 2006. An assessment of the drought tolerance of *Fraxinus* genotypes for urban landscape plantings. *Urban Forestry and Urban Greening*, 5:17–27.

Pincetl, S., Gillespie, T., Pataki, D. E., Saatchi, S., Saphores, J. D., 2013. Urban tree planting programs, function or fashion? Los Angeles and urban tree planting campaigns. *GeoJournal*, 78: 475–493.

Popko, J., Hänsch, R., Mendel, R. R., Polle, A., Teichmann, T., 2010. The role of abscisic acid and auxin in the response of poplar to abiotic stress. *Plant Biology*, 12: 242–258.

Rennenberg, H., Loreto, F., Polle, A., Brilli, F., Fares, S., Beniwal, R. S., Gessler, A., 2006. Physiological responses of forest trees to heat and drought. *Plant Biology*, 8: 556–571.

Rice, K., Matzner, S., Byer, W., Brown, J., 2004. Patterns of tree dieback in Queensland, Australia: the importance of drought stress and the role of resistance to cavitation. *Oecologia*, 139: 190–198.

Rozendaal, D. M. A., Hurtado, V. H., Poorter, L., 2006. Plasticity in leaf traits of 38 tropical tree species in response to light; relationships with light demand and adult stature. *Functional Ecology*, 20: 207–216.

Sack, L., Grubb, P. J., Maranon, T., 2003. The functional morphology of juvenile plants tolerant of strong summer drought in shaded forest understories in southern Spain. *Plant Ecology*, 168: 139–163.

Schaer, C., Vidale, P. L., Luethi, D., Frei, C., Haeberli, C., Liniger, M. A., Appenzeller, C., 2004. Variability in European summer heatwaves. *Nature*, 427: 332–336.

Schrader, S. M., Wise, R. R., Wacholtz, W. F., Ort, D. R., Sharkey, T. D., 2004. Thylakoid membrane responses to moderately high leaf temperature in Pima cotton. *Plant Cell Environment*, 27: 725–735.

Seneviratne, S. I., Corti,T., Davin, E. L., Hirschi, M., Jaeger, E. B., Lehner, I., Orlowsky, B., Teuling, A. J., 2010. Investigating soil moisture climate interactions in a changing climate: a review. *Earth Science Review*, 99: 125–161.

Sharp, R. E., Poroyko, V., Hejlek, L. G., Spollen, W. G., Springer, G. K., Bohnert, H. J., Nguyen, H. T., 2004. Root growth maintenance during water deficits: physiology to functional genomics. *Journal of Experimental Biology*, 55: 2343–2351.

Sjöman, H., Hirons, A. D., Bassuk, N. L., 2015. Urban forest resilience through tree selection – variation in drought tolerance in *Acer*. *Urban Forestry and Urban Greening*, 14: 858–865.

Tattini, M., Remorini, D., Pinelli, P., Agati, G., Saracini, E., Traversi, M. L., Massai, R., 2006. Morpho-anatomical, physiological and biochemical adjustments in response to root zone salinity stress and high solar radiation in two Mediterranean evergreen shrubs, *Myrtus communis* and *Pistacia lentiscus*. *New Phytologist*, 170: 779–794.

Wertin, T. M., McGuire, M. A., Teskey, R. O., 2001. Higher growth temperatures decreased net carbon assimilation and biomass accumulation of northern red oak seedlings near the southern limit of the species range. *Tree Physiology*, 31(12): 1277–1288.

Xiong, L., Schumaker, K. S., Zhu, J. K., 2002. Cell signaling during cold, drought, and salt stress. *Plant Cell*, 14: S165–S183.

Xu, S., He, X., Chen, W., Huang, Y., Zhao, Y., Li, B., 2015. Differential sensitivity of four urban tree species to elevated O_3. *Urban Forestry and Urban Greening*, 14: 1166–1173.

Yamada, M., Morishita, H., Urano, K., Shiozaki, N., Yamaguchi-Shinozaki, K., Shinozaki K., Yoshiba, Y., 2005. Effects of free proline accumulation in petunias under drought stress. *Journal of Experimental Botany*, 56: 1971–1985.

Zaefyzadeh, M., Quliyev, R. A., Babayeva, S. M., Abbasov, M. A., 2009. The effect of the interaction between genotypes and drought stress on the superoxide dismutase and chlorophyll content in durum wheat landraces. *Turkish Journal of Biology*, 33: 1–7.

Zimmerman, E. M., Jull, L. G., 2006. Sodium chloride injury on buds of *Acer platanoides*, *Tilia cordata* and *Viburnum lantana*. *Journal of Arboriculture*, 32: 45–53.

18

BIOTIC FACTORS

Pests and diseases

Michael J. Raupp and Paolo Gonthier

Introduction

Arthropod pests including insects and mites and plant pathogens including viruses, bacteria, and especially fungi severely affect survival, growth and aesthetics of amenity trees in urban environments. Factors such as the loss of plant biodiversity and reductions in the diversity and abundance of natural enemies and antagonists predispose urban forests to pest and disease outbreaks and catastrophic tree loss. The introduction of non-native plants can disrupt ecosystem processes with varying and sometimes contradictory results. Some non-native plants are not consumed by native herbivores in the new range. The result is fewer consumers at higher trophic levels and simplification of predator communities in urban areas. When non-native plants are introduced with their associated non-native insects and mites, eruptive outbreaks of pests can occur in the absence of important natural enemies left behind in the aboriginal range. Moreover, non-native pathogens that arrive on a plant from one region may spread to new hosts in the new geographic range. Non-native plants lacking an evolutionary history with insects and pathogens in a new geographic location may lack defenses and succumb to indigenous pests and diseases. Similar and calamitous tree loss occurs when non-native arthropod pests and pathogens are introduced to evolutionarily naïve plant hosts in a new geographic region. Classic examples of non-native pests and pathogens devastating trees in a new location include emerald ash borer, hemlock woolly adelgid, chestnut blight and Dutch elm disease. Other features of the built environment including impervious surfaces and elevated temperatures stress plants by lowering their defenses or increasing their nutritional value and predisposing them to attack by insects, mites and pathogens. Several anthropogenic inputs including ozone, nitrogen, and de-icing salts are associated with urban infrastructures. These too may increase the susceptibility of trees to attack by biotic agents. In this chapter we discuss several features of built environments that threaten the vitality and resilience of urban forests. We also deal with globally important pests and diseases in urban environments. Arthropod pests treated in the chapter encompass lethal borers, foliar pests, and sucking arthropods including scale insects, lace bugs, and spider mites. Specific arthropod case studies include oak processionary moth, *Thaumetopoea processionea*, emerald ash borer, *Agrilus planipennis*, and horse chestnut leafminer, *Cameraria ohridella*. Diseases encompass root rots and wood decays, cankers including the canker stain of plane trees caused by *Ceratocystis platani* and ash dieback associated with *Hymenoscyphus fraxineus*, vascular diseases, including Dutch elm disease and oak wilt caused by *Ophiostoma novo-ulmi* and *Ceratocystis fagacearum*, respectively, as well as the most important anthracnose and foliar diseases of oaks, planes, maples and horse chestnuts. Pests and diseases are described in their significance, impact and diagnostic characters. Integrated management strategies and tactics are discussed.

Pests

Factors of urban forests affecting outbreaks of insects and mites

Herbivorous insects and mites often attain much greater densities and cause greater amounts of injury to trees and shrubs in urban environments compared to those found on woody plants in natural forests (Raupp et al., 2010, 2012). In recent reviews of arthropod outbreaks in built environments, Raupp et al. (2010, 2012) discussed several key features contributing to insect and mite outbreaks. Here we present a summary of several of these factors and discuss mechanisms underlying outbreaks of insects and mites.

Low street tree diversity and catastrophic tree loss

Lack of plant diversity seriously compromises the sustainability of the urban forest when trees and shrubs confront new pests for which they lack coevolved defenses (Gandhi and Herms, 2010; Raupp et al., 2012). The catastrophic loss of elms to Dutch elm disease sounded a call for greater floristic diversity in urban forests; however, Raupp et al. (2006) found species and cultivars of *Acer* and *Fraxinus* had largely supplanted elms as the dominant genera of street trees in North America thereby setting the stage for catastrophic tree losses with the arrival of Asian longhorned beetle, *Anoplophora glabripennis*, and the emerald ash borer, *Agrilus planipennis*. The domination of urban landscapes by a few species or genera of woody plants predisposes cities to catastrophic loss due to pests.

Loss of top-down regulation in simplified habitats

Regulation of arthropod populations has been broadly categorized as either top-down meaning population control by predators, parasites, or pathogens; bottom-up meaning limitations imposed by the plant on the pest; or a combination of both forces. Urban habitats sometimes have reduced floristic diversity and complexity (Raupp et al., 2010, 2012). This may be accompanied by reductions in the richness and abundance of natural enemies (Raupp et al. 2010, 2012; Martinson and Raupp, 2013). However, urban habitats with greater diversity of plant material and more layers (e.g. trees, shrubs, groundcovers) of vegetation are known to support greater numbers of species and greater densities of natural enemies, especially generalist predators that may aid in suppressing outbreaks of insect pests (Shrewsbury and Raupp, 2006). These generalist predators likely play an important role in limiting pest outbreaks not only in diverse natural landscapes, but also in diverse human-altered ones.

Non-native plants and non-native insects

Non-native plants

Non-native plants are widely used in urban landscapes (Berghardt et al., 2009). The addition of non-native plants to urban landscapes can affect population dynamics of pests in several ways. Herbivorous insects with narrow host ranges and specialized feeding habits like Lepidoptera may not recognize non-native plants as food (Berghardt et al., 2009). In turn this can reduce the abundance of caterpillars on non-native plants. A paucity of prey in landscapes dominated by non-native plants could result in fewer predators and loss of top-down regulation of pests in these alien-dominated landscapes (Berghardt et al., 2009). This problem is exacerbated when non-native pests accompany their host plant into the new realm where coevolved natural enemies may be absent (Raupp et al., 2010). In North America the non-native azalea lace bug, *Stephanitis pyrioides,* is a classic example of a non-native pest that now outbreaks perennially on their non-native hosts in the invaded range (Shrewsbury and Raupp, 2006).

An additional problem arises when non-native plants enter a new biotic realm. Pests enjoy what has been termed 'defense free space' due to lack of a shared coevolutionary history with plants in the invaded range (Gandhi and Herms, 2010). Without a long standing association with a pest, plants may lack evolved defenses and be more susceptible to attack by novel pests. A notable example of this is seen in the high degree of susceptibility of North American ash trees to the emerald ash borer, *Agrilus planipennis,* a native of Asia, and the relative resistance of Asian ash trees to this pest (Herms and McCullough, 2014). Another prominent example is hemlock woolly adelgid, *Adelges tsugae,* which attacks and kills eastern North American hemlocks while Asian hemlocks are much more resistant (Gandhi and Herms, 2010). A mirror image of this relationship is seen in the high degree of susceptibility of Eurasian birches to the bronze birch borer, *Agrilus anxius,* a native of North America, and the strong resistance of North American birches to this pest (Nielsen et al., 2011).

Non-native insects

Invasions by non-native insects and mites result in direct and indirect disruption to ecological processes and economic losses in urban landscapes amounting to billions of dollars annually in the United States due to costs associated with detecting and eradicating pests, protecting and removing trees, and lost property values (Gandhi and Herms, 2010; Aukema et al., 2011). The following three vignettes describe the significance and impact, diagnosis, life cycle, and management of three major invasive insect pests of urban forests. More encyclopedic guides to the biology and management of insect and mite pests on woody landscape plants include works by Johnson and Lyon (1991), and Alford (2012).

THAUMETOPOEA PROCESSIONEA, OAK PROCESSIONARY MOTH

Significance and impact: A major defoliator of several species of *Quercus,* caterpillars of this moth also attack *Fagus, Carpinus, Corylus, Betula* and *Castanea* from Sweden to southern Europe. In 2005 it was discovered in the United Kingdom. In addition to its pest status as a defoliator, it poses a major health risk to people exposed to urticating hairs found on larger caterpillars. These toxin-laced hairs cause dermatitis, severe rashes, eye irritations, and respiratory problems (Alford, 2012; Forestry Commission of the United Kingdom,

Figure 18.1 In addition to being an important defoliator of hardwoods, irritating hairs of oak processionary moth caterpillars, *Thaumetopoea processionea*, pose a major health risk

Source: Glynn Percival

2015) (Figure 18.1). In southern Europe where the moth is native, top-down pressure from indigenous natural enemies likely works in concert with environmental factors to prevent outbreaks (Forestry Commission of the United Kingdom, 2015).

Diagnosis: Eggs are deposited on small branches in symmetrical rows called plaques. Plaques are gray in color and about 20 mm in length. Larvae are brownish with darker brown heads and hairy. Older larvae have a grayish caste and are cloaked in long silvery hairs and may attain a length of 40 mm. They are gregarious throughout their larval stages and larvae move from place to place in a long head-to-tail procession. In summer they build nests of white silk on the trunks and branches of trees. Within these tents caterpillars, frass, and pupae can be found in summer months. Defoliation by early instars results in loss of leaf margins, but late instar larvae skeletonize leave behind only veins. Adults have a wingspan of about 30 mm and are gray moths with dark gray bands running across the wings (Alford, 2012; Forestry Commission of the United Kingdom, 2015).

Life cycle: The life cycle of processionary moth consists of one generation each year. Eggs are the overwintering stage. In the United Kingdom egg masses hatch and young caterpillars begin feeding in March and April. Caterpillars consume foliage from April to June and pass through six larval instars after which time they pupate in silken nests. Adults emerge in July and can be found into September. Adults fly and mate at night and females may lay 100–200 eggs. Eggs are deposited in masses on the bark of the tree where they remain dormant until hatching the following spring (Alford, 2012; Forestry Commission of the United Kingdom, 2015).

Management: The Forestry Commission of the United Kingdom (2015) recommends that individual homeowners contact professional arborists to treat and remove caterpillars due to health risks. Monitoring trees for egg masses and old nests in the autumn, winter, and early spring may help guide interventions before larvae hatch and begin feeding. Removal of overwintering egg masses may help reduce local populations. Pheromone traps are available and can assist in locating new infestations by trapping males in late summer. Nests containing caterpillars and pupa can be removed by professionals to reduce health risks to local populations. Many biologically-based, organic, and reduced risk insecticides are available for controlling caterpillars.

AGRILUS PLANIPENNIS, EMERALD ASH BORER

Significance and impact: The native range of emerald ash borer includes northeastern China, the Korean peninsula, and eastern Russia (Herms and McCullough, 2014). The spread of emerald ash borer from eastern to western Russia threatens forests in Europe. Tens of millions of ash trees in North America have been killed by emerald ash borer (Herms and McCullough, 2014) (Figure 18.2). In the United States Aukema et al. (2011) estimated the annual losses and costs measured in terms of federal government expenditures, local government expenditures, household expenditures, residential property value losses and timber value losses to forest landowners to be $1.6 billion annually. Due to the rapid loss of forest canopy, emerald ash borer is a major disruptor of ecological processes and services including nutrient cycling and plant succession in natural forests and cooling, carbon sequestration, water infiltration, and pollution mitigation in cities where ash trees comprise much of the urban forest (Herms and McCullough, 2014). In North America *Fraxinus* has a

Figure 18.2 Emerald ash borer, *Agrilus planipennis*, has killed millions of ash trees in natural forests and built landscapes in North America

Source: Paula Shrewsbury

rich biota of more than 280 associated arthropods, 43 of which are monophagous which are imperiled by emerald ash borer (Herms and McCullough, 2014). In North America a lack of bottom-up factors such as evolved resistance in ashes and top-down mortality related to fewer specialist parasitoids likely contribute to outbreaks that are less common in the native range in Asia.

Diagnosis: Symptoms of emerald ash borer infestation include dieback and thinning of the canopy, vertical bark cracks on the bole, and epicormic shoots at the base of infested trees. Woodpeckers are adept at finding emerald ash borer larva, and elevated wood pecker activity on ash trees is indicative of an infestation. Adults exiting trees create diagnostic D-shaped holes 2–3 mm in diameter. Adults are slender, elongate beetles, iridescent green about 13 mm long. Larvae create serpentine galleries beneath bark as they bore through cambium and phloem. Larvae lack legs and are creamy white with indistinct heads. They can be 25–30 mm long when fully grown.

Life cycle: Emerald ash borer in North America has one annual generation throughout much of its range but in some locations and under varying conditions of tree health, development may take two years. Adults emerge in late spring coincident with the blooming of *Robinia pseudoacacia*. They live for 3–6 weeks after emerging from the tree and feed on ash foliage prior to laying eggs in bark crevices or under bark flaps. On average each female lays 40–70 eggs. Eggs hatch after about two weeks and tiny larvae bore though the bark and feed on meristematic and vascular tissue beneath. Larvae have four instars and feed throughout the summer and early autumn overwinter beneath the bark of infested trees. Pupation occurs the spring following deposition of eggs (Herms and McCullough, 2014) (Figure 18.3).

Figure 18.3 Emerald ash borer larvae, *Agrilus planipennis*, consume phloem and cambium, effectively girdling and killing trees. Adults (insect shown one-third actual size relative to larvae) feed on ash foliage causing little damage

Source: Michael Raupp

Management: Herms and McCullough (2014) stress the importance of early detection in managing emerald ash borer. Due to the propensity of beetles to colonize upper canopies of trees, new infestations are difficult to detect. A variety of traps baited with host volatiles and related compounds have been developed to help delineate infestations. Attempts to eradicate isolated infestations of emerald ash borer by government agencies failed in Maryland, USA, largely due to the high dispersal capability of adult beetles and the inability to limit human transit of infested wood and wooden products (Herms and McCullough, 2014). Raupp et al. (2006) noted the importance of diversifying urban forests that are now overstocked with *Fraxinus* as a way to mitigate catastrophic canopy loss. Ashes differ dramatically in resistance to emerald ash borer. Those of Asian provenance including Manchurian ash, *F. mandshurica*, are much more resistant than North American ashes including *F. americana*, *F. pennsylvanica* and *F. nigra*. Blue ash, *F. quadrangulata*, appears to be the most resistant species of North American ash and could be a source of resistance in breeding programs (Herms and McCullough, 2014). The recent discovery of white fringe tree, *Chionanthus virginicus*, as a host for emerald ash borer in North America is disturbing. Several parasitic wasps imported from Asia have been released and established in many states but their impact on population dynamics of emerald ash borer is equivocal (Herms and McCullough, 2014). While intervention in native forests remains problematic, protection of individual trees with insecticides is feasible. Systemic insecticides such as the neonicotinoids imidacloprid and dinotefuron can be applied through the soil, through the bark as a spray, or injected into the vascular system. Biologically based systemics including emamectin benzoate and azadirachtin have also proven effective in controlling emerald ash borer as preventatives and curatives (Herms and McCullough, 2014). Many cities and municipalities are taking an integrated approach to managing emerald ash borer. The first step is to create a tree inventory to establish the location, size, and value of the ash resource. Interventions include removing low value and unthrifty trees, girdling trees to attract beetles and removing trees before beetles emerge, and progressively treating trees with insecticides over several years to distribute costs through time. Simulations demonstrate that integrated approaches can preserve the many benefits of valuable urban forest canopies while limiting costs of intervention (Herms and McCullough, 2014).

CAMERARIA OHRIDELLA, HORSE CHESTNUT LEAFMINER

Significance and impact: The origins of horse chestnut leafminer are unclear but it now occupies several countries across Europe (Tilbury and Evans, 2003). The widespread use of horse chestnut trees, *Aesculus hippocastanum*, and the rapid spread of the leafminer have elevated this pest to major status (Tilbury and Evans, 2003; Percival et al., 2011). In addition to attacking *A. hippocastanum*, Tilbury and Evans (2003) list other important landscape trees including *Aesculus pavia*, *Acer platanoides* and *A. pseudoplatanus* are occasional hosts of this pest. Percival et al. (2011) demonstrated that horse chestnut trees infested by leafminers had lower photosynthesis and suffered reductions in stem extension, root carbohydrate concentration and twig starch content compared to trees protected from leafminers with insecticides. Percival et al. (2011) warn that long-term repeated defoliations related to horse chestnut leafminer infestations threaten reproductive capacity and long-term persistence of horse chestnut in the invaded range.

Diagnosis: Horse chestnut leafminer feed on parenchyma cells between the upper and lower epidermal surfaces creating serpentine mines that may coalesce. Fresh galleries are light in color but as tissues die, mines turn brown giving the foliage a scorched appearance (Figures 18.4 and 18.5). In heavy infestations leaves are shed prematurely in summer. Larvae are legless caterpillars, yellowish in color with distinct head and mouthparts at the anterior

Figure 18.4 In heavy infestations entire canopies will appear scorched with almost every leaf infested by horse chestnut leafminer *Cameraria ohridella*

Source: Glynn Percival

Figure 18.5 Individual leaves will contain several mines caused by larvae of horse chestnut leafminer *Cameraria ohridella*

Source: Glynn Percival

end. Adults are small moths with a body length of 5 mm. Wing color is orange/brown with bands of white and black running across the wings. Eggs laid on the upper leaf surface are tiny (< 1 mm), oval, and yellow. Pupae are found in silken cocoons within the mine (Tilbury and Evans, 2003; Alford, 2012).

Life cycle: Conditions of weather and climate affect the number of generations each year. The average number of generations in western Europe is three but in hot, dry locations up to five generations develop (Tilbury and Evans, 2003). Horse chestnut leafminers overwinter as pupae; in England adults appear in April and can be found throughout the growing season. Eggs are deposited along leaf veins on the upper leaf surface from May to August and individual leaves may house several hundred eggs. Eggs hatch in 2–3 weeks and larvae enter leaves and begin mining. Larvae complete five instars in about a month and pupate in a silken cocoon within the mine. Summer pupation lasts about two weeks but overwintering pupae live in fallen leaves on the ground for 6–7 months (Tilbury and Evans, 2003).

Management: Several species of parasitic wasps in the families Eulophidae, Pteromalidae, Eupelmidae, Brachonidae, and Ichneumonidae are known to attack immature stages of horse chestnut leafminer but associated mortality is generally low. Part of the explanation for this loss of top-down regulation is poor seasonal synchronization between the phenology of some parasites with that of their hosts (Grabenweger, 2004). Cultural control involving the removal of fallen leaves beneath infested trees, composting leaves, or burying them with soil or other plant material in autumn and winter may help eliminate overwintering pupae and reduce adults colonizing trees the following spring (Tilbury and Evans, 2003). Trunk injections of systemic insecticides including imidacloprid have proven highly efficacious in reducing populations of horse chestnut leafminer on individual trees (Percival et al., 2011).

Impervious surfaces, heat islands, and water stress

Impervious surfaces fragment green spaces, reduce plant density, elevate temperatures, and reduce water infiltration affecting the growth and development of plants and the ecological relationships of insects and mites attacking them (Raupp et al., 2012; Dale et al., 2016). Spider mites are among the most common taxa of eruptive herbivores in cities where elevated temperatures dramatically reduce generation times and elevate their fecundity (Raupp et al., 2012 and references therein). Scale insects also respond to warmer temperatures in cities with greater abundance and elevated fecundity (Dale et al., 2016 and references therein).

By reducing water infiltration, impervious surfaces create water stress on trees and shrubs in cities (Raupp et al., 2012; Dale et al., 2016 and references therein). Speight et al. (1998) found densities of horse chestnut scale, *Pulvinaria regalis*, elevated on trees surrounded by impervious surfaces and concluded that reduced water and nutrient availability resulted in stress that favored the scale. Despite conflicting results of the relationship between water-stress and chewing and sucking insects, there is general agreement that elevated levels of plant stress favor egregious and often lethal boring insects, in particular bark beetles and other species of cambium feeders (Raupp et al., 2012). Of trees commonly planted in urban settings, water stress has been shown to reduce resistance of ash to clearwing borers, bark beetles, and roundheaded borers (Martinson et al., 2014) and eucalypts to the roundheaded borers (Hanks et al., 1999).

Anthropogenic inputs – pollutants, nutrients, and pesticides

Several anthropogenic inputs including ozone, nitrogen, and de-icing salts are associated with cities and urban infrastructure and may occur at higher levels in built areas than in nearby rural ones (Raupp et al., 2010, 2012). Atmospheric ozone is thought to generally reduce the resistance of plants to insect attack. Herms et al. (1996) found that increased levels of ozone increased the palatability of poplar leaves for several species on Lepidoptera. Elevated nitrogen deposition and higher ozone levels from the combustion of fossil fuels in southern California, USA, in combination with drought stress were linked to outbreaks of bark beetles (Jones et al., 2004). In Munich, viburnums exposed to atmospheric pollutants expressed higher levels of foliar nitrogen which supported elevated populations of aphids (Bolsinger and Flückinger, 1987). Elevated concentrations of minerals in soil from the use of de-icing salts in several European cities have been linked to reduced resistance of plants with corresponding increases in populations of aphids along motorways (Braun and Flückiger, 1984).

A commonly held belief that fertilization can enhance plant resistance to attack by insects and mites helps explain the widespread application of fertilizers in arboriculture. However, Herms (2002) overwhelming demonstrated that fertilization generally increases the susceptibility of woody plants to defoliators including caterpillars, sawflies, and leaf beetles; sap-sucking arthropods including aphids, adelgids, lace bugs, psyllids, plant bugs, scales, and spider mites; and wood boring insects.

Insecticides are important tools for arborists to mitigate attack by pests. However, applications of insecticides have been shown to disrupt ecosystem processes at several spatial scales ranging from residential communities to individual trees (Raupp et al., 2010, 2012). Often these disruptions result in secondary pest outbreaks due to the elimination of natural enemies or attenuation of their top-down suppressive activities. Some classes of insecticides such as neonicotinoids have been shown to suppress plant defenses and thereby increase plant quality. This in turn can elevate fecundity of spider mites and cause outbreaks on several species of woody plants (Raupp et al., 2010, 2012 and references therein).

Diseases

A vast majority of tree diseases are caused by biotic agents, including viruses, phytoplasmas, bacteria, fungi, fungal-like organisms (e.g. oomycetes), parasitic plants and nematodes. Biotic disease agents are infectious, whether on their own or by means of vectors, and are thus transmissible from diseased to healthy trees. While relevant tree diseases may be caused by pathogens belonging to all groups listed above (e.g. elm yellows caused by the *Ca. Phytoplasma ulmi*, bacterial leaf scorch caused by *Xylella fastidiosa*, sudden oak death caused by the oomycete *Phytophthora ramorum*), fungi rank first as agents of significant tree diseases (Tainter and Baker, 1996).

Outbreaks with destructive effects are associated either with the emergence of native pathogens due to environmental and cultural reasons, including tree management, or can occur when pathogen or host introductions lead to novel host-pathogen interactions. Although invasiveness of tree pathogens has been reported to occur in association with ecological traits and with factors other than high virulence of pathogens (Gonthier et al., 2014), the lack of co-evolution between hosts and pathogens and the high susceptibility of native hosts to introduced pathogens are regarded as the major forces driving the invasions (Parker and Gilbert, 2004). The climate change may also be responsible for an increase of epidemics, including in urban environment (Tubby and Webber, 2010).

In this chapter, the most important classes of diseases and pathogens affecting trees in urban settings are described in their significance, impact and diagnostic characters. The most effective management strategies and tactics, including integrated disease management programs, are reported based on the biology of pathogens.

Factors affecting outbreak of infectious diseases in the urban environment

Tree diseases often attain higher incidences and severities in urban environments compared to natural ecosystems. In urban settings, trees are often planted in suboptimal conditions, with little area for root expansion (Tubby and Webber, 2010) (see Chapters 17 and 19 of this book). This, along with the occurrence of soil compaction and the presence of impervious surfaces, may exacerbate the effects of drought, thereby weakening trees and predisposing them to the attack of secondary, opportunistic pathogens (Tubby and Webber, 2010). For the same reasons, heat island effects and air pollution, which are characteristic of urban areas, may favor pathogen outbreaks. Generally, the vegetative status of trees affects more significantly necrotrophic pathogens than biotrophic ones. It is not surprising that certain groups of pathogens, including canker/dieback pathogens, are more likely to be positively associated with host stress, particularly drought stress, than other groups, encompassing most foliar pathogens (Tubby and Webber, 2010). Incidentally, most necrotrophic pathogens may be strictly dependent on or may be favored by the presence of wounds for gaining entry into the trees, including pruning wounds and mechanical injuries.

In addition to the low floristic diversity, which may favor epidemics of specialized pathogens, the reduced spacing among trees in urban landscapes may also play a role in disease outbreaks. Short distance between trees may result in higher likelihoods of transmission of pathogens spreading through root grafts (e.g. viruses, phytoplasma, vascular pathogens) or contacts (e.g. root rots).

Root rots and wood decay

Significance and impact: Root rots and wood decay are caused by lignicolous basidiomycetes or, less frequently, ascomycetes fungi destroying the plant cell wall (Tainter and Baker, 1996). They may be classified either as white rot or brown rot agents, depending on the component of the plant cell wall they are able to decompose, i.e. lignin or cellulose, respectively. While only a few cases of intercontinental introduction of wood-destroying fungi have been reported so far, some of them involve urban settings and parks (Coetzee et al., 2001; Gonthier et al., 2014). In general, however, root rots and wood decay are recognized as major forces of native ecosystems providing resources for a variety of living organisms, thus enhancing biodiversity (Lonsdale et al., 2008). Although the concept of dead wood ecology encompasses both forests and urban parks, the loss of wood mechanical properties caused by wood decay may predispose trees to the risk of failures, resulting in significant damages of property and/or tragic injuries, which is a pivotal aspect in urban environment.

In addition to hazardous situations, root rots and wood decay fungi may also determine aesthetic damages by causing stem or branch cankers, tree decline and death (Figures 18.6–18.7). Such damages are more often associated with fungi colonizing the root system and the sapwood rather than with those affecting exclusively the heartwood (Gonthier, 2010; Vasaitis, 2013).

Diagnosis: The type of rot (e.g. white vs. brown) could aid in the diagnosis but generally it is not an exhaustive trait for the identification of wood decay fungi. Diagnosis is traditionally based on the inspection of trees for the presence of fungal fruiting bodies and on their identification by using mycological keys (Bernicchia, 2005; Gonthier and Nicolotti, 2007). However, a recent study conducted in both forest and urban sites clearly indicates that, on average, inspection of trees for the presence of fruiting bodies underestimates more than 90 percent of trees infected by wood decay fungi (Giordano et al., 2015), making visual inspection an inefficient diagnostic approach. Furthermore, fruiting bodies of some species

Figure 18.6 Oak infected by the wood decay agent *Inonotus dryadeus*. Fungal fruiting body and decayed wood at the base of the tree

Source: Paolo Gonthier

Figure 18.7 Canker caused by the wood decay fungus *Inonotus obliquus* on birch. The conk of the fungus is present

Source: Paolo Gonthier

are short-lived (e.g. *Phaeolus schweinitzii*) or may be found only in particular periods of the year (e.g. *Armillaria* spp.); thus, the timing of diagnosis is also important. Most of root rots and wood decay fungi can be easily cultured from decayed wood and therefore they could be identified through the use of appropriate keys, yet such approach is difficult and time consuming (Nicolotti et al., 2010).

A number of molecular tools are now available for the identification of the most important wood decay fungi (reviewed by Nicolotti et al., 2010). Polymerase chain reaction (PCR) assays with taxon-specific primers have been developed for the early detection of the most important and widespread root rot and wood decay fungi of both broadleaves and conifers (Guglielmo et al., 2007, 2008; Gonthier et al., 2015). Such assays provide reliable diagnostics starting from both fungal pure cultures and wood samples (e.g. pieces of wood, cores, sawdust), especially when standardized sampling approaches are used (Guglielmo et al., 2010). Technology based on electronic nose is also promising, though not yet completely reliable for all host-pathogen combinations (Baietto et al., 2013).

Biology: The biology of wood decay fungi in living trees has been previously reviewed (Vasaitis, 2013). In general, primary infections occur by means of airborne spores through wounds, including pruning wound or injuries on roots, tree collar and stem. Some of these fungi may also operate a secondary, vegetative spread, allowing for the expansion of individuals established through primary infection. Depending on the pathogen, this expansion may occur through root grafts or contacts, leading to a tree-to-tree contagion, or by free growth of the fungus in the soil through rhizomorphs (e.g. *Armillaria* spp.) or mycelial cords (Gonthier, 2010; Vasaitis, 2013). Besides these general schemes, there is an increasing body of literature indicating that latent phases in symptomless tissues (i.e. endophytic phases) are possible (Vasaitis, 2013).

Management: Strategies and tactics to fight against the most important root rot and wood decay fungi have been previously reviewed (Tainter and Baker, 1996). The knowledge on the relative importance of primary vs. secondary infection is important for management purposes. Pathogens like *Armillaria* spp., *Heterobasidion* spp., etc. are able to spread secondarily. When this happens, a carry-over of the pathogen into new generations may occur, and therefore stump removal may be recommended.

To minimize primary infections, mechanical injuries should be avoided as much as possible. Shaping of irregular mechanical wounds by cutting loose pieces of knocked off bark has been suggested to enhance callus formation and wound closure (Vasaitis, 2013). Since large size pruning wounds may take years to occlude, pruning of large diameter branches should be avoided or performed only when strictly necessary. Based on literature review, wound dressing has been reported as generally ineffective in preventing infections by wood decay fungi (Clark and Matheny, 2010).

Vascular diseases: Dutch elm disease and oak wilt

Significance and impact: Dutch elm disease (DED) and oak wilt (OW) are major and destructive fungal diseases of elms (*Ulmus* spp.) and oaks (*Quercus* spp.), respectively.

Ophiostoma ulmi and *O. novo-ulmi* were responsible for two different epidemics of DED. The former species appeared around 1910 in northwestern Europe and subsequently spread throughout Europe and North America. The latter species, displaying higher virulence than the former one, was first reported in the UK in the early 1970s, and is now separated into two subspecies distributed in Eurasia and North America: *Ophiostoma novo-ulmi* subsp. *novo-ulmi* and *O. novo-ulmi* subsp. *americana*. The origin and the patterns of introduction and invasion of DED pathogens were previously reviewed (Kirisits, 2013). *Ophiostoma ulmi* caused considerable losses in elm populations: in about 50 years, 10–20 percent of the elm population was estimated to have died. In Europe and parts of Asia, *O. novo-ulmi* killed the majority of mature elm trees (Kirisits, 2013).

OW is caused by *Ceratocystis fagacearum* and is a major cause of oak mortality in many locations of the eastern and southern USA (Harrington, 2013). While loss of oak timber can be economically detrimental, losses of amenity trees are of even greater economic importance (Harrington, 2013). Members of the red oak group, such as northern red oak (*Q. rubra*) and northern pin oak (*Q. ellipsoidalis*), are highly susceptible, but some members of the white oak group can also be affected (Harrington, 2013). Whereas the disease is present only in North America, the susceptibility of European and Asian oaks and chestnuts (*Castanea* spp.) has been demonstrated, suggesting that *C. fagacearum* could be a threat if accidentally introduced in Eurasia. It should be noted that the fungus is a quarantine-regulated pathogen in Europe.

Diagnosis: Being true vascular diseases, DED and OW result in a black to brown discoloration of one or several growth rings (i.e. functional sapwood), which is visible in cross sections of wilted twigs and branches (Harrington, 2013; Kirisits, 2013). DED-affected leaves wilt, turn to yellow brown and finally drop. If the development of symptoms is acute, leaves may remain attached to the twigs for a long time. Dead tips bend downwards in a hook-like manner (reviewed by Kirisits, 2013). The distribution of symptoms on the crown also depends upon the way of disease transmission (see below). OW may also be diagnosed by mat production under the bark of recently dead trees, which are generally clustered-distributed forming infection centers (Harrington, 2013).

Biology: DED and OW pathogens are able to infect healthy trees by spreading from diseased trees through the vascular systems of grafted roots (Tainter and Baker, 1996). Functional root grafts could play a prominent role in disease transmission when trees of sufficient age and size are involved, and especially in the case of OW (Harrington, 2013). In addition to spreading locally through root grafts, DED and OW pathogens are also disseminated, though in a different way and with a different efficiency, by insect vectors.

All vectors of DED are bark beetles mostly belonging to the genus *Scolytus* which breed in stressed or recently dead trees and disseminate the pathogen propagules exozoically and endozoically (Kirisits, 2013). Disease transmission to healthy trees occurs during maturation feeding of beetles, taking place in crotches of thin twigs or at the base of leaves on newly formed shoots in the tree crown (Kirisits, 2013). When disease transmission is mediated by the feeding activity of bark beetles, external symptoms are initially restricted to one or a few branches in the upper or outer parts of the crown, rather than being generalized to large portions or to the whole crown, as it generally occurs following infections through root grafts. Although the efficiency to act as pathogen vectors has been reported to vary depending on the size of beetles, elm bark beetles are generally regarded as efficient vectors of DED (Kirisits, 2013).

In the case of OW, pathogen association with sap beetles of the Family Nitidulidae is strong, but clearly less efficient for disease transmission than *Scolytus* for DED (Harrington, 2013). *Ceratocystis fagacearum* produces mats under the bark of recently killed oaks, especially red oaks. Once the bark cracks, mats become attractive for nitidulids, some species of which (e.g. *Colopterus truncatus*, *Carpophilus* spp.) can carry spores of the pathogen to fresh wounds (Harrington, 2013). Pruning of amenity trees and other fresh injuries may lead to new infections especially in the springtime, when sporulating mats and nitidulids carrying the fungus are most abundant (Harrington, 2013).

Management: Management strategies and tactics to control DED and OW have been previously reviewed (Harrington, 2013; Kirisits, 2013). These include the adoption of planting distances large enough to avoid root graft transmission, should the diseases occur in the future, and sanitation campaigns aimed at destroying dead and dying trees, branches, stumps, etc. and any infected material which bark beetles utilize for breeding or which may be attractive for sap beetles. In the case of DED, if trees are already colonized by bark beetles, sanitation measures have to be completed before juvenile insects emerge (Tainter and Baker, 1996). Debarking can prevent insect breeding and can kill 'white stages' (larvae and pupae) of bark beetles if present. Sanitation campaigns can be incorporated into general shade tree maintenance programs.

Soil applications or injections/infusions of systemic fungicides into the functional xylem of trees have been proved effective to prevent rather than to cure infections (Tainter and Baker, 1996). Thiabendazole has been used against DED and propiconazole against both DED and OW. In the case of DED, treatments may be more effective if combined with therapeutic pruning of diseased branches. It is recommended to treat pruning and surgery wounds to decrease their attractivity to the insects and to prevent infections by *C. fagacearum* (Tainter and Baker, 1996). Any type of paint (latex, oil-based, spray-on, brush-on, or wound dressing) will suffice. Disinfecting felling and pruning equipment used for preventive, sanitation and curative treatments with 70 percent alcohol or sodium hypochlorite is also recommended (Tainter and Baker, 1996).

While a biological treatment (i.e. Dutch Trig) based on the principle of cross protection is available for DED, the transmission of OW may be prevented by digging trenches to delimit infection from healthy trees. Equipment and methods to conduct efficient soil trenching have been previously described (Harrington, 2013). Integrated disease management against vector-transmitted vascular diseases also rely on a number of interventions aimed at controlling directly or indirectly insect vector populations.

Canker/dieback diseases: canker stain of plane trees and ash dieback

Significance and impact: Canker stain of plane trees (CSP) and ash dieback (AD) currently stand among the most destructive fungal diseases affecting woody ornamentals. CSP is a lethal disease on plane trees (*Platanus* spp.) caused by *Ceratocystis platani*, an ascomycete native to eastern US and accidentally introduced into southern Europe in the middle of the last century (Harrington, 2013). Once introduced, the fungus spread throughout Italy, Switzerland, southern France and Greece, with devastating effects on London plane tree (*P. × hybrida*) and oriental plane tree (*P. orientalis*). Hundreds of thousands of trees were killed in Europe by the pathogen, which is also very infectious, making CSP probably the most harmful and damaging tree disease in parks and urban settings.

AD is a lethal disease caused by the ascomycete *Hymenoscyphus fraxineus* on *Fraxinus excelsior* and *F. angustifolia* in Europe (Gross et al., 2014). The fungus, often known with the name of its asexual stage *Chalara fraxinea*, is probably native to East Asia but was introduced into central Europe, where AD first appeared. Subsequently, the pathogen spread epidemically throughout the entire distribution range of host trees (Gross et al., 2014). The impact of AD is tremendous, hence it has been hypothesized that the combined effects of AD and emerald ash borer could seriously threaten the survival of ashes in Europe.

Diagnosis: CSP can be recognized by a rapid branch and crown dieback preceded by a shortening of the internodes, leaf wilting and chlorosis, and leaf fall. It may take only a few months or years from onset of infection for a tree to die. The fungus, which is commonly associated with sapwood becoming dark or brown stained, can attack the cambium resulting in the formation of cankers (Figure 18.8). Two characteristic features of CSP cankers are that

Figure 18.8 Canker and internal symptoms caused by *Ceratocystis platani* on London plane tree

Source: Paolo Gonthier

they lack or have very little callus growth at the canker margins and that the bark covering cankers dries, cracks and changes its color. Sprouts may develop below cankers. Detailed guidelines for laboratory diagnosis are available as well.

Early symptoms of AD include small necrotic spots on stem and branches, which then enlarge in necrotic lesions. Cankers on branches generally appear, as well as wilting, leaf fall and death of top of the crown. Isolation of *H. fraxineus* from necrotic lesions on the bark is pivotal for diagnosis, as previously reported (Gross et al., 2014). Molecular typing through PCR-based assays is also very useful to distinguish *H. fraxineus* from the native, yet non-pathogenic species *H. albidus* (Gross et al., 2014). The two fungal species can also be distinguished based on some characters of fruiting bodies, that appear as macroscopic whitish apothecia on infected tissues (Gross et al., 2014).

Biology: C. platani colonizes a tree by moving systemically and rapidly through the non-living vessels. Therefore, functional root grafts among adjacent trees represent important pathways of disease transmission. However, the most important infection courts are human-caused wounds, including pruning wounds or other wounds of even small size due to mechanical injuries (Harrington, 2013). Infectious inoculum is represented by spores or colonized sawdust. Mechanical transmission of *C. platani* through pruning saws, climbing ropes, earth moving machinery, etc. has been reported as pivotal for the epidemics of CSP (Harrington, 2013).

Sporulating fruiting bodies of *H. fraxineus* develop during the summer on leaf debris of the previous year (Gross et al., 2014). Leaves become infected by means of windblown spores.

Subsequently, the pathogen can spread along the leaf veins and can occasionally colonize across the junction between petiole and stem, thus initiating a necrotic lesion on the stem (Gross et al., 2014). The asexual stage of the fungus develops in autumn and winter on ash petioles in the litter, but asexual spores are not infectious (Gross et al., 2014).

Management: C. *platani* is a quarantine pathogen in Europe for which prescriptions for disease control are regulated at the country level. Control may be achieved with the adoption of planting distances large enough to avoid root graft transmission, as described above for vascular diseases. The prompt removal of dead or symptomatic trees along with the two neighboring plane trees on either side is recommended. Sawdust generated during cuttings should be collected in tarps and trees should not be sawn on windy days. The stump and the root system should be removed whenever possible. Pruning should be avoided or conducted in the coldest months while fungicide dressing for the protection of wounds should be used. For instance, thiophanate-methyl is recommended for that purpose. Pruning tools and other tools should be disinfected after operating on a tree and before operating on the next. Breeding efforts have led to the selection of a hybrid cultivar (i.e. 'Vallis Clausa') which is resistant to CSP (Vigouroux and Olivier, 2004), while displaying good levels of resistance to other pest and diseases, including the anthracnose of plane trees caused by *Apiognomonia veneta* (Harrington, 2013).

There are currently no effective strategies to control AD, although various physical and chemical approaches have been proposed to reduce the risk of infection, including the removal or treatment of plant debris with fungicides, the prevention of movement of infected plant material, the use of disinfectants to treat contaminated footwear, clothing and equipment. The application of endotherapic products through trunk injection has been shown promising to manage the disease (Dal Maso et al., 2014).

Foliar and shoot diseases

Significance and impact: F. and shoot diseases encompass a broad range of infectious diseases potentially reducing the ornamental value of trees. Their significance depends on the size of the epidemic area, the severity of leaf and shoot infection, tree age, the time of infection in the growing season, and the longevity of disease, which may occur in one or a few successive seasons (Kowalski, 2013). Damage is greater if the disease occurs over two or more successive years (Kowalski, 2013). It has been reported that low severity of disease, appearing as a few spots on leaves or local infection of single leaves, leads to minor damage to a tree; while higher disease severity resulting from advanced colonization of leaves and shoots causes reduction in photosynthesis and plant growth and, occasionally, premature defoliations or dieback of various parts of the tree (Kowalski, 2013). High disease severity associated with extreme environmental conditions may sporadically result in the death of trees, regardless of their age (Kowalski, 2013). Finally, foliar and shoot diseases can weaken host trees, thus favoring the attack of secondary pathogens and pests (Tainter and Baker, 1996).

Relevant damages in urban settings are caused by a number of fungal diseases, including the anthracnose of plane tree (Figure 18.9), the anthracnose of horse chestnut, *Aesculus hippocastanum*, caused by *Guignardia aesculi* (Figure 18.10) and the tar spot of maples (*Acer* spp.) caused by *Rhytisma acerinum*. In addition, powdery mildews of plane trees and oaks caused by *Erysiphe platani* and *E. alphitoides*, respectively, can be locally important, especially in nurseries or where associated with young trees. The widely grown boxwood (*Buxus* spp.) is susceptible to a damaging blight caused by *Cylindrocladium buxicola*.

Diagnosis: D. is based on the observation of symptoms of the disease in the field, commonly followed by laboratory examinations. Keys for the diagnosis and description of symptoms

Figure 18.9 Symptoms of the anthracnose of plane tree caused by *Apiognomonia veneta*: (a) foliar symptoms; (b) canker

Source: Paolo Gonthier

Figure 18.10 Symptoms of the anthracnose caused by *Guignardia aesculi* on horse chestnut

Source: Paolo Gonthier

of the main foliar and shoot diseases have been previously published (Tainter and Baker, 1996; Sinclair and Lyon, 2005). Occasionally, disease symptoms are so characteristic that they may allow the identification of the causal agent directly in the field (Kowalski, 2013). This may occur for pathogens characterized by high degrees of host specificity and causing a specific symptomatology, e.g. tar spot of maples, powdery mildew of plane trees, etc. Many pathogens of leaves and shoots develop sexual or asexual fruiting bodies having diagnostic relevance on the surface of infected tissues. If fruiting bodies are absent on leaves, induction of fructification by incubating samples in moist conditions (moist chambers) or isolation of fungi on artificial media is necessary (Kowalski, 2013). The most common symptoms of foliar diseases occurring on ornamental and shade trees are necrotic leaf spots, which vary in color and size depending on the disease. For some anthracnose diseases, including the anthracnose of plane trees, symptoms develop not only on leaves, but also on shoots, which may display cankers or turn dark before dying. In the case of powdery mildews, infected leaves and young shoots display a whitish-gray powdery looking coating on their surface, which is formed of vegetative and sporulating hyphae.

Biology: The large majority of foliar pathogens infect new leaves in the springtime through spores produced on fallen leaves, after overwintering on the ground (Tainter and Baker, 1996; Kowalski, 2013). Pathogens attacking shoots (i.e. anthracnose agents) may overwinter in cankers and dead plant tissues. Spore discharge usually increases during intensive rainfall and decreases during droughts (Kowalski, 2013). Spores are mainly dispersed by wind or rain and splashing water. With some exceptions (e.g. powdery mildews), infections and disease severity increase with increasing leaf wetness duration.

Many pathogenic ascomycetes, after infecting plant tissues, produce asexual stages liberating asexual spores. These may infect other leaves and shoots. Due to the asexual stages and depending on the fungal species and environmental conditions, one to several cycles of infection may occur during the growing season.

Management: The removal of fallen leaves in the spring before the development of new leaves is pivotal to prevent infections (Kowalski, 2013). To reduce the source of inoculum of pathogens that overwinter in cankers and dead twigs, symptomatic parts should be pruned and completely removed by spring (Kowalski, 2013). Cultural practices (e.g. pruning) aimed at reducing the level of relative air humidity in the crown and free moisture on leaves play a significant role in preventing new infections. Application of preventative fungicides on the crown may reduce the level of infection or totally prevent infections. Systemic fungicides applied to the soil, foliage or through injection/infusion can have therapeutic effects on foliar diseases. It should be noted that chemicals can be applied to a given host against a certain pathogen only if it is approved for that use, as indicated in the label and technical sheet of the product. The label also reports a number of details on how to prepare and apply the fungicide. As a general rule, it is important to alternate fungicides with different modes of action to delay the development of pathogen strains resistant to fungicides (Kowalski, 2013).

Acknowledgments

Both authors contributed equally to this chapter – Michael Raupp was lead author of the 'pests' section and Paolo Gonthier was lead author of the 'diseases' section.

Michael Raupp thanks Drs Daniel Herms and Paula Shrewsbury whose authorships on previous publications contributed greatly to the ideas discussed. We also thank two anonymous reviewers for helpful comments on a previous draft. Michael Raupp thanks Dr Glynn Percival for providing images for Figures 18.1, 18.4 and 18.5, and Dr Paula Shrewsbury for Figure 18.2. Michael Raupp's work is supported by the McIntire-Stennis project (MD-ENTM-0416) and grants from ISA Tree Fund and USDA–NIFA.

Paolo Gonthier dedicates this work to the memory of his colleague and friend Giovanni Nicolotti.

References

Alford, D. A. (2012) *Pests of Ornamental Trees, Shrubs and Flowers*, 2nd edn, Manson Publishing. London.

Aukema, J. E., Leung, B., Kovacs, K., Chivers, C., Britton, K. O., Englin, J. et al. (2011) 'Economic impacts of non-native forest insects in the continental United States', *PLoSONE* vol 6, no 9, e24587, doi:10.1371/journal.pone.0024587.

Baietto, M., Pozzi, L., Wilson, A. D., Bassi, D. (2013) 'Evaluation of a portable MOS electronic nose to detect root rots in shade tree species', *Computers and Electronics in Agriculture*, vol 96, pp. 117–125.

Berghardt, K., Tallamy, D. W., Shriver, G. (2009) 'Impact of native plants on bird and butterfly biodiversity in suburban landscapes', *Conservation Biology*, vol 23, pp. 219–224.

Bernicchia, A. (2005) *Polyporaceae s.l., Fungi Europaei,* Candusso, Alassio, Italy.

Bolsinger, M., Flückiger, W. (1987) 'Enhanced aphid infestation at motorways: the role of ambient air pollution', *Entomologia Experimentalis et Applicata*, vol 45, pp. 237–243.

Braun, S., Flückiger, W. (1984) 'Increased population of aphid *Aphis pomi* at a motorway, part 2: The effect of drought and deicing salt', *Environmental Pollution Series A, Ecological and Biological*, vol 36, pp. 261–270.

Clark, J. R., Matheny, N. (2010) 'The research foundation to tree pruning: a review of the literature', *Arboriculture and Urban Forestry*, vol 36, pp. 110–120.

Coetzee, M. P., Wingfield, B. D., Harrington, T. C., Steimel, J., Coutinho, T. A., Wingfield, M. J. (2001) 'The root rot fungus *Armillaria mellea* introduced into South Africa by early Dutch settlers', *Molecular Ecology*, vol 10, pp. 387–396.

Dale, A. G., Youngsteadt, E., Frank, S. D. (2016) 'Forecasting the effects of heat and pests on urban trees: impervious surface thresholds and the pace to plant technique', *Arboriculture and Urban Forestry*, vol 42, no 3, pp. 181–191.

Dal Maso, E., Cocking, J., Montecchio, L. (2014) 'Efficacy tests on commercial fungicides against ash dieback *in vitro* and by trunk injection', *Urban Forestry and Urban Greening*, vol 13, pp. 697–703.

Forestry Commission of the United Kingdom (2015) 'Oak processionary moth (*Thaumetopoea processionea*)', retrieved from www.forestry.gov.uk/oakprocessionarymoth#lifecycle (accessed December 15, 2015).

Gandhi, K. J. K., Herms, D. A. (2010) 'Direct and indirect effects of invasive insect herbivores on ecological processes and interactions in forests of eastern North America', *Biological Invasions*, vol 12, pp. 389–405.

Giordano, L., Sillo, F., Guglielmo, F., Gonthier, P. (2015) 'Comparing visual inspection of trees and molecular analysis of internal wood tissues for the diagnosis of wood decay fungi', *Forestry*, vol 88, pp. 465–470.

Gonthier, P. (2010) 'Controlling root and butt rot diseases in alpine European forests', in A. Arya, A. E. Perelló (eds), *Management of Fungal Plant Pathogens*, CAB International, Wallingford, UK.

Gonthier, P., Nicolotti, G. (2007) 'A field key to identify common wood decay fungal species on standing trees', *Arboriculture and Urban Forestry*, vol 33, pp. 410–420.

Gonthier, P., Guglielmo, F., Sillo, F., Giordano, L., Garbelotto, M. (2015) 'A molecular diagnostic assay for the detection and identification of wood decay fungi of conifers', *Forest Pathology*, vol 45, pp. 89–101.

Gonthier, P., Anselmi, N., Capretti, P., Bussotti, F., Feducci, M., Giordano, L. et al. (2014) 'An integrated approach to control the introduced forest pathogen *Heterobasidion irregulare* in Europe', *Forestry*, vol 87, pp. 471–481.

Grabenweger, G. (2004) 'Poor control of the horse chestnut leafminer, *Cameraria ohridella* (Lepidoptera: Gracillariidae), by native European parasitoids: a synchronisation problem', *European Journal of Entomology*, vol 101, pp. 189–192.

Gross, A., Holdenrieder, O., Pautasso, M., Queloz, V., Sieber, S. N. (2014) '*Hymenoscyphus pseudoalbidus*, the causal agent of European ash dieback', *Molecular Plant Pathology*, vol 15, pp. 5–21.

Guglielmo, F., Gonthier, P., Garbelotto, M., Nicolotti, G. (2010) 'Optimization of sampling procedures for DNA-based diagnosis of wood decay fungi in standing trees', *Letters in Applied Microbiology*, vol 51, pp. 90–97.

Guglielmo, F., Gonthier, P., Garbelotto, M., Nicolotti, G. (2008) 'A PCR-based method for the identification of important wood rotting fungal taxa within *Ganoderma*, *Inonotus* s.l. and *Phellinus* s.l.', *FEMS Microbiology Letters*, vol 282, pp. 228–237.

Guglielmo, F., Bergemann, S.E., Gonthier, P., Nicolotti, G., Garbelotto, M. (2007) 'A multiplex PCR-based method for the detection and early identification of wood rotting fungi in standing trees', *Journal of Applied Microbiology*, vol 103, pp. 1490–1507.

Hanks, L.M., Paine, T. D., Millar, J. G., Campbell, C. D., Schuch, U. K. (1999) 'Water relations of host trees and resistance to the phloem-boring beetle *Phoracantha semipunctata* F. (Coleoptera: Cerambycidae)', *Oecologia*, vol 119, pp. 400–407.

Harrington, T. C. (2013) 'Ceratocystis diseases', in P. Gonthier, G. Nicolotti (eds), *Infectious Forest Diseases*, CAB International, Wallingford, UK.

Herms, D. A. (2002) 'Effects of fertilization on insect resistance of woody ornamental plants: reassessing an entrenched paradigm', *Environmental Entomology*, vol 31, pp. 923–933.

Herms, D. A., McCullough, D. G. (2014) 'Emerald ash borer invasion of North America: history, biology, ecology, impacts, and management', *Annual Review of Entomology*, vol 59, pp.13–30.

Herms, D. A., Mattson, W. J., Karowe, D. N., Coleman, M. D., Trier, T. M., Birr, B. A., Isebrands, J. G. (1996) 'Variable performance of outbreak defoliators on aspen clones exposed to elevated CO_2 and O_3', in J. Hom, R. Birdsey, K. O'Brian (eds), *Proceedings of the 1995 North Global Change Program*, Pittsburg, Pennsylvania; USDA Forest Service General Technical Report NE-214, pp. 43–55.

Johnson, W. T., Lyon, H. R. (1991) *Insects that Feed on Trees and Shrubs*, 2nd edn, Cornell University Press, Ithaca, NY.

Jones, M. E., Paine, T. D., Fenn, M. E., Poth, M. A. (2004) 'Influence of ozone and nitrogen deposition on bark beetle activity under drought conditions', *Forest Ecology and Management*, vol 200, pp. 67–76.

Kirisits, T. (2013) 'Dutch elm disease and other Ophiostoma diseases', in P. Gonthier, G. Nicolotti (eds), *Infectious Forest Diseases*, CAB International, Wallingford, UK.

Kowalski, T. (2013) 'Foliar diseases of broadleaved trees', in P. Gonthier, G. Nicolotti (eds), *Infectious Forest Diseases*, CAB International, Wallingford, UK.

Lonsdale, D., Pautasso, M., Holdenrieder, O. (2008) 'Wood-decaying fungi in the forest: conservation needs and management options', *European Journal of Forest Research*, vol 127, pp. 1–22.

Martinson, H. M., Raupp, M. J. (2013) 'A meta-analysis of the effects of urbanization on ground beetle communities', *Ecosphere*, vol 4, no 5, p. 60.

Martinson, H., Sargent, C. S., Raupp, M. J. (2014) 'Tree water stress and insect geographic origin influence patterns of herbivory in green (*Fraxinus pennsylvanica*) and Manchurian (*F. mandshurica*) ash', *Arboriculture and Urban Forestry*, vol 40, no 6, pp. 332–344.

Nicolotti, G., Gonthier, P., Guglielmo, F. (2010) 'Advances in detection and identification of wood rotting fungi in timber and standing trees', in Y. Gherbawy, K. Voigt (eds), *Molecular Identification of Fungi*, Springer Verlag, Berlin.

Nielsen, D. G., Muilenburg, V. L., Herms, D. A. (2011) 'Interspecific variation in resistance of Asian, European, and North American birches (*Betula* spp.) to bronze birch borer (Coleoptera: Buprestidae)', *Environmental Entomology*, vol 40, pp. 648–653.

Parker, I. M., Gilbert, G. S. (2004) 'The evolutionary ecology of novel plant-pathogen interactions', *Annual Review of Ecology, Evolution and Systematics*, vol 35, pp. 675–700.

Percival, G. C., Barrow I., Novissa K., Keary I., Pennington, P. (2011) 'The impact of horse chestnut leaf miner (*Cameraria ohridella* Deschka and Dimić; HCLM) on vitality, growth and reproduction of *Aesculus hippocastanum* L.', *Urban Forestry and Urban Greening*, vol 10, pp.11–17.

Raupp, M. J., Buckelew Cumming, A., Raupp, E. C. (2006) 'Street tree diversity in eastern North America and its potential for tree loss to exotic pests', *Journal of Arboriculture*, vol 32, pp. 297–304.

Raupp, M. J., Shrewsbury, P. M., Herms, D. A. (2010) 'Ecology of herbivorous arthropods in urban landscapes', *Annual Review of Entomology*, vol 55, pp. 19–38.

Raupp, M. J., Shrewsbury, P. M., Herms, D. A. (2012) 'Disasters by design: outbreaks along urban gradients', in P. Barbosa, D. K Letourneau, A. A. Agrawal (eds), *Insect Outbreaks – Revisited*, John Wiley & Sons, Chichester, UK, pp. 313–340.

Shrewsbury, P. M., Raupp, M. J. (2006) 'Do top-down or bottom-up forces determine *Stephanitis pyrioides* abundance in urban landscapes?', *Ecological Applications*, vol 16, pp. 262–272.

Sinclair, W. A., Lyon, H. H. (2005) *Diseases of Trees and Shrubs*, 2nd edn, Cornell University Press, Ithaca, NY.

Speight, M. R., Hails, R. S., Gilbert, M., Foggo, M. (1998) 'Horse chestnut scale (*Pulvinaria regalis*) (Homoptera: Coccidae) and urban host tree environment', *Ecology*, vol 79, pp. 1503–1513.

Tainter, F. H., Baker, F. A. (1996) *Principles of Forest Pathology*, John Wiley & Sons, Hoboken, NJ.

Tilbury, C., Evans, H. (2003) 'Horse chestnut leaf miner, *Cameraria ohridella* Desch. & Dem. (Lepidoptera: Gracillariidae)', *Exotic Pest Alert. Forestry Commission*, retrieved from www.forestry.gov. uk/pdf/Horsechestnut.pdf/$FILE/Horsechestnut.pdf (accessed December 31, 2015).

Tubby, K. V., Webber, J. F. (2010) 'Pests and diseases threatening urban trees under a changing climate', *Forestry*, vol 83, pp. 451–459.

Vasaitis, R. (2013) 'Heart rots, sap rots and canker rots', in P. Gonthier, G. Nicolotti (eds), *Infectious Forest Diseases*, CAB International, Wallingford, UK.

Vigouroux, A., Olivier, R. (2004) 'First hybrid plane trees to show resistance against canker stain (*Ceratocystis fimbriata* f.sp. *platani*)', *Forest Pathology*, vol 34, pp. 307–319.

19

CONSTRAINTS TO URBAN TREES AND THEIR REMEDIES IN THE BUILT ENVIRONMENT

C. Y. Jim

Introduction

Low-density urban areas have ample above- and below-ground room for trees. This study is focused on high-density or compact urban areas where the competition for space is intense, and where the grey infrastructure prevails to suppress the urban green infrastructure (UGI) (Urban et al., 1988; McPherson et al., 2001). Under these challenging circumstances, trees can grow physically restricted and physiologically stressed (Jim, 1998; Jim and Chen, 2008; Jim and Zhang, 2013).

The critical issues include allocating sufficient plantable space at appropriate locations in the tight urban fabric, designing planting sites for optimal tree growth, ensuring adequate soil volume and quality, choosing suitable species to match site opportunities and limitations, acquiring high-quality planting materials, planting new trees correctly, caring intensively for young trees to ensure successful establishment, and maintaining mature trees in the long term. Every segment in the chain should be strong to realize the multivariate roles expected of urban forests (Jim, 1993, 2013). The notable departure from natural conditions necessitates tremendous efforts and resources to succeed.

To learn from nature, the limited space resource can be shared resourcefully among disparate stakeholders. Close juxtaposition of trees and competitors demands high-quality, tailor-made and site-specific design to mitigate the conflicts. Tree-friendly planting sites should meet the expectations of both trees and people. Whereas the right tree for the right site is often advocated, the right site for the right tree can be equally pertinent. The design can be firmly based on assessing the limitations, anticipating the problems and finding innovative, visionary and lasting solutions. The ultimate goals are extending the usable life span of urban trees, raising their ecosystem, amenity and aesthetic values, improving tree safety, reducing management burden, and improving benefit-cost ratio of urban-forest programmes.

This chapter approaches the research question from the perspective of an urban-tree constraint typology. The inherent soil limitations could dampen root development and tree performance. The subterranean constraints with respect to soil quantity and quality could confine root spread. The subaerial environment may impose headroom and lateral room obstructions, and unfavourable atmospheric conditions can dampen growth. Measures can be adopted to overcome soil volume restrictions. Assessing the limitations due to improper tree management wraps up the study. The identified problems are accompanied by ameliorative or preventive measures. The subterranean constraints due to soil composition and quality, reported in Chapter 20, will not be included here.

Subterranean constraints: soil volume

Confined soil depth and spread

Urban planting sites often suffer from inadequate availability of rootable soil. Soil volume is commonly limited laterally in linear planting areas. The ideal site has soil spread about two to three times of crown diameter at maturity (Perry, 1994), which is rarely fulfilled in urban areas. Lateral soil confinement is attributed to narrow road space, underground utilities, adjacent building foundation, and adjoining heavily compacted soil. The extreme confinement of small tree pits with <2 m^3 soil volume has been described as tea-cup or pot-bound. Moreover, the bottom and sides may be blocked by concrete in the form of a submerged container.

The optimal soil depth for trees is 1 m, which can be reduced for small trees. Effective soil depth (ESD) denotes the rootable portion. Various subsurface restrictions can reduce ESD, including rocky base, boulder, buried paving, utility installation, building foundation, and basement roof slab (Figure 19.1). Internal soil-material organization may shrink ESD, such as compacted layer, clay pan, gravel layer, coarse construction-rubble layer, and stratified layering. Occasionally, concrete baffles are installed in soil as physical root barriers. Soluble chemicals suppressing root growth constitute a special form of volume limitation.

With insufficient soil volume, root spread is curtailed. Besides inadequate capture of water and nutrients, tree anchorage may be compromised. Tree growth will be retarded and stunted, resulting in poor health and premature decline. In planting-site design, sufficient soil volume should be accompanied by high-quality soil. Poor site soil can be ameliorated with amendments and treatments. Soil too poor to be improved can be replaced by a fabricated soil mix based on a specification. If site conditions cannot permit sufficient soil volume, species with a large final size should not be planted.

Soil compaction as volume reduction

Compaction can reduce soil volume due to internal material re-organization. The applied pressure may crumble or deform soil aggregates to reduce bulk volume and pack to a higher bulk density. The inter- and intra-aggregate macro-pores could collapse in response to the applied pressure to become meso-pores and micro-pores. At the microscopic scale, soil porosity for root growth, water transmission, moisture storage and aeration is reduced. The densely packed soil is unfavourable to soil organisms, including many decomposers necessary for nutrient release from organic matter and nutrient cycling. As an undesirable geometric transformation, compaction is tantamount to internal soil volume reduction which is detrimental to root development and tree health.

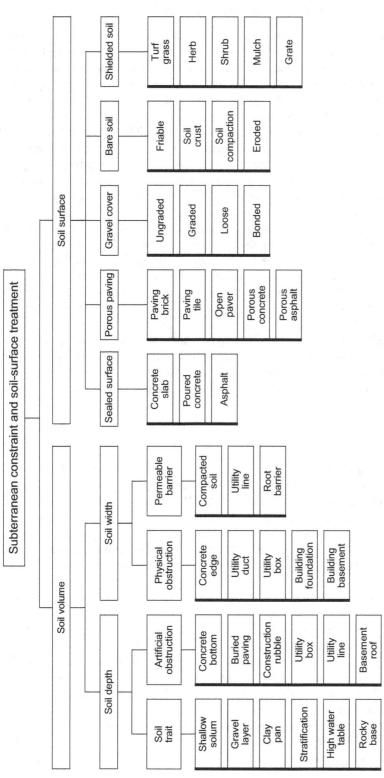

Figure 19.1 Synoptic portrayal of multiple subterranean constraints and soil-surface treatments in relation to tree growth in dense urban areas

Source: C. Y. Jim

Soil sealing as volume limitation

Urban soil is often covered and sealed by impermeable paving materials such as pour concrete or asphalt. Paving slabs or unit pavers allow limited permeability at the joints, which tend to degrade with time. Surface sealing stops water and air infiltration into the soil, and obstructs dissipation of harmful gases from the soil. The soil atmosphere has less oxygen and more carbon dioxide (Viswanathan et al., 2011). The sealed soil is deprived of natural nutrient supplies from organic litter decomposition and rainwater.

Soil moisture replenishment and evaporation are curtailed. The disrupted hydrological cycle has implications on urban stormwater management. Groundwater recharge will be reduced which may lower the groundwater table. With little evaporative cooling, soil temperature may be elevated on hot sunny days to harm tree growth. Soil organism activities are suppressed. Metabolic wastes and other undesirable substances cannot be washed away from the rooting zone by percolating water.

With degraded essential soil properties and processes, the sealed soil is unfavourable to root growth, corresponding to trimming the effective soil volume. The natural topsoil (O and A horizons) below the impermeable paving is usually stripped. Overall, the sealed soil has smaller root mass and less rigorous root functions. The affected trees suffer chronically from water and nutrient deficiency, and slow and stunted growth with implications on tree stability and safety. Young trees may not establish successfully and weak young or mature ones may decline and die.

Soil not covered by paving may also suffer from sealing. Exposed soil not protected by organic litter and living vegetation can be sealed by two processes. Rainsplash hitting the bare soil can induce soil crust formation to seal the soil surface. Trampling by human foot traffic can compact and seal surface soil. Such deleterious changes may bring impacts similar to impermeable paving.

Soil volume fulfilment

Large soil areas can facilitate vegetative growth and reproductive output (McConnaughay and Bazzaz, 1991). Large planting sites with open soil can notably improve tree growth rate, vitality and performance uniformity (Bühler et al., 2007). Conversely, limited open soil areas can suppress tree growth (Sanders and Grabosky, 2014).

If 1 m soil depth can be provided, the key consideration in *Soil volume provision* (SVP) can boil down to meeting *Soil area provision* (SAP). Two methods could estimate SAP, namely *Drip line diameter* (DLD) and *Critical root radius* (CRR = 18 × DBH in cm) (Pornosky, 2003). CRR usually defines a larger soil area than DLD. It also demarcates the area for soil improvement. To future-proof SAP, the reckoning should use expected final dimensions of the species, or the biological potential size.

Based mainly on moisture requirements, the minimum soil volume rule-of-thumb is 0.3–0.6 m^3 for 1 m^2 of crown projection area (Lindsey and Bassuk, 1991). To facilitate computation, using the upper limit this can be translated into about 1.8 m^3 of soil volume for 1 m crown diameter, and the width and length of SAP of different configurations in relation to crown diameter are plotted in Figure 19.2. The SAP dimensional guideline offers the preferred choice where space is available.

In practice, SAP shape can be flexibly adjusted. The concept of SAP compensation involves making up inadequacy in the short axis by extension in the long axis (Figure 19.3). Thus it can be satisfied as a square plot or an elongated narrow corridor of varying width

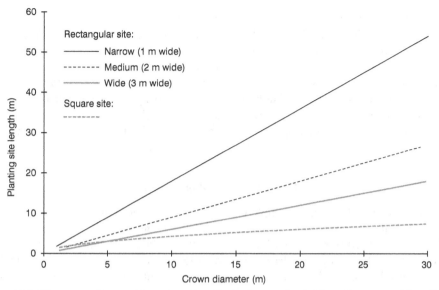

Figure 19.2 The crown diameter is proposed as the basis to estimate the soil area provision (SAR) expressed as planting site length for four site configurations, namely three rectangular and one square with varying site widths. A standard soil depth of 1 m is recommended

Source: C. Y. Jim

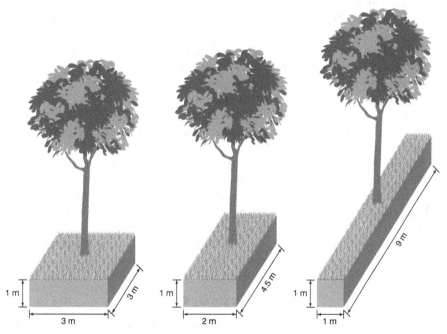

Figure 19.3 Three-dimensional depiction of the width and length of a sample planting site of 9 m³ soil volume and 9 m² soil area, with soil depth at a standard of 1 m. The soil volume can be achieved by a square or rectangular area of different widths. The soil volume of a rectangular site can be contiguous with and shared with adjacent trees

Source: C. Y. Jim

and length. In a narrow roadside site, using a linear SAP is a compromise due to limited root anchorage perpendicular to the road alignment which lowers wind resistance in the same orientation. However, it denotes an improvement over the conventional small square tree pit which is even worse in terms of windfirm characteristics.

Soil depth can be improved by removing or treating restrictions (Figure 19.4). Concrete bottom, buried paving, abandoned utilities, boulder and rocky base can be extracted, and live utilities can be relocated. Usable soil depth can be increased by loosening compacted layer, mixing stratified layers, lowering groundwater table by subsurface drainage, creating soil mound or raised planter to compensate for the lack of soil depth.

Innovative soil-volume design

SAP can be enlarged through multiple measures. The enlarged part can be contiguous to the original site, or linked to an off-site green patch via a *soil connector* (also known as root path or root channel) (Urban et al., 1988) (Figure 19.4). The soil volume can be shared with adjacent trees. The multiple modes of SAP expansion are depicted in Figure 19.5, in which the columns denote site dimension, and the rows site configuration. Moving from left to right represents progressive increase in site size, and moving downwards increase in soil continuity. The improvements are explained by rows:

1   *Pit-isolated:* Small and isolated pits are expanded by elongation along the long axis, and by widening with the footway portion covered by suspended or supported porous paving to preserve the walking surface.
2   *Pit-connected:* Small pits can break out to reach a nearby green patch with soil connectors, which can be a buried soil-filled pipe. Elongated pits can be connected similarly. The widened pits can be connected with a wider pipe. Soil sharing by adjacent trees can happen in the soil connectors and may extend into tree pits. Besides pipes, a *soil corridor* or *soil trench* covered by paving can be installed. A miniature version using vertically placed plastic drainage sheet can be adopted (Gilman, undated).
3   *Strip:* A narrow tree strip can be widened to cover the entire footway width, with the extended part covered by suspended or supported porous paving to preserve the walking surface.
4   *Peninsula:* Small pits can be extended beyond the kerb into the carriageway in the form of planting peninsulas. Instead of a small peninsular for each pit, a rectangular strip can be created and extended beyond the kerb as a wide peninsula. The peninsulas can be covered by porous paving to support vehicular traffic to maintain the usable carriageway surface.
5   *Off-site patch:* The pits or strip can be linked to a nearby green patch via soil connectors to increase the rootable soil volume. The soil connectors can be laid perpendicular or angled to facilitate root extension.

Innovative paving design

The conventional footway design with a firmly compacted sub-base to support the paving is inimical to root growth. The innovative paving does not need to rest directly on soil. Instead, piers and beams can support the suspended paving (Figure 19.4). Where appropriate, a cantilever system can be used. The soil can avoid compaction and remain friable. An alternative suspended paving uses a strong three-dimensional plastic framework known as soil cells to support paving slabs (Urban, 2009). Loose soil can fill the spaces in the open cells.

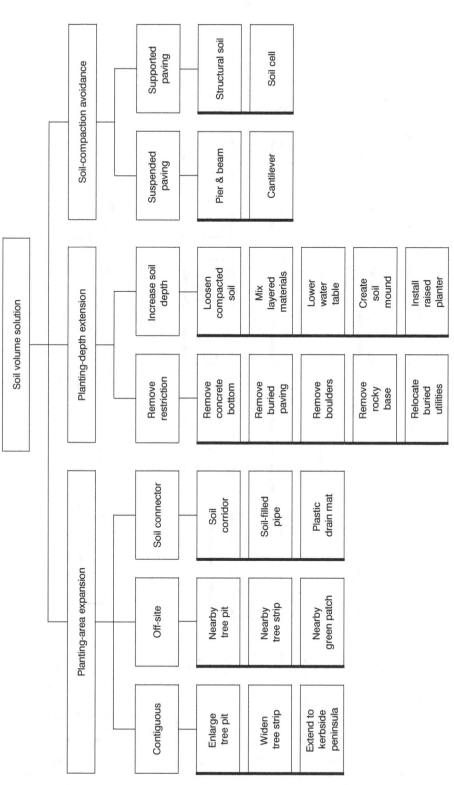

Figure 19.4 The limitations in soil volume can be resolved by three groups of solutions, namely planting-area expansion, planting-depth extension, and soil-compaction avoidance

Source: C. Y. Jim

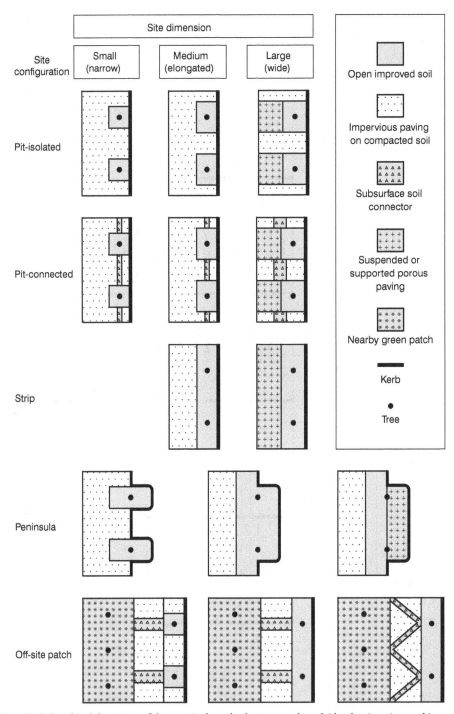

Figure 19.5 Graphical depiction of the practical methods to expand roadside planting sites and improve soil conditions for tree growth. The sites can be enlarged, connected and linked to nearby green patches. The soil can be covered by porous suspended or supported paving

Source: C. Y. Jim

The paving can rest directly on structural soil (Figure 19.4), which is composed of coarse irregular stone aggregates forming a continuous matrix with high load-bearing capacity (see Chapter 22 for details on innovative paving design). The interstitial spaces between the packed aggregates are partly filled with fine earth to support root growth. The Cornell University Structural Soil (Grabosky et al., 2009) and Amsterdam Tree Soil are fine examples (Couenberg, 1994).

A gravel sub-base below the paving can discourage root growth and reduce pavement cracking and heaving (Gilman, 2006). A raised deck can be constructed above rootable soil or exposed roots to provide a walking surface without harming or cutting roots.

Soil surface treatment

The soil surface can receive treatments to improve conditions for tree growth (Figure 19.1). The treatments can prevent compaction by pedestrian traffic, soil erosion by wind or running water and dust problems, and avoid rubbish accumulation.

Porous paving has sufficient macropores to permit water and air to infiltrate quickly to feed the soil lying under it (Volder et al., 2009), and provide a load-bearing, firm and safe surface for pedestrian traffic. Soil moisture can evaporate with little restriction. It can be installed where a walkable footway is needed above rootable soil (Figure 19.5). It is a key component of low-impact development (LID) or sustainable drainage system (SuDS). Commonly used granular materials include loose unbonded graded gravel, and graded gravel bonded with a durable adhesive substance such as synthetic resin or cement. Porous (no-fine) concrete and porous asphalt can be used. Concrete unit pavers of different size, shape and aperture proportion can allow water and air to reach the underlying soil. Grass pavers, grids and geocells made of plastic or polymer can provide load-bearing reinforcement for turf or gravel areas.

For localized surface treatment around the tree pit, an iron grate can cover the pit and the soil surface can be filled with pea or graded gravel to facilitate infiltration and aeration. The gravel can be bonded by epoxy resin to provide a firm walkable surface, in which case the grate can be omitted. Groundcover herb, turfgrass or shrubs can be planted to shield and protect the soil. A layer of organic mulch can serve a similar purpose.

Subterranean constraints: roots

Trenching and excavation damage

In cities where most utility lines are buried at shallow depth under the footway, repeated trenching can incur massive root losses and is a major cause of street tree decline. The reduced root plate may jeopardize tree stability and increase uprooting risk. The loss of absorbing fine roots may trigger induced drought and related health consequences (Fini et al., 2013). The affected trees may become stressed, stunted, unstable and hazardous. If trenching occurs frequently, the weakened tree would not be able to recover and rebuild the root system, leading to long-term decline. The roots of some trees can effectively compartmentalize fungal decay invasions around wounds, hence root severance due to trenching may not induce decay spread (Watson, 2008). In planting-site design, the utility zone can be separated from the dedicated tree zone to avoid conflicts. Alternatively, a utility duct or utility tunnel can separate the two conflicting users of limited underground space.

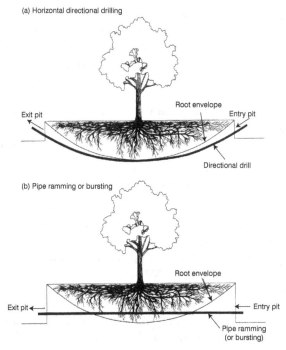

Figure 19.6 In installing underground pipes or cables, the trenchless technique can be used instead of conventional trenching which is extremely destructive to tree roots. (a) It can be implemented by installing a micro-tunnel under the root envelope using a precision directional drilling machine to accommodate the new utility line. (b) An existing pipe lying within the root envelope can be rammed or burst and a new pipe can be inserted in the same conduit so that trenching can be avoided

Source: Jim (2003)

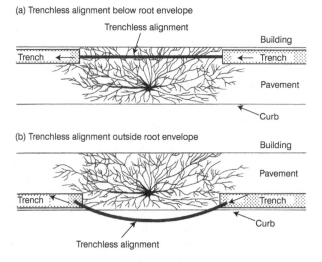

Figure 19.7 The trenchless alignment for underground utility installation can be (a) a straight line under the tree envelope, and (b) a curved line bending away from the rooting area to further reduce impact on root-system integrity

Source: Jim (2003)

In repairing existing pipes and cables, the no-dig, trenchless, directional drilling, micro-tunnelling pipe bursting or pipe rejuvenation technology can be employed to minimize root damages (Jim, 2003). Drilling a micro-tunnel below the root envelope of a tree or ramming an existing pipe to insert a new pipe in the same conduit can avoid trenching within the dripline area (Figure 19.6 and 19.7). The micro-tunnel does need to follow a straight alignment; it could curve downwards or sideways to minimize conflicts with tree roots. The mature technology, including controlled directional drilling and small footprint machines that can be accommodated in narrow street sites, should encourage adoption to reduce massive trenching damages of tree roots.

Trenching should be forbidden near large and outstanding trees. In installing new utility lines, the trench alignment can take a detour or adopt a trenchless method to avoid intruding into the root protection zone (RPZ). Two main approaches have been proposed to define the RPZ, respectively the entire area enclosed by the dripline, and multiples of trunk diameter to determine the RPZ radius. Instead of using a fixed factor, Matheny and Clark (1998) proposed varying it according to tree size and species tolerance to root damage, ranging from 6 to 18 times of DBH.

Pavement damage by tree roots

The soil below conventional impermeable paving often suffers from three physical constraints, namely compaction, surface sealing and inadequate soil volume. Root growth is restricted under the circumstances. As a response to the confinement and obstacles, roots tend to penetrate the interface just below the pavement where more air and water are available. Subsequent root thickening can cause pavement cracking, heaving, rutting and shifting to induce the tripping hazard and expensive repairs (Smiley, 2008).

Some species with large final size can develop thick root flare connecting to large structural roots protruding conspicuously above the soil surface (North et al., 2015). Some species may develop sizeable plank buttress roots. The tree pit or strip may not be large enough to accommodate them. Adjoining paving is prone to damage. Such conflicts will linger and aggravate with time.

In repairing root-damaged pavements, the roots in question are routinely cut. Disregarding the cause of the conflict and tree growth needs, the erroneous approach aims squarely at restoring the hardscape to the pre-damage paving state. Such repairs could not provide long-term solutions, as the paving damage may recur. Moreover, the destructive repairs can predispose trees to premature decline and collapse.

The small planting sites with a tiny rootable soil volume can straitjacket root-system development. Root growth into the surrounding compacted soil is difficult and sparse. Small trees can tolerate, but large trees will soon outgrow the tiny tea-cup soil volume and display distress symptoms. The victim trees will demonstrate sluggish growth, decline and instability.

Damage of underground utilities by tree roots

Tree roots are known to enter underground pipes to cause damage (Randrup et al., 2001). Roots seek water, nutrients and oxygen, hence cracks in pipes carrying potable water, stormwater or sewage are susceptible to exploration by tree roots. The irrigation and drainage pipes at the planting site can also be affected. Especially vulnerable are the old pipes or conduits with degraded shell and cracks. The conflicts are intense where trees and utilities are juxtaposed in the restricted roadside zone.

Moisture leakage from pipes attracts roots and stimulates their growth, which may grow preferentially into the wetted zone around the leak. They may have opportunistic entry into the pipe through the cracks. The pre-requisite for pipe intrusion is the presence of cracks offering moisture and portals for roots to venture into the enclosed conduit. Initially penetrated by small roots, small cracks can be enlarged by secondary root thickening to block the pipe flow and incur expensive repairs and replacements. Leaking old pipe systems offer an insidious and inadvertent supplier of water to suppress water-deficit stress and enhance tree growth. Replacement by new pipes may stress trees which have developed dependence on such supplementary water supply.

The conflict could be avoided by separation of the tree planting zone from the underground utility zone (explained in the 'Trenching and excavation damage' section above). Where they cannot be clearly separated, concrete baffles can be installed at critical conflict locations near large trees with aggressive roots. Flexible root barriers could be wrapped around the vulnerable pipe sections to inhibit root intrusion. In new planting sites, species with strong aggressive root systems and large final dimensions can be avoided. A sufficient buffer distance between trees and pipes could forestall conflict. Strong and durable pipes properly installed with good workmanship can reduce cracking or leakage. Precautions could avoid pipe damage by other causes which provide the pre-condition for tree-root conflicts. For intractable cases, pipe replacement has to be adopted.

Subaerial constraints

Confined subaerial space and tree growth

Subaerial limitation to tree growth is a relativity concept at cramped planting sites, especially at roadside and confined spaces (Jim, 1993, 2013). The key issue is whether a tree fits a site, and vice versa, whether a site fits a tree. The constraints can be firstly evaluated by the affected tree parts, namely root flare together with large surface framework roots, trunk, branch and crown. Secondly, they can be divided into three groups based on the constraint position in relation to the tree parts, namely headroom, lateral room, and omni-direction (Figure 19.8).

Headroom constraints affect the crown in two ways. The crown top requires room for upward growth. It can encounter physical obstruction of adjoining buildings and appurtenances, advertisement signs, aerial utilities, flyovers or footbridges. In compact urban areas, the common cantilevered building awning covers most of the pavement width to impose a headroom restriction. The crown-base height should conform to road traffic and safety requirements. Sufficient headroom clearance should be maintained for unimpeded pedestrian and vehicular flow.

Lateral room constraints affect all four tree parts. Different physical obstructions can block tree growth, such as adjacent buildings and appurtenances, traffic signs, lamp posts and existing trees. The tree-planting paraphernalia including tree guard, tree grate, hard edge of tree strip or pit, and the kerb, can impose girdling damage. Various road traffic clearance guidelines demand unobstructed sightline and visibility splay to traffic lights and signage, and to pedestrians at road junctions. Street lights are expected to be free from branch and foliage blockage. Physical clearance is expected at road crossings and bus and tram stops, lay-by bays, driveways and fire hydrants. Shop owners may not want shielding of shop fronts and signage. Home owners may complain about blocking natural light and ventilation, and allowing burglar access.

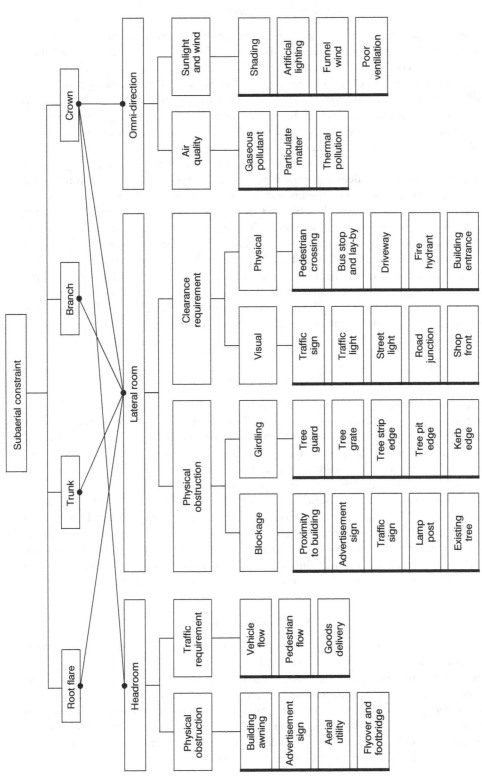

Figure 19.8 Synoptic portrayal of multiple subaerial constraints besetting tree growth in dense urban areas

Source: C. Y. Jim

Matching site and tree geometry

The common pitfall is planting trees with large final dimensions at small or narrow sites, thus predisposing them to future assured incongruity. Such a scenario may incur a double jeopardy, as the confined above-ground realm is often accompanied by equally confined subsurface domain. A heavy management burden is created by short-sighted species selection. The victimized tree has to be heavily and regularly pruned, bringing the collaterals of poor health, instability and hazard. Incorrect branch cutting would aggravate the harms. For compact cities where large planting sites are rare, the matching can ensure that large sites are filled by large trees rather than a cluster of small trees.

Most conflicts in cramped spaces can be forestalled by matching site and tree geometry (Jim, 1997). The remedies demand a refined precision matching between three-dimensional site geometry above and below the ground, and three-dimensional anticipated final tree configuration. Tree species can be classified by key final dimensional criteria, including height, crown spread and trunk diameter. To optimize species selection, a systematic database on species final dimensions and configuration can be established. Only species that can be accommodated by the site should be planted. This recommendation may sound a truism, which unfortunately has been ignored too often. The species selection step of the urban tree management regime should be accorded its due attention and respect. The shoe-fitting analogy (fitting shoes to one's feet) could offer the guiding principle. A beautiful tree that does not fit a site should be considered a misfit, if not an agony, an eyesore, a burden and a risk.

Aerial utility impacts

In cities adopting the aerial utility convention, clearance has to be kept to satisfy safety and assured-supply requirements. Large trees planted below overhead lines are often aggressively and repeatedly lopped or topped, and subject to crown truncation, incision and tunnelling. Trees suffer chronically from a disfigured tree form, and multiple wounds on lopped branches are vulnerable to wood-decay fungal invasion. The weak, unbalanced and unstable tree structure will increase tree risk, and reduce amenity and ecosystem services.

At narrow roadsides, aerial utilities and trees compete for the confined land strip. Large trees should not be planted directly below the lines. Species with small final dimensions can avoid the spatial contest for the same aerial room. Existing conflicts could be abated by replacing with trees of small final size. If there is sufficient room along the utility corridor, either the lines or the trees can be shifted sideways to remove the spatial overlap. Pruning methods could be refined to reduce impacts on tree structure and health.

Air pollution effects

The air quality, sunlight and wind conditions may affect crown growth in the omni-directional sense (Figure 19.8). Urban areas with poor air quality, including gaseous, particulate and thermal aspects, can stress trees. Low air permeability in street canyons and confined sites can hinder pollutant dispersion and aggravate impacts on trees. Trees can serve as air-cleansing and scrubbing agents. They can absorb gaseous pollutants. Particulates can deposit on leaves and branches by gravity or wind advection to literally filter the air. Rainfall can wash the deposits down to soil. If the pollutant concentration exceeds the tolerance limit, especially for vulnerable species, tree growth will suffer.

Heavy shading and street canyons

The verticality and high coverage of the urban fabric, especially high-rise buildings in narrow street canyons, substantially trim the diurnal duration and intensity of sunshine (Figure 19.8). Long and wide elevated roads present a special case of casting heavy shade below the structures. The photosynthetic efficiency and productivity of trees are depressed. The early morning and late afternoon sunlight with a low incident angle is more affected. In natural forest, trees shade each other. In dense urban areas, trees are less shaded by neighbour trees but more shaded by artificial structures (Compagnon, 2004). Shading by buildings involves blockage of sunlight, whereas adjacent trees allow dappled sunlight to filter through the leaves.

At sites with pre-existing shading problem, planting species that cannot tolerate poor light conditions lead to weak trees with low amenity value (see Chapter 22 for a list of shade-tolerant species). Heavy shading introduced after planting by new adjacent buildings can cause tree decline. The phototropism response may alter branch growth orientation and tree form to degrade tree structural balance and symmetry.

Site assessment can inform species selection. The sky view factor (SVF) of the planting site and estimated duration and intensity of solar radiation in different seasons can be ascertained. SVF can be measured from a 180-degree fisheye photograph, or computed from a three-dimensional digital elevation model. Street orientation exerts influence on diurnal solar path and seasonal changes in apparent solar angle. The latitudinal position is a key determinant of solar energy input and solar angle. Reflected sunlight in the street canyon can offer some but not sufficient relief. The effect of artificial light on trees is yet to be clarified by research.

Narrow street canyons can channel air movement and promote air flow along road alignment to create funnel wind and improve ventilation. Depending on the interactions between wind regime and urban fabric, some streets may suffer from physical blockage of wind by tall and contiguous buildings to restrict ventilation and allow air pollutants to stagnate. Low-quality pruning, especially preferential removal of low branches, can raise the centre of gravity and suppress the damping effect to render trees more susceptible to wind damage (Spatz and Theckes, 2013).

Urban heat island effect and climate change

The urban fabric absorbs and stores a considerable amount of solar energy as sensible heat, and human activities and machineries generate anthropogenic heat. Nature uses solar energy beneficially to cool by evapotranspiration. Humans damage nature and introduce artificial surfaces and materials with high specific heat capacity, which absorbs solar energy detrimentally to cause warming. The elevated urban temperature in comparison with surrounding countryside leads to the urban heat island (UHI) effect. The prevalence of impervious cover in compact urban areas would accentuate the UHI to induce hot day and hot night phenomena.

Elevated air temperature can raise evapotranspiration rates to induce water deficit in trees. The warmth can lengthen the growing season and promote growth. Climate change can upset tree physiology, phenology of leaf fall and emergence, deciduous habit, flowering and fruiting. Opportunities and frequency of successful pollination and dispersal could be reduced due to harmful effect on the population and vitality of commensal pollinators and dispersal agents. With presence of ozone and nitrogen oxides in city air, high temperature can intensify smog formation to harm trees.

Intra-city urban forests and green spaces are known to create cool islands. High tree density and open soil can bring effective cooling by absorption of latent heat of vaporization.

Large greenspace can generate significant cooling, with spillover benefits to adjacent areas especially on the leeward (downwind) side. Trees are benefactors to as well as beneficiaries of the cool-island effect. A city with a high greening rate can lower city temperature, mitigate UHI effect and offer cost-effective climate-change adaptation. Cities that are compact or undergoing densification can take actions to enhance the green-and-blue urban green infrastructure to enhance natural cooling.

Tree-management constraints

Improper and excessive pruning

The intense conflicts between trees and urban infrastructure are often resolved by drastic branch or tree removal. Heavily and repeatedly pruned trees leave many sizeable cutting wounds. The plight is aggravated by improper pruning. Topping and branch tipping, considered as crime against nature (Shigo, 1991), are liberally applied to force trees to conform to outgrown, shrinking or degraded growing space. The recommended wound size should not exceed 5 cm and 10 cm diameter for species with respectively weak and strong compartmentalization capability. Other common unprofessional tree practices include flushed cut, bark tearing, leaving stubs, and excessive pruning of more than 30 per cent of foliage in one year.

The disfigured and enfeebled trees are unable to re-establish a normal tree form. The large wounds are susceptible to invasion by wood decay fungi to cause unsafe trees. The loss of photosynthetic capability and exhaustion of food reserve can further weaken the trees. After topping, crown rebuilding by elongated and expanded epicormics branches creates unsafe attachment to old wood. Heavy epicormic branches attached at or near branch-cutting wounds, which tend to be weakened by decay, are susceptible to breakage. Overall, the short-term pseudo-solutions can lead to long-term decline and risk.

Vandalism and inadvertent injuries

Intentional tree damage by people, especially young trees, is a common problem in dense built-up areas. This unnatural stress contributes to high mortality and suboptimal performance especially in newly planted trees, and imposes a heavy drain on urban-forestry resource. Vandalism is believed to be an expression of casual unruly behaviour rather than systematic and strong tree-dislike mentality. Boredom and lack of appreciation of trees are proposed as predisposing factors. Its occurrence is not too related to the presence or otherwise of protective structures such as tree guards. Larger trees with thicker trunks are less damaged (Richardson and Shackleton, 2014). Newly planted trees situated in areas with lower social-economic status and with high unemployment rate have higher mortality (Nowak et al., 1990). However, the problem can be related to planting and maintenance techniques rather than vandalism (Gilbertson and Bradshaw, 1985). It is worth exploring how improvement in species selection, planting technique and design may help to reduce vandalism and its impact.

Conclusion: precision and visionary arboriculture

Tree growth in dense built-up areas is prone to diverse stresses which are expressed at high frequency and intensity. Trees in old towns, town centres, compact and densifying urban areas are vulnerable to varied abiotic and biotic stress syndromes. They include natural stresses accentuated by human inputs, and human-induced stresses per se. They impose acute and chronic constraints on tree growth to depress tree performance, raise tree mortality, and increase tree management burden. Under the circumstances, the high-cost tree inputs disappointingly usher low-quality products. Much effort and resources invested in urban-forestry programmes are wasted.

Studies in dense urban precincts have thrown light on the plight encountered by trees growing in confined and stressful sites. Recent advances in arboricultural research have offered new knowledge and concepts which can be translated into practical techniques to improve urban tree planting and management. Understanding of the underlying factors and processes leading to the plethora of limitations can inform development of pragmatic alternatives, recommendations and solutions. The entire spectrum of concerns in the cradle-to-grave urban tree package deserves to be optimized to improve tree health and survival in tight urban sites. Urban tree programmes should aim squarely at sustainable, trouble-free, low-maintenance, tree-root-friendly, and conflict-avoidance landscape design.

The quest for stronger, prettier, more functional and more durable trees that can thrive in the urban landscape involves holistic attention to every stage of the urban tree life-cycle. Old and inadequate sites can be ameliorated and new sites designed from scratch to forestall the constraints. It demands the attention of a host of professionals, especially the synergistic and seamless teamwork of arboriculture (trees) and engineering (urban infrastructure). It entails timely application of science to policy and practice. Beginning with site selection and design, at every step the spirit and letters of visionary and precision arboriculture can be earnestly espoused to engender a successful and sustainable delivery. To realize the plan, ingrained but inappropriate attitudes and actions have to be substituted by enlightened ones based firmly on research.

References

Bühler, O., Kristoffersen, P., Larsen, S. U., 2007. Growth of street trees in Copenhagen with emphasis on the effect of different establishment concepts. *Arboriculture and Urban Forestry*, 33(5): 330–337.

Compagnon, R., 2004. Solar and daylight availability in the urban fabric. *Energy and Buildings*, 36: 321–328

Couenberg, E. A. M., 1994. Amsterdam tree soil. In G. W. Watson, D. Neely (eds), *The Landscape Below Ground*. International Society of Arboriculture, Savoy, IL, pp. 24–33.

Fini, A., Ferrini, F., Frangi, P., Piatti, R., Amoros, G., 2013. Effects of root severance by excavation on growth, physiology and uprooting resistance of two urban tree species. *Acta Horticulturae*, 990: 487–494.

Gilbertson, P., Bradshaw, A. D., 1985. Tree survival in cities: The extent and nature of the problem. *Arboricultural Journal*, 9: 131–142.

Gilman, E., 2006. Deflecting roots near sidewalks. *Arboriculture and Urban Forestry*, 32(51): 18–23.

Gilman, E. F., undated. Urban design to accommodate trees: Sidewalk solutions. Department of Environmental Horticulture, University of Florida, Gainesville, FL. Retrieved from http://hort.ifas.ufl.edu/woody/planting (accessed 20 December 2015).

Grabosky, J., Haffner, E., Bassuk, N., 2009. Plant available moisture in stone-soil media for use under pavement while allowing urban tree root growth. *Arboriculture and Urban Forestry*, 35(5): 271–278.

Jim, C. Y., 1993. Trees and high-density urban development: Opportunities out of constraints. *Habitat International*, 17: 13–29.

Jim, C. Y., 1997. Roadside trees in urban Hong Kong: Tree size and growth space. *Arboricultural Journal*, 21: 73–88.

Jim, C. Y., 1998. Impacts of intensive urbanization on trees in Hong Kong. *Environmental Conservation*, 25: 146–159.

Jim, C. Y., 2003. Protection of urban trees from trenching damage in compact city environments. *Cities*, 20: 87–94.

Jim, C. Y., 2013. Sustainable urban greening strategies for compact cities in developing and developed economies. *Urban Ecosystems*, 16: 741–761.

Jim, C. Y., Chen, S. S., 2008. Assessing natural and cultural determinants of urban forest quality in Nanjing (China). *Physical Geography*, 29: 455–473.

Jim, C. Y., Zhang, A., 2013. Defect-disorder and risk assessment of heritage trees in urban Hong Kong. *Urban Forestry and Urban Greening*, 12: 585–596.

Lindsey, P., Bassuk, N., 1991. Specifying soil volumes to meet the water needs of mature urban street trees and trees in containers. *Journal of Arboriculture*, 17(6): 141–149.

Matheny, N. P., Clark, J. R., 1998. *Trees and Development: A Technical Guide to Preservation of Trees During Land Development*. International Society of Arboriculture, Champaign, IL.

McConnaughay, K. D. M., Bazzaz, F. A., 1991. Is physical space a soil resource? *Ecology*, 72: 94–103.

McPherson, E. G., Costello, L. R., Burger, D. W., 2001. Space wars: Can trees win the battle with infrastructure? *Arborist News*, 10(3): 21–14.

North, E. A., Johnson, G. R., Burk, T. E., 2015. Trunk flare diameter predictions as an infrastructure planning tool to reduce tree and sidewalk conflicts. *Urban Forestry and Urban Greening*, 14: 65–71.

Nowak, D. J., McBride, J. R., Beatty, R. A., 1990. Newly planted street tree growth and mortality. *Journal of Arboriculture*, 16(5): 124–129.

Perry, T. O., 1994. Size, design, and management of tree planting sites. In G. W. Watson, D. Neely (eds), *The Landscape Below Ground*. International Society of Arboriculture, Savoy, IL, pp. 3–23.

Pornosky, J. D. (coordinating author), 2003. *Urban Tree Risk Management: A Community Guide to Program Design and Implementation*. USAD Forest Service Northeastern Area, St. Paul, MN.

Randrup, T., McPherson, E., Costello, L., 2001. Tree root intrusion in sewer systems: Review of extent and costs. *Journal of Infrastructure Systems*, 7(1): 26–31.

Richardson, E., Shackleton, C. M., 2014. The extent and perceptions of vandalism as a cause of street tree damage in small towns in the Eastern Cape, South Africa. *Urban Forestry and Urban Greening*, 13: 425–432.

Sanders, J. R., Grabosky, J. C., 2014. 20 years later: Does reduced soil area change overall tree growth? *Urban Forestry and Urban Greening*, 13: 295–303.

Shigo, A. L., 1991. *Modern Arboriculture: A Systems Approach to the Care of Trees and their Associates*. Shigo and Trees Associates, Durham, NH.

Smiley, E. T., 2008. Comparison of methods to reduce sidewalk damage from tree roots. *Arboriculture and Urban Forestry*, 34(3): 179–183.

Smiley, E. T., Calfee, L., Fraedrich, B. R., Smiley, E. J., 2006. Comparison of structural and noncompacted soils for trees surrounded by pavement. *Arboriculture and Urban Forestry*, 32(4): 164–169.

Spatz, H., Theckes, B., 2013. Oscillation damping in trees. *Plant Science*, 207: 66–71.

Urban, J., 2009. An alternative to structural soil for urban trees and rains water management. In G. W. Watson, L. Costello, B. Scharenbroch, E. Gilman (eds), *The Landscape Below Ground III*. International Society of Arboriculture, Champaign, IL, pp. 301–305.

Urban, J., Sievert, R., Patterson, J., 1988. Trees and space: A blueprint for tomorrow. *American Forests*, 94(7/8): 58–74.

Viswanathan, B., Volder, A., Watson, W. T., Aitkenhead-Peterson, J. A., 2011. Impervious and pervious pavements increase soil CO_2 concentrations and reduce root production of American sweetgum (*Liquidambar styraciflua*). *Urban Forestry and Urban Greening*, 10: 133–139.

Volder, A., Watson, W. T., Viswanathan, B., 2009. Potential use of pervious concrete for maintaining existing mature trees during and after urban development. *Urban Forestry and Urban Greening*, 8: 249–256.

Watson, G., 2008. Discoloration and decay in severed tree roots. *Arboriculture and Urban Forestry*, 34(4): 260–264.

PART V

Planting sites
ANALYSIS AND MODIFICATION

20

SITE ASSESSMENT

The key to sustainable urban landscape establishment

Nina Bassuk

What is site assessment in the context of the urban forest?

A site assessment is a thorough and detailed evaluation of landscape site conditions in order to understand limitations or opportunities for successful plant growth and to maximize the potential for sustainable design.

In the most favorable situation, site assessment occurs prior to the design process, when it is possible to take advantage of favorable conditions and modify limitations in the landscape. When site assessment occurs after designs have progressed and even after construction has begun, opportunities for maximizing ecosystem benefits may be lost, such as when soil and plant health are not protected prior to construction. When site assessment does not occur at all, the successful realization of the landscape may be diminished or it may fail outright.

With all of the best of design intentions, anyone trying to plant trees or create a landscape in an urban context needs to have a firm understanding of environmental resources that allow trees to grow to their envisioned size. It is only through efforts that recognize and provide for the needs of plants, that we will gain the benefits for which they were planted in the first place – whether for shade, pollution reduction, reduced storm water runoff, increased property values, erosion control, habitats for wildlife, windbreaks, blocking undesirable views, creating parks for recreation and providing a link between our increasingly urban existence and the natural world.

Most of the places we live in whether rural, suburban or urban have been significantly impacted by human activities whether it is by building a house, creating roads, or laying pavement for a sidewalk or parking lot. These alterations are unquestionably more numerous in an inner city area than a rural village, yet they are fundamentally the same in impact.

We need to approach planting activities in human impacted landscapes as rigorously as we engineer the urban environment into which we place them. No one would think of building a house without adhering to sound construction principles from a solid foundation through sturdy walls and a non-leaky roof. Yet, with trees in the urban context, we seem to feel that they will take care of themselves. The urban landscape has modified and disturbed urban soils through their removal and replacement, re-grading, compaction, cutting and filling and contamination in the process of creating buildings, roads and pavements. All these impacts profoundly change the physical, chemical and biological nature of soil, the very substrate that trees depend on for so many of their vital resources.

Back to basics

The most basic questions asked during a site assessment relate to the acquisition of vital resources for plant growth: water, light, oxygen and carbon dioxide, nutrients and appropriate temperatures.

Where is the water coming from? Is the plant getting too little water due to soil restrictions or too much water due to lack of drainage or a high water table – or perhaps both alternating at different times of the year? Is the soil so dense that roots cannot grow thereby limiting their ability to take up water in a restricted soil volume. Are there enough available nutrients in the soil? Changes in soil pH can make nutrients more or less available to be taken up by plant roots. In urban environments because of debris from human development incorporated into the soil, pH can vary significantly. Is there enough oxygen? All parts of the plants, shoots and roots, need oxygen. If soil drainage is impaired and all soil pores are filled with water, oxygen will be limited and roots may die. Or, if a gas leak displaces oxygen in the soil, plant roots will also suffer. Lack of oxygen in the root zone is one of the quickest killers of vegetation.

Is there enough light for plant growth? Light is the driving force of photosynthesis. Many plants have evolved to take advantage of varying levels of light; however, trees for the most part, being the tallest plants in the landscape require full sun estimated at about 4 hours of sun/day in the growing season (Craul, 1999). Some smaller trees will tolerate lower levels of light. Urban environments with tall buildings that cause false horizons can limit the amount of direct sunlight that trees receive.

Carbon dioxide is rarely a limiting factor in the urban environment or elsewhere. It is the essential gas necessary for the production of carbohydrates during photosynthesis. If plants experience water deficits, their stomata close to prevent greater water loss. Low light levels will also close stomata. When this occurs carbon dioxide is prevented from entering the leaf and photosynthesis is reduced.

Are temperatures appropriate for plant growth? Of course it is critical to know what climate zone you are in. Winter hardiness and summer heat are important determinants of plant distribution and they can be exaggerated in urban environments such as when plants are grown in containers above grade, exposing roots for more extreme variation in temperature. At the other extreme, urban areas are found to create 'heat islands' which can increase temperatures several degrees from the surrounding countryside (Bornstein, 1968). More important are the microclimate effects of building facades, increasing the reflected and radiating heat from car tops and asphalt that can cause trees to lose water faster or in extreme cases directly damage leaves. If trees were in restricted soil conditions so that the increased demand for water caused by the leaves losing water so rapidly could not be met by increased water uptake from the roots, tree growth or survival might suffer.

A process of establishment

With a good understanding of site conditions that promote or limit opportunities for planting, it is now possible to proceed in the plant establishment process. After site assessment, we can then ask about what plants are best adapted to the site. Aside from the basic environmental factors that should be assessed at the site, other constraints should be factored in too. Are particular diseases and insects prevalent in the area, so that certain susceptible plants should not be planted? Are there physical barriers to plant growth underground or above ground such as underground structures, signage, light poles or building overhangs? Is the proposed planting site close to a tall building that may create a rain shadow and prevent natural rainfall from hitting the ground? Legal factors such as rights-of-way, historic districts, legal site restrictions on planting and prohibited planting lists should also be assessed before planting can occur.

Matching up the site constraints or opportunities to plant requirements is the important next step in plant establishment. However, in some cases it is important to recognize when site limitations make the plant choices so restricted (or non-existent) that sites need to be modified before successful planting can occur. Generally speaking, the more one can modify the site towards ideal conditions, the greater the number of plants that can be planted successfully. However, site modification is often labor intensive, especially when it comes to soil remediation. Costs involved in moving or amending soils may be prohibitive. If plant selection alone can match the site conditions, then it should be used.

Assessing the big picture: looking at site context

Landscape development takes place at a variety of scales. The principles of site assessment need to be able to be applied to reflect this.

It is important to obtain accurate maps and records of the surrounding context of the area to be designed and planted. Knowledge of prior use of the site would be informative in case soil contaminants might have been used. Maps should include features that may affect planting site development such as floodplain areas, proximity to streams, lakes, shorelines and wetlands. Variable water tables and seasonal flooding may affect the proposed planting site. There may be an opportunity to reduce degradation, restore watershed function and to increase ecosystem benefits through the restoration of these natural features.

Precipitation

It is essential to know what the annual and seasonal precipitation is in the area. Every option should be explored to reduce the need for irrigation beyond the plant establishment stage, which is generally accepted to be three years from planting (Sustainable SITES Initiative, 2014). If possible, the potential for capturing and using rainwater or other safe gray water should be investigated. In arid climates, plant growth is severely restricted by the lack of water. The principles of xeriscaping should be used when addressing these sites (design with water conservation in mind, improve the soil to increase water holding capacity, use appropriate drought tolerant plants, group plants with similar water needs, reduce areas with turf grass or select turf alternatives, use mulch to conserve water and irrigate efficiently) (Texas Agricultural Extension Service, 2000). In any climate it is important to get local experts involved who are familiar with the patterns of seasonal rainfall and temperature.

Temperature

Other contextual site conditions are temperature during the winter and summer seasons. Average minimum winter temperatures are important in order to choose hardy plant material, while heat load during the summer may reduce still other plants from being successful at the site. One example is the United States Department of Agriculture winter hardiness map, which has mapped minimum winter temperatures over a 20-year period for the United States (US Department of Agriculture, 2012). On the other end of the spectrum, the number of days when the temperature is greater than 30°C has been used as an indicator of heat stress. At greater than 30°C, some plants have shown to suffer cellular damage (American Horticulture Society, 1997). There are plant hardiness maps created for most European,

Asian and some African countries as well. It is important to learn about the wide context of climate patterns before concentrating on microclimates at the specific site to be planted.

Existing vegetation

It is important to identify vegetation existing on site and evaluate their health status and suitability to remain. There may be endangered or rare plants whose existence should be taken into consideration. Some countries provide legal protection for various threatened ecological communities. Identifying and evaluating existing plants will provide some measure of which plants can do well on site, although if modification of soils through grading and compaction is expected to occur, resulting altered drainage and increased root impedance will reduce the health of the soil and the palette of plants that can be expected to do well there. There may also be legal restrictions on where plants can be placed and the height of proposed plantings. Railroads have easements restricting the proximity and height of plants that line the tracks, as do some highways. It is also important to identify any invasive plants and remove them before any new planting occurs.

Soil

Soil health is probably the most critical component of the future success of any planned landscape. It is useful to acquire soil maps developed from a soil survey that may provide information about prevalent soil textures, organic matter, pH, depth and drainage over a large scale. As often as not, soils in urban areas are not mapped due to the high degree of soil disturbance. These soils are often designated as urban soils with little information provided.

Human use

Expected human use on site is an important feature to consider. The urban forest is essentially a forest that exists between the built fabric of human development. It is essential to understand the potential uses and conflicts that may arise between urban development and urban vegetation sharing the same space. Some questions to ask are:

- Will people be invited into new landscape?
- Will there be significant paving for walkways and cars on site?
- How does the new landscape respond to the presence of bicycles and other nearby modes of transportation?
- If there is a building on site, will plants potentially block windows or signage?
- Will there be reflected heat from building facades and pavement?
- Will there be interference with above- and belowground utilities, such as lights, water, gas and sanitary sewer infrastructure.

Assessing the site at the micro scale

Assessing and protecting vegetation and soil

An accurate scale map of the proposed landscape site is important as a place to record site information such as light and shadow patterns and soil characteristics as well as any barriers to planting such as utilities, windows and signage.

Once this is completed, the next task would be to inventory vegetation on site. Information should include:

- species identification;
- size;
- growth rate and general health of trees and shrubs;
- insect or disease infestation;
- potential invasive plants and particularly valuable plants; and
- if there are intact native plant communities, they should be identified, mapped, and marked for protection.

When trees are present, a thorough risk assessment should be done to determine which trees need pruning or removal prior to site construction. If trees are to remain on site, a protection plan should be developed to protect above- and belowground parts of the tree. Root and tree protection zones can be delineated by taking the tree trunk diameter at breast height (DBH) 1.3 meters from the base of the tree in centimeters. Multiply the DBH by 0.134 meters to determine the minimum radius of the tree and root protection zones on all sides (Trowbridge and Bassuk, 2004). A sturdy chain link fence or other recyclable barrier should be erected to protect the enclosed area. When trees are in close proximity, it is best to protect them as a group. Many other countries have developed tree protection standards during development. A notable example is the Australian Standard for the Protection of Trees on Development Sites (Australian Standards, 2009).

Construction staging plan

The most destructive activities on the site of a new landscape are the effects of staging construction activities. Soil compaction due to operating and parking vehicles on site can have a large impact on soil health due to severe soil compaction All parties involved in the design and construction process should agree on a plan for soil protection. This involves designating parking routes for traffic and storage of materials. It makes sense to limit vehicular movement and storage of materials to areas that will not be required to support plant growth. Such areas might be future parking lots, driveways and other paved areas (Figure 20.1).

Above-ground conditions at the micro scale

Exposure, reflected heat and wind

Knowing the site's solar exposure is critical for the choice of appropriate plant material. For plants that require full sun, we would expect to have at least 4 hours of direct sun during the growing season (Craul, 1999). Many plants have adapted to partially shaded or heavily shaded conditions; however, most large trees require full sun.

Tall buildings can create false solar horizons, so assessing solar access with proposed or existing building heights should factor in the hours of sunlight expected during the growing season.

Reflected and radiating heat can play a significant role for plants in the urban environment. Building facades, asphalt, concrete and parked cars have been shown to raise air temperature as much as 12°C higher near trees on a sunny 30°C August day in New York City. Surface

---

1. Site location_____

2. Site description_____

3. Climate

A. Winter cold and summer heat extremes

B. Microclimate factors: reflected heat load, frost pocket, wind

C. Sunlight levels: full sun (4 hrs. or more), partial sun, shade

4. Soil factors

A. Range of ph levels _____ (note actual readings on sketch)

B. Texture: clayey, loamy, sandy

C. Compaction levels: severely compacted, moderately compacted, uncompacted

D. Drainage characteristics: presence of mottled soil, low-lying topography, presence of indicator plants
Percolation test results (in./hr.)(cm/hr.)
Poorly drained (< 4")(<10cm)
Moderately drained (4"–8")(10–20cm)
Excessively drained (> 8")(>20cm)

E. Other soil considerations: indications of soil layer disturbance, evidence of recent construction,
Presence of construction debris, noxious weeds present, evidence of excessive salt usage, erosion of soil
Evident, evidence of soil contamination.

5. Structural factors

A. Limitations to aboveground space: overhead wires, lights and signage, proximity to buildings/structures:

B. Limitations to belowground space: utilities marked and noted on sketch

Approximate rooting volume for site: length: __ width: __ depth

6. Visual assessment of existing plants: species, size, growth rate, and visual assessment

---

Figure 20.1 A summary site assessment checklist to help evaluate site conditions at the proposed planting site

temperatures of car tops and asphalt peaked at 55°C while the temperature of east-facing brick building facades peaked at 43°C during the same day (Bassuk and Whitlow, 1987). While the heat island effect is well known to cause higher temperatures around the 2–4°C range over a large urban area, on a micro scale, air temperatures can be much greater than that, particularly on a sunny day (Bornstein, 1968).

Greater air temperatures act as the driving force for increased transpiration, causing the plants to lose water faster than if they were growing in a more moderate temperature. This has direct relevance for plant selection. More drought-tolerant trees tend to fare better in inner city and parking lot environments.

Particularly windy environments typically have a minor effect on plant growth, sometimes causing more water to be lost by tree leaves by increasing the atmospheric demand, although wind especially near water bodies can significantly cool the air, reducing atmospheric demand. Where winds are regularly greater than 50 mph, tree limb breakage is more prevalent. Where this is likely, smaller trees or groupings of shrubs and trees may be better choices. High winds may cause decreased human comfort. If this is the case, using windbreaks created by trees and shrubs branched to the ground can provide significant wind attenuation (Trowbridge and Mudrak, 1988).

Below-ground conditions: focus on soils

Of the six basic resources required for plant growth, soils provide, store or affect four of them. Soils store water, provide nutrients, allow oxygen to diffuse to plant roots, and affect root zone temperature due to its large or small mass. It is not surprising that many researchers have suggested that soil issues play the greatest role in determining a successful or failing urban landscape (Patterson, 1977).

In urban or human impacted landscapes, soil may be disturbed in several ways that affect resource capture. Notably, Craul listed the eight ways in which urban soils may be expected to be impaired (Craul, 1992).

1 Urban soils have vertical variability caused by cutting and filling that reflects human development.
2 Soil structure has been changed so that the crumbly aggregation of soil particles is crushed into an undifferentiated mass with high bulk density restricting root growth.
3 Urban soils often have an impervious crust that sheds water. Compaction compounded with lack of surface vegetation gives rise to this condition.
4 Soil pH may be changed due to contaminants and from runoff from built surfaces.
5 Compaction destroys the macro pores in the soil, causing impeded drainage.
6 Urban soils often have interrupted nutrient and organic matter cycling due to removal of leaves and other vegetation.
7 Urban soils often contain discarded human debris from earlier development. Evidence of building materials, foundations, utilities and other rubble remain in urban soil.
8 Urban soils are often separated from the large mass of the earth when they are placed on rooftops and planters. These relatively small masses of soil can heat and cool much faster than a large earth mass. These rapidly changing temperature extremes can challenge plants that have adapted to more modest changes.

Although these conditions can be expected in urban areas, they may or may not be present. It is important to test the soil to determine what if any limitations are present.

Soil characteristics are often separated into physical, chemical and biological components. The function of a soil, urban or not, depends on the interplay among these factors. 'Soil health' is a relatively new concept that integrates the physical, chemical and biological aspects of soil that allows it to be a medium for supporting plant life and a provider of ecosystem services. Studies have been conducted to determine minimum thresholds of soil quality indicators that are responsive to management and correlate with plant growth and health (Scharenbroch and Catania, 2012; Schindelbeck et al., 2008). The concept of soil health is important because it integrates many aspects of soil science and provides a conceptual framework that can be understood by a range of audiences from homeowners to professional scientists. It also creates quantifiable standards that are easy to interpret and can be used to develop management plans to address specific diminished or damaged components of the soil. Repeat measurement provides a feedback mechanism to monitor quality changes over time. Traditionally, soil testing focused principally on chemical aspects of soils. Physical and biological factors also play significant roles in a soil's ability to support landscape plantings and buffer environmental impacts.

Soil texture and structure

Soil texture refers to the percentage of sand, silt and clay particles in the soil while structure is defined by the amount and stability of aggregation of primary particles into peds or crumbs usually with the aid of organic matter. Texture and structure most directly affect soil characteristics by affecting drainage. The pore size and interconnectedness of pores determines the rate of drainage of water and the water holding capacity of the soil. If a soil drains poorly, oxygen will be limited in the root zone and only those plants adapted to wet soils will do well. If water drains very rapidly, water may be limited for plant growth. Ideally you would like to see adequate drainage, greater than 10 centimetres per hour, and good water holding capacity. The dryness of the soil along with expected precipitation will determine the types and range of plants that can be grown on a site (Trowbridge and Bassuk, 2004).

It is relatively easy to measure the texture and structure of a soil.

Texture may be measured in the field using a 'texture by feel' method (Figure 20.2), which is moderately accurate for determining general sandy, silty or clayey characteristics (Thien, 1997). For greater accuracy soil texture is determined in the lab where a known dry weight of soil is mixed with a 3 percent soap solution to suspend and separate particles. Sand settles out first and can be collected on an appropriately sized sieve. If further information of the sizes of sand is desired, the sand fraction can be shaken over a series of nested sieves of decreasing size. The weight of all sand fractions is then measured. After the sand is removed, silt and clay are re-suspended on the soapy solution by stirring, after which the silt is allowed to settle for 2 hours. The still suspended clay is then poured off and the silt is dried and weighed. Percentage of sand is determined by dividing the weight of sand by the total weight of the soil. The same procedure is done for determining the percentage of silt. Percentage of clay equals 100 percent minus percentage sand minus percentage silt (Gugino et al., 2009). A soil with more clay typically has greater cation exchange capacity and thus a greater ability to hold nutrients. The distribution of pore sizes will affect drainage and air movement, water infiltration and water retention. The aeration of the soil will also affect soil microorganisms, which affect nutrient availability.

Although soil disturbance doesn't affect soil texture, soil structure, or the aggregation of sand, silt and clay into larger crumbs, can be highly affected by soil disturbance. A well-aggregated soil will generally have good water retention and good drainage. However, aggregates are easily destroyed by soil compaction, especially when the soil is wet and particles within the aggregates have little resistance. Soil protection during construction is largely the practice of reducing the effects of compaction on existing soil and the destruction of soil structure. Aggregate stability can be measured in a lab. A known weight of soil made up of particles between 0.25 and 2.0 mm is layered on a 0.25 mm sieve. A rainfall simulator then drops water for five minutes at a force equivalent to a heavy thunderstorm. Soil crumbs that have slaked or de-aggregated with the force of water represent unstable aggregates. The soil that falls through the sieve is collected, dried and weighed, as is the soil that remains on top of the sieve. The percentage of stable aggregates can be calculated by dividing the weight of aggregates that didn't pass through the 0.25 mm sieve by the total soil. When a soil has about 50 percent or greater aggregates, it is considered a well-aggregated soil (Gugino et al., 2009).

Soil compaction

Soil compaction is probably the most ubiquitous condition found in urban soils.

Compacted soil has poor aggregate stability, poor drainage, increased root growth impedance, and diminished biological activity due to poor soil aeration.

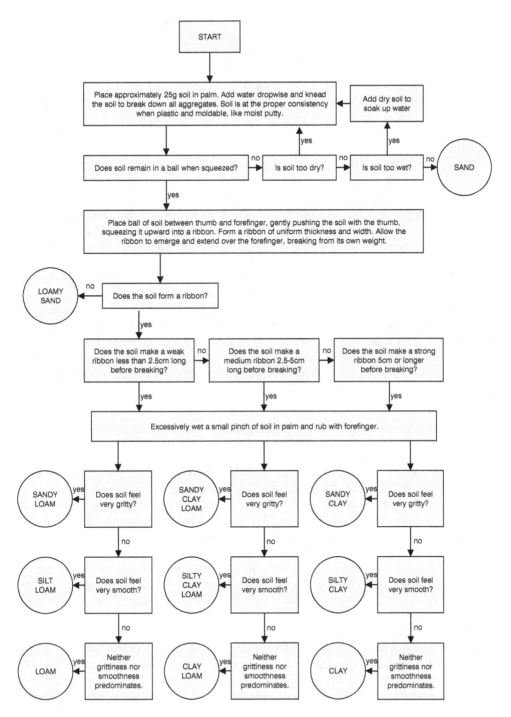

Figure 20.2 Soil texture by feel field method

Source: after Thien (1979)

It is critical that soil compaction is measured on the proposed site. While plants may be selected that are adapted to poorly draining soils, droughty soils and various pH ranges, no plant grows well in conditions where root impedance is high. Soil compaction can be measured in several ways. The most common way is to measure the soil's dry density or bulk density. Bulk density is a measure of the weight of an undisturbed soil sample (g) divided by its volume (cc).

There are well-documented densities after which root growth will not occur. These thresholds differ depending on soil texture. For a clayey soil, root impedance occurs at the bulk density of 1.4 g/cc. For a sandy soil, root impedance occurs at 1.65 g/cc. For pure sand, impedance occurs at 1.7 g/cc (Daddow and Warrington, 1983).

Although these measures are considered thresholds for root impedance, the lower the density, the greater the root growth, down to a point where soil density is so loose that root growth is no longer affected, approximately at 1.0 g/cc.

Bulk density is most often measured using an undisturbed soil sampler. A known volume of soil is carefully extracted from the soil, dried and weighed. The dry weight of the soil is divided by the volume of the sampler. It is useful to take the sample where root growth is expected to occur below any turf or mulch layer.

Another way of determining soil density is by measuring penetration resistance with the soil penetrometer. A penetrometer is an instrument that measures the force it takes to push it through soil. A useful field instrument measures force as pounds per square inch (psi). Above 300 psi, no root growth occurs. As the penetrometer is pushed into the ground, there is a point where resistance exceeds 300psi. The area above that point can be considered to be the depth of useable soil. Soil moisture is an important variable when measuring penetration resistance, the greater the soil moisture the less the resistance. When comparing sites, soil moisture should be comparable (Gugino et al., 2009).

Soil pH

The acidy or alkalinity of the soil (soil pH) is a simple but important test as part of a site assessment. Soil pH largely determines the availability of the macro and micro essential nutrients in the soil. The pH scale runs from 0 to 14, with 7 being neutral.

Most soils may have a pH between 5.0 and 8.0. Since pH is measured logarithmically, the difference between 5.0 and 6.0 is tenfold, but the difference between 5.0 and 7.0 is a thousand-fold. The range where most nutrients have the greatest availability is between 6.5 and 7.0. However, many plants have a wide range of pH adaptability into the alkaline range while some have narrow ranges of adaptability in the acidic range. Since urban environments often have a more alkaline soil pH because of leaching from building materials such as concrete and asphalt, it is useful to find good plant references that evaluate plant growth beyond the ideal range of 6.5–7.0.

Many plants have a wide range of adaptability and plants should be chosen that are adapted to the existing soil pH. Changing soil pH in the landscape is a difficult endeavor. The reasons for a soil pH are based most likely on its parent rock material. Limestone parent rock will tend to be alkaline while granite parent material will be more acidic. Limestone amendments are used to raise soil pH and sulfur is used to lower pH. It is easier to raise the pH than it is to lower it; however, any soil amendment used to change pH would have to be carefully monitored over time, and additions of the amendment added on a regular basis. Moreover, as a perennial landscape is not regularly tilled, amendments could only be placed on top of the soil after planting. It is much more sustainable to use plants that are adapted to the pH on site than trying to change it (Craul, 1992).

There are numerous colorimetric or electrode-based pH tests. Often a robust field kit will provide the site designer useful information without having to wait for laboratory results.

Soil organic matter

Soil organic matter plays an important role in plant health. It can improve water storage and decrease compaction, improve soil aggregation, act as a food source of beneficial soil microorganisms that help to release nutrients, reduce some soil borne diseases and detoxify harmful substances.

Soil available water holding capacity

Soils play an essential role in storing water. The texture and structure of the soil, which determine soil pore size as well as its organic matter content and soil depth, largely determine a soil's water storage capacity. Sandy soils generally have a lower available water holding capacity because sand macro pores allow water to drain rapidly after saturation and adhesive forces that attract water to solid surfaces are less due to a large percentage of these macro pores. In clayey soils, soil drains slowly after saturation and more water is held in the very small pores due to high forces of adhesion. Although clayey soils hold more water, they also hold more water at 1500 kPa due to the very high adhesive forces. It is important to know the available water holding capacity in contrast to the total water holding capacity to determine how much water will be available for plant growth (Gugino et al., 2009).

Infiltration, percolation and water table

Soil water storage is important for plant growth and also for storm water runoff reduction. Storm water attenuation and storage are valuable ecosystem benefits that urban soils play in largely paved environments. Because of soil compaction and a lack of organic matter in urban soils, the soil surface may become sealed due to primary particle de-aggregation and high bulk density. If this happens, water infiltration is very slow and will sheet drain over the surface to the lowest point.

Landscape grading has a significant effect of surface water flow. If the landscape is flat and water does not move to a low point, the surface soil will be saturated with all surface pores filled with water and no aeration. There are several infiltrometers that measure water movement into the soil surface.

Percolation is a measure of water movement within the soil profile.

It is a composite measure of texture, structure, compaction, topography, and layering of different materials and soils that affect water movement. As such, it is an important part of any site assessment

Because the urban soil is often highly disturbed and may be layered, percolation tests may need to be done at several depths to understand the full depth of drainage and evaluate the effect of hidden restrictive layers. For tree growth, good drainage (greater than 10 cm/hour) should reach a depth of 1 meter (Trowbridge and Bassuk, 2004). Shrub and herbaceous plant growth may not require such a deep well-drained profile. Percolation tests are conducted by first removing the sod or other vegetation from the test site. A hole approximately 30–90 cm deep and half as wide is dug into the soil. Depths will vary based on desired plant material. Sequentially deeper holes may be dug to test if hidden compacted layers and different soil textures are suspected. Water is poured into the hole and allowed to drain once or twice in

Figure 20.3 Soil percolation testing in the field

order to saturate the sides of the hole and reduce or eliminate the sideways movement of water through soil adhesive forces. As water will move in all directions against gravity when the soil is not saturated, doing a percolation test in very wet soils reduces the effects of lateral water movement that could overestimate the downward movement of water (Figure 20.3).

During a percolation test, water may be encountered that rises into the hole from the bottom or side. This may be due to an underground spring, stream or nearby water body. High water tables are critical to locate if one is suspected. If this is the case, raising the soil grade on top of the high water table is often the only way to grow a wide variety of plants. Conversely, if a persistent wet area is discovered, wet tolerant plants may be used in that area.

Soil chemistry

Soils are the primary source of nutrients for plant growth. The nutrient availability of a soil depends on the soil pH and texture, cation exchange capacity (CEC) and microorganisms that break down organic forms of essential elements to inorganic forms that can be readily taken up.

The macronutrients that plant need are:

- nitrogen;
- phosphorus;
- potassium;
- sulfur;
- calcium; and
- magnesium.

Just as critical but needed in reduced amounts are the essential micronutrients:

* iron;
* manganese;
* copper;
* zinc;
* molybdenum;
* boron; and
* chorine.

Cation exchange capacity (CEC) is the ability of soils to adsorb cations of important nutrients such as calcium, magnesium, and iron among others. As nutrients break down to simple ionic forms, a soil with a high cation exchange capacity will hold onto these ions until roots take them up in the soil solution. Nutrients can leach downward beyond the root zone in soils with low CEC. Soils that hold a negative charge, such as those with clay or organic matter, have a higher CEC. Sandy soils with low organic matter have a lower CEC. It is useful to get a soil chemical analysis from a laboratory that is used to making soil recommendations for your area. For perennial landscapes, nutrients will have greater availability when the organic matter and soil physical properties are optimal.

Soil contamination

Based on the prior use of the site, soil contaminants may be present. Testing for heavy metals is essential, especially lead and cadmium that may be on site after industrial development, and need to be recognized if people will be using the site. Other petrochemicals may also be present dependent on prior site use. Many countries have regulations that require soil testing and remediation or removal from sites that are being developed for recreation. It is important to check with environmental protection agencies before site work commences. Another potentially toxic material is aluminum, which is naturally occurring in the soil. At low pH (pH < 5.0), aluminum can become available to plant roots and will cause toxic effects. This has been shown to occur when the development site is near old coal mine spoil. In some of these instances where pH is extremely low, all nutrients will have low availability and soil may have to be removed and replaced or regular liming may need to occur to increase the pH.

Salt (NaCl) is another substance that can cause plant damage. This can occur because of ocean spray or due to de-icing salts used in northern climates. Salt spray may directly damage leaves and dormant buds through desiccation. Salt in the soil moves into the plant root on a diffusion gradient. The root cells have semi-permeable membranes that allow water and some dissolved salts to move into the root. The root generally has a greater concentration of nutrients so water moves in from the soil to try to equalize the diffusion gradient. When there is a high concentration of salts in the soil due to fertilization or deposition of de-icing salts, water can actually move out of the root into the soil and the concentration gradient is reversed. The plant will experience water deficits, or chemical drought. Sometimes this is called fertilizer or de-icing salt burn. Research has shown that the timing of salt deposition on plants is critical. Salt deposition in the autumn when the soil is still warm (>7.5°C) and the plants are taking up water, and in the spring when soils are warming up and water uptake increases are the two worst times to see plant toxicity (Headley and Bassuk, 1991). Since de-icing salt is very soluble, it will leach out below the root zone with increased rain after plant growth stops in the fall or before plant growth starts in the spring. As long as the soil drains well, flooding can remove and reduce salt levels in the soil. Soluble salts can be measured in soil and compost using an electrical conductivity (EC) meter.

Soil biological activity

Soil microorganism activity plays a large role in making nutrients available and reducing soil borne disease. There is greater interest in measuring biological activity in the soil. Two related tests for microorganism activity are *active carbon* and *potentially mineralizable nitrogen* (PMN). Active carbon is that fraction of the soil organic matter that can act as a food source for microorganisms. PMN is an indicator of the ability of the soil microbial community to mineralize nitrogen, which is tied up in complex forms and converted into the plant-available forms, nitrate and ammonium.

A root health assessment is a direct visual measurement of root growth as it may be affected by root fungi pathogens such as *Fusarium, Pythium, Rhizoctonia* and *Thielaviopsis*.

Soil samples are planted with a fungicide-treated snap bean seed and allowed to grow for 4 weeks under supplemental light or in a greenhouse. Plants are removed from their containers and roots washed in running water.

Roots are rated on a scale of 1 to 9, where 1 indicates white coarse healthy roots, and 9 is 50–75% hypocotyl and roots severely symptomatic and at an advance stage of decay (Gugino et al., 2009).

Figure 20.4 Utility conflict with tree growth

Planting space

When surveying a planting site it is important to determine the planting 'envelope' both above and below ground. Below ground there may be barriers to root growth such as when pavement is laid and purposeful soil compaction is necessary for load bearing. Because engineering specifications can call for a high level of density, roots will not be able to grow unless various load bearing soils are used (Grabosky and Bassuk, 1995). These engineered soils are an added expense. If possible do not plant within or directly next to pavement. Also, other underground obstructions such as utility vaults, conduits and underground basements should be avoided.

Above ground, signage, lights and utilities should be avoided by planting appropriately sized plants (Figure 20.4). In the most challenging situations where numerous environmental stresses and design constraints overlap, a combination of site modification strategies and appropriate plant selection provides the best result. With what we know now we can come up with enough creative solutions using all the strategies at our disposal to achieve a greener urban environment.

References

American Horticulture Society (1997) 'Heat Zone Map', retrieved from www.ahs.org/gardening-resources/gardening-maps/heat-zone-map.

Australian Standards (2009) 'Protection of Trees on Development Sites', Australian Standard AS 4970–2009, Australian Standards, Sydney.

Bassuk, N. and Whitlow, T. (1987) 'Environmental Stress in Street Trees', *Acta Hortic.* 195, 49–58.

Bornstein, R. D. (1968) 'Observations of the Urban Heat Island Effect in New York City', *Journal of Applied Meteorology* 7(4): 575–582.

Craul, P. J. (1992) *Urban Soil in Landscape Design*, John Wiley & Sons, New York.

Craul, P. J. (1999) *Urban Soils: Applications and Practices*, John Wiley & Sons, New York.

Daddow, R. L. and Warrington, G. E. (1983) *Growth-Limiting Bulk Densities as Influenced by Soil Texture*, WSDG-TN-00005, Watershed Systems Development Group, USDA Forest Service, Washington, DC.

Grabosky, J. and Bassuk, N. L. (1995) 'A New Urban Tree Soil to Safely Increase Rooting Volumes Under Sidewalks', *Journal of Arboriculture* 21(4): 187–201.

Gugino, B. K., Idowu, O. J., Schindelbeck, R. R., van Es, H. M., Wolfe, D. W., Moebius-Clune, B. N., Thies, J. E. and Abowi, G. S. (2009) *Cornell Soil Health Assessment Training Manual*, edition 2.0, Cornell University, Ithaca, NY, retrieved from http://soilhealth.cals.cornell.edu.

Headley, D. B. and Bassuk, N. L. (199). 'Effect of Time of Application of Sodium Chloride in the Dormant Season on Selected Tree Species', *Journal of Environmental Horticulture* 9(3): 160–164.

Patterson, J. C. (1977). 'Soil Compaction: Effects on Urban Vegetation', *Journal of Arboriculture* 3(9): 161–166.

Scharenbroch, B. C. and Catania, M. (2012) 'Soil Quality Attributes as Indicators of Urban Tree Performance', *Arboriculture and Urban Forestry* 38: 214–228.

Schindelbeck, R. R., van Es, H. M., Abawi, G. S., Wolfe, D. W., Whitlow, T. L., Gugino, B. K., Idowu, O. J. and Moebius-Clune, B. N. (2008) 'Comprehensive Assessment of Soil Quality for Landscape and Urban Management', *Landscape and Urban Planning* 88: 73–80.

Souch, C. A. and Souch, C. (1993) 'The Effect of Trees on Summertime Below Canopy Urban Climates: A Case Study in Bloomington, Indiana', *Journal of Arboriculture* 19: 303–312.

Sustainable SITES Initiative (2014) 'SITES Reference Guide, Rating System and Scorecard', retrieved from www.sustainablesites.org/resources.

Texas Agricultural Extension Service (2000) 'Landscape Water Conservation: Xeriscape 2000', retrieved from http://aggie-horticulture.tamu.edu/extension/xeriscape/xeriscape.html.

Thien, S. J. (1979) 'Flow Diagram for Teaching Texture by Feel Analysis', *Journal of Agronomic Education* 8: 54–55.

Trowbridge, P. J. and Bassuk, N. L. (2004) *Trees in the Urban Landscape: Site Assessment, Design, and Installation*, John Wiley & Sons, New York.

Trowbridge, P. J. and Mudrak, L. (1988) *Landscape Forms for Mitigating Winds on Shoreline Sites*, National Endowment for the Arts and Department of Landscape Architecture, Cornell University, Ithaca, NY.

US Department of Agriculture (2012) 'USDA Plant Hardiness Zone Map', retrieved from http://planthardiness.ars.usda.gov.

21

IMPROVING SOIL QUALITY FOR URBAN FORESTS

Susan Downing Day and J. Roger Harris

Introduction

Soils are a complex and ever-changing mix of minerals, organic material both living and dead, water, and gases. Their chemistry, physical structure, and biological processes are intricately entwined to create a vibrant support system for terrestrial life. Soil quality can make or break urban tree plantings, and poor soil management is a contributing or deciding factor for a host of urban forest maladies from pest predation to decline associated with construction damage.

Soils are literally the foundation of the urban forest system – both supporting vegetation and directly mediating a generous suite of ecosystem services. Many ecosystem services associated with urban forests, such as storm water management and carbon (C) sequestration, cannot be fully understood or quantified without consideration of soils. Consider, for example, the movement of storm water through the soil profile. The presence of tree roots greatly increases the rate of water movement by creating paths for gravitational flow (Bartens et al., 2008; Johnson and Lehmann, 2006). Yet soil characteristics and quality affect where these roots and their associated paths run, and soil management itself can have dramatic effects on soil permeability (Chen et al., 2014). It is, in fact, impossible to accurately answer questions about how any specific tree planting or tree preservation effort will affect ecosystem service provision without considering the soil. And, this dilemma is exacerbated if projections about future benefits are to be made, since soil quality and configuration has dramatic effects on tree growth (Watson et al., 2014; Day and Bassuk, 1994; Layman et al., 2016; Day and Amateis, 2011). Viewing trees and soils as part of an integrated system offers considerable advantages in both the management and planning of both urban forests and the ecosystem services we anticipate they may provide.

The importance of soils in supporting urban vegetation has resulted in a variety of practices aimed at creating better soils. The use of manufactured, blended, and engineered soils, for example, has exploded in recent years (Sloan et al., 2012), generating many specialized mixes – for example, CU soil (Grabosky et al., 1996), Amsterdam tree soil (Couenberg, 1994; Rahman et al., 2011), bioretention mixes (Prince George's County, 2007; Denman et al., 2006) – as well as 'recycled' or blended soil mixes used as topsoil replacements. *In situ* soil remediation techniques, such as soil profile rebuilding (Day et al., 2012; Layman et al., 2016) are also garnering interest. Yet what is the *ideal soil*? Although it is often said that the ideal soil is 50 percent mineral solids, 25 percent air, and 25 percent water, this conception of an ideal soil is firmly rooted in an agronomic view point. A soil that is ideal depends upon the objectives of a site, and becomes as much a philosophical question as a scientific one. During the past decade, the Sustainable Sites Initiative (SITES), a voluntary sustainability certification system for landscapes, tackled this question as they formulated their voluntary certification system for sustainable landscapes. Ultimately SITES embraced a holistic yet anthropocentric approach that used sustaining humankind as the ultimate benchmark, but recognizing the relation of sites, and site soils to a broad concept of environmental sustainability (Windhager et al., 2010). This necessitates an understanding of ecoregion and the context of existing or surrounding natural soils and their relation to climatic norms and indigenous plant and animal communities. This does not mean that the ideal soil is a recreation of a pre-urbanized soil. Urban dwellers make many demands upon the land that are not relevant in rural areas. For example, supporting foot or vehicular traffic is certainly one soil function that is in much greater demand in cities. Thus, balancing seemingly conflicting objectives is a hallmark of urban soil management. There is unquestionably no such thing as a one-size-fits-all soil, and objectives for any given soil decision demands serious, thoughtful consideration.

Soils clearly matter, yet the poor quality or even complete absence of soils in many urban areas leads many to dismiss the value of urban soils or be very quick to haul them away and replace them with a more amenable growing medium. Yet if we consider soils and vegetation – urban forests in particular – as a system, it quickly becomes apparent that soil management is not simply a linear process where soil deficiencies are identified and remediated. Instead soil management must respond to an array of interacting environmental factors (both existing and future) and design objectives. This chapter sets forth an iterative environmental design approach to soils management that takes into account land use and soils history and considers a broad array of soil management strategies to achieve ecological and environmental goals. This process is inherently context sensitive and requires broad appreciation of not only soil science and urban forestry, but also of local conditions from the site to the ecoregion scale. There is no one answer for the 'best' way to manage urban soils, but the concepts presented here aim to inform a decision-making process that makes transparent the consequences and trade-offs inherent in soil decisions.

Relation of urban soils to natural and agricultural soils

It is a widely held belief that all urban soils are either absent (as occurs when underground infrastructure has replaced soil as a support system or for rooftop plantings) or so disturbed as to have little value. The urban soil archetype is absent O and A horizons and graded and compacted B and C horizons liberally interspersed with buried trash, construction debris, and contaminants. Throw in a little elevated pH stemming from the ubiquity of concrete and other calcareous construction materials and you have urban soil – only fit for growing extremely adaptable species such as American elm (*Ulmus americana*) or London plane (*Platanus ×acerifolia*). Such soils are widespread, occurring around the world as a consequence of the homogenizing activity of humans.

For this reason, soil surveys are of limited utility in urban areas and site specific assessments must be made (see Chapter 20 of this volume). Yet naturals soil or elements of natural soils are ubiquitous in cities and the original soils may exert more influence on soil properties, especially soil texture, than many anticipate (Pouyat et al., 2007; Effland and Pouyat, 1997), although soil C, heavy metals, compaction, pH, potassium and phosphorus in surface layers are strongly influenced by disturbance and site management. Furthermore, soil taxonomy is advancing to better include urban soils, albeit slowly, and some urban soil surveys are now available. What then is the role of soil surveys in understanding urban soils for urban forests? Soil surveys work at a different scale than site assessments. In rural lands, soil forming factors can help interpolate soil characteristics across a landscape. In urban areas, many disturbances and management decisions occur at scales that will never be captured by surveys. Nonetheless, soil surveys should not be ignored in the site assessment as they can provide insight into characteristics such as soil texture, deep drainage, and parent material, especially where site disturbance is less extreme. But surveys can never substitute for on-site measurements. Furthermore, if site objectives include elements of ecological restoration, then understanding regional soils is essential.

Urban landscapes are highly managed systems, yet are very different from managed agricultural soil systems. Agricultural soils are managed to promote healthy and unrestrained plant growth in a sustainable way. Tilth is of utmost importance so practices that avoid compaction (minimum tillage, avoidance of traffic when soil is wet) are embraced and organic matter content is maintained by amendments or periodically incorporating green manure grown in a crop rotation system. Regular soil analysis dictates when soil pH should be adjusted with lime (to raise pH) or sulfur (to lower pH) amendments. Why can't we manage our urban systems in a similar way? The obstacles are many. In city landscapes we are dealing with perennial crops (trees) that are planted among infrastructure that prevents annual wholesale manipulation. Furthermore, urban soil has likely been manipulated for hardscape construction and not plant growth. Finally, goals for urban sites are wide ranging and complex. Unlike agricultural systems, maximizing short-term biomass accumulation is unlikely to be a priority. Instead, it may be necessary to balance such disparate goals as long-term biomass accumulation (i.e., support large, long-lived trees) and pedestrian safety at a bus stop. The multi-purpose nature of urban sites means that soil management must be a part of a holistic planning, design, and construction process.

Understanding pre-existing natural soils, soil surveys, and land use history can help interpret results of the detailed site-specific soil assessment. This comprehensive soils assessment describes what one has available to work with. The next part of the equation for improving soil quality is determining soil improvement goals. Articulating these goals in terms of the desired environmental outcomes can help create benchmarks for the soil improvement and management plan. For example, is restoring native soil characteristics an objective for a site in order to support a particular plant community? Is pollutant export a concern because of the site's connection to sensitive waterways or to groundwater? The soils will be a part of a complex ecological system that includes not just soils, but plants, humans, and many other organisms. Thus an understanding of these elements as well as regional soils and climates needs to inform such decisions.

The genesis and consequences of 'disturbed urban soil'

There are many types of disturbance affecting the quality of urban soils and every site will differ, hence the need for multiple site visits and soil testing in the site assessment. Nonetheless, understanding some of the common types of disturbance and their ramifications for soil quality is useful and may contribute to protecting soil from such damage in the first place. Many soil characteristics are both fragile and valuable, and an appreciation for the contribution of soils to urban sustainability can contribute to a more thoughtful consideration of the environmental and economic trade-offs inherent in soil management decisions. A holistic approach to managing urban land demands that objectives not be limited to supporting a given tree over a specified period of time – the contributions of high quality soil are considerably more far reaching.

Types of soil damage

Topsoil removal

Scraping away of surface soils (generally O and A horizons) is inevitable during most land development, yet even when carefully stockpiled, this process is highly damaging to soil organic matter content and structure. Soils are scraped to create stable bases for vehicular traffic, to allow grading and contouring of the land, to remove undesirable vegetation, and to support structures. It would seem to be common sense that if carefully removed and replaced, soil quality would remain intact. However, even short-term (< 1 month) stockpiling and replacement of A horizon soil can result in the loss of 35 percent of soil organic C, including 47 percent of the mineral-bound C, normally considered to be very stable (Chen et al., 2013). Stockpiling also reduces the proportion of macroaggregates, even when soil is handled at the ideal moisture content for protecting aggregates and is not tilled (Chen et al., 2014). This shift in aggregate size distribution has serious consequences for soil drainage and susceptibility to compaction, both of which will impact performance. The more physical manipulation the removed soil receives (e.g. tilling, screening, blending) the greater these effects will likely be.

Grading and compaction

Compacted soil and its close associate, poor drainage, are extremely common on urban sites. Virtually any site development, whether it is new urban development or repurposing of existing urban space, will require grading of some type. For example, grading creates stable or level surfaces for buildings or pavement, prevents surface water collection, and ensures safe inclines for roads. Yet grading mixes or buries soil horizons and nearly always requires many passes of heavy equipment to accomplish. Even grading performed when soil is not overly wet results in serious soil compaction. In the absence of grading, foot or vehicular traffic will still compact all but the sandiest soils. Compacted soils impede root growth and water movement, two processes that are absolutely essential in a productive soil. To understand why compaction is so damaging to soils, it helps to keep two things in mind. First, compaction crushes soil aggregates, the relatively stable clumps of soil mineral particles that give fine-textured soils the ability to drain. These aggregates can be crushed in a moment, but take a minimum of 5 to 15 years to reform even with significant intervention (Chen et al., 2014; Wick et al., 2009). Aggregates are critical to healthy soil. They allow soils to perform two seemingly contradictory functions essential to plant growth: hold water and nutrients while at the same time allowing water and air movement. Aggregates can reduce soil erosion and help stabilize carbon stores, and thus contribute to a variety of ecosystem services. The second factor to keep in mind is that although any degree of soil compaction reduces root exploration (Day and Amateis, 2011; Day et al., 2000), heavily compacted soils are extremely resistant to penetration by tree roots and will not only prevent tree establishment and growth, but are less likely to regain the characteristics of healthy soil precisely because they exclude these agents of change, namely roots (Day et al., 2010b; Day and Bassuk, 1994). As roots grow and die they create pathways, introduce organic matter, and produce exudates that foster microbial activity and other processes.

Alteration of soil pH

Soil pH can change during urbanization as previously buried horizons are revealed and building materials are introduced. These can result in extremely low pHs, as when acid sulfate soils are exposed to the surface (an uncommon but dramatic occurrence), or extremely high pHs where concrete and other calcareous building materials create a reservoir of alkalinity. The types of pH alteration will vary by region and building practices; however, elevated pH is probably the most common difficulty. Gravel and rock have myriad roles in building – base courses for pavement, envelopes for drains, stabilizing temporary access roads, drainage sumps, etc. However, where limestone is the most readily available stone, careless strewing of gravel throughout developed sites creates a long-term soil contamination problem. Soil pHs in such conditions may easily rise to 8 or even higher with no possibility of significantly lowering them without removing the offending construction debris. Many tree species experience micronutrient deficiencies in such soils. Even moderate pH elevations (> 6.5) are problematic for some commonly planted pH-sensitive urban tree species (e.g. willow oak, *Quercus phellos*).

Other types

There are many other types of soil damage that may be found on urban sites – contamination from heavy metals or other pollutants, buried organic horizons, buried pavement or other impervious horizons, landfill seepage, etc. However, the three described above are likely the most common and the most preventable. Sensitive design, management, and remediation can significantly curb the influence of these factors on urban forests and other green infrastructure.

Soil recovery

Building a natural soil through reintroduction of the prerequisite parts is frequently desirable, but it's worth keeping in mind that natural soils and their many positive benefits for plant growth took an extended time to develop. Nonetheless, soils can recover over time and can be 'helped along' in this process (see 'Strategies and tools for rehabilitating soils' below). Even without active soil management, ordinary urban landscapes effect significant changes in soil quality over time. For example, older landscapes may have significantly greater C stores, lower bulk densities, and greater nitrogen availability (Scharenbroch et al., 2005). Soil formation occurs under the influence of the vegetative communities present and the landscape management practices employed. Such changes even occur in engineered substrates, such as those found on green roofs (Schrader and Böning, 2006) and are tied to the physical, chemical, and biological processes associated with plant growth. Root exploration and root turnover, for example, will strongly influence soil characteristics over time. If soil conditions restrict the possibility of root penetration and other biological phenomena, then changes in soil quality will be likewise retarded. Consequently, soil interventions may have both an immediate effect on soil characteristics and a long-term effect tied to the plant communities these soils support. Thus it is possible for urban soils to recover some or all of their pre-development characteristics, but such recovery may well not occur due to irreversible or nearly irreversible damage, or lack of conditions (such as the presence of roots, for example) that facilitate recovery.

Soil as a sustainable resource

Soil supports plants, without which humans and other species would cease to exist. Soil can thus be viewed as an essential resource – a resource that must be sustained to allow life on earth to continue. Furthermore, soils facilitate a host of ecosystem services that must also be sustained to assure continuation of our species. Consequently, although not the subject of this chapter, sustainable site design emphasizes protecting soil resources. For example, SITES adopted the sustainability paradigm of ensuring ecosystem services will continue to be provided to future generations. Soil conservation is therefore an important element of the practices required by SITES (Sustainable Sites Initiative, 2014), and this rating system considers protection of healthy soils as more desirable than restoration or remediation. An approach that considers soil management for urban forests through this type of sustainability lens is a complicated undertaking in the disruptive environment of urban sites, but one that many consider worth incorporating in the planning, design, and management of urban greenspaces. Although the archetypal 'disturbed urban soil' is ubiquitous and results from the widespread manipulation of soils to facilitate building and development, there are many opportunities to prevent or mitigate this disturbance.

'Soil conscious' environmental design

In its broadest sense, environmental design incorporates environmental parameters into the planning and design of anything from a cafeteria to a factory to a park to a city. In the realm of urban forestry, environmental design occurs at multiple scales and incorporates the give and take between any landscape and its environment. For example, if heavy pedestrian use will be part of the environment, then soils must be managed or placed to avoid or resist compaction from foot traffic. If a site is near a sensitive waterway, then eliminating nutrient export may be a site requirement. At a larger scale, urban forests are part of a regional ecosystem and their management can affect everything from water quality to species composition in surrounding areas. Countless other objectives relating to aesthetics, transportation, emergency preparedness, social issues, or economics, to name a few, may also come into play. Balancing all these competing needs and interactions is what makes planning, designing, and managing urban greenspaces so interesting and so challenging.

Soils are still viewed by many as a structural engineering element rather than a living part of the environment and may get short shrift during the planning and design process, other than their manipulation for engineering purposes. Even when soils are considered as a growing medium, decisions may be dominated by the linear process of identifying what vegetation and/or engineering function is desired, and attempting to build a soil with a list of physical, chemical, and biological characteristics that will meet those needs. To some extent, however, such a linear approach divorces soils considerations from the environmental design process.

Design is a complex and iterative problem-solving process and methods and philosophies for approaching design problems have been developed in many disciplines. Recurring elements in the design process include the articulation of the problem to be solved, identifying constraints and objectives, fostering creative solutions, and information gathering. Our objective is not to propose a particular design process, but to highlight areas where soils considerations can and should play a role in the environmental design process. These include proposing an ecosystem service-based ranking of the desirability or value of soil conditions and considering trade-offs in ecosystem service provision provided by soils. It's important to note that these are aids in the decision-making process rather than indications of the best solution on a given site. Indeed, the least desirable soil from an ecosystem service perspective may be the best solution for a given portion of an urban forest planting because it is the only way to fulfil other design goals (e.g. supporting a parking lot). Yet formulating a ranking is useful in the design process because it helps articulate trade-offs that are made in the broader environmental arena which might otherwise be easily forgotten. A possible ranking is presented in Table 21.1.

Table 21.1 Ranking of soil desirability based on likely ecosystem service provision, either direct or indirect. Rankings might change based on overall site objectives, local ecological variables, environmental conditions, or design philosophy

Ranking	Soil type	Notable ecosystem service considerations
Most desirable	Nearly undisturbed natural soil with intact native plant community	Support native biota, biodiversity, intact O horizon can enhance infiltration
	Mostly undisturbed natural soil	May support native biota
	Productive agricultural soil	Support healthy vegetation, may have excessive P
	Soils on older urbanized land that was exposed to minimal grading and supports healthy plants or large trees	Large trees may be present, may have considerable soil C sequestered
	Rehabilitated urban soils that support healthy trees or other vegetation	Support rapid tree growth, enhance permeability
	Manufactured, blended, reconstituted, or recycled soils (made primarily from natural soil components)	Reduced soil structure may lead to increased water or nutrient inputs. May damage off-site land if soil was harvested from intact sites
	Soils engineered for one primary essential function (e.g., structural support, P removal, or others)	Variable, depends upon trade-offs and materials used in mix
	Highly compacted, graded soils that cannot support healthy vegetation	Restricts plant growth and generates storm water runoff
Least desirable	Heavily contaminated soils or brownfields	Can inhibit plant growth and reduce utility for consumption, other disturbance likely

   Regardless of the constraints of a particular site, incorporating soil stewardship into the environmental design process will help maintain a focus on the long-term trade-offs that are being made in a wide assortment of land use planning decisions or project designs.

Strategies and tools for rehabilitating urban soils

In spite of careful planning and protection, scraped and compacted urban soils are common and must be dealt with in order to successfully grow trees and provide basic ecosystem services such as storm water interception. The terms 'rehabilitation' or 'remediation' indicate a return of a soil to service (i.e. some ecosystem services that were lost when the soil was damaged will be restored), while 'restoration' implies a goal of returning soil to a pre-development or pre-disturbance condition. Perfect restoration is rarely, if ever, possible.

Indeed, even identifying pre-disturbance conditions may be impossible in extremely old human settlements. More importantly, there will be many instances when true restoration is not compatible with the current site objectives. For example, if current site objectives include producing vegetables in a community garden and the pre-development land was a marsh, true soil restoration is not a compatible goal. However, more modest rehabilitation goals, such as reducing soil compaction or improving drainage, are useful in a wide variety of applications. Here we will not attempt to provide a comprehensive list of soil rehabilitation techniques, but will instead highlight principles and important considerations. We will use soil profile rebuilding, a technique that we have studied intensively, as an example.

A holistic approach to soil rehabilitation

In addition to immediate landscape or environmental requirements, soil rehabilitation must set in place conditions that will function in the future as well. A holistic approach to soil rehabilitation will consider how the landscape will evolve over many decades or longer. The soil and plant system is not a static building material, but a growing, living, and changing part of a broader ecosystem. An appreciation of how soils change over time leads to looking at any given soil rehabilitation practice as setting the stage for a future tableau of events rather than a one-time soil fix. Thus a holistic approach considers the physical, chemical, and biological elements in a soil and accounts for how these will change and interact over time. When possible, soil management that takes advantage of these impending changes (e.g. facilitating root exploration of soil to contribute to future C sequestration) rather than trying to stave off change (e.g. keeping sediment out of a bioinfiltration system to prevent clogging) offers more promise for long-term sustainability. This, of course, requires considerable plant and soils knowledge; however, there are certain considerations that will facilitate this process.

Soil pH and contaminants

Strategies for 'correcting' pH and removing soil contaminants, such as heavy metals, are difficult to implement and may not always produce satisfactory results. Prevention is the best cure in the case of elevated pHs that stem from building materials. In recent years, some project managers are limiting gravel distribution around construction sites, using granite instead of limestone gravel, and protecting soils with plastic sheeting around concrete mixing areas or around concrete pours to reduce contamination with pH-raising substances. Where limestone gravel and building rubble are well distributed throughout the soil or concrete foundations and pavements are ubiquitous, lowering pH with sulfur may not be feasible. The reserve alkalinity in the soil will likely require that sulfur applications be repeated annually. Even if this were economically feasible, treating the soil that is in contact with tree root systems may be difficult since the roots of trees may be in inaccessible areas (under pavement) or well below the soil surface. In areas where the soil will be accessible, the majority of pH-raising substances can be removed, and the pH adjustment required is not large, applications of sulfur may be effective at restoring soil pH. Although the elevated pH scenario is perhaps the most common, soil pH should always be tested and other chemical imbalances that might require consultation with a soil scientist should not be discounted.

Urban soils may be contaminated with heavy metals (i.e., Pb, Cd, Zn, Ni and Cu). Lead (Pb), in particular, is ubiquitous in older cities, especially near roadways and older buildings where exterior lead paint was used. Former industrial sites may also have high levels of heavy metals. A detailed analysis of heavy metal remediation is outside the scope of this chapter, however it is critical to be aware of the possibility of heavy metal contamination. Heavy metals are toxic to both plants and humans. Heavy metals interfere with cell division and elongation in roots, resulting in compact, highly branched root systems (Kahle, 1993; Hagemeyer and Breckle, 2002) not unlike those observed in compacted soil. Management focuses primarily on reducing human exposure, however, especially exposure of children to Pb, the most common heavy metal in urban soils. If land use history suggests that heavy metal contamination is likely, testing is warranted. High levels of heavy metals should be addressed with a comprehensive management plan that may entail removing soil, limiting soil redistribution, making metals less bioavailable through organic matter additions or pH adjustments, extracting contaminants through hyperaccumulating plants, or sealing soils to prevent human exposure.

Soil depth

Perhaps because of familiarity with agronomic and turf systems that emphasize surface manipulation of soils down to plow depth, or simply because we cannot readily observe deeper soil regions, soil below the first 10 or 15 cm is often an unexploited resource. These deeper soil regions can be severely compacted during grading and therefore become highly resistant to root penetration (Day et al., 2000). Tree roots require a minimum of 60 cm of useable soil and can greatly benefit from and explore considerably deeper soils. To emphasize the importance of soil depth to tree growth, urban foresters frequently reference soil volume. Both surface area and depth affect tree growth – however, once an adequate depth is achieved, increasing useable soil depth may offer diminishing returns (Day and Amateis, 2011). Nonetheless, when an additional zone of soil is opened up for root exploration, as when a compacted subsurface layer is broken up, increases in tree growth can result (Layman et al., 2016). Additionally, this soil can provide ecosystem services such as C sequestration or storm water transmission and storage. Because rooting depth is influenced by both genetics and environment (Day et al., 2010a), a soil that can be penetrated by roots can also lead to deeper root systems, potentially contributing to drought resilience. Finally, once the significant disruption of land development is complete, deeper soil regions are more protected against some kinds of disturbance, allowing them to effectively store soil organic C (Lorenz et al., 2011), for example, or retain permeability.

In spite of the importance of soil depth, however, many soil improvement techniques focus on the surface. Ordinary tillage is unlikely to penetrate more than 15 cm below the surface, and incorporating significant organic amendments at depth can be challenging. Furthermore, drastic alteration of soil physical properties in confined areas (such as an individual planting location) can result in the accumulation of surface water – sometimes described as a bathtub or teacup effect. These challenges can be overcome by focusing on the site, rather than tree planting locations, to ensure that water can drain freely, and using subsoiling techniques such as ripping or onsite mixing with a backhoe bucket for deep incorporation of amendments.

The challenge and opportunity of fine-textured soils

Fine-textured soils, those that have significant silt and clay components, are among the most productive soils in the world, yet can also be among the most susceptible to degradation from urbanization. Soil aggregates – the clumps of soil particles held together by various binding agents including fungal hyphae and various sticky substances produced by roots and bacteria – can be considered the most critical feature of these soils, and facilitating the formation and persistence of stable aggregates thus becomes the overarching goal of soil managers to maintain productivity. Soil aggregates allow free air and water movement through fine-textured soils that would otherwise have very low permeability. This movement of air and water occurs even while the increased surface area of these fine particles increases nutrient and water retention. Furthermore, aggregates protect soil organic matter from microbial decomposition, creating more stable soil organic C stocks. In short, aggregates allow plants to have the best of both worlds: adequate drainage and adequate water and nutrient retention. However, aggregates are easily lost to compaction or soil manipulation (screening, grading or tillage) while overly wet or dry and are very slow to recover (Table 21.2). Managing trade-offs related to aggregation can be especially difficult. If site soils will be exposed to pedestrian or vehicular traffic, well aggregated fine-textured soils will be ill equipped to resist compaction unless protected in some fashion. Sand soils may be more resistant to compaction, but at a considerable cost in additional water and nutrient requirements. The severity of these conflicts will depend upon climate and degree of compactive force the soil will be exposed to. Nonetheless a fine-textured soil with a highly developed crumb structure and significant organic matter component will support abundant plant life. Long-term consequences on soil aggregation should be incorporated into any soil management decision.

Organic matter additions

Organic matter additions are a tried and true way of improving soil productivity, and organic matter management is the focus of much agricultural soil management. Organic matter serves two primary purposes:

1 to immediately aid in water and nutrient retention; and
2 contribute to future soil aggregation (Table 21.2).

Table 21.2 Activities that degrade or contribute to soil aggregation. Note that aggregate degradation occurs very rapidly, while formation of aggregates may take many years or even decades

Activity	Effect on soil aggregates
Grading (especially when wet)	Degradation of aggregates
Screening (especially when wet)	Degradation of aggregates
Tilling (especially when wet)	Degradation of aggregates
Organic amendments	May contribute to aggregation
High sand content	Reduces aggregation
Clay content	Contributes to aggregation
Surface exposed (to rainfall impact or foot traffic)	Degradation of aggregates
Surface covered with organic material (mulch, leaf litter)	Contributes to aggregation
Compaction (for any purpose)	Degradation of aggregates
Root growth	Contributes to aggregation
Microbial activity	Contributes to aggregation

However, organic matter can also contribute to a wide variety of other soil health attributes and ecosystem services, whether directly or indirectly. In urban settings, high quality, stable compost that has been tested for organic matter content, salts, nutrients, etc. is more readily available than ever before for use as a soil amendment. However, if composts have high C/N ratios, incorporation into the soil can lead to N deficiencies. Salt contents can also be unexpectedly high and only composts with electrical conductivity below 4 dS/m should be used. Large amounts of compost and repeated applications are not necessary to effect change. In the context of long-term urban tree plantings, organic matter content can build significantly over time and may affect soil physical properties in addition to chemical and biological properties. Consequently, soil rehabilitation should consider the organic matter additions and subtractions that will take place over time. Organic amendments, leaf litter, and root turnover all contribute to organic matter content, while decomposition decreases organic matter content when C is returned to the atmosphere. Furthermore, the rapidity of organic matter cycling varies. Warm moist soils with adequate oxygen will foster more rapid decomposition. Aggregates protect organic matter from microbial access, slowing the C removal process. The rapidity of root turnover surprises many people. Many roots only survive a few weeks and a majority of fine roots perish each year. Perhaps as a consequence, even under pavement, where soils would appear to be insulated from organic inputs, soil C stocks may be similar to those in greenspaces (Edmondson et al., 2012).

Soil profile rebuilding

Soil profile rebuilding is a rehabilitation technique for compacted urban soils, particularly those with deeper compacted layers that result from grading. It illustrates some of the concepts described above, in particular those related to long-term changes in soil properties that can be instigated by soil management. Briefly, the technique is applied to the entire area of compacted soil and includes these steps:

1 10 cm of compost is spread over the soil surface;
2 a backhoe bucket subsoiling technique is used to break up the compacted soil layers and incorporate compost to at least 60 cm;
3 stockpiled topsoil (or other topsoil if stockpiled soil is not available) is returned to the site and tilled; and
4 trees or shrubs are planted to allow for full occupation of the site by roots (full details can be found in the specification; Day et al., 2012).

The organic matter, in the form of compost, serves the immediate role of keeping penetrable channels open in the soil to allow for root penetration. Roots supply organic matter to the soil through turnover and exudation, thus ensuring the continued availability of organic matter over time. Thus, both in the short term and long term, organic matter that can help build soil aggregates is available. The physical breaking of compacted layers allows rapid water movement through previously nearly impermeable layers (Chen et al., 2014) providing an ecosystem service that often carries significant economic value. The immediate reduction in compaction contributes to rapid tree growth (Layman et al., 2016), enhancing both soil improvement and magnifying ecosystem service provision. The use of existing soil with minimal physical manipulation (e.g. no screening) helps conserve soil aggregates and reduces fossil fuel use. Reliance on aggregate formation reduces nutrient and water inputs required over the life of the landscape, and the focus on soil improvement at depth may have drought resilience benefits.

Working around existing trees

Soil improvement around existing trees is challenging because virtually all soil manipulations will damage some roots. To address this problem, soil managers try to minimize root disruption by treating small segments of soil at a time and sometimes spreading treatments out over a number of years. Treatments include radial trenching, where narrow trenches radiating out from the trunk are filled with uncompacted soil (Day et al., 1995), and vertical mulching where numerous holes are augered within the tree root zone and filled with loose soil, mulch, or compost (Pittenger and Stamen, 1990). More recently, air excavation tools have been used to incorporate organic matter around roots (Fite et al., 2011), but caution must still be exercised as the high pressure air stream is damaging to roots of some species (Kosola et al., 2007).

Surface treatments

Soil management cannot be considered complete in urban settings without considering surface treatments. Bare compacted soil generates significant runoff and sediment (Day and Mitchell, 2015). Surface treatments likely have a profound effect on surface infiltration of both storm water and irrigation, but have been little studied to date. The huge variety of mulches in use today vary in their properties and durability. Organic mulches, such as shredded barks, tend to add organic material to the soil over time, while inorganic mulches do not. Some mulches may distribute load somewhat, decreasing opportunities for soil compaction. However, the depths required to prevent soil compaction (as much as 30 cm) may be too extreme for many landscape situations. Surface covers may be pavements, or understory plants, or a host of other options. While in the past these have been thought of as primarily aesthetic elements or a part of maintenance regimes, we are now becoming more aware of the environmental consequences of these choices, including their effects on soil sustainability.

Evaluating effects and environmental consequences of soil rehabilitation

Success in soil improvement will, of course, depend upon one's goals. For example, is the goal to restore conditions to pre-development status or to maximize a particular ecosystem service or suite of services? The same tools and benchmarks used in the site assessment process can be deployed post-improvement to evaluate success. Unfortunately, expected changes in soil properties are not always immediately evident. For example, lowering pH with sulfur is a microbially mediated process and takes weeks or months to occur. Likewise, incorporating compost will immediately lower soil bulk density, but longer-lasting effects may not be measurable for a year or more after treatments are completed. Thus evaluation is best achieved by confirming that management has been performed as specified (for example, extracting soil samples to confirm depth of compost incorporation) followed by longer term monitoring focusing on desired environmental outcomes, rather than just instantaneous soil quality metrics. The environmental consequences of soil rehabilitation are numerous. At a minimum, long-term effects on plant growth and diversity, water quality, storm water interception, C sequestration, and climate change resilience should be incorporated into planning and design. However, it may be difficult to reliably estimate the magnitude of these services at a given site.

Conclusions

Soils are the foundation of urban forests and green infrastructure. Soil management practices range from complete neglect and indifference, to excavation and replacement of custom blended and manufactured soils on a massive scale. Unfortunately, neither extreme seems likely to result in soils that provide the broadest possible spectrum of ecosystem services. Although improving soil quality as part of an integrated system holds promise for creating truly sustainable urban ecosystems, it is not an easy task and requires significant specialized expertise. Soil scientists have traditionally operated in agronomic settings. Nonetheless, over the past few decades there has been unprecedented attention given to urban soils and their unique challenges. As we learn to bridge the communication divide between professions traditionally concerned with urban development (civil engineering, architecture, etc.) and those traditionally concerned with rural systems such as agriculture and forestry, new varieties of professional expertise are emerging, including in the area of urban soils.

References

Bartens, J., Day, S. D., Harris, J. R., Dove, J. E. and Wynn, T. M. (2008) 'Can urban tree roots improve infiltration through compacted subsoils for stormwater management?', *Journal of Environmental Quality*, vol 37, no 6, pp. 2048–2057.

Chen, Y., Day, S. D., Wick, A. F. and McGuire, K. J. (2014) 'Influence of urban land development and subsequent soil rehabilitation on soil aggregates, carbon, and hydraulic conductivity', *Science of the Total Environment*, vol 494–495, pp. 329–336.

Chen, Y., Day, S. D., Wick, A. F., Strahm, B. D., Wiseman, P. E. and Daniels, W. L. (2013) 'Changes in soil carbon pools and microbial biomass from urban land development and subsequent post-development soil rehabilitation', *Soil Biology and Biochemistry*, vol 66, pp. 38–44.

Couenberg, E. A. M. (1994) 'Amsterdam tree soil', in Watson, G. W. and Neely, D. (eds) *The Landscape Below Ground: Proceedings of an International Workshop on Tree Root Development in Urban Soils*, International Society of Arboriculture, Champaign, IL.

Day, S. D. and Amateis, R. L. (2011) 'Predicting canopy and trunk cross-sectional area of silver linden (*Tilia tomentosa*) in confined planting cutouts', *Urban Forestry and Urban Greening*, vol 10, no 4, pp. 317–322.

Day, S. D. and Bassuk, N. L. (1994) 'A review of the effects of soil compaction and amelioration treatments on landscape trees', *Journal of Arboriculture*, vol 20, no 1, pp. 9–17.

Day, S. D. and Mitchell, D. K. (2015) ' A new perspective on opportunities for stormwater mitigation through soil management in ordinary urban landscapes', *Watershed Science Bulletin*, January.

Day, S. D., Bassuk, N. L. and van Es, H. (1995) 'Effects of four compaction remediation methods for landscape trees on soil aeration, mechanical impedance and tree establishment', *Journal of Environmental Horticulture*, vol 13, no 2, pp. 64–71.

Day, S. D., Seiler, J. R. and Persaud, N. (2000) 'A comparison of root growth dynamics of silver maple and flowering dogwood in compacted soil at differing soil water contents', *Tree Physiology*, vol 20, no 4, pp. 257–263.

Day, S. D., Wiseman, P. E., Dickinson, S. B. and Harris, J. R. (2010a) 'Contemporary concepts of root system architecture of urban trees', *Arboriculture and Urban Forestry*, vol 36, no 4, pp. 149–159.

Day, S. D., Wiseman, P. E., Dickinson, S. B. and Harris, J. R. (2010b) 'Tree root ecology in the urban environment and implications for a sustainable rhizosphere', *Arboriculture and Urban Forestry*, vol 36, no 5, pp. 193–205.

Day, S. D., Layman, R. M., Chen, Y., Rolf, K., Harris, J. R., Daniels, W. L., Wiseman, P. E., McGuire, K. J., Strahm, B. D., Wick, A. F. and Mauzy, B. (2012) *Soil Profile Rebuilding Specification*, Virginia Tech, Blacksburg, VA.

Denman, L., May, P. and Breen, P. (2006) 'An investigation of the potential to use street trees and their root zone soils to remove nitrogen from urban stormwater', *Australian Journal of Water Resources*, vol 10, no 3, pp. 303–311.

Edmondson, J. L., Davies, Z. G., McHugh, N., Gaston, K. J. and Leake, J. R. (2012) 'Organic carbon hidden in urban ecosystems', *Scientific Reports*, vol 2, pp. 963.

Effland, W. R. and Pouyat, R. V. (1997) 'The genesis, classification, and mapping of soils in urban areas', *Urban Ecosystems*, vol 1, no 4, pp. 217–228.

Fite, K., Smiley, E. T., McIntyre, J. and Wells, C. E. (2011) 'Evaluation of a soil decompaction and amendment process for urban trees', *Arboriculture and Urban Forestry*, vol 37, no 6, pp. 293–300.

Grabosky, J., Bassuk, N. and Van Es, H. (1996) 'Testing of structural urban tree soil materials for use under pavement to increase street tree rooting volumes', *Journal of Arboriculture*, vol 22, pp. 255–263.

Hagemeyer, J. and Breckle, S. (2002) 'Trace element stress in roots', in Waisel, Y., Eshel, A. and Kafkafi, U. (eds) *Plant Roots: The Hidden Half* (3rd edn), Marcel Dekker, New York.

Johnson, M. S. and Lehmann, J. (2006) 'Double-funneling of trees: Stemflow and root-induced preferential flow', *Ecoscience*, vol 13, no 3, pp. 324–333.

Kahle, H. (1993) 'Response of roots of trees to heavy metals', *Environmental and Experimental Botany*, vol 33, no 1, pp. 99–119.

Kosola, K. R., Workmaster, B. A. A., Busse, J. S. and Gilman, J. H. (2007) 'Sampling damage to tree fine roots: Comparing air excavation and hydropneumatic elutriation', *HortScience*, vol 42, pp. 728–731.

Layman, R. M., Day, S. D., Mitchell, D. K., Chen, Y., Harris, J. R. and Daniels, W. L. (2016) 'Below ground matters: Urban soil rehabilitation increases tree canopy and speeds establishment', *Urban Forestry and Urban Greening*, vol 16, pp. 25–35.

Lorenz, K., Lal, R. and Shipitalo, M. J. (2011) 'Stabilized soil organic carbon pools in subsoils under forest are potential sinks for atmospheric CO_2', *Forest Science*, vol 57, no 1, pp. 19–25.

Pittenger, D. and Stamen, T. (1990) 'Effectiveness of methods used to reduce harmful effects of compacted soil around landscape trees', *Journal of Arboriculture*, vol 16, no 3, pp. 55–57.

Pouyat, R., Yesilonis, I., Russell-Anelli, J. and Neerchal, N. (2007) 'Soil chemical and physical properties that differentiate urban land-use and cover types', *Soil Science Society of America Journal*, vol 71, no 3, pp. 1010–1019.

Prince George's County (2007) *Bioretention Manual*, Department of Environmental Resources, The Prince George's County, Maryland.

Rahman, M., Smith, J., Stringer, P. and Ennos, A. (2011) 'Effect of rooting conditions on the growth and cooling ability of *Pyrus calleryana*', *Urban Forestry and Urban Greening*, vol 10, no 3, pp. 185–192.

Scharenbroch, B. C., Lloyd, J. E. and Johnson-Maynard, J. L. (2005) 'Distinguishing urban soils with physical, chemical, and biological properties', *Pedobiologia*, vol 49, no 4, pp. 283–296.

Schrader, S. and Böning, M. (2006) 'Soil formation on green roofs and its contribution to urban biodiversity with emphasis on Collembolans', *Pedobiologia*, vol 50, no 4, pp. 347–356.

Sloan, J. J., Ampim, P. A. Y., Basta, N. T. and Scott, R. (2012) 'Addressing the need for soil blends and amendments for the highly modified urban landscape', *Soil Science Society of America Journal*, vol 76, no 4, pp. 1133–1141.

Sustainable Sites Initiative (2014) *SITES v2 Rating System*, US Green Building Council, Washington, DC.

Watson, G. W., Hewitt, A., Custic, M. and Lo, M. (2014) 'The management of tree root systems in urban and suburban settings: A review of soil influence on root growth', *Arboriculture and Urban Forestry*, vol 40, no 4, pp. 193–217.

Wick, A. F., Ingram, L. J. and Stahl, P. D. (2009) 'Aggregate and organic matter dynamics in reclaimed soils as indicated by stable carbon isotopes', *Soil Biology and Biochemistry*, vol 41, no 2, pp. 201–209.

Windhager, S., Steiner, F., Simmons, M. T. and Heymann, D. (2010) 'Toward ecosystem services as a basis for design', *Landscape Journal*, vol 29, no 2, pp. 107–123.

22

DESIGN OPTIONS TO INTEGRATE URBAN TREE ROOT ZONES AND PAVEMENT SUPPORT WITHIN A SHARED SOIL VOLUME

Jason Grabosky and Nina Bassuk

The problem

Urban trees face a wide range of environmental challenges, the most significant of which is the scarcity of soil suitable for root growth (see Chapter 21). Often, soil compaction serves as the leading impediment for tree establishment. So even if there is soil present it is not available for root colonization and exploitation for growth support. While many of the problems urban trees face can be mitigated by planting species that are tolerant of a given challenge, there are relatively few tree species that can thrive within compacted soils which are prevalent throughout urban and suburban landscapes. Soil volumes are shared by grey and green infrastructure. Given a lack of space to expand and provide separate volumes for trees, there are integrating design solutions for tree soil volumes and pavement support.

In urban soils that are not covered by pavement, it is possible to break-up, amend or replace compacted soils to make them more conducive to root growth. However, where soils are covered by pavement, the needs of the tree come in direct opposition to pavement design requirements for a highly compacted supporting base on which to construct pavement. Pavement must be laid on well-draining compacted bases so that the pavement will not subside, or otherwise prematurely require replacement.

As a result, soils that must support pavement are often too dense for root growth. Pavement systems consist of a series of layers; lower layers support the materials placed in layers above. To design and support a safe, durable pavement system, soils are compacted to provide an increased load-bearing capacity, providing cost-effective support for the comparatively expensive wearing surface of the pavement. In general, sites which are expected to support pavement surfaces are compacted to a peak density as determined by a moisture-density curve from a standardized or expected field compaction effort level (based on the machinery to be used in construction). Alternatively, soils can be compacted to a specific penetration resistance value. Either way, compaction benchmarks provide a more uniform soil support behavior for pavement, foundations or other structures.

Most construction standards require soil compaction beyond what is considered desirable for tree root colonization, in all but predominantly sand and aggregate-based soils. To avoid confusion, the term 'aggregate' in this conversation refers to the larger granular or stone components in the mixture design, since this term is then further divided for rounded, smooth gravels and for angular, crushed stone particles. This highlights the importance to understand that when speaking to structural engineers or agricultural disciplines, the same word can have multiple disciplinary definitions. Care is needed to be specific and clear in all communication between all people involved in the process of urban design and installation.

It is no surprise then that trees surrounded by pavement (porous or sealed surface) are associated with slower, weaker, or poor tree establishment and growth. Access to soil volume has been shown to influence early tree survivorship, canopy dimension in 20 years and maximum trunk diameter in urban landscapes (Lu et al., 2010; Sanders and Grabosky, 2014; Sanders et al., 2013; Grabosky and Gilman, 2004). Urban tree canopy environmental benefit models suggest greater benefits are conferred by larger trees in the urban landscape (McPherson et al., 1994; Nowak et al., 2013). Reduced canopy volumes and shortened lifespans thus increase costs for urban tree management, while negatively impacting the ecosystem benefits conferred by tree canopy.

If tree roots can escape at the interface between the pavement and compacted soil or the soil-base layer interface to grow into non-compacted or more viable soil volumes away from the pavement zone, the roots near the base of the tree may radially expand to provide the vascular system connecting the proliferating distant roots to the trunk and canopy. In this 'zone of rapid taper' for tree roots (Eis, 1974), pavement surfaces may be displaced and cause pavement failure.

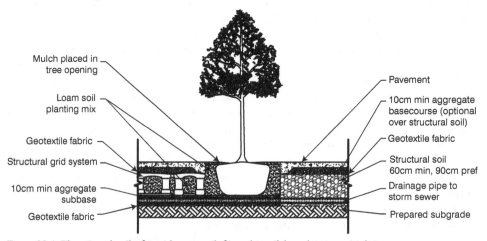

Figure 22.1 Elevation detail of a grid system (left) and a soil-based system (right)

In this detail the planting hole is filled with soil, common to the soil in the grid system, and either in common with the designed soil or as a separate soil zone in the designed soil detail. In the grid system, the root zone occurs below a decking surface supported by a column structure. In the designed soil system, the base course depth can vary with proximity to the tree if such separation is felt to protect from pavement cracking over time. A geotextile separates the base material from the rooting soil below, in part to provide separation of root displacements from the wearing surface layer.

Faced with the challenge of meeting the competing needs for pavement support and soil access for tree growth in a common soil volume, methods to integrate the needs in a common design would be a rational response. To be successful in integrating pavement and vegetation, pavement surfaces must have structural systems in place for support, which could be a physical structure, or a plan for load-bearing root zone soil. For this discussion, the responses will be divided into two generalized approaches centered on either *soil design* or *structural isolation of function* such as with structural grids or conduits.

Pavement as a layered system

It is important to understand how pavement is generally designed as a multiple-layer system, occasionally referenced as a 'pavement section design'. Pavement section design establishes a layered system to support and translate loads from expected traffic demands. The top layer which most people see is defined as the wearing surface. This wearing surface is supported by various layers of materials as base layers, which ultimately rest on the pre-existing soil occurring at the grading elevation (the 'subgrade') for the placement of entire layered system. This design can be a series of specified mineral layers, or a support structure resting on a compacted foundational layer of pre-existing soil. The wearing surface can be placed as a monolithic slab of solid or porous material or as preformed unit pavers. The jointing materials used to lock unit paver layers into place may or may not allow air and water infiltration, but represent an important opportunity. For the biotic aspects of the below-pavement tree root zone, excellent air and water exchange is desired. Porous pavement surfaces are available, but as found in a series of studies in New Zealand, they do not work alone for tree establishment (Morganroth and Visser, 2011) since the compaction of the layers to support the wearing surface can limit the rooting zone suitability.

There are instances wherein a pavement surface is placed directly on non-compacted ground, particularly in warmer climates without the concern of freezing temperature. The result can be a very short service life of some pavement surfaces which become long term management problems.

Designed soils as a base layer or below the base (defined as 'sub-base') as a load supporting rooting layer can provide root colonization and rapid water infiltration above the compacted sub grade. Additionally, the choice of wearing surface influences the tolerances for settlement and ability to bear load without breaking or subsiding. As such, the choice of wearing surface influences what materials and associated material behaviors (such as settlement and internal drainage) need to be verified as acceptable when considering specific design solutions.

Soil design response for tree pavement integrated systems

Soil design approaches can be generalized into sand-based and aggregate-based mixtures. The overall approach of designed soil systems is the development of a structural matrix or skeleton, created by larger mineral particles such as sand or an aggregate (gravel or broken rock) capable of being compacted to provide acceptable bearing capacity for pavement surface support. The level of needed compaction and bearing strength informs the choice between sands or larger aggregates. Soils, or horticultural media components, are blended into the larger sized material, meant to be suspended in a non-compacted form between the aggregates, forming a continuous interstitial rooting resource. The matrix needs to form prior to compacting the soil between the large particles to make matrix-forming contact. Use of too much soil holds the aggregates apart, and compromises engineering behavior. The fundamental question is how to design a mixture that will provide air and water in balance with root penetrability for colonization and perennial growth after compacting it for load-bearing support. Selection of aggregate, soil, and mixture ratio are considered within the context of the pavement section design needs for a given pavement and traffic support situation.

Skeletal soil systems are naturally occurring, capable of supporting forest vegetation and are characterized by a stone-dominated profile with a lack of layering. In these systems, soil occurs within the stone matrix. Indeed, designed soil systems mimic, exaggerate and selectively reproduce aspects of the natural system to meet the tree biotic capacity and the pavement support needs in a mutually acceptable manner with reproducible design documentation.

Selection of aggregate materials for forming the structural matrix

There have been various aggregates used; from lava-rock and heat-expanded slates/shales to granitic or limestone based systems (Grabosky and Bassuk, 2016; Grabosky, 2015; Liesecke and Heidger, 2000; Arnold, 1993). Aggregates need to be tested for durability for consistent behavior over time and expected loading. Narrow particle size distribution (diameter maximum: diameter minimum of 2:1), and angular crushed aggregates (at least 3 sheared faces) afford a larger void volume after compaction for soil as compared to wide size distributions (nesting of aggregates) and gravels (rounded aggregate) (Shergold, 1953). Additionally, cubical versus prismatic shapes ought to provide higher void volumes which can accommodate more soil in the final mixture design. Generally speaking, local standards for definition and use of acceptable aggregates can be found in geotechnical engineering or pavement engineering standards. When developing an aggregate-soil design, the pH of the aggregate will have an influence on the rooting zone chemistry, particularly in a limestone-based system. Granitic aggregates, when available and cost-effective, would be preferred over a high-pH aggregate. When a high pH aggregate is used, appropriate plant selection is important to avoid nutrient deficiencies. Fortunately many trees can grow well in a high pH soil up 8.0–8.2.

Expanded slates, shales and other porous aggregates have been used. These materials can provide an increased porosity, however the magnitude of the contribution is often poorly defined. As in any aggregate choice, expanded and porous materials have to be carefully considered and tested, and as a manufactured product, quality assurance and consistency in product is required, particularly with aggregate soundness or resistance to crushing. In some vitrification and heat expansion processes for slates and shales, and in some volcanic derivatives, elevated pH is a concern. Sound specifications and component testing can provide protections from high pH contamination to secure a successful tree planting after installation. Being porous, aggregate soundness may limit suitability for use in supporting load in some pavement instances, particularly in heavy traffic loading. Additionally, in climates where freezing and thawing occur, durability testing for ice development and for salt impacts need to be specifically addressed. There is a consideration of cost versus the gain in porosity, but the materials have been used with success, and there is a measurable benefit in the lower weight of expanded aggregates for situations above built structures.

Aspects of soil component selection

The soil component used in aggregate-based mixtures are biased toward clay-loams or loamy clays with a minimum of 20 percent clay by weight to provide moisture retention and cation exchange capacity (minimum 10 meq) within the total stone-soil system. In zones where ice lens formation and displacement (frost heave susceptibility) is considered, the soil component is adjusted to ensure *total* silt levels remain under 12 percent by weight, as silt content is the primary driver of soil frost heave susceptibility (Berg and Johnson, 1983). While soil pH is an important early consideration, the pH of the aggregate used and the influence of the pavement wearing-surface pH, such as concrete, ultimately influence the system pH over time. As such, it is often the case that high pH-tolerant tree species are preferred for planting in pavement systems.

Many wearing surfaces do not respond well to direct contact with organic materials in support layers (in the chemistry at the soil-concrete interface). Changes to the aggregate matrix brought by decomposition over time can also result in settlements under traffic load, which may negatively impact the wearing surface through cracking and displacement. In practice, no maxima have been established, but 1 percent by weight seems to be a convenient first approximation, with an understanding that over time the root system will raise the organic content with natural root colonization and turnover. Organic fractions within the soil components need to be minimal, less than 5 percent by weight, as volume shrinkage of organic materials in decomposition may negatively influence load-bearing stability of the pavement when under load. If there is only 20 percent soil by weight in the designed mixture, again we reduce to around 1 percent.

Stone to soil ratios in mixing design

Mixing ratios are more sensitive to engineering load-bearing demands with the matrix than they are to plant response (Grabosky et al., 1996). The preferred mix ratios vary with the regionally available aggregates and soils as much as they do with defined traffic support needs. Matrix porosity can be estimated by compacting the aggregate component into a known volume to help estimate soil component targets in systems when the soil is moved into the system after placement, such as in Danish systems (Kristoffersen, 1998). If developing a pre-mixed and delivered mixture, this estimate is not a reliable value. This is because for equivalent effort, soil in the mixture causes the aggregate matrix to form differently during compaction (Grabosky, 2015).

In the simplest context for pre-mixed, delivered blends, the mixing ratio would need to develop a compactable aggregate-soil system with at least 80 percent of the aggregate by weight in the sand or larger fractions as a first estimate within a mix-design process. The mixing ratio could be adjusted by the expected compaction, loading and acceptable settlement of the project. In such designed soil systems, the larger aggregate replaces those aspects of sand in the total mixture, so while there is a low percentage of clay and loam in the total system the textural triangle moves from sand-silt-clay to aggregate-silt-clay in total. From this vantage point, it is more intuitive that a mixture of 80–82 percent by weight crushed stone (1.25–2.5 cm) and clay-loam soil was found to behave like a clean, well-graded silty sand or gravel in compacted permeability tests with adequate plant-available water-holding capacity after compaction (Grabosky et al., 2009). A long-term working installation of 2.54 cm stone and clay loam soil has yet to show visibly reduced growth of trees after 15 years without supplemental irrigation in New York City (Grabosky and Bassuk, 2016).

Engineering design and placement considerations

In general, the engineering design considerations (and consequences in failure) are more sensitive than plant requirements when considering a shared soil volume. The level of compaction and load-bearing support from a safety consideration is governed by five main factors: the amount of loading, the frequency of load, the choice of wearing surface, the level of acceptable surface displacement and the acceptable surface settlement to prevent pavement failure. Design engineers can provide the traffic loading expectations and acceptable settlement levels to define pavement success. The forester can advocate specific wearing surface choices for water and aeration, and communicate the importance of post-compaction porosity. Together, the two can design a soil system that provides support while providing for the root growth needs of the tree. Many of the larger aggregate-based designed soils assume full construction compaction (593 kJ m^{-3} standard compaction effort up to 2693 kJ m^{-3} as a modified compaction effort, reflective of the equipment used in the specific construction). The compaction effort required in some pavement design situations or within a nationally accepted construction practice can yield most sand-based systems limited for root penetration. In those situations, the choices reduce to definition of the aggregate component to larger sizes of gravel or crushed stone aggregate. In a 1.25–2.5 cm crushed aggregate-based mixtures, successful root growth has been observed in short-term testing at compacted densities in excess of 1.9 Mg m^{-3} resulting from a standard compaction effort which can support high levels of traffic load (Grabosky et al., 1996).

Since soil design approaches are mixtures of components, there is a choice between mixing on-site or mixing off-site and delivering. In some cases, the aggregate is placed and compacted in the excavated planting zone and soil is then moved into the compacted matrix by slurry or by vibration (Kristofferson, 1998). The amount of soil incorporated in case study trials has ranged from very little (one site with zero soil but inclusion of biochar) to adding soil until lack of infiltration upon sequential slurry incorporation. Knowing the porosity in the compacted matrix can provide a rapid method to develop an estimated amount of soil incorporation potential. The estimate, coupled with observations from prior successful installations, provide an opportunity for method optimization and reproduction of success. Because it is difficult to add heavy soil to already existing aggregates, a sandy soil is generally used in these instances. Unfortunately, the water and nutrient holding capacity will be lower in these sites than in those where a clayey soil is mixed with aggregates.

Other approaches employ a mix design to add the soil prior to placement and compaction, whether on-site or off-site (Grabosky et al., 1996; Couenberg, 1994). In these systems, stability of the mixture in transit, placement and compaction become important to consider. With larger aggregates and an exaggerated gap-grading between aggregate and soil, there is a risk that the stone and soil will separate during transit and placement, which has been an argument for a stabilizing 'tackifier' agent such as hydrogel in some mixtures. While hydrogels or other binding agents have been proposed to aid mixture stability, their influence over time after installation is yet to be tested in all but short-term container and small sidewalk experimental profiles (Grabosky et al., 1996, 2001). Even if these binders lose their effectiveness over time, their main use is to maintain uniformity of the mixed soil and aggregate components during transportation and subsequent compacting into place. To increase quality control regarding mixing ratios and consistency, there is an advantage to mix components off site prior to placement within a roadway materials manufacturing plant.

Planting design considerations

Compacted structural soils have been observed to provide an available water holding capacity between 8 percent and 11 percent depending on the level of compaction and choice of soil component. Some mixes might provide higher levels of moisture, but published studies are limited. Structural soil volume calculations often use a water holding capacity of 8 percent to be on the conservative side (Grabosky et al., 2009). This water holding capacity is equivalent to loamy sand or sandy loam, suggesting that there is little need to increase soil volumes to compensate for the aggregate nature of the material. Indeed, one can look at the transition from sand to a larger particle (within reason) as an exaggeration of particle size class. This conceptual explanation is consistent with both the laboratory testing and long-term field observation (Grabosky and Bassuk, 2016). However, those studies represent only 19 years of focused observation. It may be useful to consider using 1.3–1.5 times more structural soil than the soil volume needed to grow an equivalent sized tree in sandy loam based on each soil's available water (discussed below). Stated, the need has not been rigorously demonstrated or proven in long-term observation.

Excavation is the second most expensive cost of tree-pavement integration following wearing surface costs. One of the richest areas for research consideration would be in the rapid assessment and remediation of the materials excavated from the planting trench to be used as the soil component of the designed system. By using the excavated materials, the condemnation of valuable soils for food production or in natural systems would be eliminated, and the potential expenses of disposal for excavated fill materials off-site would be minimized. There are many logistic and physical considerations to be developed in processing the excavated fill to separate out what portions are included in the desired mix design or to work the mix design for the specific excavated soil before suggesting this is feasible in practice. However, since the materials ultimately are covered in a highly alkaline pavement surface, this approach with potentially contaminated material seems to make common sense. Indeed, the largest challenge in this area would be the separation of aggregates and sands from the excavated fill, and the challenge of redeveloping the soil ecology as deemed necessary from soil testing.

Structural isolation of function: grid systems and conduit approaches

An alternative to using designed soils is to rely on a grid system to support pavement. These designed structures generally resemble a modular grid system of reinforced structural plastics. These systems are designed for both tree establishment and for storm water capture below surface grade, deployed to function as a bridge decking over the water storage volume or planting zone. In this context, we can break the systems into three categories:

1 Those with an integrated connection to soils; root zone within the deck-supporting grid.
2 Those segmented with a separation of air between the water storage in the deck-supporting grid and the root zone above the grid deck.
3 Those where the water storage in the deck-supporting grid is adjacent and passively drained to a growing zone to act as a reserve for water between rainfall events.

For category 1 systems, the supporting deck pillars are filled with a soil, and the pavement wearing surface is installed over the bridge decking. Openings in the deck allow for tree planting into the underlying soil. Many systems provide a method for free air exchange with an air space between the supported pavement/decking, and the soil surface within the pillar system. Rectangular and hexagonal systems exist on the market, and several systems are stackable to provide depth and spread options to provide root zone capacity.

Drainage through the system as a placed, disturbed, soil is needed. Grid systems rest on compacted subgrade or base layer as a foundation for stability, and deep drainage is not always possible. As such, there are likely limits on soil specifications for grid systems to ensure internal drainage by soil texture, since the transport and placement of soils can have negative impacts on soil structure during construction. Use of filling soils in the range of 70–90 percent sand by weight, would provide an estimated 9–14 percent v/v plant available moisture in the soil volume, can provide needed drainage with careful handling and installation. Interestingly, it places them rather close to the soil design approaches in mineral nutrient behavior. There are no published research studies explicitly discussing or comparing soil specification, selection and plant response for use in these systems, however there is a rich literature on general soil specification which can be directly adopted for the purpose after considering the drainage aspects. While these systems also assume the wearing surface and excavation costs in common with soil-based approaches, there are opportunities for soil improvement and replacement when using the grid system. Native soils should be considered if they can provide the needed drainage properties within the constructed system.

Category 2 modular grid systems can be designed for stormwater capture and pavement support which can support a soil volume and small trees on the decking material. The soil system is separated from the water storage below the deck since the root zone occurs above the deck. This affords very high water storage porosity (in excess of 90% by volume) of the cells to hold water. The water cell products are not specifically designed for tree root zone application, but they have been used successfully in the field after engineering approval on a limited number of sites. The systems are already marketed as supporting grass/turf surfaces.

The basic argument is that if the decking can support machine and traffic loads, it can support an equivalent level of static loading from a planting zone. It is reasonable to assume that the systems can support a limited amount of soil and vegetation as a standing load. Limited soil relegates tree planting to smaller tree species. A pumping system could be deployed for irrigation of the landscape planting above. Finally, in category 3, downward or lateral drainage could be used to move water into adjacent soils used to support vegetation at the parking lot edge, where a non-compacted soil profile would not need to support pavement but provide edge habitat for parking areas and pedestrian spaces.

Beyond the questions of soil type recommendations for use in grid systems incorporating soil, areas of research opportunity include:

- The role of plastic contamination when systems are excavated in future tree-pavement work.
- The role of habitats provided in the air gap between the pavement decking layer and the soil.
- The ability to alter and recover the soil materials developed in the excavation of the planting zone (previously described in the soil-based section).
- Implications of root intrusion into and against cell support pillars and stacking joints, firstly, whether it can happen as it does with other jointed pipes/systems in the root zone.
- Optimization of maximum grid depth for root colonization within placed-soil profiles.

Conduits

Conduits can and have been used to either create radiating spokes of accessible soil under pavement or for use as connecting bridges between limited soil volumes and larger volumes separated by a pavement section. Conduit systems have been proposed and installed; however there is no published study definitively showing the level and success of root colonization, in part from the cost of experimentation and refusal to harvest working systems due to high tree quality in response to systems which were originally experiments. The systems have been used in Osnabruck, Germany, and discussed in construction documents in the US (Grabosky, 2001). There are also installations where aggregate-based compaction-resistant soils have been placed under pavement creating a 'safe passage' for roots to grow into adjacent soil (Bassuk et al., 2015).

Details of soil characteristics of fill materials for use in conduits have not been fully developed in the literature or in published case studies. Conduits can be used for the provision of 'encapsulated' soil volume, which would suggest a higher quality soil resource for long-term root colonization and occupation. Alternatively, the conduit could be used to bridge from one zone (such as a trench parallel to the street) under pavement to another soil resource (such as a lawn). To establish the connection by the conduit, there is an incentive to have a sandier material to encourage rapid growth through the tube to the next soil volume. Since roots have been demonstrated to actively target nutrient and moisture patches, the 'lower quality' materials would encourage targeted growth across the conduit 'bridge'. Crafting systems to direct the roots to the conduits, the optimal placement geometry, and development of shrouds to allow partial conduit filling without soil cavitation into the end sections once below grade are simple challenges to surmount if the choice to deploy a conduit approach is accepted.

Finally, there are installations where the grid system is supported and surrounded by a designed soil system. Grid systems offer a fixed geometry, which can be used to cluster or track infrastructure lines. Construction and excavation can often present root zone/pavement support openings which do not lend themselves to grid systems in terms of excavation depth uniformity or excavation shape uniformity. In those cases, the irregular geometry of an excavation is normalized by designed soils to provide a base footing or a surrounding fill for the structural grids.

Soil volume calculation and root colonization

Systems used to integrate pavement and tree root zone capacity need to estimate the volume of soil needed for tree growth with consideration of the intensity and budget for tree culture and care. In general, the more soil provided, the lower the management intensity needed for water and mineral nutrition; to a logical threshold of diminishing return for added soil volume. The presumed goal would be to provide a soil volume carrying capacity scaled to the vegetation resource needed to provide the expected design functions and tree canopy sizes associated with a successful tree establishment over time.

Modeling water budgets based on expected tree canopy size, leaf area index and transpiration rates can yield estimators for required soil volumes based on water demand estimation and plant available water holding capacity (DeGaetano, 2000; Kopinga, 1998; Lindsey and Bassuk, 1991). For these models, the volume is developed from considering the water use of the proposed canopy volume, the plant available water within the designed tree-pavement root zone and the longest period between rainfall events. Soil volume will necessarily depend on the local climate variables that affect transpiration coupled to the available water holding capacity of the soil system chosen. For example, using published transpiration rate data on *Tilia tomentosa* and canopy character from the literature, it can be determined that it is possible for 13 trees with a 6 meter canopy diameter to de-water through transpiration a 0.4 hectare parking lot surface over ten days after a 7.6 cm/24 hour rainfall event, if the runoff was captured and moved to the root zone using either a sandy loam-filled grid or a designed aggregate-based soil as discussed earlier. Of course, this de-watering service is not in place until the tree grows to designed size. Other vegetation would need to be in place as the trees grow to provide transpiration demand to use the captured stormwater.

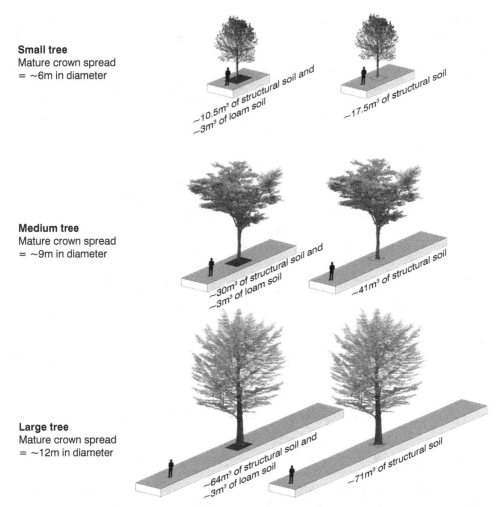

Small tree
Mature crown spread
= ~6m in diameter

~10.5m³ of structural soil and
~3m³ of loam soil

~17.5m³ of structural soil

Medium tree
Mature crown spread
= ~9m in diameter

~30m³ of structural soil and
~3m³ of loam soil

~41m³ of structural soil

Large tree
Mature crown spread
= ~12m in diameter

~64m³ of structural soil and
~3m³ of loam soil

~71m³ of structural soil

Figure 22.2 Examples of soil volume estimation using a transpiration model on soil plant water holding capacity. In this plan, 0.61 m³ soil per m² of tree crown projection at planned design size was used (Lindsey and Bassuk, 1991, 1992). Trees growing in designed soil systems in areas that normally use irrigation to grow trees should also provide low volume drip irrigation in designed soil installations

One important consideration in such design modeling is the use of tree volumes at the mature designed size. If the tree is successful and the pavement has a useful service life of 25 years, then a fairly large tree would be in place at the time of pavement replacement. Thus, one might scale the soil volume in tree-pavement design to the expected soil protection area or volume deemed normal in construction protection of 25-year-old trees of the species within the system design. One could also plan for a two-pavement cycle, or a 50-year system plan, at which point the canopy dimensions become rather stable over time, and thus soil volume calculations potentially reach their maximum limit.

While the preferred depth is variable, there is every reason to expect the depth of the root zone would be limited or somehow defined by other buried infrastructure zones such as sewerage, water or gas line layers. In some cases the existing water table or bedrock may define the maximum reasonable depth for the root zone. One meter of soil depth with varying width and length based on the expected size of the tree canopy and local precipitation and summer heat would be optimal, although several projects have been successful using soil as shallow as 60 cm.

Dimensions for the development of a supporting root system for a tree over 25 years of pavement life should inform the geometry for root zone design in windy situations. As a surrogate, many countries have minimum root plate dimension standards for tree stability in construction protection which could be used to predict the preferred situation when the pavement needs to be resurfaced. Perennial root systems compete for soil mineral and water resources, with the resulting system architecture reflecting resource exploitation potentials in scramble competition and exploitation efficiencies in 'space-filling' to capture water resources through lateral root growth branch proliferation. With this in mind, extraction of water and mineral resources in aggregate-soil based systems can be described as a root system trained into a forced soil resource exploitation efficiency within matrix voids while scrambling through the matrix with little impediment after matrix compaction.

Consequences of root colonization and growth under pavement

Encouraging tree root growth under pavement presents certain challenges. In most tree root system architectural models, there is a zone of rapid taper as roots move from close to the trunk connection and radially out into the surrounding soil. Since all roots ultimately represent a vascular system coalescing into a common single trunk in most street trees, it makes sense the roots near the trunk will have larger diameters in cross section. As roots explore soils, there must be a displacement of the soil to accommodate radial root growth in perennial root tissues. The thicker roots must displace greater volumes of soil, thus accommodations in any soil design approach are needed while they could be avoided in the grid system and conduit approaches with gaps between the top of the soil volume within the structure and the support structure (decking or top of conduit) for the pavement.

Pavement openings should allow buttress root formation with an assumption of robust tree trunk growth. This would increase tree stability while reducing sidewalk lifting. In practical design, allowances in grid structures and separation depths (base thickness separating wearing surface and designed root zone material) need to work with the natural root flare and rapid taper aspects of normal tree root system development.

In the grid systems filled with soil, perennial root growth suggests there is a question of load transfer to pillars adjacent to roots. As roots tend to follow smooth surfaces like pipes and pillars in soil due to the lesser resistance in the interface between dissimilar materials, root-pillar interactions raise a reasonable question. When connection joints between layers to develop greater root zone depth exist, there is more than enough evidence of root intrusion in pipes to suggest research might be warranted in long-term root-structure interface responses (Östberg et al., 2012).

In soil design approaches, it is important to remember that pavement design most often considers loading from the surface downward, not from loading upward as a function of root growth displacements and in reaction to external loading aspects such as wind lifting a root as the tree canopy sways. In installations in Miami, Florida, after hurricane events, trees in designed soils did not overturn or warrant pavement failure, but no data is available. Research in Singapore on short-term growth experiments suggests there are reasons to further develop this area of inquiry (Rahardjo et al., 2009). While wheel loading is already partially accommodated in normal design with or without roots, information to realistically model load transmissions into root systems and the translation upward to the wearing surface is a worthy research topic (Grabosky and Gucunski, 2011; Grabosky et al., 2011).

In the consideration of root growth displacement, it is clear that a pavement section can be designed. While the simple solution could be a very thick wearing surface, this is financially impractical as the pavement is the most expensive layer in the system. Alternatively, using a base layer to separate the wearing surface from the root zone as a subbase is a cost effective method. Early testing suggests root growth can be allowed 15–20 cm below the surface without tensile loading from displacement cracking an asphaltic concrete surface. Using the base layer to separate the wearing surface from the root zone to allow replacement of wearing surfaces without cutting roots, which can promote longevity in pavement-tree systems. The base material would likely be a clean aggregate consistent with local use under pavement or under specialized porous pavement systems.

Repairs and consequences in re-excavation

The intention of all of the systems is to provide tree life expectancies which exceed a pavement surface service life (estimated as 25 years). Of course trees may fail for many reasons over time. The larger expenses in replacement are those with pavement removal/ replacement and with the excavation for trunk and root system. Structural systems may have some challenges in excavation, since once the grids are removed they are often damaged or destroyed and cannot be used again. Moreover, plastics mixed in soil can pose issues of soil pollution in some countries. As is the case with all of the materials, oversight of the installing contractors and assurance for materials meeting the design specification need to be in place, enforced and monitored on-site, especially for construction professionals new to the methods or techniques. There are a growing number of case-study research evidences to warrant this reminder common to all construction operations.

Final considerations for integrated tree-pavement systems

Trees are a capital investment, and the use of integrative tree-pavement approaches to ensure a viable urban canopy represents an investment cost for improved urban environments. As an investment, there is wisdom in careful selection of tree species to be used in any given design. This becomes even more important within a context of climate change, since the urban environment often intensifies abiotic environmental stresses on urban trees. The climate in many temperate climate cities is changing at a rate that falls within the lifespan of most trees as individuals once established in their respective environment. Within a site assessment process, important filters for plant selection based on site variables should consider the current situation and the expected situation in 30 or more years. If the expectation is tree establishment success and the design values for the tree are envisioned from a mature canopy, then this manner of thinking is needed to accommodate environmental change and the pragmatic realities of the urban paved site. Indeed, the demands of tree establishment and growth in cities make a clear and compelling case to develop selection criteria for environmental fitness and stress tolerances over aesthetic novelty.

Regardless of the system chosen for use, urban tree plantings in temperate climates with seasonal freezing temperatures need to consider salt from snow and ice removal. Additionally, such systems need to consider water and ice formation both in terms of frost penetration and in frost lens development which can displace soil with the surrounding infrastructure. Positive drainage is common to all systems where root zone colonization is limited to a designed soil volume. It is assumed that water has a place to be intercepted or provided for the needs of the tree, but drainage is required to prevent system flooding. Salt can both enter and leave a system during snow melt and in spring rainfall events. As such, moisture and drainage come as an associated concern in freezing and salt, and selecting tree species with salt tolerance would be a common selection criterion.

Selecting tree species which are tolerant of elevated pH would be a prudent consideration. The level of this tree species selection filter as a priority would be influenced by the building materials and soils common to the area of the project. Very often, the presumption would be to select species capable of tolerating elevated soil pH. Because of its well-drained nature, trees that prefer well-drained soils likely do best in structural soil.

In summation, while all alternatives for tree-pavement systems have opportunities for research to develop further understanding, both soil approaches and structural isolation approaches are poised to provide safe and durable solutions over time as user-driven technologies.

References

Arnold H. 1993. *Trees in Urban Design*, 2nd edn. New York: Van Nostrand Reinhold.

Bassuk N, Denig B R, Haffner T, Grabosky J, Trowbridge P. 2015. *CU-Structural Soil: A Comprehensive Guide*. Ithaca, NY: Cornell University. Retrieved from www.hort.cornell.edu/uhi/outreach/pdfs/CU-Structural%20Soil%20-%20A%20Comprehensive%20Guide.pdf.

Berg R, Johnson T. 1983. *Revised Procedure for Pavement Design under Seasonal Forest Conditions*. Special Report 83-27. Hanover, NH: US Corps of Engineers Cold Regions Research and Engineering Laboratory.

Couenberg E. 1994. Amsterdam tree soil. In G Watson, D Neeley (eds) *The Landscape Below Ground: Proceedings on an International Workshop on Tree Root Development in Urban Soils*, pp. 24–33. Savoy, IL: Morton Arboretum, International Society of Arboriculture.

DeGaetano A T. 2000. Specification of soil volume and irrigation frequency for urban tree containers using climate data. *Journal of Arboriculture* 26(3): 142–151.

Eis, S. 1974. Root system morphology of western hemlock, western red cedar and Douglas fir. *Canadian Journal of Forest Research* 4: 28–38.

Grabosky J. 2001. Chapter 13: Trees and urban construction. In P Lancaster (ed.) *Construction in Cities*, pp. 157–191. Boca Raton, FL: CRC Press.

Grabosky J. 2015. Establishing a common method from which to compare soil systems designed for both tree growth and pavement support. *Soil Science* 180(4/5): 207–213.

Grabosky J, Bassuk N. 2016. Seventeen years' growth of street trees in structural soil compared with a tree lawn in New York City. *Urban Forestry and Urban Greening* 16: 103–109.

Grabosky J, Gilman E. 2004. Measurement and prediction of tree growth reduction from tree planting space design in established parking lots. *Journal of Arboriculture* 30(3): 154–164.

Grabosky J, Gucunski N. 2011. A method for simulation of upward root growth pressure in compacted sand. *Arboriculture and Urban Forestry* 37(1): 27–34.

Grabosky J, Bassuk N, Haffner T. 2009. Plant available moisture behavior in stone-soil media for use under pavement while allowing urban tree root growth. *Arboriculture and Urban Forestry* 35(5): 271–277.

Grabosky J, Bassuk B, van Es H. 1996. Further testing of rigid urban tree soil materials for use under pavement to increase street tree rooting volumes. *Journal of Arboriculture* 22(6): 255–263.

Grabosky J C, Smiley E T, Dahle G A. 2011. Observed symmetry and force of *Plantanus* × *acerifolia* (Ait.) Willd. roots occurring between foam layers under pavement. *Arboriculture and Urban Forestry* 37(1): 35–40.

Grabosky J, Bassuk N, Irwin L, van Es H. 2001. Shoot and root growth of three tree species in sidewalk profiles. *Journal of Environmental Horticulture* 19(4) 206–211.

Kopinga J. 1998. Evaporation and water requirements of amenity trees with regard to the construction of a planting site. In G Watson, D Neeley (eds) *The Landscape Below Ground II: Proceedings on an International Workshop on Tree Root Development in Urban Soils*, pp. 233–245. Savoy, IL: Morton Arboretum, International Society of Arboriculture.

Kristofferson P. 1998. Designing urban sub-bases to support trees. *Journal of Arboriculture* 24(3): 121–126.

Liesecke H J, Heidger C. 2000. Substrate für Bäume in Stadtstraßen. *Stadt und Grön* 7: 463–470.

Lindsey P, Bassuk N. 1991. Specifying soil volume to meet the water needs of mature urban street trees and trees in containers. *Journal of Arboriculture* 17: 141–149.

Lindsey P, Bassuk N. 1992. Redesigning the urban forest from the ground below: A new approach to specifying adequate soil volumes for street trees. *Arboricultural Journal* 16: 25–39.

Lu J W T, Svendsen E S, Campbell L K, Greenfeld J, Braden J, King K L, Falxa-Raymond N. 2010. Biological, social and urban design factors affecting young street tree mortality in New York City. *Cities and the Environment* 3(1): article 5.

McPherson E G, Nowak, D J, Rowntree R A. 1994. *Chicago Urban Forest Ecosystem: Results of the Chicago Urban Forest Climate Project*. Gen. Tech. Rep. NE-186. US Department of Agriculture, Northeastern Forest Experiment Station.

Morgenroth J, Visser R. (2011) Above-ground growth response of *Platanus orientalis* to porous pavements. *Arboriculture and Urban Forestry* 37(1): 1–6.

Nowak D J, Hoehn R, Bodine A R, Crane D E, Dwyer J F, Bonnewell V, Watson G. 2013. *Urban Trees and Forests of the Chicago Region*. Resour. Bull. NRS-84. Newtown Square, PA: US Department of Agriculture, Forest Service, Northern Research Station.

Östberg J, Martinsson M, Stål Ö, Fransson A M. 2012. Risk of root intrusion by tree and shrub species into sewer pipes in Swedish urban areas. *Urban Forestry and Urban Greening* 11: 65–71.

Rahardjo H, Harnas F R, Leong E C, Tan P Y, Fong Y K, Sim E K. 2009. Tree stability in an improved soil to withstand wind loading. *Urban Forestry and Urban Greening* 8(4): 237–247.

Sanders J, Grabosky J, Cowie P. 2013. Establishing maximum size expectations for urban trees with regard to designed space. *Arboriculture and Urban Forestry* 39(2): 68–73.

Sanders J R, Grabosky J C. 2014. 20 years later: Does reduced soil area change overall tree growth? *Urban Forestry and Urban Greening* 13: 295–303.

Shergold F A. 1953. The percentage voids in compacted gravel as a measure of its angularity. *Magazine of Road Research* 13: 3–10.

PART VI

Selection of planting material, planting techniques and establishment

23

CRITERIA IN THE SELECTION OF URBAN TREES FOR TEMPERATE URBAN ENVIRONMENTS

Henrik Sjöman, Andrew Hirons and Johanna Deak Sjöman

Trees in urban environments are likely to play an increasingly significant role in improving the quality of life in our towns and cities. However, the ecosystem services bestowed on communities in the future by trees are reliant on the appropriate selection of trees for planting. Unfortunately, it is common practice to inadequately evaluate the site and select from a narrow range of well-known species. This chapter presents an approach that could be used to more robustly select trees for urban green infrastructure projects. If these guidelines are adopted, a more resilient urban tree population will be secured for future generations.

Selection criteria

Since the intention when choosing tree species is usually that they will stand on the site for many years, perhaps even in excess of 100 years, it is critical to choose the best possible plant material for the specific site and climate. Based on the knowledge that many ecosystem services increase with tree size (e.g. Gómez-Muñoz et al., 2010), choosing tree species that are capable of reaching maturity on the site must be a priority. Site preparation, in the form of planting pit construction, and initial care of the trees are similarly vital for successful establishment. These latter aspects are handled in Chapters 25 and 26 of this book; this chapter deals with factors that are important to consider when choosing the right tree for the right place.

When choosing suitable plant material for a site, it is important to consider not only the soil conditions, but also factors such as temperature, wind conditions, rainfall, light availability, pollution and the potential for vandalism. Other characteristics may also be important in the selection process if the intention is to use the trees or tree plantations for one or more technical functions, such as reducing strong, cold winds in the urban landscape or contributing to stormwater management. This requires plant characteristics such as good physical resistance to strong winds or an ability to tolerate wet and periodically anaerobic soil conditions. However, in many green infrastructure projects in Europe, Asia and North America, aesthetic considerations have dominated over functional and bio-ecological selection criteria. Focusing solely on aesthetic qualities when choosing plants is seldom a successful concept since these qualities are never fully expressed if the plant material is

Figure 23.1 Many recurring maintenance measures such as continual irrigation and fertilisation or frequent pruning can be avoided by choosing the right plant species and planting bed for the site and situation. This picture shows sycamores (*Acer pseudoplatanus*) growing in a paved parking lot where the planting bed and species are both unsuitable, resulting in very poor trees that do not enhance the identity or function of the site

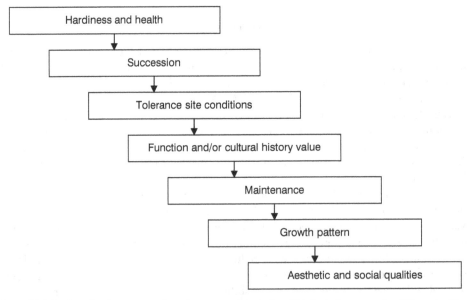

Figure 23.2 In the selection process for urban trees, technical and biological aspects should be prioritised over aesthetic values. This compilation with ranked aspects that should be included in the selection process shows that aspects such as hardiness and health together with succession, site tolerance, capacity for delivering ecosystem services, maintenance demands and growth patterns have higher priority over aesthetic and social qualities

unsuitable for the site (Figure 23.1). For example, many maples will never deliver promised autumn colours if the leaves experience significant drought stress.

To maximise the likelihood of tree species success on a particular site, a number of criteria must be considered (Figure 23.2). These criteria are presented in a hierarchical format as establishment success is governed principally by biological criteria (hardiness, succession, stress tolerance). Functional, aesthetic and social criteria for selection will be subordinate as these are fundamentally dependent on good tree performance on any given site.

Hardiness and health

The most important criteria used to determine tree selection choices are the cold hardiness and health potential of the plant material. This is largely informed by the genetic background of the tree, particularly their provenance or ecotype. These refer to the geographical site from which the seed material originates and the specific habitats that parental lineages have adapted to. A suitable ecotype of a species leads to an inherently compatible seasonal rhythm of development between the tree and the environmental conditions at the intended growing site (Hunter and Lechowicz, 1992). Failing to take this factor into consideration in tree selection can result in under-performing trees that have elevated vulnerability to other abiotic and biotic threats. Damage arising from insufficient plant hardiness at a particular site can also critically impair normal development of the species in the long term. For example, southern ecotypes of silver birch (*Betula pendula*) are less tolerant to late frosts than northern populations and commonly show sparse growth when planted in sites exposed to frequent spring frosts, especially late in the spring. Similarly, when planting Norway maple (*Acer platanoides*) in cold sites, southern ecotypes may be more prone to sunscald than northern ones (Sjöman and Slagstedt, 2015). Poorly matched hardiness can also compromise tree performance on a site because of its impact on key physiological processes such as photosynthesis and respiration (Ghannoum and Way, 2011).

Today there is relatively good knowledge in Europe and North America about the hardiness of different tree species, ecotypes and cultivars. Well-developed climate zone maps are available showing the conditions in different localities and categorising plants according to the climate zones in which they can be successfully grown. Unfortunately, this information is restricted to well-known species; published information on less common species is scarce. However, local arboreta and botanical gardens can act as an important resource, both in providing information on plant hardiness and in supplying suitable genetic material for a particular region. Another valuable but underutilised approach is the critical evaluation of current urban tree populations, as this can also often reveal the presence of rare trees that have potential for wider utilisation. New collections of wild plants may also be required in order to obtain hardier material. For example, the tulip tree (*Liriodendron tulipifera*) from eastern North America is represented in many of the major tree nurseries in Europe by provenances that are too southern to be suitable for planting in northern Europe. New collections of this species from more northerly latitudes would ensure its suitability for use in northern Europe. Furthermore, among the major tree nurseries of Europe and North America (and to some extent in arboreta and botanical gardens), many species are represented only by a very limited genetic material, collected at one or a few locations within their natural range. New collections of a wider range of genetic material are likely to provide genotypes that have a greater use-potential in parts of the world previously thought to be too cold for the species.

With regards to health, it is important to continually keep up to date with information about severe disease threats and pest problems associated with different tree species, in order to avoid investing in trees that will not be sustainable in the long term. It is also important to make a distinction between diseases and pests that primarily attack weakened trees and those that attack healthy trees growing in good conditions (Sinclair and Lyon, 2005). Since it is difficult to anticipate which pests and diseases will affect our trees in the future it is vital to create resilient and diverse urban tree populations with a broad range of species and genera (for more information, see Barker, 1975; Grey and Deneke, 1986; Moll, 1989; Santamour, 1990; Sjöman et al., 2012; Kendal et al., 2014; and Chapter 18, this volume).

Succession: learning from nature

The concept of succession can be described in simplified terms as 'the change in the species distribution from a three-dimensional perspective of a place over time' (Picket et al., 2013). Today this concept is seldom included in the selection process for urban trees, but the successional status of the species can be of critical importance for ensuring tree establishment and early development – the phase that often determines the long-term growth and survival of the tree.

An urban square or courtyard built on concrete foundations has few similarities with a mature forest environment. Therefore, late successional forest species find it much more difficult to establish in a warm square or enclosed courtyard, where the conditions resemble a much earlier stage of succession. However, planting pioneer species in narrow, heavily shaded urban canyons will expose the trees to light levels that favour more shade-tolerant late successional species. Attempting to identify the phase of succession currently represented by an intended planting site can help in the search for plant material that possesses naturally developed strategies to deal with these conditions (Figure 23.3). It can also help in anticipating the initial maintenance measures needed, depending on whether a pioneer or late successional species is chosen (see below).

Figure 23.3 Considering natural growing conditions in the search for the right tree for the right place can provide a good understanding of how well it is suited for the specific growing site. Trees that naturally occur as late successional trees in cool, moist forest environments will find establishment and development problematic in a warm and periodically dry urban square environment. Trees that grow naturally in warm, dry conditions early in the succession are much better suited to this environment, as they have well-developed strategies for coping with the conditions

One approach in the search for appropriate trees for a particular site is to try to match a natural habitat with the intended planting site. Paved urban environments, such as urban plazas, are represented in nature by warm mountain slopes with limited soil volume at an early phase of succession. This means that species such as black pine (*Pinus nigra*), sessile oak (*Quercus petraea*), goldenrain tree (*Koelreuteria paniculata*), mahaleb cherry (*Prunus mahaleb*), Russian olive (*Elaeagnus angustifolia*) and manna ash (*Fraxinus ornus*) are suitable trees, as they naturally occur in similar conditions and have developed strategies for coping with these conditions. However, if the planting site is 'improved' by ameliorating the rooting environment with structural soil, the site can be comparable to a natural scree slope with rooting conditions that provide good access to oxygen and a relatively large soil volume to hold water and nutrient resources. A number of species grow in this type of scree slope environment and display very good long-term development in their natural environment. If the urban planting site is fully exposed to the sun, pioneer species are most suitable since they can cope with the open, exposed site with high evapotranspiration. Examples of such species are Italian alder (*Alnus cordata*), Hungarian oak (*Quercus frainetto*), Turkey oak (*Q. cerris*), Swedish whitebeam (*Sorbus intermedia*), Sargent's cherry (*Prunus sargentii*), zelkova (*Zelkova serrata*), field maple (*Acer campestre*), Japanese tree lilac (*Syringa reticulata*), ginkgo (*Ginkgo biloba*) and European hackberry (*Celtis australis*). Where the urban square planting site is shaded by neighbouring buildings for part of the day, late successional species that occur naturally on scree slopes may be more suitable. These include hop hornbeam (*Ostrya* spp.), hornbeam (*Carpinus* spp.), elm (*Ulmus* spp.) and silver lime (*Tilia tomentosa*), which can cope with the soil conditions and lower light quality, as the shady conditions also create a cooler, more humid site with lower evapotranspiration.

Even on rich parkland sites, it is important to consider succession. On open sites where it may be desirable to create a windbreak, pioneer species from cool, rich forests should be selected, as they possess developmental strategies that facilitate rapid establishment. Examples of such species are silver maple (*Acer saccharinum*), poplar (*Populus* spp.), many willow (*Salix* spp.), silver birch (*Betula pendula*), alder (*Alnus* spp.) and Russian olive (*Elaeagnus angustifolia*). Where established trees already exist on parkland, their mature crowns modify microenvironmental conditions influencing the light and humidity levels on the site. These planting locations represent a later phase in forest succession and favour species such as: western hemlock (*Tsuga heterophylla*), fir (*Abies* spp.), sycamore (*Acer pseudoplatanus*), small-leaved lime (*Tilia cordata*), beech (*Fagus* spp.), yew (*Taxus* spp.) and western red cedar (*Thuja plicata*).

Table 23.1 presents a summary of pioneer and late successional species with potential for amenity landscapes. Pioneer species can cope with fluctuations in microclimate, and high weed competition, but they require high light levels. Late successional species are suitable for more mature environments with a rather stable microclimate, a higher degree of shade and lower weed pressure.

Site tolerance

Having considered hardiness and health and determined the phase of succession represented by the site, the next most important aspect to consider is the tolerance of the species to the growing site. In their natural habitats, trees have developed different strategies and characteristics to enable them to compete for resources within the constraints of a particular climate (Lamberts et al., 2008). Knowledge of plant strategies employed by different species to cope with specific growing conditions is essential to identify suitable trees for a sustainable planting scheme. While it may be tempting to generalise plant selection decisions, it

Table 23.1 Summary of tree species that can grow successfully in early successional pioneer growing conditions and late successional conditions

Species for pioneer growing conditions	Species for late successional conditions
Acer campestre	*Abies amabilis*
Acer × freemanii	*Abies grandis*
Acer negundo	*Abies homolepis*
Acer rubrum	*Abies nordmanniana*
Acer saccharinum	*Abies sibirica*
Alnus cordata	*Acer circinatum*
Alnus glutinosa	*Acer griseum*
Alnus × spaethii	*Acer·heldreichii* ssp. *trautvetteri*
Betula nigra	*Acer pensylvanicum*
Betula pendula	*Acer platanoides*
Betula utilis	*Acer pseudoplatanus*
Catalpa bignonioides	*Acer rubrum*
Catalpa speciosa	*Acer rufinerve*
Celtis australis	*Acer tataricum*
Celtis occidentalis	*Acer tataricum* ssp. *ginnala*
Cercis canadensis	*Acer tegmentosum*
Cercis siliquastrum	*Acer × zoechense*
Elaeagnus angustifolia	*Aesculus flava*
Fraxinus americana	*Aesculus hippocastanum*
Fraxinus angustifolia	*Amelanchier lamarckii*
Fraxinus ornus	*Betula alleghaniensis*
Fraxinus pensylvanicum	*Betula lenta*
Gleditsia triacanthos	*Carpinus betulus*
Juglans nigra	*Carya ovata*
Juniperis communis	*Cercidiphyllum japonicum*
Juniperus virginiana	*Chamaecyparis lawsoniana*
Koelreuteria paniculata	*Cornus controversa*
Larix × eurolepis	*Cornus kousa*
Larix sibirica	*Cornus mas*
Liquidambar styraciflua	*Crataegus monogyna*
Maackia amurensis	*Cryptomeria japonica*
Maclura pomifera	*Fagus grandifolia*
Metasequoia glyptostroboides	*Fagus orientalis*
Pinus heldreichii	*Fagus sylvatica*
Pinus nigra	*Ilex aquifolium*
Pinus sylvestris	*Magnolia kobus*
Paulownia tomentosa	*Magnolia obovata*
Populus alba	*Magnolia tripetala*
Populus balsamifera	*Ostrya carpinifolia*
Populus × berolinensis	*Ostrya caroliniana*
Populus × canadensis	*Picea abies*
Populus laurifolia	*Picea omorika*
Populus nigra	*Pinus cembra*
Populus simonii	*Pinus peuce*
Populus trichocarpa	*Pinus × schwerinii*
Prunus avium	*Pinus sibirica*
Prunus cerasifera	*Prunus padus*
Prunus maackii	*Sciadopitys verticillata*

Species for pioneer growing conditions	Species for late successional conditions
Prunus mahaleb	*Sorbus aucuparia*
Prunus padus	*Sorbus torminalis*
Prunus sargentii	*Stewartia pseudocamelia*
Prunus virginiana	*Taxus baccata*
Prunus × yedoensis	*Thuja plicata*
Pterocarya fraxinifolia	*Thujopsis dolabrata*
Pterocarya × rehderiana	*Tilia cordata*
Pterocarya rhoifolia	*Tilia × europaea*
Quercus castaneifolia	*Tilia platyphyllos*
Quercus cerris	*Tilia tomentosa*
Quercus coccinea	*Tsuga canadensis*
Quercus frainetto	*Tsuga heterophylla*
Quercus macranthera	
Quercus palustris	
Quercus robur	
Quercus rubra	
Robinia pseudoacacia	
Salix alba	
Salix fragilis	
Salix pentandra	
Salix × sepulcralis	
Sorbus hybrida	
Sorbus latifolia	
Styphnolobium japonicum	
Syringa reticulata	
Ulmus 'New Horizon'	
Ulmus 'Rebona'	
Zelkova serrata	

is important to pay attention to the precise site in question. Choosing a tree species for one street environment does not mean that it is suitable for all other streets with different climates and growing conditions. The aim should always be to consider all key factors likely to influence plant growth at a specific site. Table 23.2 provides a series of questions that should be asked to help resolve the precise nature of the planting site and the fundamental characteristics that are pertinent to species selection.

'Fitness for site'

When site conditions have been identified according to the criteria in Table 23.2, the next step is to identify tree species capable of withstanding the constraints imposed by the planting site. This requires understanding of the ecological strategies, mechanisms, and traits employed by trees in their natural habitats to cope with the inherent constraints of their native environment (Reich, 2014). Identifying and selecting trees based on this ecological knowledge can greatly aid establishment and help ensure the delivery of ecosystem services to the site.

It is not possible to provide a total review of all trees and their unique strategies for coping with different stress regimes in this chapter. However, it is possible to demonstrate the value of using plant traits and strategies to aid plant selection decisions. For example, drought stress is one of the most serious forms of abiotic stress for urban trees as a consequence of the sub-

Table 23.2 Important questions to ask when evaluating site characteristics prior to choosing suitable tree species for a planting site

Planting pit	• Is the soil predominantly dry or wet?
	• Will the site require irrigation?
	• Are there high fluctuations in soil moisture throughout the year, and if so, what is the magnitude of these fluctuations?
	• How fertile is the soil?
	• Is there a risk of water logging and root hypoxia?
	• Is the soil well aerated?
	• Can rainwater and meltwater be diverted to the planting bed?
Space	• How much rootable soil volume is there?
	• How much space is available above ground?
	• Is there underground infrastructure that must be considered?
	• How close to buildings will trees be planted and what future conflicts may arise?
Wear	• How is the space in which the trees are growing to be used?
	• Will the park or square be used for concerts or markets?
	• Is it an area with many children and will the trees be used for climbing?
	• Is there a risk of work-related damage to trunks?
	• Is there a risk of digging damage to roots?
	• Is there a risk of snow sliding from roofs onto tree crowns and causing damage?
Pollution	• What soil and air pollutants is the site exposed to?
	• How exposed is the site to road and atmospheric salt?
Wind	• What are the wind conditions like?
Light	• What is the level of shade at the planting site?
	• Is artificial lighting present at the site and, if so, could it affect the plant material?
Micro-climate	• Is the site exposed to wind, or sheltered?
	• Will paved surfaces increase the local temperature?
	• Is the site characterized by high or low humidity?

optimal rooting volumes, impermeable surfaces and the urban heat island effect (Sieghardt et al., 2005). This profound abiotic stress is likely to become more severe according to predicted future climate scenarios of warmer and drier summers in many temperate parts of the world (Allen, 2010). Therefore, drought stress is used here as an example of how to proceed in plant selection work in order to find the right tree for the site based on its capacity to cope with dry conditions.

In nature, trees have evolved a broad continuum of strategies to cope with warm and periodically dry conditions. These strategies (described in detail in Chapter 17) arise from a combination of drought avoidance and/or drought tolerance traits evolved by different species in response to water deficits of different frequencies and duration (e.g. Levitt, 1980; Kozlowski and Pallardy, 2002; Bacelar et al., 2012).

Avoidance of plant water shortage can be mediated either by maximising water acquisition or by reducing water use. Acquisition of water can be facilitated by increased hydraulic conductivity (Sack and Holbrook, 2006), deep rooting to regions of the soil profile with available water (Canadell et al., 1996) or increased potential absorptive area through root proliferation (McCully, 1999). Reducing water use through early stomatal closure also helps to postpone drought-related damage to stems (e.g. embolism) and maintain the water status of the leaf. This involves the dynamic control of stomata provided by both hydraulic and non-hydraulic mechanisms (Augé et al., 2000).

Reducing crown leaf area in response to drought, commonly known as drought-deciduousness, is another effective strategy to reduce transpiration, particularly in species with a low leaf mass per area (Fini et al., 2013). Further resistance to water loss can be associated with morphological characteristics such as leaf hairs (trichomes) and an abundance of epicuticular waxes in some species (see Figure 23.4).

Figure 23.4 Examples of avoidance strategies. Upper picture: Silver lime (*Tilia tomentosa*) which has invested in producing hairs under the leaves in order to reduce evapotranspiration. Lower picture: Katsura trees (*Cecidiphyllum japonicum*), which after a period of drought in a paved environment (left) have reduced their foliage in order to limit evapotranspiration compared with trees in a large planting bed (right) with good root space and thus better access to water

Drought tolerance traits allow the tree to maintain physiological function for longer during the drying cycle. This is advantageous on urban sites with restricted soil volume and depth that are prone to water deficits. The water potential at turgor loss point (Ψ_{p0}) is a trait of significant interest as it represents a quantifiable measure of physiological drought tolerance. Simple observations of drought tolerance based on ecological data such as climate and survival of tree species can be misleading: some species may avoid physiological drought through deep rooting, but are not very tolerant to drought in the absence of access to water deep in the soil profile. In urban environments, deep rooting is often perturbed as a result of physical barriers and/or root loss during transplantation. Therefore, physiological drought tolerance is likely to be a useful characteristic to assess for potential 'urban tree' species, as it quantifies leaf and plant drought tolerance directly: a more negative Ψ_{p0} extends the range of leaf water potential (Ψ_{leaf}) at which the leaf remains turgid and maintains function. Thus, the Ψ_{p0} value quantifies an important characteristic used by trees to 'tolerate' rather than 'avoid' drought.

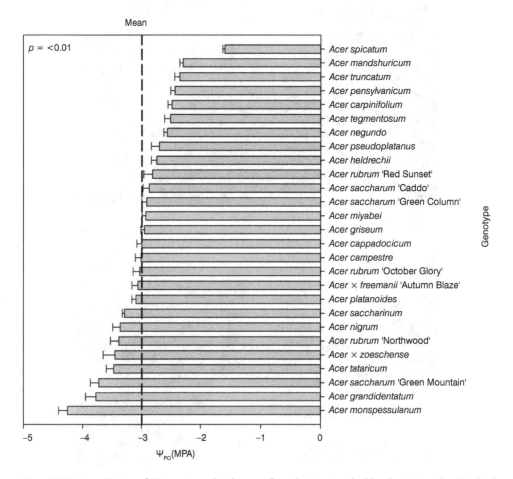

Figure 23.5 Compilation of 27 species and cultivars of maple (*Acer*) ranked by their Ψ_{p0} value. Dashed line represents the mean for all genotypes. Bars show SE and $p < 0.001$ indicates that highly significant differences occur across the dataset

Source: Sjöman et al. (2015)

In a study designed to reveal the range of species' drought tolerance within a single genera, Sjöman et al. (2015a) presented a compilation of 27 genotypes of maple (*Acer*) in which their Ψ_{p0} was evaluated (Figure 23.5). These data revealed that it is not only possible to categorise which species are drought tolerant and which are drought sensitive, but also to quantify how tolerant or sensitive they are. As can be seen in Figure 23.5, those species that originate from a warm, dry habitat, such as Amur maple (*Acer tataricum*) which comes from the steppe forests of eastern Romania, are among the species that can handle a more negative turgor pressure and thereby have the capacity to maintain turgor at low soil water potentials often found in paved urban environments. However, species listed in the upper part of Figure 23.5 do not have such an inherent ability to cope with low soil-water availability and rapidly develop symptoms of drought stress. Species with this predisposition usually originate from cool, moist forest systems, generally as understorey trees where they seldom experience dry conditions and have not needed to develop a high drought tolerance. Examples of such species are striped maple (*Acer pensylvanicum*) and mountain maple (*Acer spicatum*), both of which are found in cool forest environments in the eastern USA, often beside streams or in other damp sites and usually within a relatively humid understorey. Consequently, species such as these should only be used in park and arboretum environments with generous soil volumes and good soil quality, as they lack the capacity to cope with the more challenging conditions such as those found in many street environments.

If tree establishment is to be successful, it is essential to assess the growing conditions on the site and determine the type of strategies and plant traits that will aid tree development on that particular site. As our understanding of these traits develops, it will be increasingly possible to use quantifiable data to match species characteristics to known (or anticipated) site conditions. This approach will also make it possible to identify both unsuitable and suitable species for particular sites. Indeed, even without complex trait analysis, it is possible to evaluate species' environmental tolerances based on their natural habitat. Table 23.3 shows examples of species that have developed a high tolerance to warm, dry environments and species that are sensitive to dry environments.

Function(s)

An important aspect to include when choosing trees is what they are expected to 'deliver'. An attractive appearance is obviously important in most planting projects, particularly how the plants can give a unique identity to the site through their flowers, fruits, autumn colours, crown form and leaf structure. If this is the only function required, it is relatively easy to supply as long as the plant material chosen is suitable for the climate and the growing site. However, in a modern city, there are many more functions associated with the city's green environments than just the aesthetic one. In the future, urban densification is likely to place even greater demands on the few green environments permitted as they will need to deliver a range of ecosystem services. Consequently, it is important to consider these demands in the selection process.

Ecosystem services provided by urban trees are described in detail in Chapters 5–12 of this book. Here, we give examples of how species selection can affect the delivery of these services.

Example 1: Selection of trees for stormwater management

Trees play a number of roles in open stormwater management. In addition to being able to take up a considerable amount of water via their roots from lateral water flows under the soil

Table 23.3 Examples of species with (left) high and (right) low tolerance to warm, dry environments

Species tolerant to warm, dry environments	Species sensitive to warm, dry environments
Acer campestre	Abies homolepis
Acer tataricum	Abies nordmanniana
Acer × zoeschense	Acer griseum
Alnus cordata	Acer heldreichii ssp. trautvetteri
Cornus mas	Acer pensylvanicum
Crataegus × lavallei	Acer pseudoplatanus
Crataegus × persimilis	Aesculus flava
Elaeagnus angustifolia	Aesculus hippocastanum
Eucommia ulmoides	Betula alleghaniensis
Fraxinus angustifolia	Betula lenta
Fraxinus ornus	Betula pendula
Fraxinus pennsylvanica	Cercidiphyllum japonicum
Ginkgo biloba	Chamaecyparis lawsoniana
Gleditsia triacanthos	Cornus controversa
Koelreuteria paniculata	Cryptomeria japonica
Morus alba	Davidia involucrata
Paulownia tomentosa	Fagus sylvatica
Pinus heldreichii	Halesia monticola
Pinus nigra	Magnolia acuminata
Platanus × hispanica	Magnolia kobus
Prunus ceracifera 'Nigra'	Magnolia obovata
Prunus × eminens 'Umbraculifera'	Magnolia salicifolia
Prunus mahaleb	Magnolia tripetala
Prunus sargentii	Picea abies
Pyrus calleryana	Picea omorika
Pyrus communis	Pinus × schwerinii
Pyrus salicifolius	Prunus padus
Quercus cerris	Pterocarya fraxinifolia
Quercus frainetto	Salix alba
Quercus macranthera	Sciadopitys verticillata
Quercus petraea	Sorbus aucuparia
Robinia pseudoacacia	Stewartia pseudocamelia
Sorbus aria	Thuja plicata
Sorbus × intermedia	Thujopsis dolabrata
Sorbus latifolia	
Sorbus × thuringiaca 'Fastigiata'	
Sorbus torminalis	
Styphnolobium japonicum	
Syringa reticulata	
Tilia tomentosa	
Ulmus 'New Horizon'	
Ulmus parvifolia	
Ulmus 'Rebona'	
Zelkova serrata	

surface, trees can also capture a large amount of precipitation on their leaves and branches through the process of interception. Much of the water intercepted by leaves and branches evaporates (depending on the intensity of precipitation), without reaching the ground and infiltrating into the soil. This is most obvious in nature-like plantings, for example in parks and recreation areas, where the foliage of large, multi-layered vegetation intercepts a greater volume than that of individual trees in a street environment. A study by Florgård and Palm (1980) showed, for example, that interception by a deciduous forest represents around 40 per cent of total precipitation, while interception by a coniferous forest can be as high as 60 per cent. This suggests that quite a large amount of rain can fall before the soil receives any substantial volume of water.

In the case of individual trees, different species have a varying capacity to intercept precipitation in different parts of the year. A research project in California, USA, showed that a large plane tree (*Platanus* × *hispanica*) has an interception rate of around 15 per cent during the winter (during a rainfall event of 21.7 mm) and around 79 per cent in summer (during a rainfall event of 20.3 mm) (Xiao and McPherson, 2002). That study also showed that interception by different tree species in the summer could differ by up to 50 per cent. Thus even an individual tree can make an important contribution to interception, although this varies depending on species and time of year. It also varies between deciduous and coniferous trees. In general, conifers have a higher leaf area index than deciduous trees, which gives them a higher interception capacity compared with many deciduous trees. Another advantage with conifers is that their interception capacity remains stable throughout the year (Figure 23.6).

Figure 23.6 The bare ground underneath the pine tree illustrates the interception by the tree crown; the exposed soil follows the outer contours of the tree crown. The lack of grass is as such the result of water deficiency rather than light deficiency

Example 2: Selection of trees for wind reduction

In warm climates, amenity trees are widely referred to as *shade* trees in recognition of this vital ecosystem service provided by large trees. However, in regions where there is not only a hot summer season but also a cold winter, the ability of trees to reduce cold winds is also important for creating a more tolerable outdoor climate for urban dwellers and reducing energy used to heat buildings chilled by cold winds.

Careful planning of the city's tree population can have a great impact on wind conditions of the local area. Tree positioning in relation to the prevailing wind and built infrastructure are key factors. However, in many urban areas opportunities to create broad shelterbelts are rare so mitigation of cold winds falls to carefully placed small groups or individual trees. Therefore, selection of species capable of fulfilling that role is crucial. In addition to mechanical strength and environmental tolerance to the site, the branching pattern of the tree, i.e. crown shape and branch density, is a critical factor in shelterbelt species. In a study by Deak Sjöman et al. (2016), the crown density in winter of 72 different tree species and genotypes was evaluated (as branch area index, BAI) (Figure 23.7). The results revealed large differences between species and genotypes. Columnar deciduous trees and conifers had the densest crown structure, while common urban tree species such as *Ginkgo biloba* and *Tilia cordata* had much lower crown densities. This suggested wide variation in species' abilities to perform as windbreaks. To further analyse the capacity of trees to reduce cold winds, in the same study, a number of climate simulations were performed with selected trees using the computer program ENVI-met 3.1 (Deak Sjöman et al., 2016). The results confirmed that columnar trees with their much denser BAI have a significantly improved capacity to reduce winds when compared to trees with a smaller BAI such as *Ginkgo biloba* and *Gleditsia triacanthos* (Figure 23.8). Importantly, this modelling demonstrated that the choice of tree is critical to achieve the best possible effect in reducing wind. A poor choice of species can lead to an outdoor climate that few will want to spend time in during winter and will allow surrounding buildings to be chilled by cold winds, leading to increased heating costs.

Maintenance

Every tree will benefit from some post-planting maintenance, however, a species' ecological status can be highly relevant to the quantity of post-planting maintenance needed to get the tree established. Pioneer species such as poplar (*Populus spp.*) and silver maple (*Acer saccharinum*) (see also Table 23.1) invest in rapid root and shoot growth and are allied to the so-called *competitor* strategists (Grime, 2001). With the correct care, these species readily establish with maintenance such as irrigation only being required for 1–2 years. Other species may take much longer to establish because a more prudent rate of growth has served them well in their natural habitats. Indeed, many of the species that are slower to establish are excellent candidates for paved sites as they are *stress-tolerator* strategists that are well adapted to resource-poor sites where slow and cautious growth is necessary to conserve the limited resources (Grime, 2001). For example, Turkish hazel (*Corylus colurna*) and sessile oak (*Quercus petraea*) are good choices for paved sites but their natural disposition for slow development means that they need 4–5 years post-planting maintenance to establish effectively. With such a rich dendroflora from widely varying habitats, it is important not to generalise too much or become highly prescriptive, but simply to recognise that post-planting maintenance plans should acknowledge the type of tree material being used. For many planting schemes, this is vital because those species that are most capable of developing on challenging urban sites are those species that require a longer

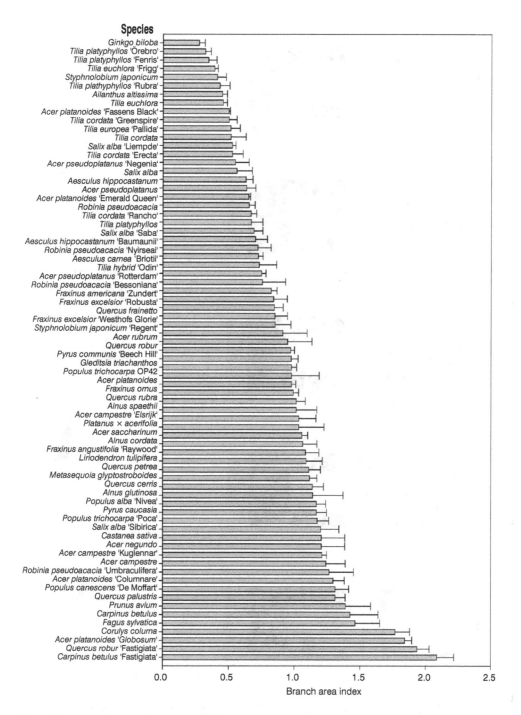

Figure 23.7 Branch area index (BAI) of different tree species and genotypes

Source: Deak Sjöman et al. (2016)

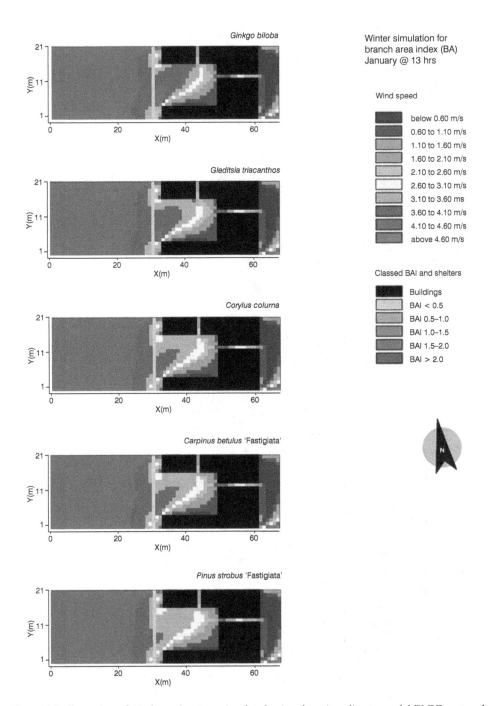

Figure 23.8 Illustration of wind speed pattern simulated using the microclimate model ENVI-met and indicating the effect on wind speed and wind pattern at 4 m height of (a) *Ginkgo biloba*, (b) *Gleditsia triacanthos*, (c) *Corylus colurna*, (d) *Carpinus betulus* 'Fastigiata' and (e) *Pinus strobus* 'Fastigiata'

Source: Deak Sjöman et al. (2016)

period of maintenance before they are established. It is also generally true that drought and heat-tolerant trees are much more tolerant of these conditions once they are established, so early maintenance procedures are closely coupled to future tree performance.

The site where the tree will grow is also important when predicting its long-term maintenance requirements. If it is a paved environment, it is important not to choose species that can create problems with their fruit. There are many examples of richly flowering ornamental crabapple trees (*Malus* spp.) being chosen as street trees, without considering the effects in autumn when the fruit falls and creates an apple sludge on pavements and cycle paths. This not only looks unattractive, but it can act as a slip hazard and attract unwanted insects. Even if the fruit does not fall to the ground but is eaten in situ by birds, this can create a problem with bird mess on pavements, cars and seats under the trees. Conflicts with tree fruits and pavements is also common when using female ginkgos (*Ginkgo biloba*) whose fruits give a very unpleasant smell when they fall down on the pavements. Another aspect associated with maintenance to consider is the leaves, petals, hairs, bud scales and other debris that falls from the tree. The large leaves of Norway maple (*Acer platanoides*) can be a problem in autumn as they can block drains and cause localised flooding. The dry leaves of common hornbeam (*Carpinus betulus*) and oak (*Quercus* spp.) do not have the same tendency to create these problems.

Site context and use may be an important consideration, particularly if the planting location is shared in a street environment where traffic flow is prioritised and green infrastructure needs to allow space for cars. By choosing species or varieties that naturally develop a straight trunk with clear horizontal branches, it is possible to avoid repeated pruning to keep the tree within the allocated space. As an example, one can consider the difference between English oak (*Quercus robur*) and swamp oak (*Quercus palustris*). Swamp oak has a tendency to develop a single straight trunk, while sessile oak tends to divide into large and wide-growing trees. In order to fit into traffic environments, sessile oak will require much more pruning work than swamp oak, which will only need crown lifting to the desired height.

Another aspect concerning crown structure and maintenance is knowing whether the selected tree species develops very acute branch angles, which can create a risk of ingrown bark between branch and trunk. This can result in the branch being less well attached to the trunk and can create a future risk. Avoiding the use of tree species/varieties that develop acute branch angles of course also reduces the need for intensive pruning to avoid these risks. Some trees such as Callery pear (*Pyrus calleryana*), silver lime (*Tilia tomentosa*) and in some cases horse chestnut (*Aesculus hippocastanum*) can develop very dense clusters or bunches of branches that are exceedingly difficult to prune, especially if the crown needs to be raised at some stage in the future (Figure 23.9).

Widespread availability of exotic species means that consideration must be given to the risk of a tree species becoming invasive. No doubt, major maintenance problems can be caused when seeds are released into near-by planting beds. However, a more substantial risk is that of invasion to natural forest environments as this may profoundly change the structure of the forest and lead to a cascade of undesirable effects. For example, Norway maple (*Acer platanoides*) has successfully spread in the eastern USA from tree plantings to natural forests, where it is having a negative effect on the natural species distribution by shading out other species. While for the more well-known and well-used species, there is good knowledge on their ability to spread; a major challenge exists in determining which new species may develop into invasive species and create large-scale maintenance problems. As a general principle, if the species is invasive in some areas of the world with a similar climate to the project site, it should never be used, while if the species has long been cultivated without becoming invasive, it can safely continue to be used.

Figure 23.9 Branching in silver lime (*Tilia tomentosa*) can often pose problems that are difficult to correct and can result in serious damage. Initial careful pruning to set the structure is essential in avoiding future problems

Growth pattern

Spatial constraints within the urban environment means that the growth pattern, or form, of the tree may be particularly relevant for species selection. Upright, narrow crowns are more suited to narrow street environments than wide-crowned species. Indeed, there are now many cultivars that are specifically selected for their predictable crown development. In the longer term, this can help reduce pruning costs and conflict with buildings and road networks. For example, seed-propagated material of small-leaved lime (*Tilia cordata*) can display huge variations in growth pattern and shape, making it difficult to anticipate future maintenance needs. By choosing a cultivar of small-leaved lime, such as *T. cordata* 'Rancho', for which the crown development pattern is known in advance, future maintenance work can be reduced and more uniform plantings can be achieved.

One should also consider how large the tree will be when fully grown. For example, the very well-known street tree pyramid hornbeam (*Carpinus betulus* 'Fastigiata') when young has a narrow, upright shape and is thus an extremely valuable tree for street environments.

Figure 23.10 When choosing the right tree for the right place, it is important to consider how large the tree will be as an adult. Pyramid hornbeam (*Carpinus betulus* 'Fastigiata') is very narrow-growing and compact when young (upper picture), making it valuable for street environments or at other sites where use of broader trees is impossible. However, it becomes very large and broad as an adult (lower picture), and is then much less suitable for street environments

However, as an adult tree it spreads out, it achieves a wide diameter and becomes too large for street environments (Figure 23.10). These adult pyramid hornbeams then have to be removed or pruned back to keep them within the space available at the site, which is unquestionably a very expensive process and a continually recurring task.

Aesthetic and social qualities

Trees have the ability to be constant and at the same time changeable over time and are thus one of the most interesting elements of urban structure. They can display interesting and varying branch work, which in winter can be enhanced by a dusting of snow. Every spring they produce a display of buds that develop into intricate and attractive leaves. In late spring and summer they display lush foliage, often accompanied by fantastic flowering, while in autumn they complete the year with a bright palette of autumn colour. The evergreen conifers have a pronounced shape that few deciduous trees can match. Conifers can also offer attractive and unique seasonal qualities, while still providing a valuable splash of green in winter. In addition to these aesthetic attributes, trees also have important social functions and values – the best trees for climbing, the trees with the best foliage for watching the sky through, the trees with the best crop of fruit, the fir or pine with the most exciting cones, etc.

Landscape design using trees is a process that provides opportunities to express a wide-ranging understanding of nature and our human relationship to it. Through design, it is possible to harness natural processes in the urban landscape to supply human needs, ecosystem services and aesthetic qualities. Therefore, trees are of critical significance to sustainable urban development. Through well-planned design, it is possible to create rooting environments capable of hosting mixed plantings of large trees that provide an aesthetically beautiful sense of place and deliver a wide range of ecosystem services to the communities where they are found. Despite the aesthetic and social qualities being ranked last in this chapter about tree selection, they are still hugely significant as the overall site design must capture the interaction between hard landscaping, buildings, people and trees if it is to achieve a truly sustainable development.

Diversification of the future urban forest

Based on data from many comprehensive tree inventories around the world, it is clear that only a few tree species and genera dominate in many cities in the northern hemisphere. For example, in New York State, USA, Norway maple (*Acer platanoides*) represents over 20 per cent of all street trees; in Helsinki, Finland, common lime (*Tilia* × *europeae*) makes up 44 per cent of all street trees; and in Beijing, China, the pagoda tree (*Styphnolobium japonicum*) comprises 25 per cent of all urban trees (Cowett and Bassuk, 2014; Sjöman et al., 2012; Yang et al., 2005). Such poor diversity poses major risks to the future of the tree population if dominant species suffer a severe disease or pest episode (see Chapter 18 for further details). Indeed, such events have already occurred: Dutch elm disease, caused by *Ophiostoma ulmi*, had serious consequences for many cities in Europe and North America with large elm (*Ulmus* spp.) populations. Today, traces of Dutch elm disease are still apparent, with some urban areas lacking large, mature trees because removed elms were not replaced. This loss of a large proportion of the tree population is a major setback in view of the many ecosystem services associated with old and large trees – and it will take a long time before new trees can replace the former elm trees, even if suitable species are used as replacements.

Pests and diseases continue to threaten our urban tree populations as global trade in goods (including plants), increases the risk of pests and diseases being spread to parts of the world where the indigenous flora have not developed defences against them. Future climate change will also affect trees by modifying growing conditions and allowing biological threats to expand into new territories. For example, a very serious threat to the tree populations in many cities is posed by the Asian longhorned beetle (*Anoplophora glabripennis*) and the citrus longhorned beetle (*Anoplophora chinensis*). A study by Sjöman et al. (2014) showed that these can attack over 140 different species of woody plants. Many of these susceptible plants are among the most widely used in many cities in the northern hemisphere. Therefore, these pests could have much more serious consequences than Dutch elm disease.

By far the best way to protect against the threats facing urban tree populations is to use as wide a diversity of species and genera as possible. It is recommended that this high diversity be evenly distributed throughout the entire city. Today it is mainly city parks, arboreta or botanical gardens that contain a high and sustainable diversity of species, genera and families. In contrast, the diversity is much lower in street environments. A recent study mapping the diversity of urban trees in Scandinavia (Sjöman et al., 2012) also analysed potential differences between park and street environments. The results confirmed that there was a wide range of species in parks, while a few species and families dominated in street environments. This finding is perhaps not surprising in light of the great differences in growing conditions that exist between parks and streets. The conditions in street environments are very challenging, with much warmer temperatures due to heat reflection from paved surfaces and buildings, water shortages as a result of rapid storm water removal, salt stress and restricted space above and below ground. Parks have much more favourable conditions, with a cooler microclimate, much more available water, little salt stress, and abundant space. With this in mind, it is perhaps not difficult to find tree species that can develop successfully in park environments. The challenge remains identifying species that are capable of establishing and performing well in street environments.

Future vista in the selection of urban trees

In many botanical gardens and arboreta around the world, there is a fantastic diversity of trees that have been growing successfully in the local climate for a long time (Figure 23.11). However, only a small fraction of this diversity is exploited by tree nurseries and made available to the urban forest community. It is, therefore, strategically important to exploit this rich diversity in public plantings and expand the knowledge of these trees in urban environments. Urban forest professionals should develop their plant knowledge, help motivate plant nurseries to start production of unusual species, and convince landscape architects to depart from the familiar species in favour of currently under-utilised species. To make this possible, it is important for the scientific community to evaluate the use-potential of new plant material so they are not planted on sites where they have no hope of establishment. Another important consideration in evaluation of future plant species is their capacity to supply different types of ecosystem services so that their future value to urban environments can be assessed.

Conclusions: selection criteria for urban trees

This chapter summarises all the different aspects that should be considered in the selection process when aiming to find the right tree for the right place. The review advocates that

Figure 23.11 Many botanical gardens contain a fantastic range of species that are never used in urban planting. This picture shows the botanical gardens in Gothenburg, which contain one of the largest tree collections in northern Europe. This may become a vital resource in future work on selecting urban trees for the region

Source: photo by Mats Havström

biological and technical aspects should always be prioritised over aesthetic aspects, since it is seldom possible to achieve attractive and long-term sustainable tree populations without considering factors such as hardiness, site tolerance, maintenance requirements and growth pattern. For the species that have been in use for a long time, there is good existing knowledge of their capacity to grow in different types of environments and their capacity to supply different types of ecosystem services. Unfortunately, it may not be possible to use many of these well-known and well-used species to the same extent in the future because of serious disease and pest problems that already exist, or their vulnerability to such threats in the future. Further, the over-representation of many species in our urban environments is likely to exacerbate the loss of canopy cover caused by future disease or pest epidemics. Thus we are facing a major challenge and must begin work on learning to cultivate under-utilised plant material. Further information on urban design and green urbanism will be presented elsewhere in this book.

References

Allen, C. D., Macalady, A. K., Chenchouni, H., Bachelet, D., McDowell, N. et al. 2010. A global overview of drought and heat-induced tree mortality reveals emerging climate change risks for forests. *Forest Ecology and Management* 259: 660–684.

Augé, R. M., Green, C. D., Stodola, A. J. W., Saxton, A. M., Olinick, J. B., Evans, R. M. 2000. Correlations of stomatal conductance with hydraulic and chemical factors in several deciduous tree species in a natural habitat. *New Phytologist* 145(3): 483–500.

Bacelar, E., Motinho-Pereira, J., Goncalves, B., Brito, C., Gomes-Laranjo, J., Ferreira, H., Correia, C. 2012. Water use strategies of plants under drought conditions. In Aroca, R. (ed.), *Plant Responses to Drought Stress*. Springer, Berlin, pp. 145–170.

Barker, P. 1975. Ordinance control of street trees. *Journal of Arboriculture* 1: 121–215.

Canadell, J., Jackson, R. B., Ehleringer, J. R., Mooney, H. A., Sala, O. E., Schulze, E. D. 1996. Maximum rooting depth of vegetation types at the global scale. *Oecologia* 108: 583–595.

Cowett, F. D., Bassuk, N. L. 2014. Statewide assessment of street trees in New York State, USA. *Urban Forestry and Urban Greening* 13: 213–220.

Deak Sjöman, J., Hirons, A. D., Sjöman, H. 2016. Branch area index of solitary trees: understanding its significance in regulating ecosystem services. *Journal of Environmental Quality* 45(1): 175–187.

Fini, A., Bellasio, C., Pollastri, S., Tattini, M., Ferrini, F. 2013. Water relations, growth, and leaf gas exchange as affected by water stress in *Jatropha curcas*. *Journal of Arid Environments* 89: 21–29.

Florgård, C., Palm, R. 1980. *Vegetation i dagvattenhantering* [*Vegetation for Stormwater Management*]. Naturvårdsverket, Stockholm.

Ghannoum, O., Way, D. A. 2011. On the role of ecological adaptation and geographic distribution in the response of trees to climate change. *Tree Physiology* 31: 1237–1276.

Gómez-Munoz, V. M., Porta-Gándara, M. A., Fernández, J. L. 2010. Effect of tree shades in urban planning in hot-arid climatic regions. *Landscape and Urban Planning* 94 (3–4): 149–157.

Grey, G. W., Deneke, F. J. 1986. *Urban Forestry*, 2nd edn. John Wiley, New York.

Grime, J. P. 2001. *Plant Strategies, Vegetation Processes and Ecosystem Properties*, 2nd edn. John Wiley & Sons, Chichester.

Hirons, A. D. 2013. Sustainable water management in the production of amenity trees: the potential role of deficit irrigation. PhD thesis, Lancaster University, UK.

Hunter, A. F., Lechowiz, M. J. 1992. Predicting the timing of budburst in temperate trees. *Journal of Applied Ecology* 29(3): 597–604.

Kendal, D., Dobbs, C., Lohr, V. I. 2014. Global patterns of diversity in the urban forest: is there evidence to support the 10/20/30 rule? *Urban Forestry and Urban Greening* 13: 411–417.

Kozlowski, T. T., Pallardy, S. G. 2002. Acclimation and adaptive responses of woody plants to environmental stresses. *The Botanical Review* 68(2): 270–334.

Lamberts, H., Stuart Chaplin III, F., Pons, T. L. 2008. *Plant Physiological Ecology*, 2nd edn. Springer, Berlin.

Levitt, J. 1980. *Responses of Plants to Environmental Stresses. Volume II. Water, Radiation, Salt and Other Stresses*. Academic Press, New York.

McCully, M. E. 1999. Roots in soil: unearthing the complexities of roots and their rhizospheres. *Annual Review of Plant Physiology and Plant Molecular Biology* 50: 695–718.

Moll, G. 1989. Improving the health of the urban forest. In Moll, G., Ebenreck, S. (eds), *A Resource Guide for Urban and Community Forests*. Island Press, Washington, DC, pp. 119–130.

Picket, S. T. A., Cadenasso, M. L., Meiners, S. J. 2013. Vegetation dynamics. In van der Maarel E., Franklin, J. (eds), *Vegetation Ecology*. Wiley-Blackwell, Chichester.

Reich, P. B. 2014. The world-wide 'fast–slow' plant economics spectrum: a traits manifesto. *Journal of Ecology* 102: 275–301.

Sack, L., Holbrook, N. M. 2006. Leaf hydraulics. *Annual Review of Plant Biology* 57: 361–381.

Santamour, F. S. 1990. Trees for urban planting: diversity, uniformity and common sense. *Proceedings of the 7th Conference of the Metropolitan Tree Improvement Alliance* 7: 57–65.

Sieghardt, M., Mursch-Radlgruber, E., Paoletti Couenberg, E., Dimitrakopoulus, A., Rego, F., Hatzistatthis, A., Randrup, T. 2005. The abiotic urban environment: Impact of urban growing conditions on urban vegetation. In Konijnendijk, C. C., Nilsson, K., Randrup, T. B., Schipperijn, J. (eds), *Urban Forests and Trees*. Springer, Berlin, pp. 281–323.

Sinclair, W. A., Lyon, H. H. 2005. *Diseases of Trees and Shrubs*. Cornell University Press, Ithaca, NY.

Sjöman, H., Slagstedt, J. 2015. Rätt träd på rätt plats [Right tree for right site]. In Sjöman, H., Slagstedt, J. (eds), *Träd i Urbana Landskap* [*Trees in Urban Landscape*]. Studentlitteratur, Lund, Sweden.

Sjöman, H., Hirons, A. D., Bassuk, N. 2015a. Urban forest resilience through tree selection: variation in drought tolerance in *Acer*. *Urban Forestry and Urban Greening* 14: 858–865.

Sjöman, H., Östberg, J., Bühler, O. 2012. Diversity and distribution of the urban tree population in ten major Nordic cities. *Urban Forestry and Urban Greening* 11: 31–39.

Sjöman, H., Östberg, J., Nilsson, J. 2014. Review of host trees for the wood-boring pests *Anoplophora glabripennis* and *Anoplophora chinensis*: An urban forest perspective. *Arboriculture and Urban Forestry* 40(3): 143–164.

Sjöman, H., Slagstedt, J., Wiström, B., Ericsson, T. 2015b. Naturen som förebild [Nature as a model]. In Sjöman, H., Slagstedt, J. (eds), *Träd i Urbana Landskap* [*Trees in Urban Landscape*]. Studentlitteratur, Lund, Sweden.

Xiao, Q., McPherson, G. 2002. Rainfall interception by Santa Monica's municipal urban forest. *Urban Ecosystems* 6: 291–302.

Yang, Y., McBride, J., Zhou, J., Sun, Z. 2005. The urban forest in Beijing and its role in air pollution reduction. *Urban Forestry and Urban Greening* 3: 65–78.

24

SELECTING NURSERY PRODUCTS

Daniel K. Struve

Introduction

Selecting trees for the urban forest is a critical consideration for the urban forester; the forester's decisions affect the long-term performance of the forest. By definition, urban trees are planted in high traffic (both people and vehicles) multiple stress sites with adjacent infrastructure (Urban, 2008). Tree failure in such sites will be likely to result in property damage and/or personal injury. Thus, only tree planting stock of the highest quality should be planted. This chapter addresses only nursery stock quality standards; it does not address species/cultivar selection criteria.

Nursery stock standards provide buyers and sellers of nursery stock with a common terminology in order to facilitate nursery stock transactions including techniques for measuring plants, specifying and stating the size of plants, determining the proper relationship between height and caliper, or height and width, and determining whether a root ball or container is large enough for a particular size and class of plant. With regard to terms, 'shall' is a requirement; 'should' is a recommendation in RASNS; the ENA standards use 'must' as a requirement.

Nursery stock quality standards focus on phenotypic characteristics with the additional requirement that nursery stock be substantially free of damaging insects and diseases. Perhaps of equal importance is the physiological condition of the nursery stock at the time of sale or transplanting and the associated stress resistance of the nursery stock. For instance, what are the physiological parameters that correlate with high establishment rates; what is an acceptable mineral nutrient or carbohydrate content or concentration in the plant material at harvest; what are the physiological effects of pre-harvest conditioning treatments; what is the acceptable level of root loss that results in sub-lethal stress levels? The absence of physiological standards is due to the lack of research-based information from which standards can be developed and the lack of rapid, low tech and inexpensive assessment methods. This paper will focus on phenotypic quality characteristics associated with American and European nursery stock quality standards. Four standards will be reviewed:

- the Revised American Standards for Nursery Stock (RASNS);
- the California Guidelines Specifications for Nursery Tree Quality;
- the Florida Grades and Standards for Nursery Plants 2015; and
- the European Nurserystock Association (ENA) Standards 2010.

Revised American Standards for Nursery Stock

The RASNS standards (ANSI Z60.1-2014) were developed to be used as a guide for nursery stock quality while acknowledging that contract specifications (if any) are the greater authority. The RASNS provide standards for most plant types from Shade and Flowering Trees to Herbaceous Perennials, Ornamental Grasses Groundcovers and Vines, as do the ENA. The RASNS and ENA standards are divided into general standards and standards for specific plant types. Both are organized into types of plants (for instance, shade or flowering tree), growth habit (upright, conical, multi-stemmed), method of production and packaging at the time of sale (field produced or containerized and balled and burlapped, in-ground fabric bag, or bare root), and the buyer's intended purpose for the plant (seedlings for conservation or restoration plantings, liners, understock, or sale to retail or commercial landscapers). Both describe the appearance of the plant material, but they do not describe the best management practices used to produce the plants.

Type 1 and 2 Shade and Flowering Trees RASNS standards will be discussed, as these are the most common plant types used in the American urban forest. The General Standards (§1.1) address the following categories: height, caliper or width sizes for a category specify the minimum acceptable size for a plant within that category (§1.1.2) with the maximum size being the smallest size of the next larger size category. Also, individual plant heights should be the average of the size category. For instance, trees within the 150 cm height category can range from 150 to 180 cm but should average at least 170 cm.

Container size specifications are described in Table1 (§1.1.3); the container class specifications are listed as trade size (note that a #1 container is not one gallon in liquid volume) as are the corresponding container volume ranges. The range of container sizes within a size category is due to lack of standardization among the American container manufacturers.

With reference to shade and flowering trees, caliper (an average of at least two trunk measurements per tree) is taken 15 cm above the soil line or substrate line for field stock and container stock for trees up to 10 cm in caliper. When describing American standards, caliper will be used for trunk size; for European standards trunk girth will be used. For trees larger than 10 cm, caliper should be taken 30 cm above the soil or substrate line. The soil or substrate line is noted by a change in color of the trunk at the soil line. If the tree is planted too deep, soil or substrate should be removed and caliper measured either at 15 cm from the root flare for trees less than 10 and at 30 cm for trees greater than 28 cm in caliper. The trunk caliper for bare root stock is measured 30 cm above the ground level or the root flare as appropriate. Plant height is measured from the ground or substrate line for field and container-grown plants, respectively.

Minimum requirements for all nursery stock (§1.3) include correct identification (genus, species and, if applicable, cultivar) and, at the time of shipment, that the nursery stock be substantially free of damaging insects and disease and in good living conditions, and be typical in habit for the species in the region of the country in which it was grown (with allowances made for specialty forms such as topiary, §1.3.2).

Co-dominate stems are allowed in Type 1 and 2 Shade and Flowering Trees only if they occur in the upper half of the crown (§1.3.3). Co-dominate stems, two similar diameter stems with common attachment point, reduce structural integrity and should be avoided if possible. However, some plants' natural growth habit (i.e. *Cladrastis kentukea*) make co-dominate stems nearly impossible to avoid. A concern with this standard is that a small sized plant transplanted into the landscape with an uncorrected co-dominate stem at an allowable height (for instance co-dominate stems occurring at 1.2 m) would be an unallowable defect when the plant's height exceeds 2.6 m, as the defect's point of origin does not change as plant height increases. Co-dominate stems can result from the stress associated with transplanting (Oleksak et al., 1997). Once induced, the defect will persist unless corrected. Corrective pruning methods shall meet the provisions of the most recent edition of Tree, Shrub and Other Woody Plant Management Standard Practices, Part 1: Pruning (ANSI A300, §1.3.2) to satisfy the minimum quality requirement.

The lowest branch height for trees to be planted on streets is not specified, but shall state the minimum height of the lowest branch or the height to which the trunk shall be free of branches while maintaining a crown-to-tree height ratio appropriate to the type of tree and in good balance with the trunk (§1.4.3). The canopy-to-tree height ratio for Type 1 trees is 2/3. The amount of clear trunk 'appropriate to the type of tree' and 'good balance' tree height is to be determined on a case-by-case basis between the seller and purchaser and should be stated in the bid specification. For instance:

- *Acer platanoides*, 10 cm girth, 3.5 to 4.5 m., trunk free of branches 1.8 m; or
- *Quercus rubra*, 8 cm, 4.5 to 5 m., lowest branch 2 m.

Bids can include specifications (§1.4.5) such as the number of times the stock has been transplanted or root pruned. However, there are no specifications for the number of times a plant should be transplanted nor the frequency of transplanting during production. The RASNS indicate that these optional practices are included to assist those in the nursery trade that appreciate the value of enhanced cultural practices. Root pruning is associated with increased root density which may increase transplant survival and speed establishment (Watson and Sydnor, 1987), but at the expense of increased production times. RASNS does not specify the number of times or the frequency that plants within a size class should be root pruned during production. Repeated root pruning may allow for smaller root balls than specified by RASNS; the smaller root ball size may be beneficial when stock is planted in sites with limited root volumes.

The RASNS addresses root ball requirements for field grown stock (§1.5). The general requirement is that ball sizes should always be of a diameter and depth to contain enough of the fibrous root system as necessary for the full recovery of the plant. This is good in principle, but a definition of 'full recovery' does not exist. Full recovery after transplant to the landscape may not be possible because the transplant site and post-transplant care are typically of poorer quality than in the nursery. Also, an appropriate 'recovery' parameter (recovery of annual shoot extension, restoration of the root system lost during harvest in the nursery or to root pruning at transplant, return to pre-transplant average leaf size or unit leaf area photosynthetic rate) or the time to full recovery have not been established. An additional consideration is the role that fibrous roots play in 'full' recovery. It is likely that fibrous roots are critical for initial water and nutrient uptake, but long term, full recovery is likely dependent on regenerating mother or pioneer roots (see Sutton and Tinus, 1983, for root terminology). Thus, the default condition is: Is there enough root system to allow the plant to survive the plant warranty period? This determination can only be made with certainty if the plant dies within the warranty period.

The trunk should be in the center of the root ball (§1.5.2) with a maximum off-center allowance of 10 percent of the root ball diameter. Root ball depth measurement should begin at the top of the root flare. If the root flare does not occur near the soil line (most commonly the result of deep planting or cultivation practices) the soil above the root flare should not be included in the soil depth measurement. Not specified is whether or not the soil above the root flare should be removed. Root ball depth varies in proportion to the root ball diameter (§1.5.3). For root balls under 50 cm in diameter, depth should be not less than 65 percent of the root ball diameter; those greater than 50 cm, depth should not be less than 60 percent of the root ball diameter. Adjustment to these recommendations can be made to accommodate soil conditions at the nursery and natural root growth patterns of the species or cultivar. The relationships between height-caliper and root ball diameter and depth are summarized in Table 24.1.

In general, the ball diameter to trunk caliper ratio decreases with plant size, from 24:1 to 10:1 (1.2 and 20 cm, respectively) as does plant height to caliper ratio, from 9:1 to 3.6:1 (using the size class range midpoint for tree height) for 1.2 and 5.5 m tall plants, respectively.

Burlap or other coverings shall completely cover the roots, be biodegradable and placed between the soil and the lacing or ball supporting the device (i.e., wire basket, §1.5.4). Wire baskets or other devices shall hold the root ball in a firm, rigid condition (§1.5.5). Wire basket 'ears' are laced to the trunk and the basket's wires crimped to maintain root ball integrity.

Table 24.1 The relationships between height and caliper and root ball diameter and depth for Type 1 shade trees

Caliper (cm)[1]/ height (m)[2]	Typical minimum height (m)[2]	Minimum root ball (cm) [2]		Acceptable container classes
		diameter	depth	
1.3/1.2	1.3	30	20	#2, #3, #5
1.6/1.5	1.6	33	22	#3, #5, #7
1.9/1.8	2.5	36	23	#5, #7, #10
2.5/2.0	3.8	41	27	#10, #15, #20
3.2	4.4	46	30	#15, #20, #25
3.8	5.1	51	34	#15, #20, #25, #45
4.4	6.4	56	37	#15, #20, #25, #45
5.1	7.6	61	40	#20, #25, #45
6.4	8.9	71	46	#25, #45, #65
7.6	10.2	81	54	#45, #65, #95/100
8.9	11.4	97	64	#65, #95/100
10.2	14.0	107	70	#95/100
11.4	17.8	122	80	
12.7	20.3	137	90	
14.0		144	96	
15.2		152	101	
17.8		178	117	
20.3		203	134	

Source: modified Table 3 of RASNS

1   English units were rounded to the nearest mm.
2   English units were rounded to the nearest cm.

The general requirement for container-grown root systems is that the plant shall have a well-established root system that reaches the container sides in order to maintain a firm root ball, but not have excessive root growth encircling the inside of the container. Root growth modifying containers (fabrics, holes, coatings or shapes) are acceptable in the trade. There are two caveats to consider when using container-grown trees, or field-grown trees originating from container-grown liners (Figure 24.1). First, no root modifying container completely controls root growth. Therefore, if root defects have been induced during production, the defect should be visible at planting so that it can be corrected before transplanting (Figure 24.2). Second, root defects can develop at all stages of nursery production. Those that occur in the early production stages and are not corrected before up-canning, can be hidden and the effects of the defect expressed many years after transplanting. It is critical that root defects be corrected at every stage during the production cycle. It is essential that the purchaser of nursery stock be familiar with the integrity of the nursery's workers and management and their cultural practices.

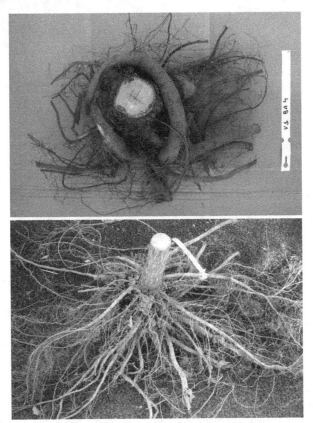

Figure 24.1 Root system of field-transplanted elm (*Ulmus*) four years after transplant (upper photo)

The plant was grown two years in a black plastic container. Root system of an elm similarly transplanted, but grown in an Air-Pot® for two years before transplanting (lower photo). Neither plant was root pruned before transplanting. Note that uncorrected root malformations, if not corrected, persist after transplanting. See Amoroso, et al. (2010) and Frangi, et al. (2016) for study details

Source: photo by Dr Alessio Fini, University of Florence

Figure 24.2 Root malformation in *Acer* × *freemanii* 'Jeffersred' Autumn Blaze plants grown in black plastic containers (left) and in an experimental root controlling container (right). Circling roots should be corrected at planting to insure the long-term health of the tree

Source: photo by Dr Pablo Jourdan, the Ohio State University

Plants grown and sold in wood boxes or in-ground fabric bags allow smaller trunk caliper-to-root ball volume ratio than for balled and burlapped plants because of reduced root loss at harvest and a more fibrous root system.

Type 1 and 2 Shade and Flowering Tree Standards (§2)

The RASNS lists two categories of shade trees: fast growing (Type 1) and slower growing (Type 2). The standards recognize growth rate differences between these two groups which are reflected in the higher average height/caliper-to-root ball diameter and depth ratios, minimum and maximum container classes and minimum acceptable in-ground fabric bag size diameter ratios of Type 1 shade trees. However, Type 2 shade tree height/caliper ratios must not be less than 2/3 of the values for Type 1 trees. Regional production differences are recognized by expressing ranges of acceptable height/caliper values; greater annual growth associated with relatively benign climates and longer growing seasons coupled with input-rich production environments, and trunk training practices typically results in nursery stock with lower caliper to height ratios.

Specifications for Type 1 and 2 shade trees shall include plant size by caliper after plants reach 2.1m in height or 2.5 cm in caliper.

Standards for Type 3, 4 (small upright and small spreading trees) and multiple trunk trees are similar to Type 1 and 2 shade trees, with adjustments made for smaller size and for minimum number of trunks for Type 4 trees. Standards for multi-stemmed trees reflect caliper differences and larger minimum ball and container sizes for multiple stems. Although not specified in the RASNS, field-grown minimum ball sizes for plants harvested during the growing season (summer-dug) should be larger than for similar sized nursery stock harvested when dormant. Also, standards for acclimating summer-dug plants are not included, but acclimation is accomplished with various combinations of foliar-applied anti-desiccants, soil-applied transplanting aids and misting under shade for various periods of times.

The California Guidelines and the Florida Grades and Standards

Other nursery standards exist, written to address regional production practices and climatic differences. Two such standards are the California Guidelines Specifications for Nursery Tree Quality and the Florida Grades and Standards for Nursery Plants.

The California Guidelines and Specifications for Nursery Tree Quality were written specifically for container-grown nursery stock. They have been incorporated into many Californian municipal specifications and modified to fit specific needs of individual municipalities, business and non-profit organizations. The intent was to help landscape professionals develop their own comprehensive and detailed specification to ensure that they obtain high quality container-grown nursery trees suited to their specific needs.

The guideline has five sections:

- proper identification;
- compliance;
- tree characteristics at the time of sale or delivery;
- moisture status;
- and visual inspection.

Tree characteristics at the time of sale or delivery is the most detailed section and goes beyond descriptions of quality to include acceptable production practices. This section has sub-sections on tree health, crown, trunk and roots. With regard to tree health, the trees should be healthy and vigorous as indicated by: relatively free from pests, with typical crown form for a young specimen of the species, typical color and appearance of leaves for the time of the year and stage of growth, with branches typical for the age and size of the species without diseased, misshaped or broken or injured branches; the trunk should be straight, vertical and free from wounds (with allowances made for proper pruning cuts) and other signs of biotic or abiotic stress; the tree height and trunk diameter should be typical for the age, species/cultivar and container size and the roots free of biotic and abiotic injuries with uniform root distribution within the substrate. The uniform root distribution requirement for container stock may be difficult to obtain. It is common for plants grown in black plastic containers to have fewer roots in the portion of the container exposed to direct sun. Sun exposure creates elevated substrate temperatures which can reach lethal levels. The detail describing tree health, which goes beyond the ANSI-A60 standards, is needed because of California's longer growing season. In California, plants can be sold year round and the long growing season promotes continuous growth and/or multiple flushes and requires repeated pruning and 'acceptance' of pruning wounds and seasonal foliage color variation. Use of the standards requires horticultural knowledge to know what constitutes 'relatively free, typical of the time of the year, stage of growth, container size and age, and uniform distribution' terminology.

The California Standards contain requirements not included in the RASNS. For instance, trees shall have a single, relatively straight central leader with a tapered trunk. The tapered trunk requirement is included because a straight central leader is developed by tying a single elongating shoot to a stake. If the shoot is tied tightly to the stake, and the tree has a full crown, the trunk may not be strong enough to hold the tree vertical once the stake is removed. Trunk taper is an indirect measure of stem strength. If the central leader is headed (often done to increase crown uniformity in strongly decurrent species/cultivars) then a new leader must be reestablished by pruning, re-staking and taping.

The standards require proper pruning cuts and branch-to-trunk caliper ratios, and branch angles. Temporary branches on trees less than 38 mm in caliper, should be present with diameters no greater than 10 mm. There is also a requirement that the root collar and large roots be free of circling and kinked roots. The guide gives an example of an acceptable root systems which shows non-circling downwardly directed roots. Although these roots are not circling, they may lead to mechanical instability due to root regeneration characteristics. Root regeneration occurs at the root tips either through elongation of existing roots or regeneration of roots at cut root tips (Stone and Shubert, 1959). Thus, if the production container was 23 cm deep, initial root regeneration will occur 23 cm below grade. If laterally growing roots develop, they will develop at least 23 cm below grade and form a root system approximating that of a plant planted too deep. The root ball periphery should be free of circling or bottom-matted roots, a laudable goal, but one difficult to obtain, even with root modifying containers. If root defects are present, they should be visible at planting and corrected. The moisture status at the time of inspection and delivery shall be moist throughout and the crown absent of moisture stress. Finally, the guidelines outline the conditions under which nursery stock can be rejected, either at inspection or delivery.

The Florida Grades and Standards for Nursery Plants are inclusive of all types of nursery crops grown in Florida. This chapter will focus on the Tree Standards, exclusive of palms. The California guidelines were developed from the Florida Grades and Standards; the Florida standards give greater detail and include production techniques and best management practices as guides for producing quality trees. The Florida Grades and Standards are based on university-industry cooperative research studies, mostly from Dr Gilman's lab at the University of Florida.

The Florida Grades and Standards were developed with the philosophy that selecting quality plants will speed establishment, reduce post-transplant maintenance costs and reduce storm damage. It is based on quality standards for trunk(s), branches, crown, leaf and root characteristics, with quality adjustments made for the degree of pruning needed to correct the defects in these three categories. The Florida grades and standards recognize that plants in lower grades can be improved with proper pruning and maintenance but correcting defects will result in increased post-transplant costs and establishment times. The Florida grades are: Florida No. 1 or 2 and Cull. There also is a Florida Fancy grade for trees of exceptional quality (Figure 24.3).

Tree quality is determined by a seven-step sequence after identifying which plant-type matrix the individual plant most closely matches. The plant type matrixes are: tall, wide (Type 1), tall narrow (Type 2) and short/wide and/or multi-trunked (Type 3). The first step is to measure trunk caliper using the RASNS methods. Three categories (trunk, crown and roots, Steps 1 to 3, respectively) with multiple criteria within each category are used to grade the trees. A Florida No. 1 tree shall have: a single stem for Type 1 and 2 trees (but not for Type 3); a branch diameter to trunk diameter ratio below 2/3 (with trunk diameter measured directly above the branch union); show few, if any, flush cuts or branch stubs on the trunk or open wounds on the trunk or major branches; a dense uniform crown, no crown necrosis, chlorotic foliage, diseases and pests; and root ball size appropriate to the trunk caliper and a root system without severe root defects. Departure from the Florida No. 1 standards are noted on a check list within each of the three major categories and the grade category

Figure 24.3 Examples of Florida Grades and Standards Trees (from left to right): Florida Fancy, Florida #1 (unpruned), Florida #1 (after pruning to correct co-dominate stems and branch clusters) and a field-grown Florida Fancy grade tree with roots uniformly regenerated from all sections of the root ball

Source: photo by Dr Ed Gilman, University of Florida

adjusted downward according to the number of defects. The point of emphasis is that trees in the lower grades will take longer to establish or more time to correct the defects.

There are seasonal and production stage differences that may affect the tree grade and these should be taken into consideration at the time of grading. For instance, in the early spring, crown density may be low or discolored because the plant is in the process of developing the spring growth flush. The foliage density may be low because the plant has just been subjected to the 'hardening off' procedure (Gilman, 2001). In the hardening off procedure a field-grown tree is dug three to six months before the final harvest, burlapped, placed in a wire basket and repositioned in the nursery row. The plant is considered hardened off when roots are regenerated and seen growing outside of the burlap. The pre-digging causes temporary root loss, plant moisture stress and reduced canopy density, but reduces post-transplant stress and speeds establishment (Figure 24.3).

The Florida Grades and Standards includes color illustrations of plants meeting the Florida No.1 grade and examples of grade-reducing defects, and goes beyond phenotypic quality characteristics by including Best Management Practices for transplanting, post-planting pruning, root ball correction, planting details and irrigation schedules.

The Florida Grades and Standards were developed for Florida's nursery industry, but they can be applied to other regions with slight modification. For instance, the Ohio State University used the standards to develop a grading check list for nursery stock purchased for campus plantings (Figure 24.4).

European Nurserystock Association Standards

European nursery quality standards were developed by the European Nurserystock Association (ENA). The technical and quality requirements have general conditions applicable to all nursery stock types and specific requirements for various types of nursery stock. Plants names are according to www.internationalplantnames.com. Plants must conform to the ENA standards; plants not meeting these standards are sold by specific agreements, with the variation(s) noted. Soil grown plants are to be lifted only when dormant. However, 'dormant' is not defined in the Definitions or Glossary sections. Tree root systems should

STEPS FOR DETERMINING THE GRADE OF A TREE Draft

Circle the Appropriate Grade

	ID Tag Number			
Step 1: Trunk Form:	Florida Fancy	Florida #1	Florida #2	Cull
Step 2: Branch Arrangement:	Florida Fancy	Florida #1	Florida #2	Cull
Step 3: Matrix Type: _____				

Step 4: Individual Tree Caliper: _____

Step 5: Crown Spread:	Florida Fancy	Florida #1	Florida #2	Cull
Step 6: Structural Uniformity:	Florida Fancy	Florida #1	Florida #2	Cull

Step 7: Lowest Grade in Steps 1, 2, 5 and 6

Step 8: If two of the Following are true reduce the grade in Step 7 by one. If more than two are true reduce the grade by two:

T F The tree with a trunk caliper larger than 1" requires a stake to hold it erect.

T F The crown is thin and sparsely foliated. Many evergreen and other trees are thin and sparsely foliated in the late winter/early spring just prior to the spring growth flush. Recently dug field-grown trees might also be thin. Do not downgrade for this.

T F More than 5% of the branches have tip die-back.

T F A) Tree height is taller than the maximum height specified in the appropriate matrix chart.

T F B) Flush cuts were made when pruning branches from the trunk

T F C) Branch stubs are left beyond the branch collar. A branch stub can be removed and not reduce the grade.

T F D) Open trunk wounds or other bark injury is evident. (Open trunk wounds **must be less than 10% of the trunk circumference** and less than 2 inches tall on Florida #1 trees. An open pruning scar on the trunk resulting from removing a branch is not considered an open trunk wound.)

T F F) More than 40% of the trunk is free of branches. (The portion of the lower trunk with shortened, temporary branches is not considered part of the canopy.) **T F.** G) More than 5% fo the leaves are chlorotic or more than 5% of the canopy exhibits damage from pests and disease infestations. Reject as Cull if significant or serious. **T F.** H) Most leaves are smaller than normal.

T F I) There is bark included between the trunk and a major lateral branch or between main trunks

T F J) Trunks and/or major branches are touching. Secondary branches on major branches may touch each other.

Consider a cull if: The branches are tip pruned:

Additional considerations of trees in containers or are already harvested if any are true consider the tree a Cull and do not accept: T F B) The root ball or container is undersized (consult proper tree matrix).

T F C) The root ball on B&B tree is not secured tightly with pins, twine or wire.

T F D) The tree is excessively root-bound.

T F E) There is evidence that one or more large roots (greater than 1/5 diameter of the trunk) were growing out of the container.

T F F) Does it have a root larger than 1/10 the diameter of the trunk circling 1/3 of the root ball.

Final Grade: _____

Figure 24.4 Check list, developed from the Florida Grades and Standards for Nursery Stock, used by The Ohio State University for selecting or rejecting nursery stock to be planted on campus

not have twisted roots close to the collar or any signs of physiological damage. All plants in containers or pots should be well rooted and plant root balls firm and solid. Specimen plants' root systems must be protected with decomposable materials (like burlap) and placed in wire baskets, wood-boxes and/or ungalvanized wire screen. Specimen trees must be regularly transplanted and correspond to the requirements of the species/cultivar. There are also handling and delivery requirements and grading requirements. Tree girth sizes are color coded for ease of identification.

Specific standards for trees are covered in §8 (container grown plants) and §9 (trees). Sections 2, 3, 4 and 5 have standards for the early stages of large girth tree production. Plants sold as liners, feathered whips or headed whips must note propagation season, pot or container size, and the number of times transplanted.

Young trees are divided into medium and vigorous growth habits. These are sold on height (measured from the soil/substrate line) to the main stem tip. Young medium-vigor tree heights range from 30 to 150 cm; vigorous young tree heights range from 40 to 180 cm. Although no girth measurements are included with the height grades, the General Requirements state that the plants must have an appropriate balance between the stem and head.

For container grown trees (§8.2.5), minimum container volumes are either 5.0 or 7.5 liters and minimum plant heights range from 100–125 or 125–150 cm for 5.0 and 7.5 liter containers, respectively. Plants are classified as whips (unbranched), feathered (a single stem with lateral branches to within 60 cm of ground level) or branched head (lateral shoot originating from the stem or a single point with few sub-lateral shoots). Plant height-to-container volumes range from 20:1 to 17:1.

Specifications for field grown trees are detailed in §9. Whips heights range from 80–150 cm for light whips and 100–250 cm whips. Feathered tree heights range from 150–300 cm. All whip types and heights are color coded. For standard trees, trunk girth is measured 100 cm above the soil line in 2 cm increments from 5 to 20 cm, in 5 cm increments from 20 to 50 cm and in 10 cm increments for trees greater than 50 cm girth. Tree specifications require that the number of times a tree has been lifted be indicated. For instance, it is typical for trees with 8–12 cm girth to have been lifted three times (Table 24.2). Transplanting must be done at least once every five years. Standard trees must have a clear straight stem, 150 cm for 6–8 cm girth trees and 180 cm for 8-10 cm girth trees. Greater clear stem height is allowed for road plantings, but the crown raising must not spoil the final shape and appearance of the tree.

The minimum bare root tree root diameter must be least four times the trunk diameter. The minimum root ball diameter to trunk girth ratio (using the smallest girth within a size class) for root balled trees ranges from 24:1 to 10:1 decreasing as the girth size increases (calculated from data in Table 24.2). No root ball depth is specified.

The ENA standards and the American standards for trees are similar, stressing quality characteristics such as straight clear trunks, acceptable crown-to-trunk ratios and root systems free from defects. The ENA standards require regular lifting; the American standards do not. The ENA and American Standards for root ball-to-trunk ratios are similar and converge on a 10:1 ratio. However, the ENA standards likely produce a more fibrous root system than does the American Standards due to repeated lifting requirements during production.

Additional quality considerations

There are other methods that can be used as quality indicators that apply equally to asexually and sexually propagated nursery stock, especially when selecting field-grown material. Try to select the most vigorous individuals within a nursery block, or within an even age planting.

Table 24.2 Trunk girth, minimum root ball sizes and the number of times transplanted for field-grown trees as required by the European Nurserystock Association Standards

Trunk girth range (cm)	Minimum diameter of root ball (cm)	Number of times transplanted
6–8	25	
8–10	30	
10–12	35	
12–14	40	3
14–16	45	3
16–18	50	3
18–20	55	3
20–25	60	4
25–30	70	4
30–35	80	4
35–40	90	5
40–45	100	5
45–50	120	5
50–60	130	6

This is especially true when selecting the first trees within the block. Typically, trees within a block are individually selected (as opposed to 'row run', where the trees are harvested from the ends of the rows and continuing until the number of trees necessary for the order is satisfied). The trees remaining after the initial harvests may have reduced vigor, relative to those trees harvested earlier, due to chronic graft incompatibility, increased pathogen load and relaxed cultural practices. Once harvesting within a block begins, fertilizer, pest control and irrigation inputs may be reduced or discontinued. Also, harvesting adjacent trees may result in root loss of the remaining trees and increased soil compaction and thus reduce the vigor of the remaining trees. In the case of sexually propagated crops, the remaining trees may also be genetically inferior. Signs of stress (early fall-color development or leaf drop, reduced leaf size and crown density that is not typical of the species/cultivar and geographical location, shortened internode length and/or reduced annual node production) can be used as indicators of relaxed management practices or inferior genetics. These factors may also contribute to reduced survival and slower establishment of large, relative to small, sized nursery stock. As stated earlier, it is good practice to purchase nursery stock from nurseries where their production practices are known and from nurseries where the managers and employees are knowledgeable of and follow best management practices.

Summary

European and American nursery stock standards provide buyers and sellers of nursery stock with a common terminology in order to facilitate nursery stock transactions. Standards include details for techniques for measuring plants, specifying and stating the size of plants, determining the proper relationship between height and caliper, or height and width, and determining whether a root ball or container is large enough for a particular size and class of plant. They also provide a common vocabulary for all members of the nursery industry. Nursery stock standards tend to raise the overall quality of nursery stock by describing the acceptable minimum size or quality within a category. The standards are subject to interpretation based on generally accepted standards for a geographical region, season of the year, plant species/cultivar, and specifications contained in a bid or contract. Thus, experience and horticultural expertise are required to interpret and defend the nursery product standards terms and conditions. Nursery standards do not specify acceptable production practices or best management practices to achieve guideline standards. Standards acknowledge that hidden defects, for instance uncorrected root defects induced in the early stages of productions, may be present. Also, detailed standards describing minimum physiological conditions for tree vigor and health have not been developed. However, despite the limitations of current standards and guidelines, they have increased nursery stock quality and communication among green industry members.

References

Amoroso, G., P. Frangi, R. Piatti, F. Ferrini, A. Fini, M. Faoro. 2010. Effect of container design on plant growth and root deformation of littleleaf linden and field elm. *HortScience* 45(12): 1824–1829.

Frangi, P., G. Amoroso, R. Piatti, E. Robbiani, A. Fini, F. Ferrini. 2016. Effect of pot type and root structure on the establishment of *Tilia cordata* and *Ulmus minor* plants after transplanting. *Acta Horticulturae* 1108: 71–76.

Gilman, E. F., 2001. Effect of nursery production method, irrigation, and inoculation with mycorrhizae-forming fungi on establishment of *Quercus virginiana*. *Journal of Arboriculture* 37: 30–39.

Oleksak, B, M. Kmetz-Gonzalez, D. K. Struve. 1997. Terminal bud cluster pruning promotes apical control in transplanted shade trees. *Journal of Arboriculture* 23(4): 147–154.

Stone, E. C., G. H. Schubert. 1959. Root regeneration by Ponderosa Pine seedlings lifted at different times of the year. *Forest Science* 5: 322–332.

Sutton, R. F., R. W. Tinus. 1983. Root and root system terminology. *Forest Science* 29(4) (Suppl.), Monograph 24.

Urban, J. 2008. *Up by Roots. Healthy Soils and Trees in the Built Environment.* Champaign, IL: International Society of Arboriculture.

Watson, G. W., T. D. Sydnor. 1987. The effect of root pruning on the root system of nursery trees. *Journal of Arboriculture* 13: 126–130.

Quality standards

California Guidelines Specifications for Nursery Tree Quality, retrieved from www.urbanforest.org/index.cfm/fuseaction/pages.page/id/424

European Nurserystock Association Standards, retrieved from http://media.wix.com/ugd/71c698_8a33d7b877964b9dba4904dba2df8328.pdf

Florida Grades and Standards for Nursery Plants 2015, retrieved from www.ufei.org/files/pubs/NurseryTreeSpecs10_09.pdf

Revised American Standards for Nursery Stock (ANSI Z60.1-2014), retrieved from www.plna.com/news/214962/Revised-American-Standard-for-Nursery-Stock-ANSI-Z60.1-now-available.htm

25

PLANTING
TECHNIQUES

Andrew K. Koeser and Robert J. Northrop

Introduction

While urban forest regeneration does occur naturally, many of the trees found in urban areas were intentionally planted. Conventional transplanting processes can be quite stressful for a tree. Field grown trees can lose a significant portion of their roots during harvest. All trees, regardless of how they were produced, can experience drought stress, mechanical damage, and exposure to extreme temperatures as they are transported, staged, and planted. These stresses and their impact on the success or failure of urban plantings have received considerable attention among researchers in arboriculture, urban forestry, and tree physiology. Their work has helped identify best practices for tree handling and installation which can reduce negative impacts to a tree's long-term growth and survival.

This chapter details some key considerations to promote successful tree planting. It includes information on minimum planting space requirements, seasonal impacts, proper tree storage and handling, planting hole excavation and backfilling, and early care practices (e.g. staking, mulching, pruning, etc.). Trees are typically sold as balled-and-burlapped, container, or bare-root nursery stock. Variations in handling and planting practices related to these production and delivery methods are noted where appropriate. This chapter also highlights planting practices that are specific to planting woody palms.

Planting space requirements

As noted in Chapter 23, there are many site and design factors that must be considered when selecting species for planting. A universal consideration that affects all trees, regardless of species and location, is the importance of sufficient planting space. Trees require adequate above- and below-ground space to fully develop the structures needed to meet species-specific requirements for photosynthesis, acquisition of water and nutrients, and structural support at maturity. Urban environments are highly-altered and often engineered landscapes. Within these landscapes, elements of infrastructure, such as neighboring trees, utilities, and signage, compete for limited available space. Identifying the space limitations for a site is fundamental to the long-term success of urban planting efforts. Beyond aspects of tree health, failing to consider the growth potential of a given species or cultivar can lead to increased levels of risk from structural failure, increase frequency of maintenance, and reduction in tree benefits.

Adequate soil volume for root growth may be more important relative to tree establishment than above-ground space. Many of the problems associated with tree establishment can be traced back to limiting soil volumes. Inadequate soil volume can limit water availability, increasing a tree's vulnerability to drought stress. This can predispose the tree to secondary insect and disease problems. As trees mature, stability can also be compromised by limiting soil conditions. Guidelines have been established based upon the mature size of the tree (Figure 25.1).

Evaluation of available above- and below-ground space should be conducted in a systematic fashion to allow identification of appropriate tree species to match site space constraints. Above-ground space can be estimated by considering:

1 Horizontal distance from buildings, paved surfaces, curbs, transportation corridors, pedestrian rights-of-way, and other large vegetation.
2 Vertical distance from overhead power and utility lines, emergency and street lights.
3 Required lines of sight for safety, road signs, and traffic signals.

Evaluating below-ground space can be difficult, but should not be avoided. Consider the tree opening, existing and prepared soil under the pavement, and adjacent open areas:

1 Review of detailed engineering plans along all transportation rights-of-way.
2 Review of detailed below-ground utility development.
3 On-site soil investigation to determine possible constraints to useable space, such as highly compacted soil layers, water tables that reach up into the active root zone, and physical barriers (e.g. construction materials).

While the base materials of roadways can effectively limit root growth, roots typically pass under sidewalks and explore soil volumes in nearby open spaces. While sidewalks may not limit soil volume, they do have an impact on planting space and can interfere with the base

Planting area guidelines			
IF YOUR SITUATION MATCHES ONE OF THE FOLLOWING:			CHOOSE THIS SIZE TREE TO FIT:
A Total planting area (lawn, island, or soil strip)	B Distance between sidewalk and curbing	C Minimum distance from pavement or wall	Maximum tree size at maturity
4.5–14.0 square meters	0.9 to 1.2 meters	0.6 meters	Small (less than 9.0 meters tall)
13.0–28.0 square meters	1.2 to 2.1 meters	1.2 meters	Medium (less than 15.0 meters tall)
More than 28.0 square meters	More than 2.1 meters	More than 1.8 meters	Large (taller than 15.0 meters tall) [*]

*In compacted or poorly drained soil, increase recommended distances and planting area size to compensate for aggressive surface root formation.

Figure 25.1 Planting space guidelines (assumes a minimum of 1 m soil depth)

Source: adapted from Gilman (2015)

of the tree. Sidewalks replacement near trees often result in root damage or removal. North et al. (2015) measured trunk flare width on four species of temperate shade trees and found they were consistently 30 percent to 50 percent wider than the measured trunk diameter at 1.4 meters (i.e. diameter at breast height). Given these findings, the authors suggested that planting widths should be wide enough to accommodate the projected trunk flare given anticipated trunk diameter (based on inventory records/experience), plus an additional 2.4 meter buffer to reduce conflicts with large tapering roots at and below the soil surface (North et al., 2015)

Impact of species and nursery production method on planting process

Once an appropriate planting site has been identified and matched with one or more suitable species, the next step is to purchase high-quality nursery stock. Nursery trees are most commonly produced as balled-and-burlapped, container, or bare-root planting stock. Within these three broad groupings, production methods can vary significantly, especially with regard to container production. Each method offers advantages and disadvantages with regard to cost, handling, planting, and early care that must be considered when selecting and planting trees. While research has shown that some methods of nursery production can outperform others in the landscape (Gilman, 2001), transplant success is generally similar for the different methods of production when proper care is given (Buckstrup and Bassuk, 2000; Anella et al., 2008; Koeser et al., 2014).

Balled-and-burlapped trees

Balled-and-burlapped trees are field-grown. When balled-and-burlapped trees are harvested in the nursery, a root ball is dug from the surrounding soil. The process leaves the roots and soil closest to the trunk intact, though as much as 95 percent of the original root system may be severed during the digging process (Gilman, 1988). The roots that remain with the tree are protected from damage and desiccation by the soil retained in the root ball, allowing for greater flexibility in planting time after harvest. However, this benefit comes at price with regard to tree handling and transport. Root balls are both heavy and fragile. In order to maintain the root ball's integrity and prevent further root damage, the root ball is wrapped in burlap and tightly bound with rope and/or metal wire.

Rope-bound root balls are most commonly associated with hand-digging. Use of rope alone to support the root ball requires mastery of a harvesting technique known as drum-lacing to create a web of rope around the root ball. Hand-digging and drum-lacing have largely given way to mechanical harvesting and the use of pre-constructed wire baskets. At harvest the wire baskets are lined with burlap. The mechanically-dug root ball is placed in the basket and the burlap is secured over the top of the ball. Rope is used to secure the basket to the base of the trunk and the basket itself may be crimped to make it fit more tightly against the root ball.

Bare-root trees

As the name implies, bare-root trees are shipped and sold with their root systems completely removed from the field soils they were grown in. This greatly reduces shipping weight, transport cost, and the effort required when handling and planting trees. One drawback to this method of nursery production is that it has traditionally limited planting to early spring as bare-root trees are harvested, cold-stored, and shipped dormant. Bare-root trees also tend to be most available in smaller caliper sizes, which may not be appropriate in areas where vandalism, lawn maintenance equipment damage, or other mechanical injury is a concern.

When harvested, bare-root trees can lose a significant portion of their root systems to undercutting. Additionally, small lateral roots may be stripped off during the lifting process (McKay, 1997). The remaining roots are typically wrapped in plastic filled with moist straw, shredded paper, hydrogel, or a similar water-holding material to prevent desiccation when transporting the trees to the planting site. Given their small size and light weight, bare-root trees may be bundled together to save spaced and packing materials.

Some species of bare-rooted trees do not readily break dormancy when removed from cold storage. When this is a concern, a process known as 'sweating' can be employed to hasten bud break (Halcomb and Fulcher, 2015). Avoiding direct sunlight, trees are placed in plastic with moist packing material to increase air temperature and humidity. Once buds begin to swell, the trees are removed and planted as normal.

If a longer planting window is desired, bare-root trees can be temporarily planted (i.e. 'heeled in') in a 35–45 cm bed of irrigated gravel (Starbuck et al., 2005). In this system, trees are supported by the gravel and watered using an automatic irrigation system. Root growth initiates while the trees are held in the gravel and this regrowth is largely preserved when the trees are pulled from the bed for final planting. Trees can be been maintained well into the growing season in this manner and typically experience similar survival rates as balled and container trees planted in comparative trials (Starbuck et al., 2005).

Container trees

Container trees can be either container-grown or containerized. As the name implies, container-grown trees are produced in containers. Seeds or cuttings are initially planted in small-volume trays, cells, or pots and replanted into bigger containers as rooting space becomes limited. In contrast, containerized trees typically start out as bare-root nursery (or sometimes balled-and-burlapped) stock, which were not sold at the start of the season. In this scenario the container serves as a temporary holding place, much like the gravel bed system noted above.

Container-grown trees offer several advantages which have made them popular among homeowners, growers, and retailers. They are shipped with their root systems intact, which can limit transplant stress. Unlike field soil, the mixes used in container production are light and drain well. This makes them a much lighter alternative to balled-and-burlapped trees, which makes transport, handling, and planting without heavy equipment considerably easier. While not as light as bare-root material, container trees offer the same flexibility in planting time enjoyed by balled-and-burlapped nursery materials. Furthermore, plastic containers are available in a wide array of sizes, which allow for greater variety in the sizes of trees available.

As with the other methods of production noted, container trees do have some notable drawbacks. Container-grown trees that are not shifted to a larger container in a timely manner will have roots that circle around the container bottom and side walls. If the latter scenario is left uncorrected and the trunk is too deep in the container, the roots may eventually girdle the tree. When purchasing a container-grown or containerized tree, the root system should be inspected to look for live roots (generally light tan or white in color) and to check for circling or girdling roots. If the latter are present, the issue can be remedied by shaving off the outer 2–3 cm of the root ball (Gilman et al., 2010). While root ball shaving does injure the tree, the process can make lasting improvements to root architecture with no noticeable impact to tree health (Gilman et al., 2010). Other methods for reducing the impact of circling roots

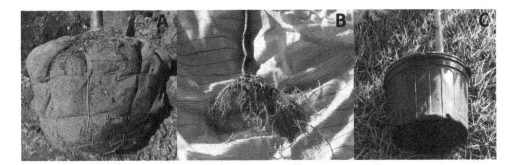

Figure 25.2 (a) Balled-and-burlapped, (b) bare-root and (c) container nursery stock

and encouraging radial root growth (i.e. scoring/slicing, butterfly pruning, and teasing out roots) have been found to be less effective (Weicherding et al., 2007).

In contrast with trees with circling roots, trees with girdling roots present should be rejected. Larger container-grown trees can have multiple layers of circling or girdling roots that correspond with the volume of the smaller containers they once inhabited over the course of production. When this is the case, shaving the outer edge of the root ball will not improve long-term survival and the tree should be rejected. Containerized trees can develop circling roots if not transplanted in a timely manner. Additionally, circling roots may be inadvertently created if bare-root trees are spun to fit their roots in the container.

Palms

Palms are either intentionally grown in a field or container nursery or harvested from forests where they grow naturally (Costonis, 1995). Unlike most woody plants, palms actually have greater transplant success when larger as they rely on stored water in their trunks to survive the water stress associated with transplanting and root disruption (Broschat, 1991). Some species, like the wild-sourced *Sabal palmetto*, experience root dieback when harvested and are transplanted with virtually no roots. For these species, removal of some or all of the fronds (in the case of *Sabal palmetto*) has been shown to increase transplanting success by reducing water stress (Broschat, 1991). For other species like *Phoenix canariensis*, *Phoenix roebelenii*, *Syagrus romanzoffiana* and *Washingtonia robusta*, frond removal and/or frond tie up offer no benefit with regard to survival under irrigated conditions (Broschat, 1994; Hodel et al., 2003, 2006). While frond removal may make transport and handling easier, it can negatively impact plant aesthetics for some time after planting (Hodel et al., 2003, 2006), which may be important for homeowner or city.

Seasonal planting considerations

Depending on the species, root regrowth can occur when soil temperatures range from 1.7 and 25°C (Richardson-Calfee and Harris, 2005). However, several studies of temperate deciduous trees have showed little root growth below 10°C (Richardson-Calfee and Harris, 2005). The seasons in which this temperature threshold is met vary by latitude, elevation, and sun exposure. Soil warming dates can often be obtained from regional agricultural agencies. Trees transplanted outside of optimal conditions may require a higher frequency of management intervention and monitoring. This is especially true for palms, particularly those species that experience root dieback at harvest. When cold-hardy palms are planted in regions where soils can cool below optimal levels, palms should be planted at the start of the warming season to facilitate root regrowth (Hodel et al., 2009). More detailed information on seasonal planting considerations can be found in Chapter 26.

To account for the root loss associated with balled-and-burlapped and bare-root production, trees have traditionally been planted while dormant – typically in early spring before bud break or in fall after leaf senescence when soil temperatures may still be high enough to support root growth. Traditionally, the environmental conditions in early spring have made it a popular time of year to transplant trees in the temperate regions of Europe and North America. However, fall is considered by some to be an optimal season for root regrowth as there is no leaf or shoot growth to compete for stored carbohydrates (Richardson-Calfee and Harris, 2005). That noted, in their review of past research, which compared differences in transplant success for spring- and fall-planted trees, Richardson-Calfee and Harris (2005) found no clear pattern in the results. This variability in findings was attributed to differences in tree species, tree size, climate, type of planting stock, and research methods employed (Richardson-Calfee and Harris, 2005). In a more recent, large-scale study of 26 urban plantings of *Quercus virginiana* (*n* = 1197), *Taxodium distichum* (*n* = 240), and *Magnolia grandiflora* (*n* = 154), Koeser et al. (2014) found no difference in caliper growth between fall- and spring-planted trees two to five years after installation.

Successful establishment of trees planted in the summer, with its longer day length and higher average temperatures, requires higher levels of available soil moisture to prevent desiccation damage. Summer-transplanted trees must support a full canopy with a reduced root mass. This often necessitates irrigation. In some regions the summer season marks the beginning of the tropical rainy season, which may reduce irrigation demand for summer-transplanted trees, though planting during spring and fall may still be preferred. Research tracking the survival of *Quercus virginiana* in Florida (United States) showed growth rates for spring-planted trees were greater than summer-planted trees, presumably because they were able to benefit from rains over the duration of the rainy season (Koeser et al., 2014). Subtropical and tropical climates have continuously warm soil and air temperatures conducive to root and shoot growth year round. In these climates, trees can be successfully established year round, especially when irrigated during periods of reduced rainfall.

Transport and handling of nursery trees

With the exception of lightweight bare-root nursery stock, care should be taken not to lift trees by their trunk alone during handling and transport. Lifting a balled-and-burlapped or container tree by its trunk with the weight of the root ball left unsupported may cause trunk or root damage. When moving balled-and-burlapped or container trees, the tree should be lifted by the root ball. For larger trees suspended by straps or chains during handling, a sling can be placed on the lower trunk to prevent the tree from rotating; however, care should be taken to make sure the weight of the root ball is fully supported.

If balled-and-burlapped or container trees are transported on an exposed flatbed trailer or truck, trees should be protected from desiccation and damage with shade cloth or a similar covering. Once delivered, trees should be inspected for damage and watered as needed to maintain root ball moisture. If not planted immediately, trees must be stored and cared for appropriately. Bare-root trees that will be planted dormant can be stored temporarily in a cool, shaded location unless sweating is required. If long-term staging is required, balled-and-burlapped trees can be 'heeled in' by surrounding their root balls with mulch and watering them to prevent desiccation. This creates a warm and moist environment that encourages root growth. These same conditions also increase the microbial activity which degrades burlap (even copper-treated burlap), twine, and other natural materials, which may make handling somewhat difficult depending on the severity of the degradation. Container trees may be heeled in as well to help reduce water loss and prevent overturning in wind. However, trees purchased in properly-sized containers can be maintained under irrigation for some time in the growing mix found within to container.

When transplanting palms, care must be taken to avoid damage to the trunk and apical meristem (Costonis, 1995; Hodel et al., 2009). For species where fronds are retained during transplanting, shade cloth should be used to prevent desiccation during open transport. If palms are not planted immediately after delivery, care should be taken to prevent desiccation. Root balls should be watered and crowns should be shaded to reduce transpiration. Palms can be heeled to maintain health when stored onsite for longer periods of time.

Planting hole excavation

Roots are a storage site for carbohydrates, regulate the uptake of minerals and water, provide the anchorage required for maintaining aboveground growth, and produce/respond to various hormonal signals. While root system depth varies with soil conditions and species (Day et al., 2010), it is often generalized as being within the upper 0.6 m of soil. Care should be taken to assure these roots stay near the surface at planting and that the surrounding soil conditions near the surface are sufficient for root development.

Trees are often planted with their root collar and structural roots placed well below soil grade. This can occur at the nursery, during installation in the landscape, or at both stages of transplanting. Deeply planted trees can have reduced establishment rates, shortened lifespans and increased defects such as girdling roots, and reduced stability (Day and Harris, 2008). Research has shown these impacts can vary by species or site conditions and may not fully be realized until much later in a tree's life (Day and Harris, 2008).

Generally, the hole should be the same depth or 2.5 to 5 cm shallower than the height of the first major root coming off the root ball (Watson, 2014). This allows for settling that may occur after installation. Planting slightly above grade (by digging the hole shallow) may also be helpful for sites with a high water table or high clay content. However, choosing tree species that most closely fit the site conditions are more important for long-term success. The width of the planting hole should be approximately 2 to 3 times the width of the root ball with the sides slowing up into a bowl shape to reduce unnecessary digging. This will loosen the surrounding soil and facilitate root growth. If the site is severely compacted, tilling or soil amendment/replacement may be required (see Chapter 21 for more information on soil amelioration). When planting in soils with a high clay content care should be taken to avoid glazing the sides of the hole as this can create a barrier for root penetration and water drainage. If glazing is noticed, the sides of the hole can be scarified to alleviate the problem.

Tree installation

Bare-root trees

Once the planting hole has been properly excavated, the tree is positioned in the center of the hole and supported as needed to remain upright for backfilling. When planting bare-root trees, kinked or damaged roots may be pruned away prior to placement in the hole.

The remaining roots should be straightened to prevent circling and arranged in a spoke-like manner. Once situated, the original fill from the hole is replaced.

Balled-and-burlapped trees

Larger caliper balled-and-burlapped trees are very heavy. When planting a balled-and-burlapped tree without the aid of heavy equipment, the root ball must be carefully rolled into the hole. Rolling the tree so the side of the root ball rides along the edge of the planting hole will ensure the tree rights itself once in place. Final adjustments with regard to position or angle should be made by pushing on the root ball itself, not by pulling or pushing on the trunk of the tree.

Wire baskets, natural burlap, and natural twine/rope can be removed or retained without detectable impact on initial or long-term above-ground growth and survival (Koeser et al., 2015; Lumis and Struger, 1988). However, partial or full removal of these materials may temporarily reduce tree stability (Koeser et al., 2015). Some root girdling can occur if treated burlap does not break down in the landscape (Kuhns, 1997). Synthetic burlap will persist for many years and should not be left in place as it can impede root growth. Any rope or twine that circles the base of the trunk should be removed regardless of whether it is synthetic or natural in origin as it can girdle the stem (Koeser et al., 2015).

If desired, partial removal of the wire and burlap can be completed after the tree is placed in the planting hole. Bolt cutters are used to remove as much of the wire as practical and the underlying burlap is either pulled down or cut away to prevent potential interference with roots in the upper part of the root ball. If full removal is preferred, the bottom of the basket can be cut away prior to placing the basket in the hole. This leaves the sides intact to support the root ball as it is moved into place. Once in the hole, the side of the wire basket can be removed and all but a small portion of the burlap underneath the root ball can be cut away.

Container trees

The vast majority of containers used in tree production are non-plantable and must be removed prior to installation. It may be difficult to remove a container from a pot-bound tree, especially when roots grow through drain holes. Care should be taken to not cause undue root damage when removing the pot – cutting the container away if necessary. As noted above, issues of circling roots should be corrected. Currently, shaving the outer roots from the root ball seems to be the most effective and efficient means of root system modification (Gilman et al., 2010).

Palms

If container grown, palms should be planted as indicated above. Field-grown palms often are shipped with plastic covering their roots/root initiation zone. This must be removed at planting.

Backfilling and initial irrigation

Rapid root development occurs in well-aerated soils (Bridel et al., 1983) and slows as root tips come into contact with denser or compacted soils (Perry, 1982). The process of digging, turning, and replacing soils when excavating and refilling the planting hole helps break up compacted or heavy soils. While trees benefit from the larger volume of aerated backfill associated with larger planting holes, amending backfill has not consistently enhanced root development and tree establishment. In the ideal soils typical of agricultural research plots,

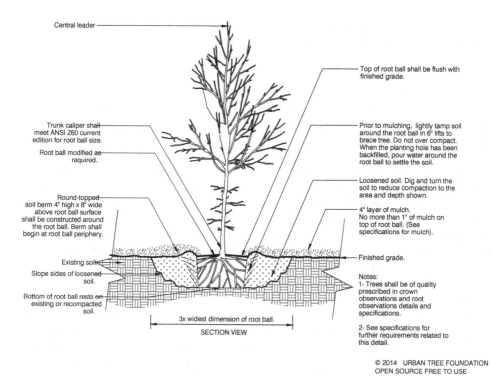

Figure 25.3 Planting detail for installing a tree in unmodified soil

Source: details by Gilman et al. (2015)

root regrowth was good regardless of whether amended or unamended soil was used to backfill the hole (Watson et al., 1992). While amending the planting backfill in tougher urban sites may offer some improvement, the root systems of mature trees are quite extensive. Proper species selection may be the best option unless the property manager is willing to conduct large-scale cultivation and/or amendment of the site.

When backfilling the planting hole, hold the tree in the desired position and add soil around the root ball, tamping lightly to help secure the tree in place. As backfill is added, water should be added to the hole to settle the soil and eliminate air pockets. When backfilling, do not cover the root ball with soil. Excess backfill soil can be removed from the site, used to create an irrigation berm, and raked into the surrounding bed or lawn.

Irrigation is the most critical cultural practice associated with the establishment of transplanted trees. Transpiration rates are often the highest immediately following transplanting (Harris and Gilman, 1993). Roots that are drought stressed have a 90 percent reduction in the rate of growth (Watson et al., 2014). Application of too much water can also lead to stress and reduced rates of root reproduction, lengthening establish periods.

Irrigation should be directed to the root ball. Application of water at slow rates to ensure percolation into the root ball is preferred. Micro-irrigation systems are very efficient and effective, capable of placing calculated amounts of water directly on the root ball at a predetermined frequency. On sites without direct access to irrigation system, portable watering devices can help ensure water is applied consistently and in sufficient amounts. Watson and Himelick (2013) suggest that irrigation frequency can be reduced once the root

Figure 25.4 Two examples of portable irrigation bag designs

system is three times the size of the original root ball, but that irrigation should not be entirely stopped until establishment.

Tree stabilization

Tree stabilization may or may not be required depending on the size of the tree planted, the nursery production method employed, and the site conditions present. Balled-and-burlapped trees planted with the root ball intact are less reliant on additional support systems (Koeser et al., 2015). In contrast, larger container trees and bare-root trees lack a heavy anchoring root mass and may need additional support on windy sites until root regrowth is sufficient to support the tree. A wide array of stabilization systems are used and sold to prevent newly planted trees from overturning. These range from large underground staples that stabilize the root ball directly to guy wires or posts (with connecting material) that attach to the trunk of the tree.

Trees increase root growth and produce taper in response to wind-induced movement. As such, the use of overly rigid stabilization systems should be avoided (Watson and Himelick, 2013). The use of wide, flexible staking materials will reduce the likelihood of trunk injury and allow for controlled movement. Wire (even wire threaded through rubber hose or some other protective material) should be avoided because it can damage trees at the point of contact. All materials should be removed as soon as the tree is stable enough to support itself to prevent girdling. This timeframe varies given site conditions and the size of the tree transplanted, but for the smaller-caliper trees commonly planted in urban forestry programs, one growing season/year is typically sufficient.

Mulching

Urban environments rarely resemble the abiotic and biotic conditions found within forest ecosystems. High and low temperatures are more extreme, wind velocity is stronger, humidity levels are lower, and the daily period of solar radiation is longer (see Chapter 18 for details). These conditions lead to increased demands for moisture which can dry

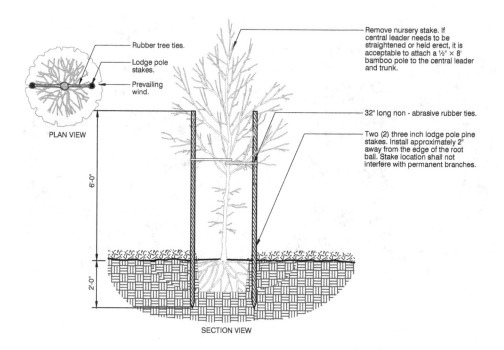

PLAN VIEW

Rubber tree ties.

Lodge pole stakes.

Prevailing wind.

Remove nursery stake. If central leader needs to be straightened or held erect, it is acceptable to attach a ½" × 8' bamboo pole to the central leader and trunk.

32" long non - abrasive rubber ties.

Two (2) three inch lodge pole pine stakes. Install approximately 2" away from the edge of the root ball. Stake location shall not interfere with permanent branches.

6'-0"

2'-0"

SECTION VIEW

Tree staking – lodge poles

© 2014 URBAN TREE FOUNDATION
OPEN SOURCE FREE TO USE

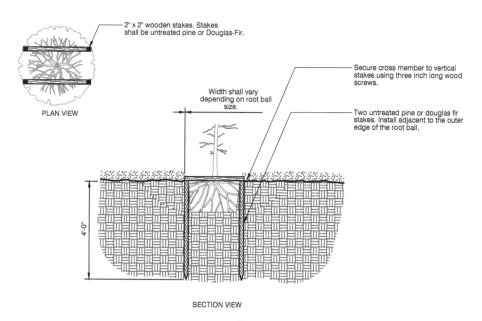

2" x 2" wooden stakes. Stakes shall be untreated pine or Douglas-Fir.

PLAN VIEW

Width shall vary depending on root ball size.

Secure cross member to vertical stakes using three inch long wood screws.

Two untreated pine or douglas fir stakes. Install adjacent to the outer edge of the root ball.

4'-0"

SECTION VIEW

Tree staking – staple

© 2014 URBAN TREE FOUNDATION
OPEN SOURCE FREE TO USE

Figure 25.5 Examples of tree stabilization options

Source: details by Gilman et al. (2015)

out soils, alter soil processes, and stress newly planted trees. Mulch, in the form of wood chips, shredded bark, pine straw, or similar material is often used to incorporate organic matter into the soil-planting site. Organic matter influences physical and chemical processes, encourages the retention of soil water, supplies energy for microorganisms, and cools the soil environment. The benefits of organic mulch with regard to soil health and tree growth have been documented in many studies (Scharenbroch, 2009). Most recently, researchers showed the use of composted mulch in urban areas increased tree height and diameter, carbon storage, and improved water use efficiency of trees (Fini et al., 2016).

An additional benefit of mulch is its ability to suppress weed and turfgrass growth (Marble et al., 2015), reducing competition for nutrients and moisture. Few trees used in urban settings developed in ecosystems dominated by sod-forming grasses. Consequently trees have a difficult time becoming established in an environment where the cultural practices support the vigor and growth of sod-forming grasses common in urban lawns. Additionally, lawn mowing and trimming equipment is a significant source of lower trunk and root injury for trees (Morgenroth et al., 2015). A ring of mulch provides a much-needed barrier to reduce wounding and re-wounding (Morgenroth et al., 2015).

Wider mulch areas provide the greatest benefit to trees as their roots extend well beyond their dripline. A 5- to 10-cm thick layer of mulch should be applied to the base of a tree – avoiding contact with the trunk and extending outward as far as practical (to the dripline if possible). Excessive mulch over the root ball can lead to the formation of stem girdling roots, the root ball being kept too wet, or repelling of rainfall during light rainfall events. Mulch should not touch the trunk. Mulch touching the trunk can lead to stem diseases, and rodent damage.

Mulch with small particle size or a wide diversity of particle sizes should be avoided since they tend to settle quickly and inhibit infiltration and gas exchange. Pine straw settling reduced its usable life after just one year, while bark and wood chunk mulches remained viable for 2 years (Duryea et al., 1999). Mulch flammability should be considered when using mulches close to homes in fire-prone locations. Care should be taken to ensure that all mulches are weed and seed free, and that its constituents have not been treated with preservatives. Additionally, some organic mulches like fresh wood chips can have a high carbon-to-nitrogen ratio and can create nitrogen-limiting conditions in the surrounding soil. Composting chips or selecting another mulch material can limit this potential concern.

Pruning

Trees were once pruned as a means of compensating for the root losses associated with transplanting. However, several studies have discredited the merits of compensatory pruning, noting it has little impact on growth when soil moisture conditions are limiting and actually decreases tree growth when conditions are ideal (Hauer, 1999). Typically pruning at planting is limited to removal of dead, diseased, or damaged limbs which should be minimal if quality stock is procured and care is taken with transport, staging, and handling.

While pruning as a means of preventing water stress is not recommended, there could be other justifications for pruning at planting. Young tree training can have lasting impacts on tree architecture while minimizing branch removal. In a municipal setting, trees may not be revisited for years after planting. It may be that young tree training at planting is an efficient means of beginning a structural pruning program for young trees. However, research is needed to see if the benefit of early training persists and outweighs potential costs with regard to tree health and survival.

Conclusion

Planting success depends a great deal on reducing any or all of the potential stresses that can inhibit establishment in the landscape. Some aspects of tree planting are dependent on the method of nursery production used. Many others are consistent regardless of stock type. With proper care and timing, urban reforestation efforts can achieve high rates of establishment. This greatly reduces economic waste and maximizes the long-term environmental potential of planting initiatives.

References

Anella, L., Hennessey, T. C., Lorenzi, E. M. et al. (2008) 'Growth of balled-and-burlapped vs bare-root trees in Oklahoma, US', *Arboriculture and Urban Forestry*, vol 34, no 3, pp. 200–203.

Bridel, R., Whitcomb, C., Appleton, B. L. et al. (1983) 'Planting techniques for tree spade-dug trees', *Journal of Arboriculture*, vol 9, no 11, pp. 282–284.

Broschat, T. K. (1991) 'Effects of leaf removal on survival of transplanted sabal palms', *Journal of Arboriculture*, vol 17, no 2, pp. 32–33.

Broschat, T. K. (1994) 'Effects of leaf removal, leaf tying, and overhead irrigation on transplanted pygmy date palms', *Journal of Arboriculture*, vol 20, no 4, pp. 210–214.

Buckstrup, M. J., Bassuk, N. L. (2000) 'Transplanting success of balled-and-burlapped versus bare-root trees in the urban landscape', *Journal of Arboriculture*, vol 26, no 6, pp. 298–308.

Costonis, C. (1995) 'Factors affecting the survival of transplanted sabal palms', *Journal of Arboriculture*, vol 26, no 6, pp. 298–308.

Day, S. D., Harris, J. R. (2008) 'Growth, survival, and root system morphology of deeply planted Corylus colurna 7 years after transplanting and the effects of root collar excavation', *Urban Forestry and Urban Greening*, vol 7, no 2, pp. 119–128.

Day, S. D., Wiseman, P. E., Dickinson, S. B. et al. (2010) 'Contemporary concepts of root system architecture of urban trees', *Arboriculture and Urban Forestry*, vol 26, no 4, pp. 149–159.

Duryea, M. L., English, R. J., Hermansen, L. A. (1999) 'A comparison of landscape mulches: Chemical, allelopathic, and decomposition properties', *Journal of Arboriculture*, vol 25, no 2, pp. 88–97.

Fini A., Degl'Innocenti, C. Ferrini, F. (2016) 'Effect of mulching with compost on growth and physiology of Ulmus "FL634" planted in an urban park', *Arboriculture and Urban Forestry*, vol. 42., no 3, pp.192–200.

Gilman, E. F. (1988) 'Tree root spread in relation to branch dripline and harvestable root ball', *HortScience*, vol 23, no 2, pp. 145–152.

Gilman, E. F. (2001) 'Effect of nursery production method, irrigation, and inoculation with mycorrhizae-forming fungi on establishment of quercus virginiana', *Journal Of Arboriculture*, 27: pp. 30–39.

Gilman, E. F. (2003) *Where are Tree Roots?* University of Florida, IFAS Extension, Gainesville, FL.

Gilman, E. F. (2015) 'Landscape plants – UF/IFAS', retrieved from http://hort.ifas.ufl.edu/woody (accessed June 9, 2016).

Gilman, E. F., Paz, M., Harchick, C. et al. (2010) 'Root ball shaving improves root systems on several tree species in containers', *Journal of Environmental Horticulture*, vol 28, no 1, pp. 13–18.

Halcomb, M., Fulcher, A. (2015) 'Sweating nursery stock to break dormancy', retrieved from www2.ca.uky.edu/HLA/Dunwell/LnrSweat.html (accessed March 18, 2015).

Harris, J. R., Gilman, E. F. (1993) 'Production method affects growth and post-transplant establishment of "East Palatka" holly', *Journal of the American Society for Horticultural Science*, vol 118, no 2, pp.194–200.

Hauer, R. J. (1999) 'Compensatory pruning ... to prune or not to prune at planting that is the question!' *Minnesota Shade Tree Advocate*, vol 2, no 2, pp. 7–8.

Hodel, D. R., Downer, A. J., Pittenger, D. R. et al. (2006) 'Effect of leaf removal and tie-up on transplanted large Mexican fan palms (*Washingtonia rubusta*)', *Palms*, vol 50, no 2, pp.76–81.

Hodel, D. R., Downer, A. J., and Pittenger, D. R. et al. (2009) 'Transplanting palms', *HortTechnology*, vol 19, no 4, pp. 686–689.

Koeser, A. K., Gilman, E. F., Paz, M. et al. (2014) 'Factors influencing urban tree planting program growth and survival in Florida, United States', *Urban Forestry and Urban Greening*, vol 13, no 4, pp. 655–661.

Hodel, D. R., Pittenger, D. R., Downder, A. J. et al. (2003) 'Effect of leaf removal and tie up on juvenile, transplanted Canary Island Date Palms (*Phoenix canariensis*) and queen palms (*Syagrus romanzoffiana*)', *Palms*, vol 47, no 4, pp. 177–184.

Koeser, A. K., Hauer, R., Edgar, J. et al. (2015) 'Impacts of wire basket retention and removal on planting time, root-ball condition, and early growth of *Acer platanoides* and *Gleditsia triacnthos* var. inermis', *Arboriculture and Urban Forestry*, vol 41, no 1, pp. 18–25.

Kuhns, M. (1997) 'Penetration of treated and untreated burlap by roots of balled-and-burlapped Norway maple', *Journal of Arboriculture*, vol 23, no 1, pp. 1–7.

Lumis, G. P., Struger, S. A. (1988) 'Root tissue development around wire-basket transplant containers', *HortScience*, vol 23, no 2, p. 401.

Marble, S. C., Koeser, A. K., Hasing, Gitta et al. (2015) 'A review of weed control practices in residential landscape planting beds: Part I – non-chemical methods', *HortScience*, vol 50, no 506, pp. 851–856.

McKay, H. B. (1997) 'A review of the effect of stresses between lifting and planting on nursery stock quality and performance', *New Forests*, vol 13, no 1, pp. 363–393.

Morgenroth, J., Santos, B., Cadwallaer, B. et al. (2015) 'Conflicts between landscape trees and lawn maintenance equipment – the first look at an urban epidemic', *Urban Forestry and Urban Greening*, vol 14, no 4, pp. 1054–1058.

North, E. A., Johnson, G. R., Burk, T. E. et al. (2015) 'Trunk flare predictions as an infrastructure planning took to reduce tree and sidewalk conflicts', *Urban Forestry and Urban Greening*, vol 14, no 1, pp. 65–71.

Perry, T. O. (1982) 'The ecology of tree roots and the practical significance thereof', *Journal of Arboriculture*, vol 9, no 8, pp. 197–211.

Richardson-Calfe, L. E., Harris, J. R. (2005) 'A review of the effects of transplant timing on landscape establishment of field-grown deciduous trees in temperate climates', *HortTechnology*, vol 15, no 1, pp. 132–135.

Scharenbroch, B. C. (2009) 'A meta-analysis of studies published in Aboriculture and Urban Forestry relating to organic materials and impacts on soil, tree, and environmental properties', *Arboriculture and Urban Forestry*, vol 35, no 5, pp. 221–231.

Starbuck, C., Struve, D. K., Mathers, H. et al. (2005) 'Bareroot and balled-and-burlaped red oak and green ash can be summer transplanted using the Missouri Gravel Bed system', *HortTechnology*, vol 15, no 1, pp. 122–127.

Watson, G. W. (2014) *Best Management Practices – Tree Planting* (2nd edn), International Society of Arboriculture, Champaign, IL.

Watson, G. W., Himelick, E. B. (2013) *The Practical Science of Planting Trees*, International Society of Arboriculture, Champaign, IL.

Watson, G. W., Hewitt, A. M., Custic, M. et al. (2014) 'The management of tree root systems in urban and suburban settings: A review of soil influence on root growth', *Arboriculture and Urban Forestry*, vol 40, no 4, pp. 193–217.

Watson, G. W., Kupkowski, G., von der Heide-Spravka, K. G. et al. (1992) 'The effect of backfill soil texture and planting hole shape on root regeneration of transplanted green ash', *Journal of Arboriculture*, vol 18, no 3, pp. 130–135.

Weicherding, P. J., Giblin, C., Gillman, J. et al. (2007) 'Mechanical root-disruption practices and their effect on circling roots of pot-bound *Tilia cordat* Mill. and *Salix alba* L. 'Niobe'', *Arboriculture and Urban Forestry*, vol 33, no 1, pp. 43–47.

26

NAVIGATING THE ESTABLISHMENT PERIOD

A critical period for new trees

J. Roger Harris and Susan Downing Day

Introduction

Urban and community forests are often reliant on planted as opposed to naturally regenerated trees for significant portions of the canopy, especially in zones of dense settlement. As such, healthy transplanted trees create the foundation of much of urban green infrastructure and make our urban areas immensely more livable. They provide cool shade on hot days, filter our air from dust and air pollution, mitigate the harmful effects of stormwater runoff, and have significant ramifications for public health (see Chapters 4–10, this volume). The fulfillment of the vision of design professionals for a thriving tree canopy that offers such ecosystem and social services is, unfortunately, a difficult process and the original promise is often unfulfilled. Successfully navigating tree establishment is a critical component of this process and requires a thorough understanding of both the planting site itself and how this planting environment, below and above ground, imposes physiological restrictions on the planted tree. A high quality tree from a proven grower is the best beginning (see Chapter 24 of this volume for details on nursery standards). In this chapter we explore the physiological foundations of tree establishment and how this can be managed and understood in urban environments to optimize planting success.

What is establishment?

A critical question on a site manager's mind is likely 'How will I know when these trees are established?' Or, more succinctly, 'What is establishment?' Restoration of normal tissue hydration is often considered a prerequisite for establishment and researchers frequently measure it as a metric for establishment progress. It is important to note that return to normal tissue hydration is only a prerequisite to establishment and that species react differently to moisture stress. Metrics that may signal completion of the establishment process include the return of pre-transplant shoot extension or trunk expansion, the replacement of roots lost at harvest, re-establishment of tree height to root spread, and recovery of leaf gas exchange. The fulfillment of these metrics can take a considerable amount of time even in favorable sites. For example Watson (1985), using a root extension rate of approximately 45 cm per year, theorized an establishment rate of one year per 2.5 cm of trunk diameter (caliper; measured 15 cm above soil line for trunk diameters < 10 cm and 30 cm above soil line for larger trees) for balled and burlapped (B&B) field grown trees in warm continental climates, such as Chicago (Dfa according to Koppen climate classification). A typical newly planted landscape-sized tree (5–8 cm caliper) would take approximately three years to fully establish, and likely longer in unfavorable sites since roots may grow more slowly. Sites in warmer climates would take less time due to the extended period of favorable soil temperatures, although warmer sites can also exacerbate risk to early desiccation. Conversely, colder sites would take longer. The implication for site managers is that newly planted trees can require close monitoring for an extended period of time, most likely several years, before they are established.

All of these metrics are useful proxies for the completion of what we really mean by the *establishment period*: the time post-transplant when a tree is not self-sustaining at its new location and is at heightened risk of failure. Put another way, a tree that fails to establish is soon dead or only persists because of ongoing, unsustainable interventions. It is in our interest to usher trees through this period of risk as quickly as possible. A longer establishment period means more costs associated with post-transplant maintenance, more risk of structural or other defects resulting from plant stress, longer exposure to secondary stressors such as diseases or insects, greater likelihood of failure (plant death), and delays in achieving the ultimate goal – increased urban tree canopy.

Speed matters: the planting environment

The rapid regeneration of a new root system is essential for the survival of a newly transplanted tree since the limited soil volume initially available for water and nutrient acquisition can make survival difficult in an uncertain environment where drought can quickly develop. Newly transplanted trees become increasingly less vulnerable to transplant failure as they become established, so in the uncertain built environment, speed matters. The most effective way to speed new root growth is to plant the tree in an environment that does not restrict growth. There are many potential site factors that can limit root growth ranging from physical impediments to soil contamination. However, compacted soil is perhaps the most common restriction in ordinary urban environments, and remediation of compacted rooting space will likely pay big dividends.

Where does this unrestricted rooting space need to be created? During the traditional establishment period, roots might be expected to extend two or three meters out from the trunk. However, when considering the lifetime of the tree we need to ask how far from the tree should we expect roots of an established tree to extend. The ratio of root system radius to trunk diameter at breast height (DBH, 1.3 m from ground) is about 38:1 on well-established trees in mesic sites (Day et al., 2010), so a tree of say 10 cm DBH would normally have roots reaching out almost 4 m from the trunk. Restricting root growth for the new transplant in this zone will very likely limit the potential growth of the tree. However, even where soil conditions are homogeneous, roots may not be uniformly distributed around the tree. In addition, tree roots are also able to navigate through soil cracks and around below-ground impediments, sometimes unexpectedly increasing the volume of explored soil in urban plantings. Nonetheless, to realize our vision of a healthy mature tree in an urban environment, a large, unrestricted rooting area is needed. Unrestricted rooting space is especially critical for newly planted trees where our goal is to promote quick establishment so that trees can thrive and ecosystem services can best be realized. Since unrestricted rooting space is so critical for fast establishment, the recommendation that 'wider is better' and digging a planting hole that is at least 2–3 times the width of the root ball is a standard planting recommendation (see also Chapter 25, this volume). Loose cultivated soil makes the best environment for new roots emerging from the transplanted root balls, allowing establishment to proceed as quickly as possible. Additional loosening of the sides of the planting pit (not the bottom) may also be beneficial. These new roots may extend to over 0.75 m beyond the root ball within the first post-transplant growing season in temperate climates, emphasizing the need for a wide zone of loose soil.

Compacted soil works against establishment in two different ways. Root penetration becomes increasingly difficult as compacted soil dries (think a brick). However, since compaction restricts drainage as soil macropores are lost, the excess water restricts oxygen diffusion down to roots. Root growth in compacted soil is therefore restricted to a window when there is enough moisture to ease root penetration resistance but not so much water to severely restrict oxygen diffusion. This window gets smaller as compaction gets worse. In other words, root growth opportunity becomes severely restricted. Site analysis and amelioration before planting will pay big dividends, but species selection for such sites is especially critical. Trees native to low, periodically flooded areas may be best adapted to compacted urban sites since they can exploit the low resistance of the wet soil while tolerating the low oxygen levels (Day et al., 2000). Decisions about these environmental and genetic factors will set the stage for the establishment process, before planting even takes place or specific plants or nursery production systems are selected.

Figure 26.1 B&B maples and lindens experiencing extensive root loss at transplant

Photo credit: Alessio Fini

Root response to transplanting

Transplanting is a unique type of disturbance in the life of a tree. Depending upon production method and planting process, roots may be cut, maintenance regimes will be disrupted, and the above and below ground environments may completely change. Root severance is a significant part of this disturbance and understanding root physiological response to severance is key to problem solving establishment issues.

A transplanted field-grown tree will start life in the new landscape with a much reduced root system whether it is B&B or bare-root (Figure 26.1). Even container-grown trees are likely to experience root loss since research supports slicing or shaving container root balls at planting (Gilman, 2015). Root growth into the backfill soil from sliced or shaved container root balls will be faster and result in a more stable tree (Gilman, 2015).

Root regrowth

Regrowth of cut roots is an essential process for tree establishment. Regrowth primarily occurs near the cut end, with numerous new roots appearing (Figure 26.2). The newly regenerating root system therefore consists of a proliferation of new roots exploring the backfill soil and beyond. Although numerous roots are produced from each severed root end, in most cases,

Figure 26.2 Root regrowth two years after severance of *Tilia* × *europaea* roots. Note that severed large conducting roots released several finer exploring and absorbing roots

Source: photo by Alessio Fini

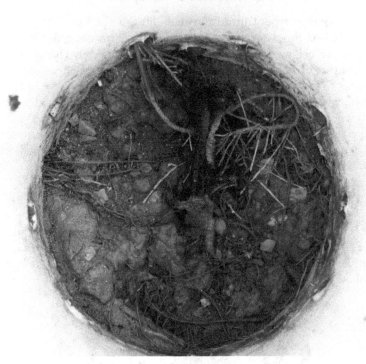

Figure 26.3 Fine roots of *Fraxinus ornus* proliferating around patches of high nutrient availability, in this case a decomposing lizard

Source: photo by Alessio Fini

one (or at most, a few) roots will become dominant within a couple of years and the remainder of the small roots will die (Watson, 1986). Therefore, many of the first roots regenerated after a tree is transplanted may serve as 'temporary' roots that grow quickly and support the newly transplanted tree until a more permanent root system is regenerated.

The rapid production of roots on newly transplanted trees serves to prevent desiccation by quickly increasing the absorptive capacity of the root system. Maintaining this new absorptive capacity while simultaneously increasing the foraging area by growing long roots would carry a hefty metabolic cost. Consequently, shedding excess roots near the root ball so that resources can be spent on longer foraging roots fits the economic model in which resources are spent where needed and withdrawn when not. Root turnover and production is a much-studied aspect of forest ecology. The dynamic system results in plasticity to exploit resource 'patches', while shedding roots when they are not needed (Figure 26.3). Richardson-Calfee et al. (2010) studied the longevity of such newly produced roots in a rhizotron study on transplanted sugar maple (*Acer saccharum* Marsh.). Although the authors predicted that root production would dominate the dynamics of first-season below-ground activity, root mortality was substantial during the first winter after transplanting. Root production mostly was limited to the growing season when leaves were present and the photosynthetic machinery was in place. Root mortality, however, occurred year-round and reached a maximum in winter for the transplanted trees and late spring for the non-transplanted trees, supporting Watson's theory regarding 'temporary' roots in transplanted trees (Watson, 1986). The root system of a transplanted tree will very likely begin to resemble that of a non-transplanted tree (minus a tap root) when it is well established.

Fostering conditions that support rapid root regrowth is therefore essential to successfully navigating the establishment period. In addition to the paramount need for a suitable soil environment, adequate water supply is essential since root regeneration is directly dependent upon water for both physiological viability and cell expansion. Other sometimes unanticipated factors that influence root regrowth are discussed below.

Species differences

Genetics plays a significant role in the rapidity of root emergence post transplanting as well as the ubiquity of viable growing points. Such favorable genetics are typically exploited as insurance against failure to establish, although at a cost to biodiversity. Because the rate of post-transplant root regeneration is closely linked to the innate capacity of a species for root regrowth, many urban forests are heavily populated with 'easy-to-transplant' species. For example, green ash (*Fraxinus pennsylvanica* Marsh.) will regenerate roots from severed root ends in as few as 17 days (Arnold and Struve, 1989), whereas northern red oak (*Quercus rubra* L.) takes at least 24 days (Struve and Rhodus, 1988). Ten weeks after transplanting, field-grown Leyland cypress (*Cupressocyparis leylandii* Dallim.) had over four times the dry mass of new roots outside root balls than laurel oak (*Quercus laurifolia* Michx.) (Harris and Gilman, 1991). Kelting et al. (1998) reported that roots were first visible outside of the root ball of red maple (*Acer rubrum* L. 'Franksred') 38 days after spring transplanting, but Harris et al. (1996) found that new roots of spring-transplanted fringe tree (*Chionanthus virginicus* L.) took at least 85 days to grow outside of root balls. Even then the amount of soil explored by the new roots was extremely limited. These differences can contribute to delays in establishment times for some species of as much as one or two additional years since weak early root growth will not safeguard against post-transplant physiological stress, resulting in long-term consequences. Consequently, ease of transplanting has had an enormous effect on the makeup of our planted urban forests. Although they may be desirable when established, difficult-to-transplant species (e.g. white oak, *Quercus alba* L.) do not comprise a significant percentage of our planted species, whereas easy-to-transplant species (e.g. sugar maple, *Acer saccharum* Marsch.) do.

Another important component of quick-to-establish transplants is the growth of intact (not severed) lateral roots. Easy-to-transplant species generally have a fibrous root system with many small-diameter roots, and chances for transplant success are enhanced by the increased numbers of intact root tips on fibrous root balls as compared with non-fibrous (coarse) root balls. Intact lateral roots initiate elongation faster than severed roots, speeding root growth into backfill soil (Ritchie and Dunlap, 1980). Intact laterals of dormant green ash begin elongation as early as seven days after placement in a warm greenhouse (Arnold and Struve, 1989). In addition, severed smaller diameter roots may regenerate root tips faster than larger diameter roots (Struve and Rhodus, 1988) and the ability of damaged roots to form new roots generally decreases with increasing diameter (Watson et al., 2014). Overall, research supports the notion that trees transplanted with a more fibrous root ball (i.e. higher proportion of small-diameter roots) will regenerate roots more quickly, speeding establishment.

Seasonal effects

Season of transplant affects tree establishment in two ways – effects of physiology and effects of environment. These two factors interact, leading many times to perplexing instances of transplant failure. First, season, or time of year, dictates specific plant growth stages (e.g. dormancy, shoot expansion, leaf drop) and consequently affects a variety of plant resources that influence the potential for quick post-transplant root system regeneration, the key to successful establishment. For example, buds of temperate zone plants are released from dormancy upon satisfaction of chilling requirements and, along with lengthening days, shoots begin elongation when spring temperatures rise. Consequently, transplanting during active shoot elongation is generally not a good idea due to the risk of desiccation and competition between roots and shoots for available carbohydrates. Even after shoots harden, transplanting when trees are inleaf results in much greater desiccation risk than when trees are dormant. Consequently, growth stage of shoots can have a major impact on root system regeneration.

The second way that season of transplant affects post-transplant establishment is that seasons have characteristic weather (e.g. soil temperature, soil moisture, humidity, wind, etc.) that affects plant growth and potential for root system regeneration. In contrast to the shoots of temperate zone hardwoods, which have a dormant period that can be overcome by chilling, roots may not exhibit an easily identified period of innate dormancy. Although root growth is linked to shoot growth by physiological signals, root growth is strongly influenced by environmental factors such as soil temperature and moisture. Thus, when considering tree establishment, we mainly focus on weather conditions that affect root growth.

Each species has different amplitudes or 'ideal' ranges of soil temperatures that are most suitable for root growth. This range is usually related to the normal climatic amplitude of temperatures in the region to which the species or ecotypes of the species are native. Local soil temperatures therefore strongly affect root growth opportunity for transplanted trees. Struve and Moser (1985) determined that as temperature increased from 10.0 to 26.1°C in the root zone of root pruned scarlet oak (*Quercus coccinea* Muenchh.) seedlings, time until new root initiation decreased, numbers of initiated new roots increased, and root elongation rate increased. Larson and Whitmore (1970) reported that little new root growth occurred in northern red oak seedlings at temperatures less than 12.8°C. Maximum root elongation occurred in Struve and Moser's research at 26°C, while no new root growth was evident below 10°C. These findings are similar to data for other temperate zone species. At 21°C, roots initiated growth 6 days following pruning, whereas at 16°C, root growth was initiated after 12 days. No elongation or initiation occurred at 10°C. Winter soil temperatures in temperate zones will therefore be limiting to root growth of most trees. Mitchell (2014) found that soil temperature in linear roadside tree planting areas in Arlington, Virginia USA, reached 12°C between 10 and 12 April and fell below 8°C between 24 and 25 November, the temperatures linked to the initiation and cessation of root growth in *Q. coccinea* (Harris et al., 1995). Periods with favorable root growth temperatures will vary in length according to local climate and species.

Planting depth

Deep planting and fill over root systems has long been recognized to hinder establishment and to affect long-term tree health. For example, reduced soil aeration in the root zone of established shade trees was attributed to compacted clay fill soil over fifty years ago (Yelenosky, 1963), and the recognition of decline in recently planted street trees with deep roots was reported by Berrang et al. (1985) over 30 years ago. It is clear from contemporary research that when conditions lower in the soil profile are less favorable than those near the surface, deep planting can indeed inhibit establishment. Arnold et al. (2007) found that planting small (9.3 l), container-grown trees as little as 7.5 cm below grade decreased survival and growth of all five species tested except *Platanus* spp. after three years in a sandy loam underlain at 15–30 cm with a hard clay pan in Texas. The clay pan in this layered soil was punctured during planting excavation, suggesting that the root systems of the deeply planted trees were partly surrounded by clay. Growth increased in some instances for trees planted above grade, suggesting that minimizing exposure to the clay hardpan may have been beneficial. In a seven-month establishment study with ~7.6 cm caliper, field-grown live oak (*Quercus virginiana* Mill.), tree growth was unaffected by planting as much as 18 cm below grade in a fine sand soil. However, deep-planted trees experienced greater water deficits than trees planted at grade when lightly irrigated after an extended dry period. This irrigation event was apparently unable to penetrate down to the deep root balls, even in sandy field soils (Gilman and Grabosky, 2004). These results suggest that the interactions of climate and soil properties at differing depths are primary factors influencing survival and growth of deep-planted trees during the establishment period. For example, if lower soil regions were highly compacted, very wet, or very dry, establishment would likely be impaired, especially in species sensitive to the particular conditions present. On the other hand, when no exacerbating conditions are present, trees may grow normally for many years. For example, in the first five years of a study with Turkish hazel (*Corylus colurna* L.), growth and establishment of trees planted both 15 and 30 cm deep were essentially identical to trees planted at grade. But after two severe flooding events, 40 percent of the most deeply planted trees died compared with no deaths among trees planted at grade or only 15 cm deep (Day and Harris, 2008). Tree establishment may also be unaffected if the species in question can tolerate the adverse soil conditions present in the lower regions of the site. For example, red maple (*Acer rubrum* L.) is a very flood-tolerant tree, while Yoshino cherry (*Prunus* × *yedoensis*) is not. When these two species were planted 30 cm below grade on a slope where drainage was poorer on the lower end of the slope, all red maples survived, while deep-planted cherries died in much greater numbers during establishment than at-grade trees (50% versus none), especially on the lower end of the slope (Wells et al., 2006). Trees planted well below grade may develop undesirable structural root patterns (Day and Harris, 2008), but these will rarely come into play during the establishment period.

Container-grown trees

Container production of landscape sized trees has increased and imposes new establishment challenges compared to field grown trees harvested B&B or bare-root. Since roots are less hardy than shoots, roots of container-grown trees must be protected in cold-winter areas during production. This limits production of large trees to warmer growing areas unless containers are protected, such as in the pot-in-pot production system where growing containers are held in 'socket' containers embedded in the ground. Although container-grown trees may be planted with complete root systems and have many other advantages to producers and landscapers (e.g. easier handling, instant effect), field-grown trees planted while still dormant are likely to establish faster and have lower mortality than container-grown trees when irrigation is less than optimal (Harris and Gilman, 1991; Gilman, 2001). This faster establishment is perhaps partly due to a stimulation of new root growth from cut root ends for field-grown trees and compensatory root growth compared with shoot growth. Shaving or slicing of container root balls, however, may achieve the same effect (Gilman et al., 2010). Poor establishment of large container-grown trees may sometimes be a result of planting during active growth when hot dry weather exacerbates desiccation, whereas field-grown trees are usually planted when dormant.

A major concern with container-grown trees as far as establishment is the distortion of roots caused by deflection of the container sidewalls during production. In trees that have

Figure 26.4 Circling roots in container-grown *Tilia cordata*

Source: photo by Piero Frangi

been held for an extended time in the container, roots often circle and the plants become 'pot bound' (Figure 26.4). Such trees will establish more slowly and there is real concern for the future stability of a maturing tree. This is of greater concern for large growing trees than for shrubs or small stature flowering trees since instability of smaller plants will not likely pose a hazard to people and infrastructure. A moderately pot-bound root ball can be remediated through physical slicing or shaving of the circling roots at planting. Roots emerging from the cut ends of the sliced roots will grow outward into the surrounding soil, much like the cut roots of a field-grown tree (Gilman et al., 2010).

Bare-root trees

Field-grown trees moved bare-root can establish quickly if roots do not desiccate during handling. Some species appear more tolerant of handling and smaller trees generally transplant and establish more easily than larger trees (Buckstrup and Bassuk, 2000). Advantages to planting trees bare-root are primarily financial. Bare-root trees are much cheaper than trees produced by other production methods because of ease of digging, storing and shipping. Many species respond well to moving bare-root. Longer root lengths are possible since weight is of little concern, and bare-root trees can potentially retain a greater proportion of the original root system. Another highly valued characteristic of bare-root production is that inspection of the entire root system is possible, and inferior root systems or defects, such as girdling roots, can be detected. An advantage of bare-root vs. container-grown plants is that the native backfill soil will surround and be in close contact with roots in the bare-root transplant, alleviating the excessive drainage of a planted container substrate. This may favor establishment time for the bare-root trees and also be an advantage over B&B trees when the B&B root ball soil is very different from the landscape soil (e.g. clay vs. sand) and water movement between the two soil types may be difficult. This also occurs for container plants that are grown in soilless substrates. The ability to inspect root systems in particular has led to the popularity of novel production systems such as the gravel bed system (Starbuck et al., 2005) and even the bare-rooting of trees produced in container or B&B systems, although this last system removes the financial benefits of bare-root production.

Water stress

Field-grown trees

Desiccation is the most common cause of failed establishment in field-grown trees since the drastic reduction of the root system makes fully maintaining post-transplant tissue hydration difficult (Figure 26.5). Furthermore, these tissue moisture deficits can be present long after transplanting. Early in the establishment period, water uptake will be limited by the absorptive capability of the much reduced transplanted root system. The limited size of the transplanted root system relative to the aboveground portion of the tree results in a reduced soil reservoir available for water absorption. Consequently, water absorption from the

Figure 26.5 Drought can kill entire plantings at once, if establishing trees are not properly managed

Source: photo by Alessio Fini

immediate root zone may occur more rapidly than water can move in from the surrounding soil, and whole tree moisture stress may then occur despite the presence of ample moisture in the surrounding soil. The more rapidly a root system is regenerated into the surrounding soil, the less moisture stress will be imposed upon the tree and the greater the chance of survival and establishment, again emphasizing the critical nature of rapid root regrowth.

Although the effects of post-transplanting soil water deficits are diminished by the reduction in leaf area that typically occurs in newly transplanted field-grown trees when leaves emerge after transplanting as well as the increased percentage of photosynthates allocated to the root system, root regeneration can still be greatly inhibited. Keeping root balls moist before as well as after planting is essential. Even one severe desiccation event can result in the restriction of water flow through root systems after the plant is rehydrated (Harris and Bassuk, 1995), leading to dieback and even, in extreme cases, plant death. Even with adequate hydration after transplanting, the removal of so much of the root system of field-grown trees inevitably slows post-transplant growth, a phenomenon known as 'transplant shock' or 'planting check'. Although initial root regeneration can rely entirely on stored carbohydrates, these reserves are quickly depleted. New root growth is needed not only to keep the transplant hydrated but to support new canopy growth that will photosynthesize and produce new carbohydrates. For most rapid establishment, the

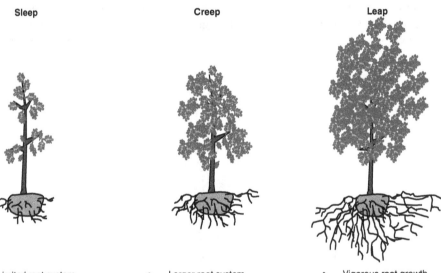

Sleep	Creep	Leap
• Limited root system	• Larger root system	• Vigorous root growth
• Wound recovery	• More intact root tips	• Near normal leaf size
• New root initiation	• Leaves larger	• Strong shoot growth
• Leaves small	• Longer shoots	• Dense canopy
• Shoots short	• Canopy more dense	• Large canopy promotes
• Thin canopy	• Larger canopy leads to	vigorous root growth
• Limited root growth	more root growth	• Root system replaced

Figure 26.6 Normal three season establishment process for a 5–8 cm caliper shade tree planted in a favorable site and well irrigated. The transplanted tree 'sleeps' during year one as the root system begins to regenerate. More substantial canopy development is noticed in year two and the tree 'creeps' along as the pre-transplant shoot:root ratio begins to reestablish. The larger root system supports strong canopy growth during year 3 and the tree appears to 'leap' as it becomes established

Source: diagram by Roger Harris

naturally reduced canopy of early deciduous transplants must not become water stressed due to insufficient irrigation since stomatal closure will result in a reduction in photosynthesis. Some light canopy thinning of evergreen trees at planting may be beneficial since the remaining leaves may be less water stressed, resulting in greater stomatal conductance and higher photosynthesis rates.

Container-grown trees

In theory, container-grown trees need not be subjected to the desiccation stress of field-grown trees since all roots are contained in the root ball. However changes in both the above- and below-ground environments subject even high quality container-grown trees to significant water stress. Container root balls can quickly become very dry following transplanting. This is mostly due to the fact that container substrate that provides favorable moisture and air conditions during production results in suboptimal water holding capacity when the container is removed at transplanting (Nelms and Spomer, 1983; Hanson et al., 2004). In other words, a good substrate for growing trees in containers is a poor substrate for transplanting and the transplanted substrate will rapidly become drier than the surrounding soil. Precision irrigation research may result in adoption of more suitable substrates that are not so excessively drained in some cases. Outdoor producers in temperate zone areas, however, will experience rainfall that may flood substrates for several days, reducing aeration in the root balls and limiting the use of more slowly draining substrates. Actively growing trees will further deplete container root balls of available moisture after transplanting. In a study of similar-aged red maple trees transplanted from different sized containers, trees in larger containers were more water stressed for the first 3 years after transplanting than trees in smaller containers and established more slowly due to the differences in root ball moisture dynamics (Gilman et al., 2013). Water stress on newly transplanted trees clearly imposes a physiological stress that will impair root regeneration and rapid establishment. To prevent such water stress, thorough and frequent irrigation of transplanted container-grown trees is therefore very critical until roots grow into the backfill soil and beyond.

How much water is needed?

As discussed above, water stress is a main deterrent to tree establishment. Newly planted trees must remain hydrated for the natural process of root regeneration and growth to proceed with optimum expediency. The most critical period is the first growing season after transplanting, with decreasing needs the following seasons. Sivyer et al. (1996) used a pan evaporation model created by others to develop a method for predicting irrigation amount and frequency for newly planted street trees and tested it on mulched, 7.5 cm caliper, B&B Callery pear (*Pyrus calleryana* Decne. 'Redspire') and river birch (*Betula nigra* L. 'Heritage') trees five months after planting. The model predicted that root balls should be saturated every 3 days with 38 liters of water. Refitting the model assumptions with actual tree measurements and adjusting the root ball soil tension point at which root balls were to be irrigated to well above the permanent wilting point, resulted in a 19-liters-every-3-days regime. The authors indicated that full saturation of the root balls was critical, and that although 19 l was the minimum volume required, more would be needed to assure thorough wetness. The take home message was that unless soaking rains occur, 20–40 liters of water every three days are needed on newly planted 7 cm caliper (trunk diameter taken 15 cm from soil) trees transplanted B&B. Witmer (2000) monitored the water status of root balls of

55 liters container-grown (approximately 5 cm caliper) red and sugar maple for three years after planting and irrigated when root balls reached a predetermined dry point. Irrigation intervals for both species (all seasons when not dormant) was 4.9, 5.5 and 6.6 days for post-transplant seasons one, two, and three, respectively. Approximately 40 liters of water was applied to each tree at each irrigation event. Naturally, irrigation needs will vary according to local climate and site conditions as well as the rapidity of root regeneration that is subject to environmental and genetic controls described earlier. Precision placement of irrigation water through a mini sprinkler or drip emitter may reduce the volume needed to something approaching 20 liters. Intervals could be lengthened during the cooler fall period and the following season as the tree becomes more established. For the first growing season, a typical landscape-sized tree (5–8 cm caliper or over 3 m tall) would probably need a minimum of 15–40 liters of water applied directly to the root ball twice a week to prevent water stress. Trees with greater leaf area will likely need more. Although irrigation would be less critical for the next season, careful monitoring is required. If trees are healthy, most should be fully established during the third growing season. Post-transplant fertilization has generally proven ineffective in speeding along the establishment process, regardless of production method or site characteristics (Harris et al., 2008).

Conclusions

A well-established tree that is fulfilling the vision of providing meaningful social and ecological services to urban landscapes is best achieved by first selecting high-quality trees from reputable growers. Site analysis and modification are keys to selecting the proper species. Compacted rooting zone soil is a very common occurrence in urban and community landscapes and is a real and often ignored limitation on establishment. An understanding of the establishment process will help guide irrigation and maintenance strategies. The establishment period varies according to the size of the transplant and the site factors that might inhibit root growth. It is imperative that root balls of newly transplanted trees are not allowed to dry out for the first season after transplanting. As roots grow beyond the root ball, irrigation can focus more on the surrounding soil. Irrigation will be needed for at least two years after transplanting for landscape-sized trees. Besides irrigation, aftercare would include removal of stakes and guys, light corrective pruning, and maintenance of mulch (if applied). Fertilization of newly planted landscape-sized trees has generally not proved to aid establishment (see also Chapter 29 about fertilization of urban trees).

Reduced canopy growth (transplant shock) can be expected even in the best of conditions until equilibrium between roots and shoots (i.e. root:shoot ratio) is restored by compensatory growth of the root system. Trees characteristically maintain a defined root:shoot ratio, so transplanted trees do not put on significant shoot growth until the pre-transplant ratio has been reestablished. As this ratio is being restored, canopies are generally thin and leaves are relatively small due to feedback limitations from the reduced root system. Above-ground growth will naturally be very slow during year one. As the root system develops, increased canopy growth will occur allowing for greater production of photosynthates. A return to pre-transplant growth is often noticed the third year after transplanting for a landscape-sized tree planted in good site conditions and well irrigated. Practitioners have traditionally called this sleep (year 1), creep (year 2), and leap (year 3) (Figure 26.6). Poor site conditions and poor irrigation management can dramatically slow post-transplant growth and the sleep and creep periods can last for several seasons. In these conditions the leap year may never materialize. Finally, even though a tree may be considered established, its growth rate may never be what it was in the excellent growing conditions of the production nursery. Moderately slow but steady growth is usually considered acceptable as long as leaves show good color with no obvious nutrient deficiency symptoms. This is a common case of the tree adapting to the resources of the new site. In other words a *functional equilibrium* has been reached and the tree is growing at a rate supported by the available resources – the tree is *established*.

References

Arnold, M. A. and Struve, D. K. (1989). 'Green ash establishment following transplant', *Journal of the American Society for Horticultural Science*, 144, 591–595.

Arnold, M. A., Mcdonald, G. V., Bryan, D. L., Denny, G. C., Watson, W. T. and Lombardini, L. (2007). 'Below-grade planting adversely affects survival and growth of tree species from five different families', *Arboriculture and Urban Forestry*, 33, 64–69.

Barton, A. J. and Walsh, C. S. (2000). 'Effect of transplanting on water relations and canopy development in *Acer*', *Journal of Environmental Horticulture*, 18, 202–206.

Berrang, P., Karnosky, D. F. and Stanton, B. J. (1985). 'Environmental factors affecting tree health in New York City', *Journal of Arboriculture*, 11, 185–189.

Buckstrup, M. J. and Bassuk, N. L. (2000). 'Transplanting success of balled-and-burlapped versus bare-root trees in the urban landscape', *Journal of Arboriculture*, 26, 298–308.

Day, S., Seiler, J. and Persaud, N. (2000). 'A comparison of root growth dynamics of silver maple and flowering dogwood in compacted soil at differing soil water contents', *Tree Physiology*, 20, 257–263.

Day, S. D. and Harris, J. R. (2008). 'Growth, survival, and root system morphology of deeply planted *Coryluys colurna* seven years after transplanting and the effects of root collar excavation', *Urban Forestry and Urban Greening*, 7, 119–128.

Day, S. D., Wiseman, P. E., Dickinson, S. B. and Harris, J. R. (2010). 'Tree root ecology in the urban environment and implications for a sustainable rhizosphere', *Arboriculture and Urban Forestry*, 36, 193–205.

Gilman, E. F. (2001). 'Effect of nursery production method, irrigation, and inoculation with mycorrhize-forming fungi on establishment of *Quercus virginiana*', *Journal of Arboriculture*, 27, 30–39.

Gilman, E. F. (2015). 'Effect of eight container types and root pruning during nursery production on root architecture', *Arboriculture and Urban Forestry*, 42, 31–45.

Gilman, E. F. and Grabosky, J. (2004). 'Mulch and planting depth affect live oak (*Quercus virginiana* Mill.) establishment', *Journal of Arboriculture*, 30, 311–317.

Gilman, E. F., Paz, M. and Harchick, C. (2010). 'Root ball shaving improves root systems on seven tree species in containers', *Journal of Environmental Horticulture*, 28, 13–18.

Gilman, E. F., Miesbauer, J., Harchick, C. and Beeson, R. C. (2013). 'Impact of tree size and container volume at planting, mulch, and irrigation on *Acer rubrum* L. growth and anchorage', *Arboriculture and Urban Forestry*, 39, 173–181.

Hanson, A. M., Harris, J. R., Wright, R. W., Niemiera, A. X. and Persaud., N. (2004). 'Water content of pine-bark growing media in a drying mineral soil', *HortScience*, 39, 591–594.

Harris, J. R. and Bassuk, N. L. (1995). 'Effect of drought and phenological stage at transplanting on root hydraulic conductivity, growth indices, and photosynthesis of Turkish hazelnut', *Journal of Environmental Horticulture*, 13, 11–14.

Harris, J. R. and Gilman, E. F. (1991). 'Production method affects growth and root regeneration of Leyland cypress, laurel oak, and slash pine', *Journal of Arboriculture*, 17, 65.

Harris, J. R., Day, S. D. and Kane, B. (2008). 'Nitrogen fertilization during planting and establishment of the urban forest: a collection of five studies', *Urban Forestry and Urban Greening*, 7, 195–206.

Harris, J. R., Knight, P. and Fanelli, J. (1996). 'Fall transplanting improves establishment of balled and burlapped fringe tree (*Chionanthus virginicus* L.)', *HortScience*, 31, 1143–1145.

Harris, J. R., Bassuk, N. L., Zobel, R. W. and Whitlow, T. H. (1995). 'Root and shoot growth periodicity of green ash, scarlet oak, Turkish hazelnut, and tree lilac', *Journal of the American Society for Horticultural Science,* 120, 211–216.

Kelting, M., Harris, J. R., Fanelli, J. and Appleton, B. (1998). 'Humate-based biostimulants affect early post-transplant root growth and sapflow of balled and burlapped red maple', *HortScience*, 33, 342–344.

Larson, M. M. and Whitmore, F. W. (1970). 'Moisture stress affects root regeneration and early growth of red oak seedlings', *Forest Science*, 16, 495–498.

Mitchell, D. K. (2014) 'Urban landscape management practices as tools for stormwater mitigation by trees and soils', MS thesis, Virginia Polytechnic and State University, Blacksburg, VA.

Nelms, L. R. and Spomer, L. A. (1983). 'Water retention of container soils transplanted into ground beds', *HortScience*, 18, 863–866.

Richardson-Calfee, L. E., Harris, J. R., Jones, R. H. and Fanelli, J. K. (2010). 'Patterns of root production and mortality during transplant establishment of landscape-sized sugar maple', *Journal of the American Society for Horticultural Science*, 135, 203–211.

Ritchie, G. and Dunlap, J. (1980). 'Root growth potential: its development and expression in forest tree seedlings', *New Zealand Journal of Forestry Science*, 10, 218–248.

Sivyer, D., Harris, J. R. and Fanelli, J. (1996). 'Evaluation of a pan evaporation model for estimating post-planting street tree irrigation requirements', *Journal of Arboriculture*, 17, 250–256.

Starbuck, C., Struve, D. K. and Mathers, H. (2005). 'Bareroot and balled-and-burlapped red oak and green ash can be summer transplanted using the Missouri gravel bed system', *HortTechnology*, 15, 122–127.

Struve, D. K. and Moser, B. C. (1985). 'Soil temperature effects on root regeneration of scarlet oak seedlings', *Research Circular (Ohio Research and Development Center)*, 2841, 12–14.

Struve, D. K. and Rhodus, W. T. (1988). 'Phenyl indole-3-thiolobutyrate increases growth of transplanted 1-0 red oak', *Canadian Journal of Forest Research*, 18, 131–134.

Watson, G. (1985). 'Tree size affects root regeneration and top growth after transplanting', *Journal of Arboriculture*, 11, 37–40.

Watson, G. W. (1986). 'Cultural practices can influence root development for better transplanting success', *Journal of Environmental Horticulture*, 4, 32–34.

Watson, G. W., Hewitt, A. M., Custic, M. and Lo, M. (2014). 'The management of tree root systems in urban and suburban settings II: a review of strategies to mitigate human impacts', *Arboriculture and Urban Forestry*, 40, 249–271.

Wells, C., Townsend, K., Caldwell, J., Ham, D., Smiley, E. T. and Sherwood, M. (2006). 'Effects of planting depth on landscape tree survival and girdling root formation', *Arboriculture and Urban Forestry*, 32, 305–311.

Witmer, R. K. (2000) 'Water use of landscape trees during production and during establishment in the landscape', PhD dissertation, Virginia Polytechnic and State University, Blacksburg, VA.

Yelenosky, G. (1963) 'Soil aeration and tree growth', *Proc. Int. Shade Tree Conf*, 1963, 16–25.

PART VII

Managing urban forests and urban trees

27

PRUNING

Brian Kane

Introduction

Pruning is the selective removal of plant parts to achieve an objective, and has been practiced in various disciplines for many years. Orchardists are interested in the effects of pruning on edible fruit production and foresters prune lower branches on the trunk to improve timber quality. This chapter focuses on pruning trees in developed landscapes, which are variously referred to as 'street', 'shade', 'ornamental', 'landscape' or, as I will use here, 'amenity' trees. Owners and managers of amenity trees usually do not expect them to provide products like timber or produce – although urban biomass production and edible landscapes are possible – but rather, appreciate them for the variety of benefits they provide (see Chapters 5–12). Pruning is arguably the most common arboricultural practice, and certainly the practice that consumers most often associate with arboriculture.

Common objectives for pruning amenity trees include: improve aesthetics, structure, or tree health; reduce the likelihood of failure; create clearance or a view; stimulate flower production; restore trees damaged by poor pruning, storms, or vandalism; and manage pests. Arborists use different types of pruning to achieve these objectives. Pruning types and techniques differ by discipline and geographically, and arboricultural standards around the world set guidelines for various aspects of pruning. Terms to describe different aspects of pruning vary by discipline and geographically. In this chapter, I will use terms from Part 1 of the American National Standards Institute A300 Standard (Anonymous, 2008). The Standard is regularly revised (including an expected revision in 2016), but previous research has primarily used terms consistent with the 2008 revision, which describes the following types of pruning: clean, espalier, pollard, raise, reduce, restoration, structural, thin, and vista. Pruning types describe one or more of the following: (i) the part of a branch to be removed, (ii) the location in the crown from which branches will be removed, and (iii) the type of cut used to remove branches. Cleaning is the removal of non-beneficial tree parts – dead, diseased, or broken branches; raising provides vertical clearance; reducing decreases height or spread; and thinning decreases the density of live branches. Structural (also called formative) pruning is pruning to establish or improve branch architecture. Types of specialty pruning include, espalier – training plants to grow in a single plane with the help of a supporting structure;

pollarding – creating a pollard head to which sprouts are frequently pruned back; restoration – pruning to correct damage to structure, form, or appearance from topping or storms; topiary – training and shearing trees into formal shapes; and vista – creating or enhancing a specific view. Utility pruning is a type of specialized pruning to 'uphold the intended usage of the … utility space while adhering to accepted tree care performance standards' (Anonymous, 2008).

In addition to recommended pruning types, inappropriate pruning types sometimes still occur. Two types of pruning that are no longer recommended are lion's tailing and topping. Recent studies have demonstrated the adverse physiological and structural consequences of topping. Adverse consequences include greater discoloration and decay (Gilman and Knox, 2005; Dahle et al., 2006a; Fini et al., 2015) – which may be due, in part, to slower wound occlusion (Fini et al., 2015), poor health (Karlovich et al., 2000; Fini et al., 2015), weakly attached branches (Dahle et al., 2006b; Fini et al., 2015), and an assessed greater likelihood of failure (Karlovich et al., 2000). Topping accentuates the temporary, post-pruning imbalance between roots and shoots and removes apical control, which stimulate short-term growth (Gilman et al., 2008b; Fini et al., 2015) at the expense of trunk growth (Fini et al., 2015). Consisting of leaves of less mass per unit area, the post-topping flush of growth can be slender (Fini et al., 2015), which, along with weak attachments (Dahle et al., 2006b; Fini et al., 2015) and greater extent of decay (Dahle et al., 2006a), increases the likelihood of failure. Topping can increase maintenance costs over time (Campanella et al., 2009), but consumers consider it as a way to reduce the likelihood of tree failure (Fazio and Krumpe, 1999). Pollarding appears to be a better way to maintain a fixed tree size since there is less decay and dieback (Gilman and Knox, 2005). Short-term growth induced by topping also occurs at the expense of long-term tree health and storage of reserve carbohydrates, leading to greater occurrence of crown dieback (Fini et al., 2015), which has also been observed in communities (Karlovich et al., 2000).

Rigorous experimental data have yet to quantify the presumed adverse effects of lion's tailing (e.g. poorly tapered and sagging branches, sunscald, poor growth). However, experimental data (Smiley and Kane, 2006; Gilman et al., 2008a) have debunked the myth that lion's tailing more effectively reduces the likelihood of failure (by reducing drag-induced bending moment) than other pruning types. In the absence of evidence to justify lion's tailing as an effective means of reducing the likelihood of failure, it seems prudent to avoid it.

Pruning cuts can be categorized in different ways. Fundamentally, they can be 'proper' or 'improper'. The latter typically refers to cuts that (i) damage or remove part, or all, of the branch collar, (ii) leave a stub beyond the branch collar, or (iii) do not occur at a node (i.e. internodal). Cuts that completely remove the branch collar are called flush cuts. There are two types of proper cuts: branch removal and reduction cuts. The former removes a higher-order branch back to its parent stem, while the latter removes the parent back to a higher-order branch of sufficient diameter – the recommendation is greater than or equal to one-third the diameter of the parent (Anonymous, 2008). Heading cuts (which reduce a parent back to a bud or daughter branch not large enough to assume the apical role, i.e., less than one-third the diameter of the parent) can be labeled proper or improper. Although reduction cuts are preferred to heading cuts, in restoration pruning and other anomalous situations, heading cuts may be the only alternative. Some references use 'heading' instead of 'internodal'.

Pruning induces physiological responses in, and has mechanical consequences for, trees. Although horticulturists and foresters have long studied the effects of pruning, the pruning types, species of interest, growing conditions, and pruning objectives are not always arboriculturally relevant. For example, foresters prune trees to increase the volume of merchantable timber, and orchardists prune trees to maximize fruit production, not to improve aesthetics or reduce risk. Arboricultural researchers and practitioners can learn from pruning studies from other disciplines, but must apply these results to amenity trees only with caution. Physiological responses (growth partitioning, carbohydrate storage, wound occlusion) and mechanical consequences (changes in loading and dynamic properties) can vary among species, type and severity of pruning, growing conditions, and age and health of the individual tree.

Physiological responses to pruning

In the short-term, pruning invigorates existing growth because it disrupts the balance between roots and shoots: the root system that sustained a certain crown size before pruning needs only to supply some fraction of the original crown after pruning. Manifestations of invigorated growth after pruning (that may vary by the type of pruning cut) include the following: (i) greater leaf size – whether this occurs for both reduced and thinned trees (Fini et al., 2015) or just reduced trees (Hipps et al., 2014) is unclear; (ii) increased growth of remaining branches that were not pruned (Findlay et al., 1997; Gilman and Grabosky, 2009); (iii) slowed growth of pruned branches (Gilman and Grabosky, 2009; Kristoffersen et al., 2010; Gilman, 2015a, 2015b); (iv) production of watersprouts, suckers (root sprouts), and epicormic branches (Grabosky and Gilman, 2007; Gilman et al., 2008b; Fini et al., 2013; Hipps et al., 2014; Fini et al., 2015) and (v) a temporary increase in the photosynthetic rate per unit leaf area. These responses facilitate the return to equilibrium of the root:shoot balance (three years after pruning, leaf area of pruned trees was similar to unpruned trees; Hipps et al., 2014), which is why most studies have shown that moderate pruning using proper cuts does not adversely affect trunk growth of young, recently-transplanted trees (Gilman and Grabosky, 2009; Fini et al., 2013, 2015; Gilman, 2015b). However, some studies have noted a modest post-pruning retardation of trunk growth rate under more severe pruning (Gilman, 2015a). Not all pruning studies have quantified the post-pruning reduction in total leaf area (some only quantified the reduction per branch), but it has usually been described as moderate (i.e. ≤30% of total leaf area). More severe pruning (>50% of total leaf area) can slow trunk growth of forest trees and substantially deplete reserve carbohydrates over time (Clair-Maczulajtys et al., 1999), although on young trees, Findlay et al. (1997) did not detect any adverse effects of removing 50 percent of shoots at the time of planting.

Pruning can alter the amount and location of reserve carbohydrates in a tree. In unpruned trees, the concentration of reserve carbohydrates is seasonally stable in large roots, and there are more reserves stored in the lower trunk (compared to the upper trunk) and in the upper crown (compared to the lower crown). However, pollarded trees store the majority of reserve carbohydrates in the pollard head, rather than at the base of the branch, in the trunk and in large roots – which reveal marked seasonal fluctuations in carbohydrate storage (Haddad et al., 1995). Stored carbohydrates affect sprouting and cambial activity, which can influence occlusion of wounds (Dujesiefken and Stobbe, 2002) and the area of discoloration (Grabosky and Gilman, 2007).

Pruning can also affect apical dominance. On amenity trees, removing the apical bud (e.g. when topping trees) increases the number of root (Gilman et al., 2008b; Fini et al., 2015) or basal sprouts (Gilman et al., 2008b) or those that originate near the cut (Fini et al., 2015). Whether reduction cuts are more likely to stimulate watersprouts compared to removal cuts has not yet been determined. Some studies have observed more watersprouts with reduction cuts (Fini et al., 2013; Hipps et al., 2014), but Fini et al. (2015) did not find any differences. Fini et al. (2015) suggested that the apical bud of the higher-order branch to which the parent was reduced can more easily replace the terminal bud that was removed by reducing the parent branch. More growth also occurs in higher-order branches arising from the new terminal branch (i.e. the lateral branch to which the original parent branch was reduced) (Grabosky and Gilman, 2007). An internodal cut (made when topping) removes the apical bud without a natural substitute available to assume the apical role and sprouting is more prolific (Fini et al., 2015). This reasoning is consistent with the observation of the number of root and basal sprouts increasing with the size of internodal cuts (Gilman et al., 2008b). Further study into the extent of sprouting induced by reduction cuts is necessary because Grabosky and Gilman (2007) observed watersprouts growing near only 53 percent reduction cuts on two *Quercus* species. Neither could they predict the occurrence of watersprouts from measurements of the reduced branch – diameter, aspect ratio, angle, or post-pruning growth rate. Sprouting also occurs in the vicinity of removal cuts (Fini et al., 2013, 2015). Findlay et al. (1997) did not measure watersprouts, but did observe that most of the post-pruning growth on retained branches was in apical extension, rather than growth of higher-order branches from the retained branch (Figure 27.1). Exposure of cuts to sunlight may also influence sprouting, since Fini et al. (2015) observed poor growth of watersprouts when sunlight was scarce.

Reduction cuts retard growth of reduced branches, which reduces aspect ratio (the ratio of the diameters of the lateral and parent branches) over time (Gilman and Grabosky, 2009; Kristoffersen et al., 2010; Gilman, 2015a, 2015b). Branches with smaller aspect ratios are more strongly attached to the trunk and, when removed, cause smaller wounds and areas

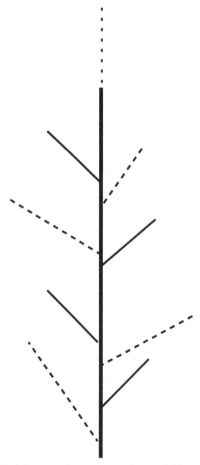

Figure 27.1 Pruned (thin, dashed lines) and remaining (thin, solid lines) lateral branches on a parent twig (thick, solid line), which grew primarily as apical extension of the twig (thin, dotted line) after pruning, rather than extension of remaining lateral branches

of discoloration (Gilman and Grabosky, 2006; Grabosky and Gilman, 2007). The latter considerations are important for street trees because branches can be retained on newly-transplanted trees to encourage good structure and subsequently removed without creating large wounds (Kristoffersen et al., 2010). Removing between 25–30 percent of the leaf area of individual branches reduced aspect ratio of branches with an initial aspect ratio of approximately 0.50 and 0.60 by about 10 percent (Kristoffersen et al., 2010; Gilman, 2015a). Pruning to retard growth of branch unions with large aspect ratios may be most effective on unions with larger aspect ratios and those growing in proximal parts of the crown. More severe pruning (50–75%) was needed to slow the growth of *Q. virginiana* branches with larger (≥0.70) aspect ratios (Gilman and Grabosky, 2009), consistent with Grabosky and Gilman's (2007) finding that growth of higher-order branches of *Q. virginiana* was inversely proportional to the diameter and aspect ratio of the reduced parent branch. Although growth of branches with initial aspect ratio of approximately 0.55 decreased over time even without pruning (Gilman, 2015b), on branches with a very large aspect ratio (>0.80), the aspect ratio of unpruned branches did not diminish over time (Gilman and Grabosky, 2009).

Pruning wounds trees, which can lead to discoloration and decay. The vast majority of pruning studies have been conducted on forest- or plantation-grown trees, but there are some studies of amenity trees. Given important differences in species, growing conditions, and types of pruning cuts, practitioners and researchers must interpret results from such work cautiously.

With respect to pruning wounds on amenity trees, research has explored the size of wounds, the rate of wound occlusion – a surrogate measure of expected decay, and the area and depth of discoloration (and, for a few studies of longer duration, decay). Specifically, researchers have investigated whether these response variables vary among different pruning cuts and species, and by branch morphology.

Evidence from temperate (Dujesiefken and Stobbe, 2002) and tropical (Ow et al., 2013) species has shown that despite faster woundwood growth on flush cuts, because they are larger, it takes longer to occlude the wound, and the area of discoloration is greater than for branch collar cuts. Flush cuts also induce greater cambial dieback, which, on larger wounds, further reduces rate of occlusion (Dujesiefken and Stobbe, 2002). For a variety of cuts (flush, branch collar, reduction, or removal), the area of discoloration or decay is proportional to the size of the cut (Dujesiefken and Stobbe, 2002; Gilman and Grabosky, 2006; Grabosky and Gilman, 2007; Ow et al., 2013). Even after ten years, many cuts larger than five centimeters in diameter were not completely occluded (Dujesiefken and Stobbe, 2002). In contrast, other pruning studies have shown that occlusion of even larger wounds can occur within three (Ow et al., 2013) or four (Neely, 1991) years. Even though both considered trees growing in a city, growing conditions in temperate (Dujesiefken and Stobbe, 2002) and tropical (Ow et al., 2013) settings are clearly different. Such differences, in combination with the lack of winter dormancy in the tropics, might explain why, for a variety of tropical species, between 10 and 50 percent of branch collar cuts were completely occluded in only 3 years. In another study in a temperate climate and 4 years after pruning, Neely (1991) reported complete occlusion of 12 and 17 percent of conventional and branch collar cuts, respectively, even though conventional cuts made substantially larger wounds. (No longer used, conventional cuts are different from flush cuts: the cut starts distal to the branch bark ridge but continues parallel to the parent stem – through the branch collar – rather than following and remaining just distal to the branch collar.) Both Dujesiefken and Stobbe (2002) and Neely (1991) tested trees in temperate climates. The disparity between their findings might be attributable to better growing conditions in an arboretum (Neely, 1991), since less vigorous trees are less able to compartmentalize wounds (Dujesiefken and Stobbe, 2002).

Direct comparison to Dujesiefken and Stobbe (2002) is not possible given different size and age of trees, size of cut, and growing conditions, but it is notable that under very similar experimental conditions (including branch diameter), Fini et al. (2013) found, in contrast to Fini et al. (2015), that 2 years after the first pruning, *removal* cuts were 96 percent occluded and *reduction* cuts were 70 percent occluded. The effect of the wide variety of factors that influence occlusion of pruning wounds is clearly still not well understood, including the effect of repeated pruning. Recent work suggests that repeated pruning reduces the rate of wound occlusion. One year after the second pruning (3 years after the first pruning), removal and reduction cuts were less than 20 percent occluded (Fini et al., 2013, 2015). Two years after the second pruning (4 years after the first pruning), cuts were, at most, 50 percent occluded (Fini et al., 2015).

Studies have also revealed a wide range of variation in wound responses among species in both tropical (Ow et al., 2013) and temperate (Neely, 1991; Dujesiefken and Stobbe, 2002; Grabosky and Gilman, 2007) climates. Since these studies were undertaken under different growing conditions with trees of varying age and size, it is not possible to compare results directly. The effect of species (and presumably other effects such as growing condition and tree age and vigor) confounds a basic trend that many previous studies have demonstrated. Under controlled experimental conditions, cut diameter predicted 62 percent of the variance in the area of discoloration for *Quercus shumardii*, but only 12 percent of the variance for *Q. virginiana* (Grabosky and Gilman, 2007).

In response to pruning wounds, Dujesiefken and Stobbe (2002) characterized genera as 'strong' and 'weak' compartmentalizers (Table 27.1). The genus *Quercus* was among the strong compartmentalizers, but Grabosky and Gilman (2007) found many significant differences in the area of discoloration between *Quercus shumardii* and *Q. virginiana*. For example, three years after pruning, *Q. shumardii* showed greater depth and area of discoloration than *Q. virginiana*. Also, the depth of discoloration was inversely proportional to branch angle for *Q. virginiana*, but not *Q. shumardii*, and the area of discoloration was proportional to the diameter of the cut for *Q. shumardii*, but only marginally so for *Q. virginiana*. Grabosky and Gilman (2007) measured the wound response to reduction cuts, which is another factor confounding a direct comparison with Dujesiefken and Stobbe's (2002) data, but it illustrates the complexity and myriad interactions that may affect a tree's response.

Table 27.1 Genera commonly considered strong and weak compartmentalizers (adapted from several sources)

Strong	Weak
Acer	Aesculus
Carpinus	Betula
Fagus	Malus
Fraxinus	Populus
Gleditsia	Prunus
Ginkgo	Salix
Pyrus	Styphnolobium (Sophora)
Quercus	
Tilia	

Data have also revealed differences in wound response for different branch morphology, but not always consistently. Three years after reduction cuts on Q. *shumardii* and Q. *virginiana* (Grabosky and Gilman, 2007), and four years after removal cuts on *Acer rubrum* (Gilman and Grabosky, 2006), the area of discoloration was proportional to the aspect ratio of the pruned branches. And while Dujesiefken and Stobbe (2002) recommended pruning branches with included bark parallel to the long axis of the parent stem to avoid cambial dieback, Gilman and Grabosky (2006) did not observe any increase in discoloration on branches with included bark when pruning perpendicular to the branch axis. The small size of removal cuts in the latter study may have facilitated wound occlusion and negated any adverse effect of cambial dieback.

Mechanical effects of pruning

Discoloration and decay are also mechanical consequences of pruning because they affect the likelihood of tree failure. Likelihood of failure is one of the key components of tree risk assessment, and pruning to reduce risk is a common pruning objective. In the following discussion, 'tree failure' includes failure of individual branches, trunk breakage, or windthrow.

Rigorous studies have focused exclusively on the effect of pruning on reducing the likelihood of failure. The likelihood of failure of a tree is related to the type and magnitude of applied load(s) and the ability of the tree (or tree part) to resist the applied load(s). The most common loads that damage trees are wind, snow, and ice. In addition to their magnitude, loads can be quantified with respect to time. The tree's response can also be quantified in this way. Wind loads (and the tree's response to them) are dynamic (i.e., the magnitude and direction change over short periods of time). Snow and ice loads increase in magnitude very slowly (as does a branch's deflection), and are thus considered static. Pruning to reduce the likelihood of failure typically concerns wind loads, and can affect several aspects of the wind-induced load as well as a tree's response to it.

Drag is the force that a structure experiences when it obstructs the flow of a fluid, such as when trees are exposed to wind. The classical equation to quantify drag (D) is:

$$D = \rho \star A \star C_D \star U^2 / 2 \tag{27.1}$$

where ρ is air density, A is area, C_D is drag coefficient, and U is wind speed. Studies have demonstrated that pruning reduces drag by reducing area (Smiley and Kane, 2006; Gilman et al., 2008a, 2015; Pavlis et al., 2008) and drag coefficient (Cao et al., 2012). Quantifying crown area presents challenges, and a more reliable surrogate measure is mass, to which drag is proportional (Kane and Smiley, 2006; Kane et al., 2008). Given this finding, it is not surprising that there is a direct correlation between pruning-induced reductions in mass and drag (Smiley and Kane, 2006; Pavlis et al., 2008). The location in the crown from which biomass is removed (assuming the same proportion of high- and low-drag elements) tends not to affect drag, which remains proportional to the mass removed.

Leaves increase area (and mass) of the crown, and also have disproportionately greater drag coefficients than branches because they are less rigid and flutter in the wind. Increases in area and drag coefficient increase drag on in-leaf trees (Smiley and Kane, 2006), and pruning-induced reduction in drag is small compared to the reduction due to leaf senescence (Smiley and Kane, 2006). In temperate climates where some trees are deciduous, pruning to reduce drag (and, consequently, likelihood of failure) may be less effective if severe wind events are more likely when trees are leafless. A better understanding of the seasonal variation in likelihood of severe loading will help gauge the effectiveness of pruning at different severities.

Severe loading events are important because Equation 27.1 shows that wind speed, which is squared, influences drag more than other factors. This effect has also been demonstrated experimentally: Gilman et al.'s (2008a) analysis of variance showed that the effect of wind speed was nearly five and ten times greater than the effects of pruning severity and type, respectively. In contrast, pruning-induced reduction in drag is linearly proportional to mass removed. This means that to compensate for a doubling of wind speed, an arborist must reduce the mass of the entire crown by 75 percent. Although tree response will vary among growing conditions, tree age, and species, in most cases such severe pruning is physiologically inadvisable. This means that there is a limit to how effectively pruning can reduce the likelihood of failure during severe loading events. Tree failure occurs under a variety of loading conditions, however, and pruning remains a useful method for reducing the likelihood of failure.

When a tree experiences drag, roots anchored in the ground provide the reaction force that prevents translational motion (i.e. sliding) of the tree relative to the ground. The lines of action of the reaction force, at the root-soil plate, and drag, in the crown, are separated by a distance, called the lever arm, that induces a bending moment on the tree. The bending moment (M) is the product of drag and the lever arm (l):

$$M = D \star l \tag{27.2}$$

and is directly proportional to bending stress (σ):

$$\sigma = M \star y / I \tag{27.3}$$

where I is the area moment of inertia of the cross-section and y is the distance from the neutral axis of the cross-section to the bark. To reduce the likelihood of the tree or an individual branch failing by bending stress, pruning can reduce the bending moment by reducing drag, the lever arm, or both.

Drag-induced bending moment increases with tree mass and wind speed (Kane and Smiley, 2006; Kane et al., 2008). Concomitantly, pruning generally reduces drag-induced bending moment (or trunk deflection, a surrogate measurement) in proportion with mass removed (Smiley and Kane, 2006; Gilman et al., 2008a; Pavlis et al., 2008). The effect is most noticeable at higher wind speeds (Smiley and Kane, 2006; Gilman et al., 2008a, 2008c). Although the location of pruned biomass does not affect drag, which is reduced in proportion with mass removed (Smiley and Kane, 2006; Gilman et al., 2008a; Pavlis et al., 2008), the location of pruned branches does affect drag-induced bending moment (Smiley and Kane, 2006; Pavlis et al., 2008) and, consequently, trunk deflection (Gilman et al., 2008a, 2008c). This occurs because bending moment is the product of drag and the lever arm on the crown, which can be taken as the distance from some height (typically, the lower trunk or root flare) and the centroid of area of the crown. Raising, which removes branches in the lower crown to achieve vertical clearance, therefore increases the lever arm. Even if drag is reduced, the increase in lever arm may keep drag-induced bending moment constant (or even increase it), or increase bending higher up on the trunk where foliage is concentrated (Gilman et al., 2008c). In contrast, reduction pruning, which shortens the crown and, consequently, the lever arm, more effectively reduces drag and drag-induced bending moment (Smiley and Kane, 2006; Pavlis et al., 2008) or trunk deflection (Gilman et al., 2008b). Per unit mass removed, reduction pruning also more effectively reduced drag-induced bending moment than raising (Pavlis et al., 2008) and thinning (Smiley and Kane, 2006; Pavlis et al., 2008) because reduction pruning disproportionately removes high-drag elements of the crown (leaves) rather than low-drag elements (rigid branches). Although Gilman et al. (2008a) did not find differences in trunk deflection between reduction and other pruning types (except thinning), this disparity might be attributed to (i) using trunk deflection instead of measuring bending moment directly – deflection is inversely proportional to the fourth power of trunk diameter, so small differences between trees can dramatically alter deflection, and (ii) not normalizing the effect of pruning type by the mass removed.

Since wind and trees interact dynamically, measuring the effect of pruning on what are effectively static wind loads (Smiley and Kane, 2006; Pavlis et al., 2008; Gilman et al., 2008a, 2008c; Cao et al., 2012) provides helpful, but insufficient results. It appears that no rigorous studies have measured the drag-induced bending moment on trees *in situ*, although Gilman et al. (2015) measured the effect of pruning on trunk strain (a surrogate measure of bending moment) in a simulated, turbulent wind at four different frequencies. Other studies have investigated the effect of different types of pruning on two important sway characteristics of amenity trees (Kane and James, 2011; James, 2014; Miesbauer et al., 2014): (i) natural frequency – the reciprocal of the time required for a single oscillation, and (ii) damping ratio, which reflects how well a swaying tree dissipates energy.

Understanding the dynamic interaction of trees and wind is important, but complicated. Trees that sway at higher frequency absorb less wind energy, which theoretically reduces the likelihood of failure. For forest- and plantation-grown trees, dynamic beam theory provides valuable insights into tree sways because such trees can be modeled as single-degree-of-freedom (SDOF) systems and sufficient empirical data facilitate comparisons with theory. In contrast, open-grown amenity trees typically have a greater proportion of their total mass in branches and frequently trunks bifurcate into two or more large upright branches (sometimes as co-dominant stems). Assuming an SDOF system for these trees is problematic and there are few empirical studies that have quantified the sway characteristics of large amenity trees (James, 2014; Kane et al., 2014). For open-grown trees, one or more large branches of similar mass can influence the sway behavior of the tree; removing such branches could induce large changes in natural frequency (James, 2014).

The effect of pruning on natural frequency and damping ratio of amenity trees has not been extensively tested. Reduction pruning tends to have a greater effect on increasing frequency than raising or thinning (Kane and James, 2011), which is consistent with the simple physical model of less slender cantilevered beam vibrating at higher frequency. Pruning did not affect damping ratio (Kane and James, 2011). On deciduous trees, the effect of leaf senescence on natural frequency and damping ratio can have an equal or greater effect than pruning at moderate severities (Kane and James, 2011; Miesbauer et al., 2014). For example, damping ratio of leafless trees was up to 75 percent less than trees in-leaf for three species (Kane and James, 2011; Miesbauer et al., 2014). Larger pruning-induced changes on frequency and damping ratio can be achieved with more severe pruning (James, 2014). The effect of leaves on frequency and damping ratio reiterates the importance of the seasonal occurrence of severe wind events with respect to whether pruning will measurably reduce the likelihood of tree failure. Testing greater pruning severities on amenity trees is also necessary to explore further the effect of different types of pruning on sway characteristics.

Limitations on pruning research and future prospects

With some exceptions (notably studies focusing on wound occlusion and discoloration), studies on the effects of pruning have mostly considered small, young trees of relatively few genera and species. And most experiments have occurred in arboreta or nurseries representing a limited range of climates and soil textures. Often, trees are tested within several years of transplanting (or to quantify the effect of pruning at transplanting); sometimes, they are fertilized and irrigated before and during experiments. While controlled conditions are necessary to minimize experimental variability, they are not representative of most growing conditions in residential communities. Even under controlled experimental conditions, (i) variability in tree growth can still be large (see, for example, growth of unpruned trees in Fig. 3 in Gilman and Grabosky, 2009); and (ii) species, even within the same genus can respond differently to the same pruning treatment (Grabosky and Gilman, 2007). Considering (i) the inherent variability in tree growth, (ii) the variety of factors that influence the response of trees to pruning (and many possible interactions between the factors), and (iii) the limited range of species and growing conditions that have been tested, practitioners should take care in broadly applying results from existing studies. Throughout this chapter, I have pointed out inconsistencies among studies and variable responses to pruning, but I believe it is very much worth repeating.

This general limitation applies to physiological and mechanical consequences of pruning. Age, species, climate, soil texture, and cultural practices affect tree growth; and size, shape, and material properties affect mechanical behavior – the former often have a non-linear effect. Growth and material properties also change as trees age. Additional studies on mature and large trees growing in residential areas are critical to understanding better the physiological and mechanical consequences of pruning. Long-term manipulative experiments are also absent from the literature on pruning amenity trees. This is understandable given funding and logistical constraints, but such studies are integral to understanding better the long-term consequences of pruning.

With respect to the focus of this chapter – how pruning affects tree structure, growth, and decay – the underlying goal of most pruning studies was to reduce the likelihood of tree failure, a common pruning objective. Examples include studies of how pruning affects wound occlusion, discoloration and decay; the growth and aspect ratio of branches; and drag, bending moment, and sway characteristics. But none have focused on how pruning affects long-term tree growth, which directly affects benefits like providing shade, storing carbon, mitigating air pollution and runoff. Results from Miesbauer et al. (2014) nicely illustrate this conundrum. Trees pruned to maintain or enhance an excurrent form had higher frequency and less biomass higher in the crown, both of which reduce the likelihood of failure. However, the increase in frequency did not last more than a growing season and pruning to maintain or enhance the excurrent form required removing 1.7 and 2.5 times as much biomass pruning one and two years, respectively, after the initial pruning. This suggests that maintaining or enhancing an excurrent form required greater inputs of time and carbon. It is not clear that these trends would apply for longer timeframes or on larger trees or other species, but it does highlight the importance of more accurately quantifying whether pruning meaningfully reduces the likelihood of failure and whether doing so adversely affects the benefits that trees provide.

Minimizing tree risk is clearly important, and pruning can influence some of the many factors that affect the likelihood of tree failure. However, of studies that have quantified the effect of pruning on actually reducing the likelihood of failure, only Duryea et al. (1996) reported a positive impact. Much more work is needed to understand whether and, if so, under what conditions pruning meaningfully reduces the likelihood of failure.

Practitioners will continue to prune trees to achieve a variety of objectives, but they encounter two important challenges when attempting to implement guidelines from research on pruning. First, there are many limitations on existing research, and for some aspects of pruning, there is not a clear consensus on the 'best' method to achieve an objective. Of the limitations, the most important may be a lack of data quantifying the long-term effects of pruning on reducing the likelihood of failure and the benefits that trees provide. Second, pruning to achieve one objective may adversely affect another objective. The ideal season to prune highlights this challenge. Pruning earlier in the summer stimulates greater storage of carbohydrates (Clair-Maczulajtys et al., 1999) and expedites wound occlusion (Perry and Hickman, 1987), but increases the likelihood of attracting beetles that vector oak wilt and Dutch elm disease. Arborists must therefore use good judgment and experience when determining pruning objectives for individual situations.

References

Anonymous. (2008). *American National Standard for Tree Care Operations: Tree, Shrub, and Other Woody Plant Management: Standard Practices (Pruning)*. Londonderry, NH: Tree Care Industry Association.

Campanella, B., Toussaint, A. and Paul, R. (2009). Mid-term economical consequences of roadside tree topping. *Urban Forestry and Urban Greening*, 8(1), 49–53.

Cao, J., Tamura, Y. and Yoshida, A. (2012). Wind tunnel study on aerodynamic characteristics of shrubby specimens of three tree species. *Urban Forestry and Urban Greening*, 11(4), 465–476.

Clair-Maczulajtys, D., Le Disquet, I. and Bory, G. (1999). Pruning stress changes in the tree physiology and their effects on the tree health. *Acta Horticulturae*, 496, 317–324.

Dahle, G. A., Holt, H. H., Chaney, W. R., Whalen, T. M., Grabosky, J., Cassens, D. L., Gazo, R., McKenzie, R. L. and Santini, J. B. (2006a). Decay patterns in silver maple (*Acer saccharinum*) trees converted from roundovers to V-trims. *Journal of Arboriculture*, 32(6), 260–264.

Dahle, G. A., Holt, H. H., Chaney, W. R., Whalen, T. M., Cassens, D. L., Gazo, R. and McKenzie, R. L. (2006b). Branch strength loss implications for silver maple (*Acer saccharinum*) converted from round-over to V-trim. *Arboriculture and Urban Forestry*, 32(4), 148–154.

Dujesiefken, D. and Stobbe, H. (2002). The Hamburg tree pruning system: A framework for pruning of individual trees. *Urban Forestry and Urban Greening*, 1(2), 75–82.

Duryea, M. L., Blakeslee, G. M. and Hubbard, W. G. (1996). Wind and trees: A survey of homeowners after hurricane Andrew. *Journal of Arboriculture*, 22, 44–50.

Fazio, J. R. and Krumpe, E. E. (1999). Underlying beliefs and attitudes about topping trees. *Journal of Arboriculture*, 25, 193–199.

Findlay, C., Last, F., Aspinall, P., Thompson, C. W. and Rudd, N. (1997). Root and shoot pruning in root-balled *Acer platanoides* L.: Effects on establishment and shoot architecture. *Arboricultural Journal*, 21(3), 215–229.

Fini, A., Ferrini, F., Frangi, P., Piatti, R., Faoro, M. and Amoroso, G. (2013). Effect of pruning time on growth, wound closure and physiology of sycamore maple (*Acer pseudoplatanus* L.). *Acta Horticulturae*, (990), 99–104.

Fini, A., Frangi, P., Faoro, M., Piatti, R., Amoroso, G. and Ferrini, F. (2015). Effects of different pruning methods on an urban tree species: A four-year-experiment scaling down from the whole tree to the chloroplasts. *Urban Forestry and Urban Greening*, 14(3), 664–674.

Gilman, E. F. (2015a). Pruning *Acer rubrum* at planting impacts structure and growth after three growing seasons. *Arboriculture and Urban Forestry*, 41(1), 11–17.

Gilman, E. F. (2015b). Pruning severity and crown position influence aspect ratio change. *Arboriculture and Urban Forestry*, 41(2), 69–74.

Gilman, E. F. and Grabosky, J. C. (2006). Branch union morphology affects decay following pruning. *Journal of Arboriculture*, 32(2), 74–79.

Gilman, E. F. and Grabosky, J. C. (2009). Growth partitioning three years following structural pruning of *Quercus virginiana*. *Arboriculture and Urban Forestry*, 35(6), 281–286.

Gilman, E. F. and Knox, G. W. (2005). Pruning type affects decay and structure of crapemyrtle. *Journal of Arboriculture*, 31(1), 48–53.

Gilman, E. F., Harchick, C., Grabosky, J. C. and Jones, S. (2008a). Effects of pruning dose and type on trunk movement tropical storm winds. *Arboriculture and Urban Forestry*, 34(1), 13–19.

Gilman, E. F., Knox, G. W. and Gomez-Zlatar, P. (2008b). Pruning method affects flowering and sprouting on crapemyrtle. *Journal of Environmental Horticulture*, 26(3), 164–170.

Gilman, E. F., Masters, F. and Grabosky, J. C. (2008c). Pruning affects tree movement in hurricane force wind. *Arboriculture and Urban Forestry*, 34(1), 20–28.

Gilman, E. F., Miesbauer, J. W. and Masters, F. J. (2015). Structural pruning effects on stem and trunk strain in wind. *Arboriculture and Urban Forestry*, 41(1), 3–10.

Grabosky, J. C. and Gilman, E. F. (2007). Response of two oak species to reduction pruning cuts. *Journal of Arboriculture*, 33(5), 360–366.

Haddad, Y., Clair-Maczulajtys, D. and Bory, G. (1995). Effects of curtain-like pruning on distribution and seasonal patterns of carbohydrate reserves in plane (*Platanus acerifolia* Wild) trees. *Tree Physiology*, 15(2), 135–140.

Hipps, N. A., Davies, M. J., Dunn, J. M., Griffiths, H. and Atkinson, C. J. (2014). Effects of two contrasting canopy manipulations on growth and water use of London plane (*Platanus* × *acerifolia*) trees. *Plant and Soil*, 382(1-2), 61–74.

James, K. R. (2014). A study of branch dynamics on an open-grown tree. *Arboriculture and Urban Forestry*, 40(3), 125–134.

Kane, B. and James, K. R. (2011). Dynamic properties of open-grown deciduous trees. *Canadian Journal of Forest Research*, 41(2), 321–330.

Kane, B. and Smiley, E. T. (2006). Drag coefficients and crown area estimation of red maple. *Canadian Journal of Forest Research*, 36(8), 1951–1958.

Kane, B., Modarres-Sadeghi, Y., James, K. R. and Reiland, M. (2014). Effects of crown structure on the sway characteristics of large decurrent trees. *Trees*, 28(1), 151–159.

Kane, B., Pavlis, M., Harris, J. R. and Seiler, J. R. (2008). Crown reconfiguration and trunk stress in deciduous trees. *Canadian Journal of Forest Research*, 38(6), 1275–1289.

Karlovich, D. A., Groninger, J. W. and Close, D. D. (2000). Tree condition associated with topping in southern Illinois communities. *Journal of Arboriculture*, 26(2), 87–91.

Kristoffersen P., Buhler O., Larsen S. U. and Randrup T. B. (2010). Growth of newly established *Tilia platyphyllos* 'Rubra' roadside trees in response to weed control and pruning. *Arboriculture and Urban Forestry*, 36(1), 35–40.

Miesbauer, J. W., Gilman, E. F. and Giurcanu, M. (2014). Effects of tree crown structure on dynamic properties of *Acer rubrum* L. 'Florida Flame'. *Arboriculture and Urban Forestry*, 40(4), 218–229.

Neely, D. (1991). Branch pruning wound closure. *Journal of Arboriculture*, 17(8), 205–208.

Ow, L. F., Ghosh, S. and Sim, E. K. (2013). Mechanical injury and occlusion: An urban, tropical perspective. *Urban Forestry and Urban Greening*, 12(2), 255–261.

Pavlis, M., Kane, B. C., Harris, J. R. and Seiler, J. R. (2008). The effects of pruning on drag and bending moments of shade trees. *Arboriculture and Urban Forestry*, 34(4), 207–215.

Perry, E. and Hickman, G. (1987). Wound closure in *Eucalyptus*. *Journal of Arboriculture*, 13(8), 201–202.

Smiley, E. T. and Kane, B. (2006). The effects of pruning type on wind loading of *Acer rubrum*. *Journal of Arboriculture*, 32(1), 33–40.

28

IRRIGATION OF URBAN TREES

Alessio Fini and Cecilia Brunetti

The water resource and its use in urban settings

The availability of fresh water is an issue as old as civilizations. Across the ages, cities thrived where the supply of potable water was abundant and collapsed in the face of drought. Today, mankind uses about 4000 km^3 water/year for domestic/industrial use, and 14,000 km^3/year for agriculture and grazing lands (Oki and Kanae, 2006). These withdrawals account for less than 10 and 30 percent, respectively, of the world's renewable fresh water resource, which indicates that, if the water cycle is maintained wisely, fresh water can cover human demand far into the future (Oki and Kanae, 2006). The uneven spatial and temporal variability of fresh water, however, results in 2.4 billion people currently living in water-scarce environments. The rapid and often irreversible changes which follow urbanization may further enhance such inequity.

Over the last century, a tremendous increase of urban population, relative to rural, has occurred: on May 23, 2007, citizens living in urban areas equaled those living in the countryside and up to 84 percent of the world's population is expected to live in cities by 2050. The fast-expanding cities and their related urban sprawl, typical of our 'Anthropocene' era, have become a huge sink of natural resources (urban settings occupy less than 3 percent of the world's emerged land but they consume about 75 percent of resources), including fresh water. For example, in 2001, the municipality of Copenhagen (Denmark) consumed 405 mm groundwater, while the 'non-Copenhagen' area of the Sjaelland island extracted 24 mm (Boegh et al., 2009).

Low-density housing is typically characterized by the presence of private gardens and, in warmer climates, swimming pools, and requires far more water compared to flats in the compact city core. In Barcelona metropolitan region, for example, where residential water use accounts for 77 percent of total water demand, indoor water use accounts for 72 percent of residential water use in the city center, and 36 percent in the low density suburbs (Domene and Saurí, 2006). This results in about 131 liters/person/day consumed in the city center, and about 178 in the suburbs (Domene and Saurí, 2006). These observations were reliably upscaled to Coastal Mediterranean Spain, where 200–600 liters/person are used every day, depending on housing density and garden characteristics (Hof and Wolf, 2014). In the United States, water used for irrigation of landscape plants and gardens accounts 40–70 percent of total water use. The highest consumptions are found in desert areas such as Phoenix, Arizona, where every day about 911.3 liters per person are used, 74 percent of which for outdoor purposes (Balling and Gober, 2007).

It has been recently observed that landscape irrigation consumes more water per unit area than pressurized agricultural irrigation (Salvador et al., 2011). There are several causes leading to high outdoor water requirements and low irrigation efficiency resulting from over-irrigation in low-density residential landscapes. First, inappropriate species selection and poor planting design increase water needs. For example, turfgrass is widely used worldwide, regardless of climate type. Turf is the most water-requiring type of vegetation (mean monthly irrigation requirement estimated over different Spanish cities was 61.0–86.5 mm), while trees and shrubs require lower amounts (37.3–55.3 mm) (Hof and Wolf, 2014). Second, water requirements vary in relation to vegetation type, species, and design, but irrigation is often adjusted to satisfy requirements of the most sensitive species. Third, landowners often do not consider water requirements information to schedule irrigation (Salvador et al., 2011). Finally, water is a relatively inelastic good and its consumption can be hardly regulated through price policies (Balling and Gober, 2007).

Global warming will further increase water use in cities: in dry and warm climates, household water use was found to increase about 0.8 percent for every 1°C of increase in air temperature (considering an average annual water use of 300 kL, 1°C warming will make 4.61 kL water be additionally required) (Balling and Gober, 2007). These projections and the co-occurring increasing competition for water between different sectors impose new challenges for urban water management, including irrigation scheduling.

Is there a real need for irrigation in urban settings?

Unlike agricultural areas, there is substantially no irrigation reference point for urban vegetation, which consists in a variety of different green spaces, including urban forests, public parks, private gardens, golf courses, street trees, green roofs and other vegetated open areas. At one end of the spectrum, communities may decide that urban vegetation should be only rain-fed, regardless of possible wilting during dry periods. At the other end, it may be desirable to irrigate for zero plant stress as for agricultural crops (Nouri et al., 2013). In between, a wide range of possibilities exist that may be selected depending on the role and type of urban vegetation, and on the environmental characteristics of the planting site. In practice, while private, institutional, and business gardens, where aesthetics is a primary requirements, are often (over-)irrigated, irrigation of public green areas, such as urban parks and street trees, is rare, challenging, and often restricted to the very first year after planting (Whitlow et al., 1992). This practice is supported by the lack of evidence about chronic drought stress in urban trees and by the belief that replacing dead trees during sudden dry spells is cheaper than performing routine or supplemental irrigation (Whitlow et al., 1992). This belief may now require re-evaluation in the light of climate change, which will likely increase the variability of rainfall and the frequency of drought episodes.

As detailed in Chapters 4 to 12 of this book, urban vegetation provides a wide range of environmental, economic, social, cultural, psychological benefits, which largely depend on health, size at maturity, leaf area, leaf gas exchange, and longevity. As several other maintenance techniques, irrigation should be targeted to improve tree health, carbon gain, transpiration, and life-span, thus increasing ecosystem services provided by the tree and reducing CO_2 emissions for replacing dead or declining plants. This is the case of newly planted trees (not discussed here; see Chapter 26), but may be worth for established plants in environments where available soil moisture is permanently or temporarily inadequate to support ETP demand from the atmosphere, also depending on species' mechanisms to cope with drought and planting site characteristics. In heavily sealed sites, in particular, irrigation may be required because most rainfall is ineffective (lost as runoff because soil infiltration is prevented by impervious pavements) and because limited rooting volume reduces soil water storage. It must be considered that isolated trees growing in cities (e.g. street trees) transpire 2–3 times more water than trees in dense stands (e.g. forest trees) and their higher transpiration, in addition to shade cast on buildings and pavements, makes urban trees 3–5 times more effective than forest trees for microclimate amelioration. If soil moisture is not limiting, a mature urban tree can transpire up to 180 liters of water per day (Pataki et al., 2011), thus dissipating up to 409 MJ of sensible heat as latent heat. On the contrary, non-irrigated trees may experience sudden or chronic drought spells, which can temporarily or permanently lower leaf gas exchange and/or plant water potential, depending on stress intensity. Mild and moderate stresses generally impose diffusional limitations to photosynthesis, due to stomatal closure and reduction in mesophyll conductance to CO_2. Severe stresses, instead, impose biochemical limitations (i.e. downregulation or impairment of Rubisco and electron transport), which may impede complete recovery of the photosynthetic apparatus even after relief from water stress (Fini et al., 2013).

Irrigation can also directly ameliorate urban microclimate and thermal comfort and can reduce the 'urban heat island' effect (Yang and Wang, 2015). In fact, irrigation alters surface heating, heat reflection, turbulent air mixing, and heat exchange between the air and the building, thus decreasing the external thermal load of buildings. During summer, irrigation reduced buildings, air, and ground temperature by 3, 4 and 6°C, respectively, and resulted in about 6 percent energy saving for building conditioning compared to the 'no irrigation' scenario (Yang and Wang, 2015). Considering that urban areas account for 67–74 percent of global energy use, a city-wide wise irrigation program may effectively reduce urban heat island and energy use during summer. On the contrary, if irrigation keeps running during cold months, this results in higher energy use and resource-use inefficiency (Yang and Wang, 2015).

How to irrigate in urban settings

The choice of irrigation method is critical for efficient watering of urban trees. Common watering methods include:

1 *Hand irrigation:* it has a high labor cost and is commonly used when installing an irrigation system is considered unfeasible. Water is distributed manually with a hose to fill the berm around the trunk flare and saturate soil in the rootzone. This may lead to deep percolation and evaporation losses, decreasing irrigation efficiency, which can be 64 percent lower than using a properly designed automatic system (Machado e Silva et al., 2014). Efficiency can be increased by using hand filled slow release watering bags (i.e. Treegator; Figure 28.1). New plantings require frequent irrigation (e.g. twice per week for the first 3 months, weekly for the rest of the first year; then every 2 weeks for the next 2 years), established trees need to be irrigated once/twice per month from May to September. On average, 55–75 liters/month and 120–150 liters/month are required for newly planted and established trees, respectively, but these values may vary consistently depending on species and site characteristics (Costello et al., 2000).

2 *Sprinklers, sprays, and microsprays:* commonly used for turf, but little effect with trees, because most irrigation water is retained in the shallowest soil layers, thus promoting evaporation and shallow rooting. Typically these systems operate at 1.5–5 bars,

consume 30–200 (sprays) to 400–1500 (sprinklers) liters of water/hour and have an efficiency around 70–75 percent.

3 *Surface drip:* very well suited for established and mature trees. Should be placed below a mulch layer or porous pavement to reduce vandalism, which otherwise can be an issue. If dripping spacing and flow rate are properly sized, efficiency can be around 95 percent. Typically operate at pressure lower than 2 bars and consume 1–6 liters/hour.

4 *Subsurface drip:* can be buried shallow (0–15 cm) or deep (20–40 cm). Works at similar pressures and water use as surface drip, and can have 99 percent efficiency because water is distributed directly in the root-zone. Recommended for new plantings, while some digging damage may account for installation near mature trees.

5 *Watering wells and trenches:* wells (vertical) or trenches (horizontal) are typically 30 cm wide and 30 cm deep; trenches can be 1.2 m long. They are filled with free draining gravel material to maximize void space and water storage when saturated. Water is delivered to the trench through an irrigation bubbler. Typically, 4–8 trenches are required for a mature tree. Their installation may harm tree roots if location of main roots is not previously assessed.

6 *Water storage devices with controlled release of water:* include a variety of water storage items, from cisterns to modified traffic barriers, which can collect and store rainwater or runoff. Depending on climate, and impervious area contributing to water storage, tanks are capable to store up to 100 mm rainfall can be required. If a large portion (i.e. > 90%) of households within a city/neighborhood is involved in water storage, irrigation can be fully sustained using stored water.

The choice of irrigation method is highly dependent on water availability, site characteristics, soil texture, and vegetation type (Haley et al., 2007). For example, while xeric gardens can be efficiently irrigated by micro-irrigation, turf is commonly watered using sprinklers, which requires about 30 percent more water (Haley et al., 2007).

Irrigation systems can be manual or automated. Automated irrigation can be time-scheduled or controlled by various technologies to match plant water needs or environmental goals. These technologies include: soil moisture sensors, soil temperature probes,

Figure 28.1 Treegator for irrigation of street trees in Copenhagen: (a) detail; (b) a non-irrigated, declining pedunculate oak (*Quercus robur*) and next to a Treegator-irrigated oak in full vegetation

evapotranspiration controllers, rain sensors (McCready and Dukes, 2011; Yang and Wang, 2015). Soil moisture sensors can be adjusted to the desired moisture threshold and bypass an irrigation event if soil moisture is higher than the selected threshold. Soil temperature probes work similarly, activating irrigation when soil temperature exceeds a desired threshold. Evapotranspiration controllers ideally irrigate according to calculated evapotranspiration of the plant, estimated through proper coefficients from reference evapotranspiration (ET_0). They can be programmed according to site and species characteristics. Rain sensors are designed to interrupt time-scheduled irrigation events after a certain depth of rain.

Automated irrigation can either improve or decrease irrigation efficiency compared to non-automated systems. Most automated systems are time-scheduled and schedule doesn't change with weather conditions and soil moisture, leading to frequent over-irrigation (up to 47% more water can be used than with non-automated systems). On the other hand, smart automated irrigation can save up to 75 percent compared to manual irrigation (Machado e Silva et al., 2014). Soil-temperature controlled systems can consume up to three times more water than non-automated systems, but are extremely effective in reducing the urban heat island effect and in improving thermal comfort (Yang and Wang, 2015). The savings arising from reduced summer cooling of buildings may largely compensate for the additional cost of irrigation water, particularly in hot climates, making this soil-temperature irrigation control extremely cost-effective (Yang and Wang, 2015). Conversely, soil-moisture-, evaporation-, and rain-controlled sensors may reduce water use for irrigation on average by 70, 60 and 30 percent, respectively, compared to time-scheduled irrigation, if soil moisture threshold and evapotranspiration are correctly estimated (McCready and Dukes, 2011).

Water holding capacity and water cycle in urban soils

Soil cannot retain an unlimited quantity of water, so knowing soil water holding capacity is a key requirement to design an efficient irrigation system. After a rain/irrigation event that saturates the soil, water fills both micropores and macropores. If soil is not compacted, water held in macropores percolates by gravity (gravitational water) within a few hours and has a limited availability to trees. After gravitational water is lost, macropores become filled with air while micropores still retain water. The soil is at field capacity (FC), which is known to maximize water availability to plants. Volumetric water content at FC may vary from 15 percent (sandy soils) to over 40 percent (heavy soils). Water in micropores is held by capillarity (capillary water) and can be absorbed by plant roots or evaporate. If not replaced by rain or irrigation, water content decreases to permanent wilting point (WP). No capillary water is left in the soil, and only a thin water layer of hygroscopic water, unavailable to plants, remains strongly attached to soil colloids. Typical volumetric soil water content at WP range from 3 percent (sand) to 15 percent (heavy soils). Water between FC and WP is available water (AW) to plants, and can be further divided into easily or not easily AW, depending on the tension required to extract it from the soil (water potential). At decreasing volumetric water content, soil water potential becomes more negative according to soil texture-driven moisture retention curves (Figure 28.2); the more soil water potential is negative, the more the plant must lower its leaf water potential to absorb water. Typical soil water potentials range from –0.01 in well-watered soils to –10 MPa in exceptionally dry soils; values lower than –0.5 MPa typically induce some degree of stomatal closure and prevent full recovery of pre-dawn water potential (depending on species, stress severity and duration) (Porporato et al., 2001). If soil moisture remains below WP, plants may experience stress and decline; if irrigation brings the soil at higher soil moisture than FC, that will result in inefficient water use because gravitational

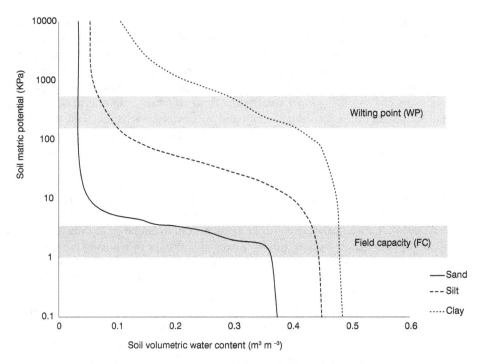

Figure 28.2 Examples of soil water retention curves for sand, silt, and clay soil textures

Note that larger surface area allows larger water retention in finer than in coarser textures. Finer textures, however, hold water more tightly. As a consequence, for example, at volumetric water content around 0.2 a plant would have plenty of available soil water in a sandy soil, but be severely drought stressed in clay

water will be lost before being absorbed by plants. Urban soils are highly heterogeneous (see Chapter 21 of this volume), therefore knowing hydraulic and physical characteristics of soil in the planting site is a key prerequisite for designing an efficient irrigation system. Volumetric water content at FC and WP, and available water can be easily assessed on soil samples of known volume using gravimetric methods (Fini et al., 2011).

The actual water content (AWC) in the root zone is the amount of available water in the soil in a given moment and can be calculated as:

$$100 \, (X - WP)/(FC - WP) \tag{28.1}$$

where: X is the actual weight of the soil sample of known volume, and FC and WP are weights of the same volume of soil at field capacity and wilting point, respectively (Fini et al., 2011). The actual water content of a given soil depends on several variables, described in Equation 28.2:

$$\Delta AWC = R + I + W - ET_L - D - R \tag{28.2}$$

where ΔAWC is the change in AWC over a given time period, R is rainfall, I is irrigation, W is capillary rise, ET_L is landscape evapotranspiration, D is deep percolation, and R is runoff.

In urban landscapes, the extensive use of paved (impervious) surfaces affects the terms of Equation 28.2: infiltration rate can be as low as 15 percent in sealed sites (compared to about 90% in unpaved, uncompacted soils), thereby greatly reducing the amount of effective rainfall while increasing runoff. Soil sealing, however, also reduces evaporation: soil moisture can therefore be higher below impervious pavements than in bare soil, if no trees are planted. For example, in Denmark, ET_L in urban landscapes was 210 mm per year, much lower than in agricultural (493 mm/y) and forest (415–628 mm/y) sites, because of soil sealing and little canopy cover (Boegh et al., 2009). With trees, instead, transpiration (which consumes far more water than evaporation; Vico et al., 2014) coupled with low infiltration rate of sealed soils may progressively deplete locally available water in the root zone, resulting in drought stress and tree hydraulic deterioration, particularly if irrigation isn't applied (Savi et al., 2015). Thus, in paved soils, the extension of the permeable area (either unpaved or covered with porous pavements) is the main determinant of irrigation needs, with frequent irrigation events required to maintain tree health in heavily sealed sites (Vico et al., 2014).

Optimizing water use in cities – irrigation efficiency and scheduling

Estimating the irrigation needs of urban green areas with mixed vegetation is extremely challenging, but it is a key requirement for getting the most benefit from urban trees while saving water. Plant water needs can be accurately estimated with several methods accurately reviewed by Nouri et al. (2013), as described below.

Soil moisture probes

Knowledge of soil moisture content is essential for precision scheduling of irrigation. Soil moisture data can be collected by several types of soil sensors, measuring either soil tension (e.g. gypsum blocks, tensiometers), or soil volumetric water content, such as time-domain-reflectometer (TDR) and frequency-domain-reflectometer (FDR). In particular, FDR probes have been successfully used for irrigation scheduling, because of relatively low cost and high accuracy even in sandy soils with low water holding capacity. To avoid misleading information, volumetric probes must be calibrated to the characteristics of the soil of the planting site – at least the moisture (v/v) at which WP and FC occur must be identified. For example, 16 percent (v/v) water content corresponds FC in a sandy soil, while being near WP in a clay soil. Calibration can be done using the gravimetric method (Machado e Silva et al., 2014), burying the moisture probes in the selected soil samples during the watering–drying processes.

A soil-moisture-based irrigation scheduling can be arranged in a bypass configuration (i.e. if soil moisture is above a threshold, the time-scheduled irrigation event is skipped) or start irrigation as soon as soil moisture declines below a moisture threshold. The moisture threshold at which the irrigation starts can be set according the aesthetics and the level of tree benefits desired: where water availability is scarce, and a certain level of stress may be tolerated on trees, irrigation events may start at very low soil available water. For example, starting irrigation at AWC as low as 7–10 percent resulted in acceptable plant quality (McCready and Dukes, 2011). Although a single sensor placed in the root zone provides valuable information, burying multiple sensors at different depths and/or distance from the trunk allows greater accuracy and resolution.

Evapotranspiration-based controllers

Measuring the total amount of water lost through evapotanspiration provides an estimation of water that should be partially or fully replaced by irrigation. In agricultural crops, crop evapotranspiration (ET_c) is easily calculated from reference evapotranspiration (ET_0) and crop coefficient (K_c) according to Equation 28.3:

$$ET_c = K_c ET_0 \qquad (28.3)$$

ET_0 can be estimated from a class-A evaporation pan or using micrometeorological equations (Allen et al., 2011). In urban landscapes, instead, the estimates of K_c are very challenging because various species of trees, shrubs, and turf are planted with different densities in planting sites with different (even contrasting) characteristics. To overcome these uncertainties, Costello et al. (2000) developed the Water Use Classification of Landscape Plants method, which estimates landscape evapotranspiration (ET_L) as function of ET_0 and a landscape coefficient (K_L):

$$ET_L = K_L ET_0 \qquad (28.4)$$

K_L is calculated as the product of three coefficients which account for plant species (K_s), planting density (K_d) and site microclimate (K_{mc}). K_s ranges from 0.1 for very low water-demanding species to 0.9 for high demanding ones (see Table 28.1). K_d ranges from 0.5–0.9 in low density stands (i.e. trees: less than 70% canopy cover; shrubs and ground cover: less than 90%) to 1.1–1.3 in high density ones (i.e. full canopy cover for all types of vegetation in multi-tiered stands). K_{mc} ranges from 0.5–0.9 in shaded, wind-sheltered sites (e.g. north side of buildings) to 1.1–1.4 where site features increase the evaporative demand (e.g. street medians, parking lots). A complete list of values for the coefficients and specification for their determination are reported in Costello et al. (2000). A similar, alternative method to calculate ET_L, the Landscape Irrigation Management Program (LIMP), has been recently developed by Snyder et al. (2015).

Lysimeters

Lysimeters are large containers filled with disturbed or undisturbed soil that use gravimetric weighing to estimate water use (Nouri et al., 2013). They are easy for inspection, can be automated, and are very sensitive to vegetation type, so that lysimeters can be used to calibrate other methods for soil moisture measurement (Todd et al., 2000). Lysimeters, however, aren't very practical to be used for scheduling irrigation in urban settings and can induce shallower rooting in trees planted in the lysimeter.

Bowen ratio-energy balance

This robust inexpensive micrometeorological technique indirectly estimates latent heat flux from a surface using measurements of net radiation, air temperature and humidity gradients, and soil heat flux (Todd et al., 2000) using Equation 28.5:

$$\lambda E = R_n - G - H \qquad (28.5)$$

where λE is latent heat flux, R_n net radiation, and H sensible heat flux, all expressed in W m^{-2}. The reliability of λE (and consequently ET) estimate depends on the accurate measurement of net radiation and soil heat flux (Allen et al., 2011), which may be complicated because of high heterogeneity of the urban environment.

Table 28.1 Species coefficients (Ks) for selected tree and shrub species. Ks may vary depending on climatic zone and plant phonological phase

Species	Ks	Species	Ks
Abies spp.	0.4–0.6	*Hedera helix*	0.4–0.6
Acacia dealbata	0.1–0.2	*Hydrangea*	0.6–0.8
Acer campestre	0.4–0.6	*Hypericum* spp.	0.4–0.6
Acer platanoides	0.5–0.7	*Ilex aquifolium*	0.4–0.5
Acer rubrum	0.6–0.8	*Juglans* spp.	0.4–0.6
Aesculus × carnea	0.4–0.6	*Juniperus* spp.	0.3–0.4
Alibizzia julibrissin	0.3–0.5	*Koelreuteria paniculata*	0.3–0.5
Araucaria araucana	0.4–0.5	*Laburnum* spp.	0.4–0.6
Arbutus unedo	0.2–0.4	*Lagerstroemia indica*	0.3–0.5
Banksia integrifolia	0.4–0.5	*Lantana camara*	0.2–0.3
Bauhinia purpurea	0.4–0.6	*Larix decidua*	0.4–0.6
Berberis spp.	0.1–0.3	*Laurus nobilis*	0.2–0.4
Betula spp.	0.7–0.9	*Ligustrum* spp.	0.3–0.5
Brachychiton populneus	0.3–0.4	*Magnolia* (deciduous)	0.4–0.6
Brachychiton × hybridus	0.4–0.6	*Magnolia grandiflora*	0.5–0.6
Buxus sempervirens	0.4–0.6	*Malus* spp.	0.4–0.6
Callistemon citrinus	0.2–0.3	*Melaleuca* spp.	0.1–0.4
Calycanthus floridus	0.4–0.6	*Metasequoia glyptostroboides*	0.7–0.9
Camellia japonica	0.6–0.8	*Morus alba*	0.4–0.6
Carpinus betulus	0.4–0.6	*Musa* spp.	0.7–0.9
Casuarina cunninghamiana	0.3–0.4	*Myrica* spp.	0.2–0.4
Catalpa speciosa	0.4–0.6	*Olea europaea*	0.1–0.3
Cedrus spp.	0.3–0.5	*Parrotia persica*	0.4–0.6
Celtis australis	0.4–0.5	*Paulownia tomentosa*	0.5–0.7
Celtis occidentalis	0.3–0.5	*Picea abies*	0.4–0.6
Celtis reticulata	0.1–0.3	*Pinus pinea*	0.2–0.4
Ceratonia siliqua	0.1–0.3	*Pinus ponderosa*	0.2–0.4
Cercidiphyllum japonicum	0.4–0.6	*Pinus sylvestris*	0.4–0.6
Cercis canadensis	0.4–0.6	*Platanus × acerifolia*	0.5–0.7
Cercis siliquastrum	0.3–0.4	*Platanus orientalis*	0.4–0.6
Chimonanthus praecox	0.4–0.6	*Podocarpus macrophyllus*	0.4–0.6
Choisya ternata	0.4–0.6	*Populus* spp.	0.6–0.8
Chorisia speciosa	0.1–0.3	*Prunus* spp. (cherry, peach, plum)	0.4–0.6
Cistus spp.	0.1–0.3	*Quercus ilex*	0.3–0.4
Citrus spp.	0.4–0.6	*Quercus robur*	0.4–0.6
Cornus spp.	0.5–0.7	*Quercus rubra*	0.4–0.6
Corylus avellana	0.4–0.6	*Quercus virginiana*	0.4–0.6
Cotynus coggygria	0.1–0.3	*Rhododendron*	0.6–0.8
Cupressus arizonica	0.1–0.2	*Schinus molle*	0.1–0.3
Cupressus sempervirens	0.1–0.3	*Sophora japonica*	0.3–0.5
Diospyros kaki	0.4–0.6	*Sorbus aucuparia*	0.4–0.6
Eucalyptus globulus	0.2–0.4	*Tamarix* spp.	0.1–0.2
Eucalyptus sideroxylon	0.2–0.4	*Taxus baccata*	0.4–0.6
Euphorbia pulcherrima	0.3–0.5	*Thuja occidentalis*	0.4–0.6
Fagus sylvatica	0.6–0.8	*Tilia* spp.	0.4–0.6
Ficus spp.	0.4–0.6	*Tipuana tipu*	0.4–0.6
Forsythia × intermedia	0.3–0.5	*Ulmus* spp.	0.4–0.6
Fraxinus spp.	0.4–0.6	*Ulmus pumila*	0.2–0.4
Ginkgo biloba	0.4–0.6	*Viburnum* spp.	0.4–0.6
Grevillea spp.	0.2–0.4	*Zelkova serrata*	0.4–0.6

Source: modified from Costello et al. (2000)

Remote sensing

Remote sensing can be used to estimate the spatio-temporal distribution of Leaf Area Index and evapotranspiration in heterogeneous stands (Nouri et al., 2013). Remote sensing methods include energy balance via sensible heat flux (using models as METRIC and SABAL) and use of vegetation indexes (e.g. NDVI) (Allen et al., 2011). The Normalized Difference Vegetation Index (NDVI), in particular, has been successfully correlated to plant biomass and physiological status (Allen et al., 2011). NDVI is based on the differential absorbance/reflectance of sunlight irradiance by leaf cells than by other inorganic materials such as water and rocks. NDVI can be calculated according to Equation 28.6:

$$NDVI = (R_{NIR} - R_{RED})/(R_{NIR} + R_{RED}) \tag{28.6}$$

where R_{NIR} and R_{RED} indicate reflectance in the near infra-red and in the red spectrum. Negative values are associated with water; values around 0 indicate barren areas; slightly positive values (0.2–0.4) indicate shrub- or grass-land; finally, values around 0.4–0.8 indicate dense canopies. The time-course of NDVI over a given area is representative of growth, transpiration and health of vegetation.

Other methods to estimate irrigation needs

Both point-based (e.g. leaf gas exchange, sap flow) and plot-based (e.g. eddy covariance) techniques can be used to estimate irrigation needs. Through gas exchange techniques it is possible to evaluate species-specific photosynthesis to stomatal conductance ratio and to schedule irrigation for minimizing the marginal water use per unit carbon gain, although more species-specific studies are required. The sap flow method estimates the water flow through the plants from heat pulse movement or heat convection in the stem. Eddy covariance measures high speed fluxes of water, heat, gas in the atmospheric boundary layer from a fixed (i.e. tower) or movable (i.e. airplane) point of measurement (Nouri et al., 2013). Eddy covariance is suitable for estimating ET from mixed vegetation, although the high heterogeneity of the urban environment and anthropogenic heat release can make it very challenging.

Regardless of the method used to estimate plant water needs, irrigation depth can be selected to either fully or partially restore ET (or FC). The latter approach, known as deficit irrigation, is based on maintaining plants in mild/very mild water stress conditions to trigger acclimation to drought, and can lead to optimum irrigation efficiency if the stress level is set properly. The more severe the stress, the fewer benefits can be expected from trees, but higher water savings can be achieved. For example, restoring 50–75 percent FC at 6 days interval, or supplying 20–80 percent of ET_0, maintain acceptable development and performance for the most common landscape species (Pittenger et al., 2008). Irrigation frequency needs to be also properly scheduled: low frequencies and high depth per event may lead to runoff and deep percolation if active soil depth is low and soil is highly sealed (Porporato et al., 2001); high frequencies and low depth may result in shallow rooting (Vico et al., 2014).

Optimizing water use in cities – garden design and species selection

Garden design and species selection have been found to affect water requirement to a greater extent than garden size (Domene and Sauri, 2006). Xeric gardens with drought tolerant species consume 40–60 percent less water than mesic gardens with turf. Gerhart et al. (2006)

showed that through the selection of trees, shrubs, and ground-covers with low water use, irrigation requirements can be lowered by 44 percent during summer months. Less intuitively, with targeted planning choices, it is possible to achieve water saving without changing the identity of the green area. For example, Litvak et al. (2014) showed that planting trees to shade turfgrass and lower turf ET reduced plot-ET by 50 percent (average during the summer period) compared to stands with turf exposed to full sunlight. Despite the contribution of trees to total plot-ET, the reduction of shaded turf ET was larger than tree transpiration, thus resulting in net water saving.

Selection of the right species, based on the microclimatic conditions of the planting site, is needed for maximizing ecosystem services. Vico et al. (2014) showed that, under well irrigated conditions, a high water requiring species (100 liters day^{-1}) has higher cooling effect than a lower requiring one (60 liters day^{-1}). Conversely, in non-irrigated, highly sealed planting sites, cooling capacity decreased by 56–60 percent in the high demanding species, while it declined by only 17–30 percent in the low demanding one. The latter species was therefore more effective for cooling rainfed-only planting site. It is difficult to provide a general moisture threshold below which ecosystem services are greatly diminished or can no longer be provided, because the effects of water stress are species-specific. Woody species have evolved a broad continuum of strategies and mechanisms to withstand drought, ranging from isohydric (i.e. maintenance of leaf water potential and volumetric leaf water content during drought) to anisohydric (i.e. maintenance of leaf physiological processes at low leaf water potential) behaviors. Early stomatal closure, leaf shedding, and deep rooting are typical isohydric traits, whereas osmotic adjustment and changes in cell wall elasticity are key anisohydric traits (see Chapter 18 for a more complete description). The mechanism adopted should not be confused with the degree of drought-resistance (e.g. species adopting anisohydric mechanisms can be either very drought resistant, as *Olea europaea*, or sensitive, as *Liquidambar styraciflua*). This is clear, for example, in Mediterranean coastal area, where species with opposite mechanisms, such as the anisohydric *Phillirea latifolia* and the isohydric *Cistus incanus,* coexist. In artificial environments, such as cities, where no plant species is really native, selecting species with mechanisms which better fit the characteristics of the planting site can result in greater benefits provided per unit of water used. For example, anisohydric species may provide larger benefits than isohydric ones in sites with short but frequent drought spells, or under deficit-irrigation conditions, because they are better able to maintain gas exchange and growth (Moreno-Gutierrez et al., 2012). Conversely, in harsher or more prolonged dry conditions, isohydric traits such as drought-deciduousness may allow greater survival rates, but often at expense of growth potential and leaf gas exchange (Moreno-Gutierrez et al., 2012). Similarly, isohydric species which rely on deep rooting, as *Ceratonia siliqua*, may provide large benefits when planted in sites with large active soil depth, whereas in shallow soils the capacity to adjust osmotically (an anisohydric trait) may be advantageous because it allows higher capacity to extract water from soils and maintain leaf turgor at decreasing leaf water potential.

Unfortunately, to date, information about the mechanisms adopted by urban trees to survive drought is limited to only a few landscape species. An early work ranked five different genera on an anisohydric to isohydric scale (Whitlow et al., 1992). *Fraxinus* was the most anisohydric genus, followed by *Ulmus, Acer, Cornus; Tilia* was the most isohydric one. More recently, the capacity to withstand drought and tolerance mechanisms have been investigated for temperate (Niinemets and Valladares, 2006), Mediterranean (Flexas et al., 2014), and tropical (Engelbrecht et al., 2003) species. Species within a genus may also largely differ in their response to drought. Comparing *Acer* genotypes, Sjöman et al. (2015) found a greater capacity to adjust osmotically and, by consequence, a greater capacity to survive drought in

A. grandidentatum and *A. monspessulanum* compared to other *Acer* species. For this reason, these species may be successfully planted in shallow soils with low water holding capacity, where less drought-tolerant species (*A. spicatum, A. mandshuricum, A. truncatum*) would struggle. Similarly, Percival et al. (2006) identified through chlorophyll fluorescence techniques the effects of drought on some *Fraxinus* species. Authors found a superior capacity to tolerate drought in European species and *F. nigra* compared to other American species. Fini et al. (2009) investigated the capacity to withstand drought in four *Tilia* species. Authors found *T. platyphyllos* to be more sensitive to drought than *T. cordata, T.* × *europaea* and, in particular, *T. tomentosa*. Also, while the latter species showed an effective isohydric strategy based on early stomatal closure and dense trichome cover, *T.* × *europaea* and *T. cordata* displayed a water spending behavior, which requires larger soil volumes. Oaks have been extensively investigated for drought tolerance, and high- (e.g. *Q. rubra, Q. palustris, Q. shumardii, Q. robur*) and low- (*Q. prinus, Q. velutina, Q. pubescens*) water-use species have been identified for American (Abrams, 1990) and European (Struve et al., 2009) oak species.

More opportunities for water-wise selection of planting material come from significant differences in drought tolerance between cultivars and ecotypes of the same species: *Acer* × *freemanii* 'Autumn Fantasy' was found to withstand drought better than 'Celebration' and 'Marmo' because of more favorable leaf area to sapwood ratio and leaf mass per area (Zwack et al., 1998). Similarly, *Acer platanoides* 'Deborah' and 'Emerald Queen' were more tolerant to drought than 'Summershade' (Fini et al., 2009). In another work, *Eucalyptus globosus* clone CN5 was more tolerant than clone ST51 because of maintenance of more favorable water relations during drought and higher carbon allocation to the root system (Costa e Silva et al., 2004).

Intra-specific differences in the capacity to withstand drought may also exist between ecotypes/accessions/provenances from different climatic conditions. Provenances of *Pinus canariensis* with greater capacity to adjust osmotically showed higher survival rates than other sources during water deficit (Lopez et al., 2009). Similarly, *Jatropha curcas* ecotypes from tropical monsoon climates were less drought tolerant than ecotypes from tropical savanna climate (Fini et al., 2013). In the light of climate change, it has been suggested that species/ecotypes which, in the planting site, are near the northern limit of their distribution range should be planted (Ghannoum and Way, 2011). This may allow a better acclimation of physiological processes to the predicted warming (Ghannoum and Way, 2011).

Optimizing irrigation efficiency – reuse of wastewater

Reuse of wastewater for irrigation is an important conservation strategy to save freshwater and improve nutrient cycling. Reusing wastewater may rise, however, concerns in authorities administering public health and may lead to long-term soil degradation because of changes in soil's physical and chemical traits and build-up of salinity and heavy metals (Morugan-Coronado et al., 2011). As a result, the water quality legislation controlling the use of reclaimed water can be quite strict in some countries (e.g. in Italy, a maximum of 0.1 colony forming unit ml^{-1} *Escherichia coli* is allowed for direct wastewater use) (Styczen et al. 2010). On the contrary, if wastewater is reused safely, several benefits may occur: a substantial increase in irrigated area without a corresponding increase in freshwater use; prevention of any potential pollution of aquatic and terrestrial ecosystems; restoration of marginal land and conversion to functional green areas; improvement of nutrient cycling within the urban ecosystem because of the richness in organic matter and/or mineral nutrients of treated wastewater.

Main qualitative characteristics to be monitored in wastewater include: content of bacteria, protozoa, and helminth; salinity; heavy metals and other pollutants (in particular, the presence of

Figure 28.3 Species of the same genus can differ in tolerance to abiotic stress, so that through experienced species selection, stress tolerance can be improved without affecting visual and aesthetic requirements. In the picture: (a) *Tilia cordata*, (b) *T. tomentosa* and (c) *T.* × *europaea*, which proved to be more drought tolerant than (d) *T. platyphyllos*

Source: Fini et al. (2009)

As, Zn, Mn, Ag, Be, Cr, Cu, Li, and Se in high concentrations prevents the reuse of wastewater for a large number of uses); concentration of organic pollutants (e.g. benzoyl, oil, tar, aldeids, phenolics). Wastewater should be considered of low quality when it has high concentrations of salts (EC > 2 dS m^{-1}, although salt damage may occur at EC > 0.75 dS m^{-1} in sensitive species) and solids in suspension (> 300 mg l^{-1}), and a high biological (BOD > 300 mg O$_2$ l^{-1}) and chemical (COD > 900 mg O$_2$ l^{-1}) oxygen demand (Morugan-Coronado et al., 2011). Upon irrigation with saline wastewater, salinity of the soil solution can rise 7-fold the salinity in irrigation water (up to 15,000 mg l^{-1} total dissolved solids in soil solution compared to 2340 mg l^{-1} in wastewater, see Gerhart et al., 2006). Soil salinity increased as a function of increasing levels of salt in irrigation supply, but soil salinity did not increase over time, indicating that an equilibrium between application rates and leaching can occur, particularly in well drained soils (Gerhart et al., 2006). Also, heavy metal such as cadmium accumulated in soil and leaves of poplars irrigated with wastewater. To reduce salt and pollution load, prior to use, wastewater is subjected to primary (phase separation), secondary (biochemical oxidation), and tertiary (microfiltration) treatments. Tertiary treatment (using a sandy filter) successfully reduced EC, BOD, COD, suspended solids, and cations compared to wastewater only subjected to primary and secondary treatments (Morugan-Coronado et al., 2011). If wastewater salinity is still higher than 2–3 dS m^{-1} after treatment, blending with freshwater (1:1) can successfully reduce local salinity hazard (Gerhart et al., 2006). In any case, when using wastewater for landscape irrigation, salinity and dissolved solids should be periodically monitored in irrigation water and in the soil solution, to avoid undesired effects on the urban ecosystem. Scientific and technical studies are required to maximize this contribution through developing water recycling opportunities and reuses particularly for green space irrigation to provide environmentally, socially and economically sustainable environments.

References

Abrams, M. D., 1990. Adaptation and responses to drought in *Quercus* species of North America. *Tree Physiology*, 7: 227–238.

Allen, R. G., Pereira, L. S., Howell, T. A., Jensen, M. E., 2011. Evapotranspiration information reporting: I. Factors governing measurement accuracy. *Agricultural Water Management*, 98: 899–920.

Balling, R. C. Jr., Gober, P., 2007. Climate variability and residential water use in the city of Phoenix, Arizona. *Journal of Applied Meteorology and Climatology*, 46: 1130–1137.

Boegh, E., Poulsen, R. N., Butts, M., Abrahamsen, P., Dellwik, E., Hansen, S., Hasager, C. B., Ibrom, A., Loerup, J.-K., Pilegaard, K., Soegaard, H., 2009. Remote sensing based evapotranspiration and runoff modeling of agricultural, forest and urban fluxes in Denmark: from field to macro-scale. *Journal of Hydrology*, 377: 300–316.

Costa e Silva, F., Shvaleva, A., Maroco, J. P., Almeida, M. H., Chaves, M. M., Pereira, J. S., 2004. Responses to water stress in two *Eucalyptus globulus* clones differing in drought tolerance. *Tree Physiology*, 24: 1165–1172.

Costello, L. R., Matheny, N. P., Clark, J. R., 2000. *A guide to estimating irrigation water needs of landscape plantings in California: the landscape coefficient method and WUCOLS III*. Sacramento, CA: University of California Cooperative Extension. Retrieved from http://water.ca.gov/wateruseefficiency/docs/wucols00.pdf.

Domene, E., Saurí, D., 2006. Urban and water consumption: influencing factors in the metropolitan region of Barcelona. *Urban Studies*, 43: 1605–1623.

Engelbrecht, B. M. J., Kursar, T. A., 2003. Comparative drought-resistance of seedlings of 28 species of co-occurring tropical woody plants. *Oecologia*, 136: 383–393.

Fini, A., Ferrini, F., Frangi, P., Amoroso, G., Piatti, R., 2009. Withholding irrigation during the establishment phase affected growth and physiology of Norway maple (*Acer platanoides*) and linden (*Tilia* spp.). *Arboriculture and Urban Forestry*, 35: 241–251.

Fini, A., Bellasio, C., Pollastri, S., Tattini, M., Ferrini, F., 2013. Water relations, growth, and leaf gas exchange as affected by water stress in *Jatropha curcas*. *Journal of Arid Environments*, 89: 21–29.

Fini, A., Frangi, P., Amoroso, G., Piatti, R., Faoro, M., Bellasio, C., Ferrini, F., 2011. Effect of controlled inoculation with specific mycorrhizal fungi from the urban environment on growth and physiology of containerized shade tree species growing under different water regimes. *Mycorrhiza*, 21: 703–719.

Flexas, J., Diaz-Espejo, A., Gago, J., Gallé, A., Galmés, J., Gulias, J., Medrano, H., 2014. Photosynthetic limitations in Mediterranean plants: a review. *Environmental and Experimental Botany*, 103: 12–23.

Gerhart, V. J., Kane, R., Glenn, E. P., 2006. Recycling industrial saline wastewater for landscape irrigation in desert areas. *Journal of Arid Environments*, 67: 473–486.

Ghannoum, O., Way, D. A., 2011. On the role of ecological adaptation and geographic distribution in the response of trees to climate change. *Tree Physiology*, 31: 1273–1276.

Haley, M. B., Dukes, M. D., Miller, G. L., 2007. Residential irrigation water use in central Florida. *Journal of Irrigation and Drainage Engineering*, 133: 427–434.

Hof, A., Wolf, N., 2014. Estimating potential outdoor water consumption in private urban landscapes by coupling high resolution image analysis, irrigation water needs and evaporation estimation in Spain. *Landscape and Urban Planning*, 123: 61–72.

Litvak, E., Bijoor, N. S., Pataki, D. E., 2014. Adding trees to irrigated turfgrass lawns may be a water-saving measure in semi-arid environments. *Ecohydrology*, 7: 1314–1330.

Lopez, R., Rodriguez-Calcerrada, J., Gil, L., 2009. Physiological and morphological response to water deficit in seedlings of five provenances of *Pinus canariensis*: potential to detect variation in drought-tolerance. *Trees*, 23: 509–519.

Machado e Silva, M. D. F., Calijuri M. L., Ferreira de Sales, F. J., Batalha de Souza, M. H., Lopes, L. S., 2014. Integration of technologies and alternative sources of water and energy to promote the sustainability of urban landscapes. *Resources, Conservation and Recycling*, 91: 71–81.

McCready, M. S., Dukes, M. D., 2011. Landscape irrigation scheduling efficiency and adequacy by various control technologies. *Agricultural Water Management*, 98: 697–704.

Moreno-Gutierrez, C., Dawson, T. E., Nicolas, E., Querejeta, J. I., 2012. Isotopes reveal contrasting water use strategies among coexisting plant species in a Mediterranean ecosystem. *New Phytologist*, 196: 489–496.

Morugan-Coronado, A., Garcia-Orenes, F., Mataix-Solera, J., Arcenegui, V., Mataix-Beneyto, J., 2011. Short-term effects of treated wastewater irrigation on Mediterranean calcareous soil. *Soil and Tillage Research*, 112: 18–26.

Niinemets, U., Valladares, F., 2006. Tolerance to shade, drought, and waterlogging of temperate northern hemisphere trees and shrubs. *Ecological Monographs*, 76: 521–547.

Nouri, H., Beecham, S., Kazemi, F., Hassanli, A. M., 2013. A review of ET measurement techniques for estimating the water requirements of urban landscape vegetation. *Urban Water Journal*, 10: 247–259.

Oki, T., Kanae, S., 2006. Global hydrological cycles and world water resources. *Science*, 313: 1068–1072.

Pataki, D. E., McCarthy, H. R., Litvak, E., Pincetl, S., 2011. Transpiration of urban forest in the Los Angeles metropolitan area. *Ecological Applications*, 21: 661–677.

Percival, G. C., Keary, I. P., Al-Habsi, S., 2006. An assessment of the drought tolerance of *Fraxinus* genotypes for urban landscape planting. *Urban Forestry and Urban Greening*, 5: 17–27.

Pittenger, D., Henry, M., Shaw, D., 2008. Water needs of landscape plants. Paper presented at UCR Turfgrass and Landscape Research Field Day, University of California, Riverside, CA.

Porporato, A., Laio, F., Ridolfi, L., Rodriguez-Iturbe, I., 2001. Plants in water-controlled ecosystems: active role in hydrologic processes and response to water stress. III. *Vegetation under stress. Advances in Water Resources*, 24: 725–744.

Salvador, R., Bautista-Capetillo, C., Playán, E., 2011. Irrigation performance in private urban landscapes: a case study in Zaragoza (Spain). *Landscape and Urban Planning*, 100: 302–311.

Savi, T., Bertuzzi, S., Branca, S., Tretiach, M., Nardini, A., 2015. Drought-induced xylem cavitation and hydraulic deterioration: risk factors for urban trees under climate change? *New Phytologist*, 205: 1106–1116.

Sjöman, H., Hirons, A. D., Bassuk, N. L., 2015. Urban forest resilience through tree selection – variation in drought tolerance in Acer. *Urban Forestry and Urban Greening*, 14: 858–865.

Snyder, R. L., Pedras, C., Montazar, A., Henry, J. M., Ackey, D., 2015. Advances in ET-based landscape irrigation management. *Agricultural Water Management*, 147: 187–197.

Struve, D. K., Ferrini, F., Fini, A., Pennati, L., 2009. Relative growth and water use of seedlings from three Italian *Quercus* species. *Arboriculture and Urban Forestry*, 35: 113–121.

Styczen, M., Poulsen, R. N., Falk, A. K., Jorgensen, G. H., 2010. Management model for decision support when applying low quality water in irrigation. *Agricultural Water Management*, 98: 472–481.

Todd, R. W., Evett, S. R., Howell, T. A., 2000. The Bowen ratio-energy balance method for estimating latent heat flux of irrigated alfalfa evaluated in semi-arid, advective environment. *Agricultural and Forest Meteorology*, 103: 335–348.

Vico, G., Revelli, R., Porporato, A., 2014. Ecohydrology of street trees: design and irrigation requirements for sustainable water use. *Ecohydrology*, 7: 508–523.

Whitlow, T. H., Bassuk, N. L., Reichert, D. L., 1992. A 3-year study of water relations of urban street trees. *Journal of Applied Ecology*, 29: 436–450.

Yang, J., Wang, Z.-H., 2015. Optimizing urban irrigation schemes for the trade-off between energy and water consumption. *Energy and Buildings*, 107: 335–344.

Zwack, J. A., Graves, W. R., Townsend, A. M., 1998. Leaf water relations and plant development of three Freeman maple cultivars subjected to drought. *Journal of the American Society for Horticultural Science*, 123: 371–375.

29

FERTILIZATION IN URBAN LANDSCAPE

Cecilia Brunetti and Alessio Fini

Nutrients in urban soils

Urban soils show high variability regarding the content and availability of different nutrients. Nitrogen is usually lower in urban soil than in agricultural soil, because soil sealing in urban areas can have negative consequences for soil fertility and long-term storage of nitrogen. Potassium is rarely deficient, whereas iron and phosphorus can be little available to plants because of insolubilization processes due to higher pH levels resulting from addition of debris (e.g. cement, plaster, etc.) in urban soils. Although organic and inorganic forms of phosphorus (P) are abundant in soils, plant phosphorus availability is restricted as it mostly occurs in insoluble forms. Furthermore, phosphate-solubilizing microorganisms (PSM) (bacterial and fungal strains) are needed for gradual P-release, but PSM are scarcely present in urban soils, so that only 0.1 per cent of the total P may be available to plants.

Nutrient use efficiency in urban environment

The nutrient use efficiency (NUE) is the amount of nutrients absorbed by plants from the soil within a certain period of time compared with the amount of nutrients supplied over the same period of time. In urban environment NUE is a crucial marker for plant growth. NUE depends on both soil properties to supply adequate levels of nutrients (the rate of organic matter decomposition and mineralization, erosion processes and soil transformation, the delivery of dissolved nutrients from the bulk soil to the root via mass flow and diffusion) and the rate of root uptake (Chapman et al., 2012) (Figure 29.1).

Urban soils have been produced in the process of urbanization and have been defined as 'a soil material having a non-agricultural, man-made surface layer more than 50 cm thick, that has been produced by mixing, filling, or by contamination of land surfaces of urban and suburban areas' (Craul, 1992) (for more information see Chapter 21). In contrast to natural soils, urban soils are particularly difficult to categorize, since their anthropogenic origin results in a huge spatial variability of soil traits.

Urban soils usually contain less than 1 per cent of organic matter and the continuous removal of leaf litter and other organic debris can further reduce inputs to the soil organic matter pool (Figure 29.1). In addition, urban litter is characterized by a high content of

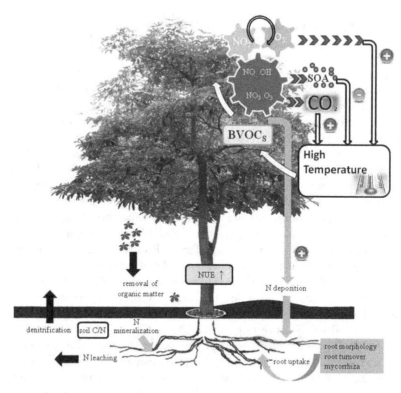

Figure 29.1 Impact of urban environment on nitrogen use efficiency and volatile organic compounds emission

In the past less attention has been given to BVOC's effects on climate assuming that the short lifetime of BVOCs would preclude their influence on climate. However, there are emerging evidences on the relationship between climate and BVOCs emission at different spatial scales (local, regional and global). BVOCs emission could be one of the most important factors inducing the formation of tropospheric ozone, which is both a key pollutant in photochemical smog events and a greenhouse gas. The formation of ozone through BVOCs depends on the presence of NOx (nitrogen oxides), which essentially derive from the combustion processes in polluted urban areas. The oxidizing effects of BVOCs on atmospheric chemistry can be multiple and can have different effects on temperature and deposition of airborne nitrogen, influencing in turn plant NUE. In the figure, grey arrows indicate processes which enhance nutrient availability for plants, whereas black arrows indicate processes with negative impact on NUE.

carbon from recalcitrant organic matter pool (passive soil carbon) and a low content of carbon with a fast turnover time (labile carbon). The low level of labile carbon, together with some urban pollutants, such as ozone and heavy metals, may in turn depress microinvertebrates biodiversity and activity, thus impairing biogeochemical cycling in the soil profile (Lorenz and Lal, 2009). In particular, N cycling is usually altered in urban and suburban areas as a consequence of several environmental factors. First, the atmospheric deposition of inorganic N (mainly NO_x) derived from fossil fuel is higher in the urban sites as compared to the rural sites. Second, the higher incidence of exotic species of decomposers (e.g. earthworms) combined with the warmer climate of the cities ('urban heat island effect') may increase the rates of N mineralization. In contrast to these two factors which increase N availability, other co-occurring processes such as leaching and denitrification deprive soil N pool, either moving N to the water table (leaching) or to the atmosphere as molecular nitrogen (N_2, denitrification), increasing the loss of nitrogen as N_2. Denitrification is triggered in waterlogged soils, a common situation in the urban stand because of soil compaction. As a consequence, the great ecological heterogeneity of urban soils enhances variability in nitrogen content and availability, making difficult the standardization of fertilization management in urban soils (Figure 29.1).

In the urban environment, soil transformation occurs faster than in undisturbed sites. Indeed the natural weathering processes play a small role in soil erosion, whereas excavations for new buildings, surface removal, accumulation of rubble and vehicular traffic accelerate the erosion processes (De Kimpe and Morel, 2000; Lorenz and Lal, 2009). In addition, because of the paved surfaces or crusts on compacted bare soils, runoff occurs more rapidly, reducing infiltration of water available for urban vegetation but increasing N leaching (Figure 29.1) as well as the release of pollutants and heavy metals from debris (Sieghardt et al., 2005). Once released in the soil, interception, mass flow and diffusion are responsible for the fluxes of dissolved nutrients towards the root surface (Chapman et al., 2012). In urban compacted soils, low porosity results in little diffusion of dissolved nutrients to the roots (Sieghardt et al., 2005). In addition, since the physical resistance of highly compacted soils impedes root penetration, tree roots tend to follow paths of low resistance, becoming shallow and sparsely branched, which decreases nutrient uptake by interception.

Mycorrhizal networks play important roles in mobilization and uptake of nutrients, mainly P, N and Zn, because of a greater degree of soil exploitation than that of fine roots alone and an enzymatic depolymerization of organic matter containing nitrogen (Karandashov and Bucher, 2005). Mycorrhizal symbioses are nearly ubiquitous in plants, nevertheless plants growing in urban environments have significantly lower mycorrhizal colonization compared to forested sited because anthropogenic activities strongly depress mycorrhizal fungi (Bainard et al., 2011) (Figure 29.1). The low mycorrhizal infection of plant roots in urban soils induces changes in rhizosphere functioning and reduces plant tolerance to some environmental stresses such as drought (Fini et al., 2011).

Global changes may affect nutrient uptake in urban environment

The kinetics of nutrient uptake by the roots are deeply influenced by the effects of global change, particularly rising CO_2 concentrations and increased temperatures (Bassirirad, 2000). At present, there is very limited information about how plant nutrition will be affected by the concomitant change of multiple environmental parameters induced by climate change.

Current data indicate that elevated CO_2 stimulates plant growth (if other factors, i.e. water, are not limiting) by increasing plant nutrient demand, at least in short term. It has been suggested that elevated CO_2 concentrations could increase root growth and exudation (Lynch and St Clair, 2004; Lukac et al., 2010) as well as stimulate symbiotic association facilitating nutrient uptake of macro- and micro-nutrients from the soil. However, Bassirirad (2000) underlined variability in root growth responses to elevated CO_2 because of a high degree of species-specific differences in root physiological characteristics. In addition, most data available in literature are based exclusively on field experiments or experiments conducted in forest ecosystems; studies on root nutrient uptake responses to high CO_2 in urban environments are still lacking.

Concomitantly to a rising atmospheric CO_2 concentration, it is estimated that the temperature of the globe might increase up to 4°C during this century. Warming is likely to affect nutrient availability and uptake by trees in different ways. The most notable are the direct effects of soil temperature on biogeochemical processes that regulate N and water availability in the soil. A number of studies have demonstrated that increases in soil temperature can directly improve foliar nutrient status by altering root transport properties for NH_4^+, NO_3^-, PO_4^{3-} and K^+. Nonetheless, the exact mechanism is still not well understood. First, higher root respiration and activity triggered by soil warming may partly explain higher uptake and transport. Second, higher soil temperatures stimulate root N uptake as NO_3^-, which is the form of atmospheric N deposition over NH_4^+ (Bassirirad, 2000). Last, soil warming can increase soil N mineralization, with beneficial effects on plant growth but also with higher possibility of nitrate leaching (Lukac et al., 2010).

It is not known whether elevated CO_2 and higher soil temperature will have an interactive effect on root nutrient uptake kinetics. The large number of possible interactions between climate change variables, edaphic constraints and differential sensitivity of species to changes in soil temperature and CO_2 make generalizations difficult.

Diagnosis of nutritional deficiency using soil and foliar analysis

Soil analysis

Before planning a fertilization schedule, soil should be analysed to assess nutrient availability or deficiencies. Main parameters to be assessed are: pH, organic matter, nitrogen, phosphorus, major cations (K^+, Ca^{2+}, Mg^{2+}) and micronutrients (B, Cl, Cu, Fe, Mn, Zn). The first step in an effective soil testing program is the collection of a sample that is representative of the area of interest. In most cases several samples are collected from the area of interest and then merged to produce one sample that is submitted to the laboratory for analysis.

Soil pH is a crucial indicator of nutrients availability in the soil. Indeed, it affects the availability of essential elements and micronutrients: in acid soils (pH < 5) Ca and Mg become unavailable; in alkaline soils (pH > 5) the availability of P, K, Fe, Mn, Cu decrease, while the solubility of toxic elements (e.g. Al) and the biological activities of microbial populations increase (Barker and Pilbeam, 2007). Soil pH is usually measured potentiometrically in a soil water/slurry following standardized protocols as reported in previous works (Kopinga and van den Burg, 1995; Jones, 2001), but it can be easily assessed in situ using a portable pH-meter.

Soil organic matter consists of two forms:

1 plant material and soil microorganism residues in various stages of decomposition;
2 humus, a very stable complex of large molecules (mainly humic acid and fulvic acid).

Plant material and microbial residues are the source of essential plant nutrients whereas humus, because of its high cation exchange capacity, impacts physical and physiochemical properties of the soil increasing a friable soil condition. Though occurring in relatively small amounts in urban soils, organic matter plays a very important role for soil structure, porosity, nutrient reserve and CEC (cation-exchange capacity), soil biological activity and water holding capacity. The commonly used organic matter determination procedures, wet oxidation (Jones, 2001) and loss-on-ignition (Kopinga and van den Burg, 1995) do not distinguish between these two forms of organic material in the soil. A more recent tool for determination of organic matter content in soils is the diffuse reflectance spectroscopy in the visible-near infrared region (Vis-NIR); this technique is faster than traditional methods because it may be used routinely to predict soil organic matter of surface and subsurface soils without extraction procedures (Fidencio et al., 2002) (Table 29.1).

Nitrogen is derived almost exclusively from the decomposition of organic matter (usually >95%), partially from atmospheric deposition, and in some trees, through symbiotic relationships with bacteria capable of acquiring N directly from the atmosphere (N fixation). The two main inorganic forms of N, NH_4^+ and NO_3^-, derived from microbial transformation of organic matter and are routinely measured in soil laboratories, as they reflect the extent of mineralization, and the forms of N taken up by plants. Furthermore, it should be considered that the excess of NO_3^- is lost because of leaching, whereas NH_4^+ remains in the soil exchange complex. The standard procedures to quantify total inorganic N or NO_3^- levels in soils have been widely reported (Jones, 2001). The accurate determination of the NO_3^- concentration in the upper 70 cm of the soil profile is critical to regulate the addition of N through fertilizers (Table 29.1). The goal is to supply plants with an adequate amount of N avoiding excessive applications that can leach into water aquifers and run off onto the surface as NO_3^-.

Plants utilize phosphorus as orthophosphates ($H_2PO_4^-$ and HPO_4^{2-}) depending on the pH of the soil. Different pools of phosphorus exist in the soil, mainly divided into inorganic phosphate and phosphorus contained in soil organic matter. Ortophospates may become available through a series of mechanisms, most of which depend on the activity of soil microrganisms: solubilization of inorganic form of P, excretion of phosphatase enzymes, mineralization of P organic matter from both the compounds containing phosphate ester bonds and the more stable organophosphonates. The net phosphorus release depends on the phosphorus concentration of the decaying residues and the activity of specific soil microbial population. There are several soil testing procedures commonly used for determining soil P (Bray P1 and P2, Olsen, Morgan, Mehlich 1 and 3), often designed for specific soil situations. The interpretation ranges for these various test methods vary considerably, and frequently the levels of extractable P do not correlate well among P soil test methods unless the soil characteristics (e.g. pH) are similar (Table 29.1). A detailed description of these protocols is reported in Jones (2001) and Pierzynski (2000). In addition, advanced spectroscopic techniques such as ^{31}P-NMR and P K-edge X-ray absorption near-edge structure spectroscopy (P-XANES) have been recently used for a fast analysis of P in organic matter. NMR and P-XANES assays allow the quantification (expressed as proportions of total P) of the different organic phosphorus species (phosphonates, phosphate monoesters and phosphate diesters) (Liu et al., 2013).

Availability of the major cations (K^+, Ca^{2+}, Mg^{2+}) in soil is determined by their water soluble forms and their exchange from the colloidal complex. The amounts of K, Ca, and Mg in the soil solution are quite small irrespective to the amounts in the exchangeable form. Hence, the quantities of these three cations extracted in most soil test procedures are simply referred to as exchangeable K, Ca, and Mg (Table 29.1). In the literature there are plenty of methods used to measure these exchangeable cations (Jones, 2001). Briefly, these methods

Table 29.1 Optimal values of the main parameters evaluated in a soil analysis

Parameter	Range	Units
Organic matter	≥ 20	g/Kg dw
Nitrogen (total inorganic N)	≥ 1	g/Kg dw
Phosphorous ($H_2PO_4^-$ and HPO_4^{2-})	10–20	mg/Kg dw
Extractable potassium (K^+)	0.2–0.4	meq/100 g dw
Extractable calcium (Ca^{2+})	5–10	meq/100 g dw
Extractable magnesium (Mg^{2+})	0.8–1.6	meq/100 g dw

use an exchange cation (e.g. NH_4^+, H^+, Na^+) in the extraction reagent in order to remove the cations on the colloidal complex. To quantitatively assess the cations, the extracted samples are generally analysed by inductively coupled plasma-atomic emission spectroscopy (ICP-AES).

The purpose of soil analysis is to chemically extract the amount of nutrient from the soil that should be proportional to that which will be available to the plant during the growing season.

Unfortunately, as mentioned above, soil nutrient analysis may not reflect the effective availability to plants, and nutrient deficiencies may be unreliably predicted from soil analysis for a few reasons. First, the availability of some elements also depends on the physical conditions of the soil (water-logging, soil hypoxia, soil compaction). Second, the uptake of some elements, especially N, depends on the size and activity of the root system. Third, spatial heterogeneity of urban soil introduces additional problems to sampling. In consequence, it is recommended to implement soil analysis with foliar analysis.

Foliar analysis

Nutritional status of plants can be assessed by foliar analysis, that should involve a careful observation of plant growth, an accurate sampling procedure and chemical analysis on leaf samples since some symptoms of nutrient deficiency are similar to damage due to pollution or pathogens. An early leaf analysis allows to detect symptomless detrimental conditions or to confirm the nature of visible toxicity preventing severe metabolic disruptions. Due to the different mobility of nutrients within the plant, leaf analysis must be carefully targeted on old versus young leaves, depending on the nutrient suspected to be deficient. For nutrients which are mobile and immobile within the plant, see Table 29.2.

The symptom of nitrogen deficiency is mainly a pale green or yellow colour of foliage combined with a restricted growth of the vegetative organs (Table 29.2). If the deficiency develops during the growth cycle, the nitrogen will be mobilized from older/basal leaves to young/apical leaves causing the former leaves to become chlorotic or even necrotized in the case of severe deficiency. Nitrogen in plant tissue can be determined by two destructive analytical procedures: Kjeldahl digestion and the Dumas technique (Benton et al., 1991). However, non-destructive and less time-consuming methods can also be used to measure plant N status. At leaf level, optical properties of pigments have been used as plant N indicators. Indeed chlorophyll content, measured through SPAD chlorophyll meter, dualex and chlorophyll a fluorescence, shows a good correlation with plant N status (Muñoz-Huerta et al., 2013).

Phosphorus deficiency reduces root growth and blooming. A typical symptom of phosphorus deficiency are small, dark-green leaves in stunted plants (Table 29.2). Because phosphorus is mobile in plants, it is transported readily from old to young leaves as deficiency occurs, and symptoms and necrosis are often observed on old leaves (Barker and Pilbeam, 2007). However, P-deficiency symptoms can be easily confused with other nutrient deficiencies and an overuse of P should be avoided because it poses risk of eutrophication of water bodies. Phosphorus in leaves can be evaluated by the Bray-II colorimeter method (Bray and Kurtz, 1945). In addition, Frydenvang et al. (2015) recently described a non-destructive method to quantitatively predict the foliar P levels using chlorophyll a fluorescence. This method allows the assessment of the bioactive P concentration by using multivariate analysis of OJIP transients.

The first symptom of K deficiency is growth retardation, which is a rather nonspecific symptom and consequently it makes it hard to early detect K deficiency (Table 29.2).

Table 29.2 Essential nutrients in plants: functions and main visual symptoms of deficiency

Nutrient element	Symbol	Functions in plants	Sufficiency range	Deficiency symptoms	Available form from soil
Nitrogen	N	Promotes shoot growth, chlorophyll formation and protein synthesis.	1–6%	Slow and stunted growth, undersized leaves, short new twig, premature fall color and leaf drop, chlorosis in lower leaves, firing of lower leaves, blossom-end rot.	NO_3^- NH_4^+
Phosphorus	P	Stimulates early root growth, blooming and seed formation.	0.2–0.5%	Intense green coloration or reddening in older leaves, leaf tips look burnt.	$H_2PO_4^-$ HPO_4^-
Potassium	K	Increases resistance to drought and disease.	1–4%	Intervenial chlorosis of older leaves, which look scorched near tips and margins.	K^+
Calcium	Ca	Improves root formation. Increases resistance to seedling diseases. Role in nitrogen absorption.	0.5–1.5%	Stunting of root systems. Blossom-end rot. Tips of new leaves are distorted or irregularly shaped.	Ca^{++}
Magnesium	Mg	Promotes chlorophyll formation and regulates uptake of other nutrients.	0.2–0.4%	Older leaves turn yellow at the edge leaving a green arrowhead shape in the center of the leaf. Leaf margins may curl.	Mg^{+++}
Sulfur	S	Structural component of some vitamins and amino acids.	0.1–0.5%	Younger leaves turn yellow with veins lighter than surrounding areas.	SO_4^-
Zinc	Zn	It is required in a large number of enzymes, affects endogenous levels of auxin and gibberellins.	5–20 ppm	Terminal leaves may be rosette, and yellowing occurs between veins of the new leaves. Reduction of internodes and a decrease in leaf size.	Zn^{++}
Iron	Fe	It is involved in the production of chlorophyll and in the functioning of many enzymes.	50–10 ppm	New upper leaves turn yellow between veins. Edges and tips of leaves may die.	Fe^{++} Fe^{+++}

Manganese	Mn	It is involved in nitrogen metabolism. It is a direct component of chlorophyll reaction centers.	10–200 ppm	Interveinal chlorosis. New upper leaves have dead spots over surface.	Mn^{++}
Molybdenum	Mo	Aids nitrogen fixation and assimilation.	0.1–0.3 ppm	Irregular leaf blade formation. Mottling of lamina in midleaves of the plant.	MoO_4^-
Copper	Cu	It is essential in several plant enzymes involved in photosynthesis.	3–7 ppm	Plant is stunted. Distortion of the younger leaves, and possible necrosis of the apical meristem.	Cu^{++}
Boron	B	Aids carbohydrate transport and cell division.	~20 ppm	Tips of the shoot dies. Stems and petioles are brittle.	H_3BO_3 $H_2BO_3^-$ HBO_3^- BO_3^- $B_4O_7^-$
Chlorine	Cl	It is essential in photosynthesis.	50–200 ppm	Chlorosis of younger leaves and wilting of the plant. It occurs seldom because Cl is found in atmosphere and rainwater.	Cl^-
Cobalt	Co	It is essential for nitrogen fixation and phosphorus uptake.	1–2.5 ppm	Reduced seed germination in dry conditions and reduced plant growth.	Co^{++}
Nickel	Ni	Grain filling, seed viability.	0.1–0.5 ppm	Reduced growth and yield.	Ni^{++} Ni^{+++}

Plants require 15 essential elements (besides carbon, hydrogen and oxygen which are supplied by the atmosphere and water) for normal growth and to complete their life cycle. These nutrients are acquired from the soil by roots and they are commonly classified as macronutrients (N, P, K, Ca, Mg, S) and micronutrients (Zn, Fe, Mn, Mo, Cu, B, Cl, Co, Ni). Macronutrients usually constitute more than 0.1 per cent dry mass, indeed they are needed in larger amounts than micronutrients. Micronutrients accumulate only in small amounts ($\leq$0.05% of dry mass) within plant tissues and are expressed in parts per million (ppm). Once incorporated by the plant, nutrients are transported to the site of their function, typically to growing tissues, but transport rate varies depending on specific nutrients. Some elements remain immobile in their original location (immobile elements), while others can be remobilized and moved to tissues with greater demand (mobile elements). Mobile nutrients are nitrogen (as NO_3^-), phosphorus (as $H_2PO_4^-$ and HPO_4^-), potassium (K), magnesium (Mg), chlorine (Cl), zinc (Zn) and molybdene (Mo). Calcium (Ca), sulphur (S), iron (Fe), boron (B) and copper (Cu) are immobile.

As K deficiency becomes more severe, K is transported from older to younger leaves and in most plant species chlorosis and necrosis appear at leaf tips and leaf margins of older K-deficient leaves. K can be determined on dry ashes (see organic matter destruction methods in Benton et al., 1991) or after extraction by acid solutions (dilute HCl or 2% CH_3COOH) using flame emission spectrophotometry or ICP-AES. More recently, Menesatti et al. (2010) developed an alternative non-destructive and cost-effective technique to predict leaves' nutritional status using a Vis-NIR portable spectrophotometer. The results obtained with this method were compared with standard chemical analysis, and the best fitting model was achieved for potassium.

Magnesium is a mobile element within the plant, therefore, similar to P and K, the symptoms of deficiency are mainly detectable in older leaves (Table 29.2). Magnesium deficiency is associated with early reductions in plant growth and higher starch accumulation in the leaves (as a consequence of decreased allocation of carbohydrates to developing sinks) followed by fading and yellowing of the tips of old leaves, which progresses intervenally toward the base, forming a typical 'V'-pattern as a result of chlorophyll degradation. These symptoms usually appear mainly on sun exposed leaves, while shaded portions of the plants remain dark green. Magnesium can be routinely quantified in leaves by atomic absorption spectrophotometry or ICP-AES after mineralization of the samples (Jones, 2001).

Calcium can be quantified with the same analytical procedures of magnesium, however, given the abundance of calcium in soils such deficiency is unusual.

Sulphur (as sulphate SO_4^-) is rarely quantified because after absorption it is quickly incorporated into plant components. Further, the results of sulphate analysis may be interpreted difficultly since many factors influence sulphate concentrations in leaves (e.g. age of plant, type of plant, status of other nutrient elements).

Iron is frequently quantified in leaf samples after the occurrence of chlorosis, which occurs on young leaves. Several protocols have been suggested to measure the 'active' form, which is the ferrous-iron (Fe^{2+}) form. A procedure involves the extraction with some chelating agents (o-phenantroline) and the quantification of Fe^{2+} by reading the transmittancy using a spectrophotometer. Other methods quantify Fe^{2+} by emission spectrophotometry after digestion of leaf samples or extraction with acids (CH_3COOH or HCl) (Jones, 2001). Recently, Paltridge et al. (2012) applied the energy-dispersive X-ray fluorescent spectrometry (EDXRF) to quantified iron in plant tissues obtaining results well-correlated with the standard protocols. This method was shown to be a reliable and economic non-destructive tool to analyse iron in plant tissue.

Management of fertilization in urban environment

The management of fertilization practices in urban landscape should aim at maintaining trees in an adequate nutritional status to provide the wide variety of benefits that trees can supply in urban settings. Fertilization protocols in urban areas should not be prescribed for increasing the yield, such as in agricultural crops and forest plantations, but to supply essential elements to plants thus avoiding nutrient deficiencies and metabolic injuries. Since assessing nutrients demand can be difficult because of the variability in urban plants (species, ages, physiological and phenological conditions) and soil characteristics (soil compaction, paved areas, shortage of organic component), a combination of soil analysis, foliage analysis and visual observation (see previous section) should always be carried out in order to choose the most appropriate fertilizer, application amounts and application rate.

When trees are planted in urban landscapes, their nutritional status depends on both nutrient management during nursery cultivation and on available nutrients in the rooting zone. In the nursery, an appropriate fertilization, which results in balanced shoot to root ratio, may enhance nutritional status and promote establishment, but excess or unbalanced fertilization (i.e. excess of nitrogen) to fasten growth in the nursery may lead to opposite results. Effectiveness of fertilization at or after planting, instead, depends on soil properties, but also on species ability to regenerate roots outside the rootball into the fertilized native soil. It is not surprising, therefore, that several studies have shown conflicting results of fertilizing plants after transplanting (Kane, 2008). Since it is not clear that fertilization at transplant promotes establishment, current tree-care recommendations suggest application of inorganic fertilizers only after soil test confirms deficiencies of specific nutrients (Ferrini and Baietto, 2006).

In addition, in order to avoid unnecessary fertilizers that could pollute water resources through leaching or runoff, nutrients should be applied to fit as precisely as possible to plants' needs. This can be achieved by choosing the most suitable type of fertilizer and the most appropriate application technique.

Fertilizers can be classified in: quick-release, slow-release and controlled-release fertilizers. Quick-release fertilizers (QRFs), are the most widely-used (and cheap) commercial fertilizers in traditional agriculture and forestry. QRFs make all nutrients available when dissolved in water, namely within a short period of time after being applied to soil. Their use should be carefully evaluated on the basis of plant growth stage and requirements. When QRFs are used at transplanting, salinity can build up in the root-zone and induce leaf burning, often followed by nutrient deficiencies if fertilization is not repeated. Also, leaching can occur if fertilization is not scheduled to meet plant needs. For these reasons, QRFs efficiency is relatively low (about 50% for N). For trees in urban areas, controlled-release fertilizers (CRFs) and slow-release fertilizers (SRFs) are of greater interest. Indeed, both in CRFs and SRFs, nutrients are slowly released to meet plant demand, to increase nutrient uptake efficiency and to reduce environmental impact (Morgan et al., 2009). CRFs consist of a soluble fertilizer core (consisting of urea alone or a complete fertilizer containing N, P, K and micronutrients) covered with water-insoluble coating. The coating material controls the rate, pattern, and duration of plant nutrient release, that can range from 3 to 16 months. CRFs have been studied to synchronize N release with N requirements pattern of crop, but studies in urban environment are still lacking. The release of nutrient from SRFs depends on soil (biological decomposition of fertilizer) and climatic conditions (mainly temperature). Based on the source, SRFs can be of synthetic or natural origin. Synthetic SRFs (e.g. ureaformaldehyde, methylene urea, urea-acetakdehyde/cyclo diurea, urea-isobutyraldehyde/isobutylidene) need multistep processes, involving dissolution and decomposition, before releasing plant-available N (Morgan et al., 2009). The multiple factors affecting the release of nutrients from synthetic SRFs are not only environmental conditions such as soil moisture, pH and temperature, but also microbial activity. Therefore, nutrients are released in a period that can range from 20 days to 18 months but that can be highly variable and difficult to accurately predict, differently from CRFs. Natural SRFs include animal manure, treated sewage, fish meal, bat guano, etc. Because of their organic nature, microbial activity is essential to release the nutrients to the plants. Organic SRFs contain both macronutrients and micronutrients, however their nutrient concentration is relatively lower than those of synthetic SRFs. SRFs and CRFs present greater advantage than QRFs because they provide nutrients over a long period of time. However, this advantage comes at an increased cost, for example the price per unit of nitrogen is almost 50 per cent greater for CRFs than for conventional urea. In addition, a low microbial population in urban soils can represent an impediment to the use of some SRFs because nutrient release can be unpredictably delayed by 6 to 8 weeks (Morgan et al., 2009). Organic SRFs can be a valid solution, both for the cost and the environmental impact, to improve soil structure and have a release of nutrients over a long period.

Fertilizers are applied using different methods:

1 *Broadcast:* fertilizers are distributed onto the surface of the soil. It is rapid but not recommended for compacted soils, paved surfaces and for trees with dense turf grass in the understory.
2 *Sub-surface:* fertilizers are applied in the soil (using liquid or dry formulation) beneath the crown of the tree. In case of street trees, holes can be drilled around the tree trunk at 50 cm from the trunk and fertilizers are applied 15–30 cm beneath the soil surface dividing the entire fertilizer dose equally among the individual holes.
3 *Foliar spray:* readily soluble fertilizers are sprayed on the leaf surface. It is used when plants suffer from specific nutrient deficiencies, such as iron or zinc. This type of delivering the fertilizers results in fast plant response but it cannot be considered a substitute for soil fertilization.
4 *Trickle-irrigation:* fertilizers are applied through an irrigation system that drips water onto the soil at very low rates (2–20 liters/hours depending on the type of soil, e.g. slowly on clay soils to avoid ponding and runoff). Water is applied close to plants, which enables wetting the whole soil profile supplying precise and continuous dosages near to the root zone.

Currently, the recommended annual application rates are of 140 kg of nitrogen per hectare for QRFs or 270 kg per hectare for SRFs. However, application rates should be carefully evaluated after soil analysis. These values represent the upper limits for an annual rate of application whose objective is to promote growth. The application rates for maintenance fertilization are lower and should be evaluated on the bases of foliar analysis.

A slow-release form of fertilizer should be used to fertilize trees so that nutrients are available throughout the season. Slow-release fertilizer formulations can be applied during the fall and winter, and the nutrients will still be available the following spring when they are most effective. However, fertilization in the early fall may reduce stem freeze resistance, so that spring fertilization is often recommended. Early spring is the best time to fertilize using QRFs because it avoids leaching during the winter, especially in heavy rainfall areas or in winter-cold regions (Benedikz et al., 2005). In addition. QRFs can be used to provide a prompt fertilization when symptoms of stress appear (e.g. chlorosis).

As a general rule, regardless of the form of the fertilizer and method used to deliver fertilizer, the objective is to make the essential elements in the fertilizers available when plants can absorb them (i.e. after sprouting and during leaf expansion).

Fertilization, irrigation, and drought tolerance

Nutrient uptake and water availability are tightly related, both because soil moisture affects nutrient availability to plant roots, and because plant nutritional status affects several plant morphological traits which can affect drought tolerance in plants. Although they should be considered with caution, because most of the works investigating the interaction between available moisture and nutritional status were performed in the nursery or in forest plantations, research findings are quite consistent across a wide range of plant species and environmental conditions (Fini et al., 2007; Ovalle et al., 2016). Using a factorial design consisting in field-grown trees of *Acer pseudoplatanus* which were only fertilized (using a complete controlled release fertilizer), only irrigated, both fertilized and irrigated, and neither fertilized or irrigated (control), Fini et al. (2007) found a clear interaction between fertilization and irrigation: 'fertilized only' maples had longer shoot growth than control trees, but this came at expense of stomatal conductance and photosynthesis, which were indeed lower than in 'control', 'irrigated only', and 'irrigated and fertilized' plants. Photoinhibition occurred during dry periods in fertilized only trees, but not in irrigated, irrigated and fertilized, and control plants, indicating a fertilizer-induced decrease in drought tolerance, as recently confirmed by Ovalle et al. (2016). Mechanisms through which fertilization may decrease drought tolerance include: larger total plant leaf area and higher shoot to root ratio, which increase plant water use to a greater extent than plant water uptake capacity; lower concentration of phenolics and secondary metabolites, which decrease plant defences against oxidative stresses (e.g. drought) and pests. Clearly, the effects of fertilization on drought tolerance depend on the type of fertilizers applied (or formulation, in complete fertilizers): N is generally associated with enhanced shoot growth (relative to root growth) and mineral N fertilization has been recommended only in good water availability conditions; P and K are, instead, associated to root growth and osmotic adjustment and much less likely to negatively affect drought tolerance. If fast growth and prompt effect are desired, coupling irrigation and fertilization is recommended, since it results in long-term positive, additive effects on growth, leaf gas exchange, and gross benefits provided by trees (Fini et al., 2007; Ovalle et al., 2016).

Fertilization, environmental stresses and BVOC emissions

In urban environments plants are rarely exposed to single stressors. Instead, they often face a combination of multiple environmental co-occurring stress factors. The ability of a plant to react to multiple stresses and survive under changing environmental conditions depends on the effectiveness of defence mechanisms. The accumulation of secondary metabolites often plays a crucial role to withstand a wide range of environmental constraints. Among secondary metabolites, BVOCs, because of their high reactivity, substantially contribute to atmospheric processes, such as ozone and secondary organic aerosol (SOA) formation, in the presence of anthropogenic nitrogen oxides (NO_x), altering the concentrations of hydroxyl radicals and inducing methane oxidation (Ghirardo et al., 2015; Calfapietra et al., 2013) (Figure 29.1). Since the effect of nutrients on plant responses to abiotic stresses is related to their role in supporting enzyme production and functioning, the study of the interactions between abiotic stresses, nutrients availability and BVOCs emission results are of particular interest. The carbon-nutrient balance theory suggests that the production of secondary metabolites is stimulated by limited nutrient resources, however the effects of fertilization on BVOCs emission remain still unclear. Results of some studies fit well with the carbon/nutrient balance hypothesis, because it was found that N fertilization treatments reduced terpene emissions (Holopainen et al., 1995). However, more recent studies support the existence of a positive correlation between nitrogen fertilization and terpene biosynthesis and emission (Ormeño and Fernandez, 2012; Blanch et al., 2007). Some works investigated also the effect of different nitrogen supply combined with water stress (Blanch et al., 2007), salt stress (Teuber et al., 2008) or ozone treatment (Carriero et al., 2016). These studies show complex species-specific responses depending on the duration and severity of the stress, since environmental stress conditions combined with different N fertilization levels have the potential to influence both BVOCs emission rates and plant growth. Apart from N availability, since some terpene precursors contain high-energy phosphate bonds, such as isopentenyl diphosphate and dimethylallyl diphosphate that are immediate precursors of isoprene, it is likely that also P availability influences isoprenoids emission rates. Blanch et al. (2012) found that even within the same species (*Pinus pinaster*) P scarcity affected differently terpene emission pattern. Since BVOCs emission is genus- and species-dependent, the contribution of BVOCs emission to the atmospheric reactivity in urban environment is related to plant diversity in urban areas. Consequently, some plant families might be more suitable than others for expanding the urban green lungs.

Further studies of landscape planning of urban areas should take into account the potential for BVOCs emission when considering how to reduce emission of O_3 precursors, and to mitigate urban air pollution, especially when planning large scale tree planting programmes.

Conclusion

Fertilization is a management technique widely consolidated in agriculture, forestry, and timber arboriculture, while information and protocols for fertilization of urban trees are still scanty, except for the nursery production stage. High variability of soil and environmental conditions across and within cities greatly affects the efficacy of fertilizer application. Despite prescribing fertilization treatment is accepted by national standards in some countries, the use of proper independent plant and soil nutrient assays is critical to enhance nutrient use efficiency while avoiding undesired effects (e.g. higher sensitivity to drought). Investigation of soil nutrient pools, organic matter quality and quantity, as well as soil microbial populations, is essential to assist landscape managers in understanding dynamic processes which govern nutrient availability or deficiency in urban soils.

References

Bainard, L. D., Klironomos, J. N., Gordon, A. M. (2011) 'The mycorrhizal status and colonization of 26 tree species growing in urban and rural environments', *Mycorrhiza,* vol 21, pp. 91–96.

Barker, A. V., Pilbeam, D. J. (2007) *Handbook of Plant Nutrition*, CRC Press, Boca Raton, FL.

Bassirirad, H. (2000). 'Kinetics of nutrient uptake by roots: responses to global change', *New Phytologist*, vol 147, pp. 155–169.

Benedikz, T., Ferrini, F., Garcia-Valdecantos, J. L., Tello, M.-L. (2005) 'Plant quality and establishment', in C. C. Konijnendijk, K. Nilsson, T. B. Randrup, J. Schipperijn (eds), *Urban Forests and Trees*, ch. 9, Springer, Berlin.

Benton, J. J., Wolf, J. B., Mills, H. A. (1991) *Plant Analysis Handbook: A Practical Sampling, Preparation, Analysis and Interpretation Guide*, Micro-Macro Publishing, Athens, GA.

Blanch, J., Peñuelas, J., Llusià, J. (2007) 'Sensitivity of terpene emissions to drought and fertilization in terpene-storing *Pinus halepensis* and non-storing *Quercus ilex*', *Physiologia Plantarum*, vol 131, pp. 211–225.

Blanch, J. S., Sampedro, L., Llusià, J., Moreira, X., Zas, R., Peñuelas, J. (2012). 'Effects of phosphorus availability and genetic variation of leaf terpene content and emission rate in *Pinus pinaster* seedlings susceptible and resistant to the pine weevil, *Hylobius abietis*', *Plant Biology*, vol 14, pp. 66–72.

Bray, B. M., Kurtz, L. T. (1945) 'Determination of total, organic and available forms of phosphorus in soils', *Soil Science,* vol 59, pp. 39–45.

Calfapietra, C., Fares, S., Manes, F., Morani, A., Sgrigna, G., Loreto, F. (2013) 'Role of biogenic volatile organic compounds (BVOC) emitted by urban trees on ozone concentration in cities: a review', *Environmental Pollution*, vol 183, pp. 71–80.

Carriero, G., Brunetti, C., Fares, S., Hayes, F., Hoshika, Y., Mills, G., Tattini, M., Paoletti, E. (2016) 'BVOC responses to realistic nitrogen fertilization and ozone exposure in silver birch', *Environmental Pollution*, vol 213, pp. 988–995.

Chapman, N., Miller, A. J., Lindsey, K., Whalley, W. R. (2012) 'Roots, water, and nutrient acquisition: let's get physical', *Trends in Plant Science*, vol 17, pp. 701–710.

Craul, P. J. (1992). *Urban Soil in Landscape Design*, John Wiley & Sons, New York.

De Kimpe, C. R., Morel, J.-L. (2000) 'Urban soil management: a growing concern', *Soil Science*, vol 165, pp. 31–40.

Ferrini, F., Baietto, M. (2006) 'Response to fertilization of different tree species in the urban environment', *Arboriculture and Urban Forestry*, vol 32, pp. 93–99.

Fidencio, P. H., Poppi, R. J., de Andrade, J. C. (2002) 'Determination of organic matter in soils using radial basis function networks and near infrared spectroscopy', *Analytica Chimica Acta*, vol 453, pp. 125–134.

Fini, A., Ferrini, F., Frangi, P., Amoroso, G. (2007) 'Growth and physiology of field grown *Acer pseudoplatanus* L. trees as influenced by irrigation and fertilization', *Proceedings of the SNA Conference in Field Production*, vol 52, pp. 51–58.

Fini, A., Frangi, P., Amoroso, G., Piatti, R., Faoro, M., Bellasio, C., Ferrini, F. (2011) 'Effect of controlled inoculation with specific mycorrhizal fungi from the urban environment on growth and physiology of containerized shade tree species growing under different water regimes', *Mycorrhiza*, vol 21, pp. 703–719.

Frydenvang, J., van Maarschalkerweerd, M., Carstensen, A., Mundus, S., Schmidt, S. B., Pedas, P. R., Laursen, K. H., Schjoerring, J. K., Husted, S. (2015) 'Sensitive detection of phosphorus deficiency in plants using chlorophyll a fluorescence', *Plant Physiology*, vol 169, pp. 353–361.

Ghirardo, A., Xie, J., Zheng, X., Wang, Y., Grote, R., Block, K., Wildt, J., Mentel, T., Kiendler-Scharr, A., Hallquist, M., Butterbach-Bahl, K., Schnitzler, J. P. (2015) 'Urban stress-induced biogenic VOC emissions impact secondary aerosol formation in Beijing', *Atmospheric Chemistry and Physics Discussions*, vol 15, pp. 23,005–23,049.

Holopainen, J. K, Rikala, R., Kainulainen, P., Oksanen, J. (1995) 'Resource partitioning to growth, storage and defence in nitrogen fertilized Scots pine and susceptibility of the seedlings to the tarnished plant bug *Lygus rugulipennis*', *New Phytologist*, vol 131, pp. 521–532.

Jones Jr, J. B. (2001) *Laboratory Guide for Conducting Soil Tests and Plant Analysis*, CRP Press, Boca Raton, FL.

Kane, B. (2008) 'Nitrogen fertilization during planting and establishment of the urban forest: a collection of five studies', *Urban Forestry and Urban Greening*, vol 7, pp. 195–206.

Karandashov, V., Bucher, M. (2005) 'Symbiotic phosphate transport in arbuscular mycorrhizas', *Trends in Plant Science*, vol 10, pp. 22–29.

Kopinga, J., van den Burg, J. (1995) 'Using soil and foliar analysis to diagnose the nutritional status of urban trees', *Journal of Arboriculture*, vol 21, pp. 17–24.

Liu, J., Yang, J., Cade-Menun, B. J., Liang, X., Hu, Y., Liu, C. W, Zhao, Y., Li, L., Shi, Y. (2013) 'Complementary phosphorus speciation in agricultural soils by sequential fractionation, solution ^{31}P nuclear magnetic resonance, and phosphorus K-edge X-ray absorption near-edge structure spectroscopy', *Journal of Environmental Quality*, vol 42, pp. 1763–1770.

Lorenz, K., Lal, R. (2009) 'Biogeochemical C and N cycles in urban soils', *Environmental International*, vol 35, pp. 1–8.

Lukac, M., Calfapietra, C., Lagomarsino, A., Loreto, F. (2010) 'Global climate change and tree nutrition: effects of elevated CO_2 and temperature', *Tree Physiology*, vol 30, pp. 1209–1220.

Lynch, J. P., St Clair, S. B. (2004) 'Mineral stress: the missing link in understanding how global climate change will affect plants in real world soils', *Field Crop Research*, vol 90, pp. 101–115.

Menesatti, P., Antonucci, F., Pallottino, F., Roccuzzo, G., Allegra, M., Stagno, F., Intrigliolo, F. (2010) 'Estimation of plant nutritional status by Vis–NIR spectrophotometric analysis on orange leaves [*Citrus sinensis* (L) Osbeck cv Tarocco]', *Biosystems Engineering*, vol 105, pp. 448–454.

Morgan, K. T., Cushman, K. E., Sato, S. (2009) 'Release mechanisms for slow- and controlled-release fertilizers and strategies for their use in vegetable production', *Horttechnology*, vol 19, pp. 10–12.

Muñoz-Huerta, R. F., Guevara-Gonzalez, R. G., Contreras-Medina, L. M., Torres-Pacheco, I., Prado-Olivarez, J., Ocampo-Velazquez, R. V. (2013) 'A review of methods for sensing the nitrogen status in plants: advantages, disadvantages and recent advances', *Sensors*, vol 13, pp. 10823–10843.

Ormeño, E., Fernandez, C. (2012) 'Effect of soil nutrient on production and diversity of volatile terpenoids from plants', *Current Bioactive Compounds*, vol 8, pp.71–79.

Ovalle, J. F., Arellano, E. C., Oliet, J. A., Becerra, P., Ginocchio, R. (2016) 'Linking nursery nutritional status and water availability post-planting under intense summer drought: the case of a South American Mediterranean tree species', *iForest-Biogeosciences and Forestry*, vol 789, pp. e1–e8.

Paltridge, N. G., Palmer, L. J., Milham, P. J., Guild, G. E., Stangoulis, J. C. (2012) 'Energy-dispersive X-ray fluorescence analysis of zinc and iron concentration in rice and pearl millet grain', *Plant and Soil*, vol 361, pp. 251–260.

Pierzynski, G. M. (2000) 'Methods of phosphorus analysis for soils, sediments, residuals, and waters', *Southern Cooperative Series Bulletin,* vol 396, pp. 102.

Sieghardt, M., Mursch-Radlgruber, E., Paoletti, E., Couenberg, E., Dimitrakopoulus, A., Rego, F., Hatzistathis, A., Randrup, T. B. (2005) 'The abiotic urban environment: impact of urban growing conditions on urban vegetation', in C. C. Konijnendijk, K. Nilsson, T. B. Randrup, J. Schipperijn (eds), *Urban Forests and Trees*, pp. 281–323, Springer, Berlin.

Teuber, M., Kreuzwieser, J., Ache, P., Polle, A., Rennenberg, H., Schnitzler, J. P. (2008) 'VOC emissions of grey poplar leaves as affected by salt stress and different N sources', *Plant Biology*, vol 10, pp. 86–96.

30

TREE BIOMECHANICS

Frank W. Telewski and Karl J. Niklas

Introduction

The goal of this chapter is to provide an overview of the basic physical principles and phenomena that influence the mechanical behavior of trees. This overview revolves around eight biomechanical features that hold true for all trees, regardless of their habitat:

1 trees sustain two general categories of mechanical forces (static loads and dynamic loads);
2 these forces are additive (stresses as well as strains are additive);
3 static loads increase slowly over time as trees increase in size (therefore, tree growth patterns can compensate for these increasing loads);
4 dynamic loads can change dramatically over short periods of time (these loads are unpredictable and therefore potentially dangerous);
5 trees generally fail as a result of dynamic loads;
6 plant tissues resist bending more than twisting (eccentric loadings are potentially dangerous);
7 the young parts of woody plants (leaves and first-year twigs) are more flexible than older parts; and
8 belowground growth generally does not keep pace with aboveground growth (which reduces a tree's safety factor as overall size increases).

Before exploring each of these eight topics, it is important to emphasize that trees grow and respond to their mechanical environment. This may seem obvious if not indeed trivial, but it is a vital fact. Fluid and mechanical engineering makes a number of assumptions that are consistently violated by anything that grows organically. For example, elementary engineering theory by and large assumes that a structure is composed of materials that are homogeneous and elastic in behavior. No animal or plant tissue is homogeneous and every tissue is viscoelastic. Engineering practice deals with structures that are fabricated to meet specific specifications for a particular work-place environment. Natural selection has no agenda. Perhaps most important is the fact that an engineered structure cannot heal itself and cannot change its size (with the exception of corrosion). Every plant has the capacity to replace damaged parts and every organism changes size as it grows, reproduces, and ultimately dies. Trees grow in response to their local environment in ways that cannot be modeled easily nor comprehended fully. Engineering theory also typically deals with small deflections, whereas trees often experience extremely large deflections. For these reasons, many of the assumptions made in engineering theory and practice are naïve in the context of tree biology. Consequently, the principles presented in this chapter can provide, at best, only guidance when predicting the mechanical behavior of trees.

As noted, the concepts presented in this chapter pertain generally to all trees, regardless of whether they grow in the tropics, deserts, temperate, alpine, boreal ecosystems, or in urban or natural environments. However, neotropical and tropical trees flourish under some conditions that would seem to set them apart. For example, many tropical trees have buttressed root systems that have been interpreted to function as anchorage stabilizers in moist or soft soils. Tropical trees tend to have greater epiphyte loads than trees growing in most other types of ecosystems. They can also have large liana loads that additionally entangle the branches of neighboring trees in ways that can affect dynamic dampening. Likewise, arboreal animals and large fruits often distinguish tropical trees, which suggests greater canopy dynamics and seasonal variations in self-loading compared to trees in other ecosystems. Finally, dense stands of trees growing in low latitudes often experience limited light availability during the early and late daylight hours. These and other features seem to set tropical tree biomechanics apart from tree biomechanics in general. Specific growing conditions and resultant growth responses vary between the different ecosystems in response to these conditions. Nevertheless, with few exceptions, as for example buttressed root systems in tropical trees and some temperate trees which present a unique morphological and biomechanical context, these and other differences are a matter of degree and not of kind. For example, the deflection of a cantilevered uniformly loaded beam (which can be used to model tree branches) depends on the load per unit length (see Equation 30.15), regardless of whether the load is solely due to self-loading or involves additional loads resulting from epiphytes or lianas in tropical trees, or ice and snow loading in temperate, boreal or alpine trees.

Finally, space precludes giving the derivations for the equations presented here. The derivations of these equations are provided by Niklas (1992) and Niklas and Spatz (2012), and in additional literature, which is cited as required.

Strength, flexibility, brittleness and viscoelasticity

To understand biomechanics, it is important to have a clear understanding of the definition of terms associated with the various parameters that are applied in the field. Some of the parameters can easily be confused, a mistake which should be avoided to better understand the relationships among them.

Strength refers to the maximum load a structure (branch, trunk, root) can withstand before it fails. As a structure is loaded, it will twist or bend. The ability of a structure to bend or twist is determined primarily by four parameters: the amount of material within the structure in the direction of the applied load (the second moment of cross sectional area, I, and the polar moment of inertia, J) and two of the structure's material properties, the elastic modulus (Young's elastic modulus, E, see Equation 30.9 below) and the shear modulus (G). The functional parameter with regard to the ability of a structure to resist bending is the product of E and I (the product is called flexural stiffness, EI, see Equation 30.11 below). The functional parameter with regard to the ability of a structure to resist twisting is the product of G and J (the product is called torsional stiffness, GJ; see Equation 30.12 below). Strength and flexibility are not related. A strong structure may be flexible or a weak structure can be ridged. What can lead to confusion with regard to strength and flexibility is the correlation between strength and flexibility in wood (Niklas and Spatz, 2013). It is only a correlation, and not cause and effect. Similarly, the forest products industry has relied on the measure of wood density as a proxy for strength (Panshin and de Zeeuw, 1980) even though density does not determine strength. These considerations are critically important when understanding the strength, elastic modulus and density of reaction wood where the relationships fail to hold.

When discussing strength and flexibility, it is important to also understand elastic and plastic deformation. When a structure is loaded, regardless of strength, it will bend or deflect from its original position. The degree of bending depends upon EI; the degree of twisting depends on GJ. As a structure is loaded, it will bend initially in a linear relationship (the first part of a stress–strain curve; Figure 30.4). This is referred to as the elastic region. As the load is released, the structure will return to its original position. However, if the load exceeds the elastic limit, the stress–strain relationship is no longer linear (non-linear behavior) entering a region known as plastic deformation. At this point the yield stress or breaking stress is reached leading to structural failure. This is the determination of the strength of the structure.

How the structure fails also depends on the nature of the material. Most living or wet (green) plant structures are ductile, the point of failure results in a plastic deformation and possibly a green stick fracture. In some cases and as is more common in dried plant material, the failure is more obvious and results in a fracture or breaking of the structure. A brittle material or structure will produce a fairly clean fracture when the yield or breaking stress is reached. A good example of a brittle material is glass. Green wood and other living plant tissues can also be brittle, although this material property is fairly rare in plants. Brittle failures have been reported in the stems and branches of oak trees which undergo very slow growth, resulting in growth rings composed mostly of large earlywood vessels (Koehler, 1933). It is also common in wood undergoing the decay process where the failure is referred to as a brash break (Panshin and de Zeeuw, 1980; Guyette and Stambaugh, 2003).

As stated in the introduction, no animal or plant tissue is homogeneous and every tissue is viscoelastic. A viscoelastic material exhibits properties of both viscosity, the ability of a fluid to resist shearing deformation and elasticity, the ability to restore deformations instantly when the level of stress drops to zero. Solid, liquid and viscoelastic materials can be differentiated based upon a quantity known as the relaxation time. This is the measure of the amount of time required to deform the molecular structure of the material to which a load is applied. An ideal elastic solid has a relaxation time of infinity and zero for a fluid. Therefore, the relaxation time cannot be measured for an ideal solid or liquid. The relaxation time for a viscoelastic material lies between these two extremes and can be measured. In a very real and mathematically precise way, the relaxation time is the ratio material's dynamic viscosity to its E (Niklas, 1992).

Viscoelastic materials experience creep, a mechanical behavior characterized by changes in the magnitude of strain under a constant level of stress (Niklas, 1992; Niklas and Spatz, 2012). In other words, under a constant load (stress), the degree of displacement (strain) will increase over time. The rate at which deformation occurs is governed by the material's viscosity. It also depends on the magnitude of the applied load and the temperature (Niklas, 1992). This becomes an important element when trying to interpret tree and branch behavior during constant static loading experienced during snow and ice storms. A branch loaded suddenly by snow or ice will undergo an immediate elastic deformation De_1 and it will continue to deform over time and may eventually fail without an increase in load. This delayed deformation has two components: a delayed elastic deformation De_2 and a plastic deformation D_p, the sum of which is creep (i.e. $De_2 + D_p$ = creep). When the load is removed (as for example when snow falls off a branch, or when ice melts rapidly), the branch will undergo a rapid elastic contraction (equal to De_1) returning it to some of its original orientation followed by a less rapid retarded elastic contraction (equal to De_2), but it will retain its plastic deformation (D_p), which may or may not be restored by reaction wood. In passing it is worth noting that the water in xylem can freeze during the winter and this causes branches bent under snow or ice loads to remain bent even when the snow or external ice is removed because of fractured and displaced internal ice. The effects of internal ice in wood are reviewed by Hogan and Niklas (2004).

Static and dynamic loadings

In biomechanics, any internally or externally applied force is referred to as a load. There are two general types of loads – static loads and dynamic loads, denoted as P_s and P_d, respectively. Static loads (also called self-loads) are the result of gravity acting on the aboveground parts of a plant or recall from above, any external load placed on a tree such as epiphytes, lianas, ice, snow or in the case of urban trees, swings or any other object hung on a tree. They are the product of mass m and the acceleration of gravity g:

$$P_s = mg \tag{30.1}$$

Although the numerical value of g differs as a function of altitude or the mass density of a rock layer below, this variation is irrelevant to dealing with tree mechanics. The conventional standard is $g = 9.80665$ m/s².

Dynamic loads are externally applied forces such as the pressure exerted by wind, the collision of stems or leaves with other stems or leaves, or the activities of animals in a canopy. The most pervasive dynamic load is wind pressure, which produces a drag force, denoted as D_f. This force is the product of air density ρ, wind speed U, projected (sail) area Sp, and a dimensionless parameter called the drag coefficient C_D:

$$D_f = 0.5 \rho U^2 Sp C_D \tag{30.2}$$

C_D is often assumed to be a constant for computational ease. However, it is not a constant for plants because leaves, branches, and stems reorient with respect to oncoming wind and thus reduce the numerical value of this 'constant' (Vogel, 1981). This reorientation also typically reduces Sp. Consequently, under some conditions, the drag force exerted by wind can be reduced even if wind speed increases (Telewski and Jaffe, 1986a). This phenomenology and the loss of leaves, twigs, and branches in strong winds can increase the safety factor of a tree (Niklas and Spatz, 2000).

Mechanical loads are additive. Therefore, the total load experienced by a tree P_t is given by the formula

$$P_t = P_s + D_f = mg + 0.5 \rho U^2 Sp C_D. \tag{30.3}$$

Because loads are vectors (i.e. they have direction and magnitude), Equation 30.3 is a vector equation.

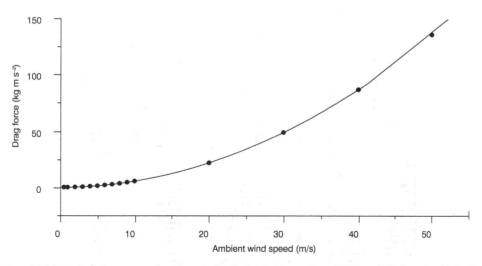

Figure 30.1 The drag force exerted on a vertical cylinder (with a radius of 0.03 m and a length of 1.5 m) subjected to different uniform wind speeds. It is important to note that vertical wind speed profiles are never uniform. Under natural conditions ambient wind speeds increase from zero at ground level to their maximum with increasing distance from ground level

With the aid of a few simplifying assumptions, Equations 30.1–30.3 can be used to evaluate whether static or dynamic loads contribute most to the total load. Consider a vertical cylinder with a radius of 0.03 m and a length (height) of 1.5 m subjected to a uniform wind speed of 11 m/s. Assuming that it is composed of wood with a density of 844 kg/m³ (which is the average density of green conifer and angiosperm wood; see Niklas and Spatz, 2010), Equation 30.1 shows that the static load equals π (0.03 m)² × 1.5 m × 844 kg/m³ × 9.80665 m/s² ≈ 35.09 kg m s⁻² or 35.09 N. With an ambient temperature of 20 °C, the density of air is 1.2041 kg/m². Assuming that the drag coefficient is approximately 1.0, Equation 30.2 shows that the drag load equals 0.5 × 1.2041 kg/m³ × (11 m/s)² × 0.06 m × 1.5 m × 1.0 ≈ 6.56 kg m s⁻² or 6.56 N. Consequently, the drag load exerted on this hypothetical cylinder by air moving at 11 m/s is a small component of the total load. This situation changes as the ambient wind speed increases because the drag force scales as the square of wind speed (Figure 30.1).

However, as we will see, what really matters are the stresses that develop at the base of the column when subjected to its own compressive load and to the stresses developed by the wind-induced drag. These stresses can be easily calculated, but this requires an understanding of other concepts and other formulas (see 'Concluding remarks' below), which need to be presented first.

Stresses, strains, and the elastic and shear moduli

A load produces stresses and strains. A stress equals a load P divided by the area A over which the load is applied, i.e. stress is a force per unit area and thus has units of kg m⁻¹ s⁻² or N/m². There are three kinds of stresses: compressional stresses, tensile stresses, and shear stresses (denoted as σ_-, σ_+, and τ, respectively):

$$\sigma_-, \sigma_+, \text{ and } \tau = P/A. \tag{30.4}$$

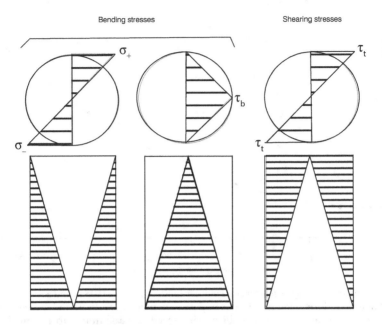

Figure 30.2 Transverse and longitudinal distributions of tensile, compressive, and shear stresses (σ_+, σ_-, and τ, respectively) in a circular solid cylinder fixed at its base and subjected to bending and twisting

The magnitudes of each stress are denoted by the stippled lines in each figure. Note that σ_+ and σ_- reach their local maximum intensities at the surface of each transection and achieve their overall maximum at the base of the cylinder, and that the bending shear stresses (τ_b) achieve their local maximum intensities at the center of each transection and achieve their overall maximum at the base of the cylinder. In contrast, torsional shear stresses (τ_t) reach their local maximum intensities at the surface of the cylinder and their overall maximum at the free end.

These stresses result when a material is subjected to pure compression, pure tension, or pure shear. They also result when a material is bent or twisted (Figure 30.2).

A strain is a dimensionless measure of deformation. It is typically expressed as a decimal fraction or as a percentage (e.g. 0.01 or 1%). Strains can be measured in different ways. For example, the Cauchy strain ε (also called the engineering strain) is the change in a reference dimension ΔL divided by the original non-deformed dimension L_o:

$$\varepsilon = \Delta L/L_o = (L - L_o)/L_o \tag{30.5}$$

The true strain ε_{true} (also called the logarithmic or Henchy strain) is calculated by integrating the incremental Cauchy strain:

$$\varepsilon_{true} = \ln(1 + \varepsilon) \tag{30.6}$$

For small strains (< 5%), Equations 30.5 and 30.6 yield similar results. For large strains, true strains should be used. Regardless of which equation is employed, it is always prudent to report how strains are calculated.

Shear strains γ are more complicated. Referring to Figure 30.3, the shear strain equals the deformation in the direction that parallels the direction of the shearing force Δx divided by the distance over which the shearing force acts z:

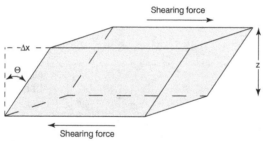

Figure 30.3 Diagram of shear strains γ developing in a small volume of an isotropic material

The shear strain equals the deformation in the direction that parallels the direction of the shearing force divided by the distance over which the shearing force acts. In this illustration the direction of the shearing force is in the x–axis and the distance over which it acts is the height of the volume, which is z. Therefore, the shear strain γ equals $\Delta x / z$, which in turn equals the tangent of Θ (see Equation 30.7)

$$\gamma = \Delta x / z = \tan \Theta. \tag{30.7}$$

Equation 30.7 is convenient and simple. However, it is not appropriate when shearing forces produce substantial gradients of deformation within a structure. In this case, the shear strain must be calculated in its differential form:

$$\gamma = \partial x / \partial z. \tag{30.8}$$

The ability of a material to resist a load is measured by its elastic modulus E (also called Young's modulus) and its shear modulus G, which are given by the quotient of the appropriate stress and its corresponding strain:

$$E = \sigma / \varepsilon \tag{30.9}$$

and

$$G = \tau / \gamma. \tag{30.10}$$

Experimentally, these moduli are calculated by measuring the slope of the linear portion of bivariate plots of stress versus strain (Figure 30.4).

Most materials are less capable of coping with shearing than with compression or tension because, for most materials, $G << E$. This fact does not imply that materials will fail more easily in shearing than in bending. It simply means that most materials shear more easily. The delamination of wood subjected to bending looks like a shearing failure, but it is not. It reflects the fact that upon bending, wood fails more easily in the longitudinal direction than in the radial direction. It is worth noting further that engineering theory typically deals with what is called *pure* tension and *pure* compression (that is, compression and tension without shearing). However, this hardly ever happens in reality. Shearing always occurs in one form or another when a material is bent, compressed, or pulled. The fact that $G << E$ does help to explain why less force is required to twist a material than to bend it. However, it is important to draw a sharp distinction between a material and a structure. A structure can fail in a variety of ways depending upon the type and direction of loading. Indeed, as we will see, tall and slender structures can bend gracefully or catastrophically (see 'Buckling' below).

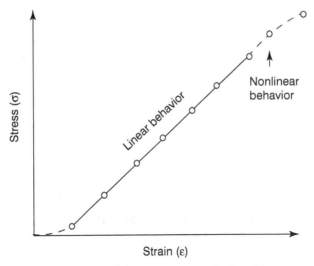

Figure 30.4 A hypothetical bivariate plot of the increasing magnitudes of the stresses and strains (σ and ε, respectively) developing within a sample of a material subjected to either tension, compression, or bending

The slope of the linear portion of the stress–strain plot equals the elastic modulus E. In this example, non-linear behavior is denoted by the dashed lines, which are curvilinear

The role of geometry

Although the ability to resist loads depends on the physical properties of a material or structure, it also depends on size, shape, and geometry (note that shape and geometry are not the same thing). Specifically, the ability to resist bending is quantified as the product of the elastic modulus E and the second moment of area denoted by I, which is called flexural stiffness. In a similar manner, the ability to resist twisting is quantified as the product of the shear modulus G and the polar moment of area J, which is called the torsional stiffness:

$$\text{Flexural stiffness} = EI \tag{30.11}$$

$$\text{Torsional stiffness} = GJ \tag{30.12}$$

Formulas for I and J for different geometries are easily obtained from the literature. A few are provided in Figure 30.5. The second moment of area and the polar moment of area quantify how the cross sectional geometry, shape, and size of an object contribute to the ability of the object to resist mechanical forces. For a circular cross section, $I = \pi r^4/4$ and $J = \pi r^4/2$, where r is radius. Notice that, in this particular case, $J = 2I$, and, second, both I and J increase dramatically as r increases. Although J is twice the numerical value of I, the ability of a material or structure to resist twisting is still much less than the ability to resist bending because, as noted, the shear modulus G of virtually every material is many times lower than that of the elastic modulus E. For example, for structural steel, $G = 79\ \text{GN/m}^2$ and $E = 200\ \text{GN/m}^2$, whereas, for Douglas fir (*Pseudotsuga menziesii*) wood at a density of 625 kg/m³, $G = 0.68\ \text{GN/m}^2$ and $E = 8.3\ \text{GN/m}^2$.

A third feature is also important. With the exception of a circular cross section, the numerical value of I depends on the plane of bending. Consider the formulas for the second moment of area of a beam with an elliptical cross section (see Figure 30.5). When the major

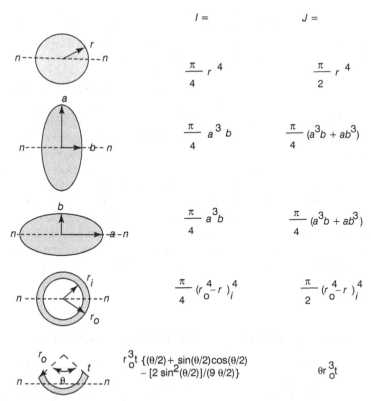

$$I =$$ $$J =$$

$$\frac{\pi}{4} r^4$$ $$\frac{\pi}{2} r^4$$

$$\frac{\pi}{4} a^3 b$$ $$\frac{\pi}{4} (a^3 b + ab^3)$$

$$\frac{\pi}{4} a^3 b$$ $$\frac{\pi}{4} (a^3 b + ab^3)$$

$$\frac{\pi}{4} (r_o^4 - r_i^4)$$ $$\frac{\pi}{2} (r_o^4 - r_i^4)$$

$$r_o^3 t \{(\theta/2) + \sin(\theta/2)\cos(\theta/2) - [2\sin^2(\theta/2)]/(9\,\theta/2)\}$$ $$\theta r_o^3 t$$

Figure 30.5 Equations for computing the second moments of area (I) and the polar moments of area (*J*) of beams, columns, etc. characterized by different transverse geometries and shapes

Note that in each case *J* > I. Also note that, with the exception of the circular cross section, the numerical values of *J* and I depend on the plane of bending (i.e. the neutral plane), which is denoted in each figure by n – – – – n. The importance of the plane of bending is shown here by comparing the formulas for the two elliptical cross sections, which have different (orthogonal) orientations with respect to the neutral plane

axis of the ellipse is oriented in the vertical direction, $I = \pi a^3 b/4$. However, when the minor axis is oriented in the vertical direction, $I = \pi ab^3/4$. Consequently, the same object can resist bending differently depending on its orientation or the direction of an applied force. This phenomenon is evident when the cross sections of cantilevered branches or leaf petioles are inspected. Many of these structures have an elliptical cross section in which the major axis is oriented vertically. This orientation helps a branch or petiole to resist the pull of gravity, but it permits sideways deflection in the wind.

Another aspect of geometry requiring consideration centers on the mechanics of a solid rod versus a hollow tube. Under bending, the maximum stresses and strains occur at the surface of the object under loading. The material in the center of the object contributes very little to stiffness or strength and therefore it can be removed to achieve lighter weight. For this reason, a less expensive tubular construction can be as strong as structures composed of solid rods. In fact, the use of tubular construction reduces the amount self-loading. This mechanical principle contributes significantly to understanding tree stability. A hollow tree is not necessarily an unstable tree. For a 2 m diameter tree trunk, we can calculate the second moment of cross sectional area (I) of a solid trunk by $I = \pi D^4/64$. We can also calculate the value of I for a 2 m diameter trunk with a 1m hollow in the center by $I = \pi/64\,(D^4 - d^4)$ where $d = 1$ m. In the case of the solid stem, $I = 0.7854$ m^4, and for the hollow trunk $I = 0.7363$ m^4. Applying this to trees, Wagener (1963) reported that a loss of half of the total trunk diameter will only result in a loss of 12.5 percent of the trunk's total strength as long as the void is centered in the stem and the wood is solid, without cracks or wounds. Wagener (1963) concluded, based on his observations of conifers, that a tree can lose up to 70 percent total wood diameter inside the bark, which translates to a 33 percent loss of strength, without affecting the safety of the tree as long as no other defects are not present. Smiley (1989) applied and modified Wagener's formulas to analyze hollow tree failures after Hurricane Hugo and concluded that strength loss of broken trees varied from 1 to 90 percent, with an average of 33 percent strength loss. Smiley et al. (2000) conclude that a strength loss of 33 percent is a maximum tolerable limit, equating to 15 percent of trunk diameter composed of a solid wood shell without any cavities that would compromise the outer shell. They go on to emphasize that a number of other factors will reduce the strength-loss threshold including wood strength, severity of stress exposure, proximity of decay to main branch unions, lean, and size and density of crown. It is important to bear in mind that these estimates do not take into account the lengths of hollow trunks. Computer simulations indicate that the susceptibility of hollow tubes to mechanical failure is extremely complex and that tube dimensions, including length, play critical roles in its determination (Spatz and Niklas, 2013).

Bending and twisting of cantilevered beams (branches)

We are now in a position to consider bending and twisting moments. A moment, denoted as M, is the product of an applied load and the lever arm over which the force acts. This is illustrated for a cantilevered beam of length L fixed at one end and loaded at its free end (Figure 30.6a). Assuming that the beam is uniform in its cross sectional area and homogenous in its composition, the bending moment at any point x from its fixed end is given by the formula

$$M = P\,(L - x). \tag{30.13}$$

Thus, the bending moment increases linearly from the point-loaded free end toward the base of the beam where $x = 0$. The maximum deflection δ_{max} of such a beam occurs at the free end. It is given by the formula

$$\delta_{max} = PL^3/3EI. \tag{30.14}$$

In the case of a uniformly loaded beam, the maximum deflection is given by the formula

$$\delta_{max} = wL^4/8EI, \tag{30.15}$$

where w is the weight of the beam per unit length.

The situation with torsion is more complex. Consider a condition called pure shear in which a beam is twisted but not bent. Referring to Figure 30.6b, the angle of twist θ results in strains γ that decrease toward the fixed end, and the circular arc between two points A and B has a length equal to $L\gamma$ and $r\theta$, where r and L are the radius and the length of the beam, respectively. Assuming that the beam is homogeneous and elastic and that the distortions are not excessive, the relationships among the shear stress τ just beneath the surface of the beam, the shear strain γ, the shear modulus G, and the geometry of the beam (r and L) are given by the formula

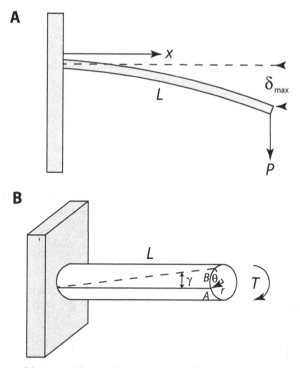

Figure 30.6 Diagrams of horizontally cantilevered beams (fixed at one end and free at the other) subjected to (a) bending and (b) twisting.

(a) The displacement of a beam (with length L subjected to a point-load P at the free end) from the vertical (δ) increases along the distance $\times$ from the fixed end toward the free end and reaches its maximum (d_{max}) the free end of the beam (P). Likewise, the maximum bending moment (M_{max}) occurs at the fixed end of the beam where $x = 0$ (see Equations 30.13–14). (b) The material elements in a beam (with length L subjected to a clockwise torque T) located along line A will be displaced along the dashed line B. The magnitude of the displacement (the strain) of these material elements (γ) increases from the fixed end toward the free end. Note that angle of twist θ is equal to the strain γ (see Equations 30.16–17)

$$\tau / \theta = G \ (r/L). \tag{30.16}$$

The total torque *T* experienced by the beam is given by the formula

$$T = GJ \ (\theta/L). \tag{30.17}$$

Buckling

Bending can result in buckling. This phenomenon results from the addition of a load that exceeds the maximum load that a column can sustain. Once this critical load is reached, the column undergoes mechanical instability that can lead to mechanical failure. Two forms of failure are possible: compressive (crushing) failure or buckling. Which of these occurs depends on the ratio of a column's radius *r* to its length *L*. The value of the quotient that characterizes the transition from one mode of failure to the other is given by the equation

$$r/L = (4/\pi)\,(\sigma_{comp}/E)^{1/2}, \tag{30.18}$$

where σ_{comp} is the maximum compressive stress a material (such as wood) can sustain. Equation 30.18 shows that very short and thick columns will undergo compressive failure, whereas tall and thin columns are more likely to bend under the applied load.

The load exerted on a tree trunk increases annually due to the addition of new leaves and branches and the loading condition of the trunk can be stylized as a point load at the top of a vertical column. The girth of a tree trunk also increases as a result of the annual addition of wood. However, different trees attain different r/L as they get older as a result of species-specific differences (e.g. *Populus* versus *Quercus*) and differences in local growing conditions (e.g. shaded habitats versus sunny habitats). Although it is conceivable that a trunk might undergo compressive (crushing) failure, a far more common response to applied loads is bending (Figure 30.7). Assuming that a slender trunk has a solid cross section, the critical load is given by the formula

$$P_{crit} = \pi^2 EI/(2L)^2, \tag{30.19}$$

where L is the length (height) of the column. The equation describes what is called Euler buckling in honor of the great mathematician Leonhard Euler (1707–1783) who derived it.

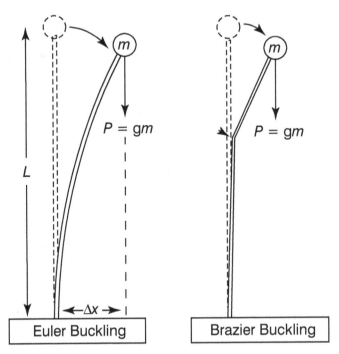

Figure 30.7 Diagrams illustrating Euler (long-wave) buckling and Brazier (short-wave) buckling in a slender vertical thin-walled tube subjected to a point-load (P, which equals the point-mass m times the acceleration of gravity g) at the tube's free end

In each case the original undeformed tube and point-mass are indicated by dashed outlines. In Euler buckling, the column deflects under the point-load and the displacement along the horizontal plane (Δx) increases along the length of the column (L). In Brazier buckling, the tube deflects from the vertical (not shown) and its circular cross sections gradually ovalize until catastrophic failure occurs at a confined location (indicated by short arrow)

Euler's equation shows that the elasticity of the material (wood) used to construct a column (trunk) determines the load bearing capacity of the column. It also shows that the critical load is directly proportional to the second moment of area of the column. Thus, the load bearing capacity of the column can be increased by using a material with a higher elastic modulus, or by maximizing the second moment of area. In general, denser materials such as woods have higher elastic moduli, but a denser material contributes more weight per unit volume, which decreases the critical load because of increased self-loading.

An alternative to increasing the elastic modulus of a column is to distribute the material as far from the principal axis of the cross section as possible without incurring local buckling. This can be seen by dividing the critical load of a hollow column by the critical load of a solid column with the same material and the same outside diameter, i.e. $P_{crit=hollow}/P_{crit=solid} = [\pi^2 EI/(2L)^2]_{hollow}/[\pi^2 EI/(2L)^2]_{solid} = 1 - (D_i/D_o)^4$, where D_i is the internal diameter and D_o is the external diameter. As D_i increases, $P_{crit=hollow}/P_{crit=solid}$ increases. Although this helps to explain why a tubular section is much more efficient than a solid section for the construction of a column, this tactic has limits because very thin tubes are susceptible to a phenomenon called Brazier buckling. This mode of failure, which is also called short-wave buckling, results when a bending load causes a cross section to ovalize. As the load increases, the ovalization of the cross section increases and eventually results in catastrophic localized crimping (Figure 30.7). Hollow stems and leaves are susceptible to this mode of failure. It should be noted that Euler buckling and Brazier buckling are interrelated because a tall slender hollow column starts to undergo Euler buckling before it experiences Brazier buckling. The critical bending moment for this case is given by the equation

$$M_{crit} = (0.99\, E\, r_o\, t^2)/(1 - v^2)^{1/2}, \tag{30.20}$$

where r_o is the outside radius, t is the thickness of the hollow column, and v is the Poisson ratio, i.e. the negative quotient of the transverse and the axial strains a material experiences. The equation holds true provided that $r_o > 10\,t$ and that the column is composed of an isotropic material (i.e. a material that mechanically responds the same regardless of the direction a load is applied). Whether a hollow trunk or branch undergoes compressive failure or Brazier buckling failure depends on many factors, including wall thickness, outside diameter, and length (a detailed treatment is provided by Spatz and Niklas, 2013).

The critical (maximum) buckling height

Euler's formula for the critical buckling load (see Equation 30.19) provides a way to calculate the critical height to which a column can be elevated before it begins to bend under its own weight. This calculation begins by noting that the critical load P_{crit} has to be some fraction or multiple γ of the weight of a column. The weight of a column equals the column's density times the acceleration of gravity times its volume. The weight of an untapered circular cylinder with radius r and length L equals $\delta g\pi r^2 L$. Therefore, the critical buckling load equals $\gamma\delta g\pi r^2 L$. Inserting this expression into Euler's formula gives

$$\gamma\delta g\pi r^2 L = \pi^2 EI/(2L)^2. \tag{30.21}$$

Solving for the critical buckling length L_{crit}, we obtain

$$L_{crit} = [\pi^2 EI/(4\gamma\delta g r^2)]^{1/3}. \tag{30.22}$$

Since the column is circular, $I = \pi r^4/4$ (see Figure 30.5). Inserting this expression into Equation 30.22 gives

$$L_{crit} = (\pi^2/16\gamma)^{1/3}(E/\delta g)^{1/3} \, r^{2/3}, \tag{30.23a}$$

or

$$L_{crit} \approx (0.85/g^{1/3})(E/\delta g)^{1/3} \, r^{2/3}. \tag{30.23b}$$

Unfortunately, we do not know the precise numerical value of γ. However, more sophisticated mathematics gives us the following relationships for columnar and conical stems:

Columnar stems $L_{crit} = 0.79 \, (E/\delta g)^{1/3} D^{2/3}$ (30.24a)

Conical stems $\quad L_{crit} = 1.25 \, (E/\delta g)^{1/3} D^{2/3},$ (30.24b)

where D is basal stem diameter. Notice that a conical (tapered) trunk provides for greater maximum height than does a cylindrical (untapered) trunk.

However, the extent to which these formulas can be used to estimate the extent to which a tree has reached its maximum height becomes problematic, particularly when we examine the assumptions underlying Equation 30.24. Tree canopies are never completely symmetrical, trunks are never perfect cylinders or cones, and they almost always experience lateral loads due to moving air. For these and other reasons, Equation 30.24 always over-estimates maximum tree height, which may help to explain why empirical observations indicate that trees growing in open habitats almost never reach their critical buckling heights. In contrast, trees growing in habitats sheltered from the wind, particularly in dense stands, can approach their critical buckling heights in part because they are sheltered from wind by their neighbors and because they do not experience a phenomenon called wind-induced thigmomorphogenesis (see next section).

Before leaving the topic of critical buckling height, it is worth noting that the geometry of a tree and the material properties of a tree are equally important when considering safety factor analyses. This is no better illustrated than by the detailed study of six baobab (*Adansonia*) tree species by Chapotin et al. (2006). Despite the considerable girth with respect to the height of mature specimens, which might give the impression that these trees have very high factors of safety, Chapotin et al. (2006) report that safety factors (based on estimated elastic buckling heights) are rather low, i.e. baobab trees are no more overbuilt than the majority of temperate or tropical trees. This feature is the result of baobab wood, which has a low elastic modulus owing to its high water content and parenchymatic volume fraction (Chapotin et al., 2006). It is also worth noting that the elastic modulus of baobab wood can change as a function of the withdrawal of water, since the elastic modulus of this wood decreases with water content. This opens up the curious possibility that the second moment of area of the trunk and the elastic modulus of its wood are inversely correlated and thus possibly compensatory which is altered by the addition or removal of water. In this context, it is interesting to return to Equation 30.11, which shows that a stem's ability to resist bending depends on *E and* on *I* such that 'fat but weak' can be just as effective as 'slender and stiff'.

Thigmomorphogenesis: responses to dynamic loadings

During their lifetimes, trees are constantly subjected to mechanical stresses resulting from the effects of bending, twisting, vibration, and swaying under the influence of wind. The earliest published observations regarding the response of trees to the dynamic loading imposed by the wind were recorded by the Greek philosopher Theophrastus in 300 BC. The first experiment to quantify the effect of wind on tree growth was conducted by Knight (1803) on apple trees which were guyed to prevent wind-induced sway and compared to free-swaying trees. Free-swaying trees did not grow as tall as the guyed trees, but exhibited increased radial growth. Subsequent researchers have confirmed that stems subjected to a uni- or bi-lateral flexing take on an eccentric or elliptical cross section with the major axis aligned parallel to the direction of force application. The increased radial growth being the result of increased cambial growth in the direction of flexing (see Telewski, 1995, 2012, 2016; Gardiner et al., 2016). The prevalence of an elliptical stem cross section is quite common in urban street trees growing adjacent to multistory buildings which restrict wind flow up or down the street in an artificial canyon (Telewski et al., 1997; Telewski, 2016). Inspection of Figure 30.5 reveals that a change from a circular cross section to an elliptical cross section results in a change in the second moment of area such that an elliptical cross section contributes more to the ability of a stem to resist bending if the major axis of the ellipse is aligned in the direction of the applied load. Jacobs (1954), employing a similar experimental design to that used by Knight (1803) 150 years earlier, provided additional insights on the effects of bending and swaying by guying the trunk of young Monterey pine trees (*Pinus radiata*) with wires approximately 20 feet (≈ 6.1 m) above ground so that trunks would sway above the attachment points of the wires but remain unperturbed below. The portion of the trunks above the guy wires was observed to grow in girth more rapidly that the portion of the restrained trunks. When the wires were removed, the lower portions of the trunks grew in girth at the same pace as the portions of trunks that were permitted to sway. These experiments indicate that wind-induced swaying stimulates the activity of the vascular cambium and results in thicker stems.

Experiments with non-woody plants indicate that mechanical perturbation produces similar results (Mitchell and Myers, 1995; Gardiner et al., 2016). Mechanically stimulated stems grow more rapidly in girth and less rapidly in length compared to stems that are not permitted to move. In addition, mechanical tests reveal that the tissues of mechanically perturbed stems have lower elastic moduli compared to tissues that are not perturbed. Collectively these experiments indicate that mechanical stimulation increases the second moment of area but decreases the elastic modulus of stem tissues. The decrease in the elastic modulus appears to be related to an increase in the cellulose microfibrillar angle within the secondary cell walls within the xylem of both conifer and angiosperms species exposed to wind (Telewski, 2016). These responses increase the contribution of geometry to resist bending and twisting but decrease the contribution made by the mechanical properties of tissues (presumably so that stems deflect more easily in the wind and thus reduce their projected area toward the oncoming wind). The decrease in the E observed in stems of trees exposed to flexing, and resulting from changes in the microfibrillar angle within the cells of the xylem have also been reported to occur in branches of wind-exposed trees. Specifically within branch junctions an increase in E and the MFA increases flexibility augmenting the damping effect of branches in windy environments (Jungikl et al., 2009).

This phenomenology is an example of thigmomorphogenesis (Jaffe, 1973), a term that refers to any growth response to mechanical perturbation (for a mechanistic review, see Telewski, 2006). It is most often studied in terms of the changes in morphology or anatomy attending the application of external forces such as loading. In addition to the reported alterations in stem growth, both in terms of a decreased rate of elongation and increased radial growth in the direction of the applied force, trees grown in windy environments also have smaller leaves resulting in reduced leaf area and smaller crowns (Telewski, 1995; Niklas, 1996). The reduction in leaf area and crown profile result in reduced profile areas reducing the overall drag imposed upon the crown (Telewski and Jaffe, 1986b). In extreme cases, the crowns of trees take on a permanent wind-blown morphology where the crown is streamlined in the direction of the prevailing wind further reducing drag (Telewski, 2012).

Thigmomorphogenesis has been reported for over 80 percent of all the species examined (Jaffe, 1973). More recently, it has been studied in terms of the molecular events preceding changes in shape and size. Braam and Davis (1990) have shown that ten to twenty minutes after mechanical stimulation by handling, rain, or wind, the mRNA levels of mouse-ear cress (*Arabidopsis*) increase up to a hundredfold. Four touch-induced (TCH) genes are involved. These genes encode for calmodulin, suggesting that calcium ions are required for the transduction of mechanical signals, thereby enabling plants to sense and respond to dynamic as well as self-imposed mechanical forces. Subsequent reviews of thigmomorphogenesis have detailed additional studies regarding the molecular events preceding developmental changes including the role of mechanosensing stretch activated ion channels and the cytoskeleton-plasma membrane-cell wall in the mechanopreception of wind and additional mechanical loads (Telewski, 2006; Chehab et al., 2009; Coutand, 2010).

Gravitropism: responses to static loadings

With few exceptions, the majority of trees respond to gravity by what is referred to as the gravitropic response; shoots being negatively gravitropic (growing upward, away from gravity), roots of seedlings, tap roots and sinker roots are positively gravitropic (growing downward, towards gravity) and branches and lateral roots are plagiotropic (growing at an oblique angle to gravity). A tree growing in the normal vertical orientation, away from the stimulus of acceleration of gravity must, as previously mentioned, support itself (self-loading). If the tree trunk is displaced with respect to the gravitational vector, it will respond gravitropically and form reaction wood to correct for the displacement or stabilize the displaced stem. In conifers, reaction wood takes the form of compression wood. In most angiosperms it takes the form of tension wood. Excellent reviews on the subject of reaction wood have been provided by Wilson and Archer (1977), Timell (1986a, 1986b, 1986c), Du and Yamamoto (2007), Gardiner et al. (2014) and Donaldson and Singh (2016).

Reaction wood is produced by the vascular cambium in response to the types of stresses (and their locations) generated by externally applied loads, among which gravity is a persistent form of load. Thus, the formation of reaction wood facilitates maintaining or controlling the posture or orientation of a tree's trunk and branches with respect to environment stimuli including gravity, light, and wind (Timell, 1986c; Moulia et al., 2006; Bastien et al., 2015). For example, compression wood forms on the side of wind-deflected branches of flag-formed conifer trees (Telewski, 2012).

In terms of tree biomechanics and gravitropism, the formation of reaction wood can impart a redirection of a displaced stem, returning it to the vertical orientation. This is accomplished by the development of internal growth strains resulting from the differentiation and maturation of reaction wood cells. In conifers, compression wood forms on the lower side of displaced stems. The tracheids of compression wood are round in cross-section, creating intracellular air spaces between cells. The secondary cell wall is greatly thickened and lignified. The cellulose microfibrils in the secondary cell walls are deposited at a higher angle, more transverse or horizontal to the orientation of the tracheid in the stem. Upon maturation, the compression wood expands axially resulting in the creation of a compressive force which pushes the stem towards the vertical, upright position.

In most angiosperms, tension wood forms on the upper side of a displaced stem. Tension wood is a diverse tissue which varies between different species and families of trees (Ruellé, 2014). Upon maturation, the tension wood shrinks creating a tensile force on the upper portion of the displaced stem resulting in a pulling action which pulls the stem towards the vertical orientation. The specialized fibers of the tension wood of some species are referred to as gelatinous fibers, so named for the gelatinous-like, jelly-like appearance of the secondary cell wall. This gelatinous layer is abbreviated G-layer and is composed mostly of highly crystalized cellulose and cellulose microfibrils deposited nearly vertical with respect to the axis of the stem. Originally thought to be a key feature of tension wood, gelatinous fibers only occur in certain species. In temperate tree species, gelatinous fibers are common in genera of the beech family (Fagaceae) including beech (*Fagus*), oak (*Quercus*) and chestnut (*Castanea*), and in the willow family (Salicaceae) including poplar (*Populus*) (Ruellé, 2014). It is worth noting that tension wood forms in porous wood angiosperms, those angiosperms with vessels (pores) in their secondary xylem for conducting water. Vessels are reduced both in frequency and in diameter in tension wood. Non-porous angiosperms produce compression wood.

With only one exception, all studies investigating the influence of mechanical loading (dynamic or static) on tree growth and development have been conducted during the growing season, which opens the question can trees respond to mechanical loads imposed during the dormant season? To address this question, Valinger et al. (1995) conducted a study with specific reference to ice and snow loading during the winter dormant period. During the winter, they displaced tree stems 30° from the vertical position imposing a static load to simulate snow, ice, or wind load. The trees were returned to the vertical position before the break of dormancy. During the subsequent growing season, growth of the treated trees was compared to non-treated controls. They reported that the trees bent during dormancy increased their radial growth compared to non-bent controls. These results indicate that trees can perceive mechanical loading during dormancy and respond in the subsequent growing season after the break of dormancy. Unfortunately, anatomical data were not provided and there was no indication if compression wood was formed as a result of the winter static displacement.

Trunk-root interactions and damping

The growth and development of woody roots are known to respond to chronic dynamic perturbations with a predictable directional component by adaptively altering cross-sectional morphology, patterns of wood deposition, and even the mechanical properties of tissues. The data reported from a number of case studies confirm that woody roots have the capacity to alter important physical and geometrical properties that are adaptive to stress and strain conditions (e.g. Knight, 1811; Coutts, 1983, 1986; Ennos, 1993, 1994; Stokes et al., 1995, 1996; Niklas, 1999a).

Dynamic loadings typically actuate roots to bend and twist. Typically, the roots on the windward side of the tree will bend upward and the roots on the leeward side will bend downward as a result of the rotational pivoting of the trunk caused by the wind (Figure 30.8). Roots oriented at an angle with respect to the direction of the oncoming wind will twist as well as bend to varying degrees. This bending and twisting can also cause the more

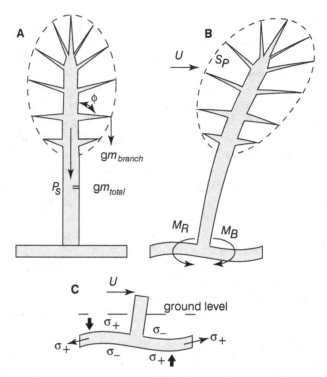

Figure 30.8 Synoptic diagram of self-loading resulting from individual branches and the total above-ground mass of a tree (a); dynamic loading resulting from wind drag (b); and trunk-root interactions under conditions of wind-induced drag forces (c)

(a) Self-loading for an individual branch with mass m_{branch} results in a bending moment M_B whose magnitude depends on tangent of the branch angle φ. In the simplest case, $M_B = g\, m_{branch}\, L \sin φ$, where L is the length of the branch. Since sin φ = 1.0 when φ = 90°, a horizontally oriented branch achieves the maximum bending moment (which occurs at its base). The maximum self load and the maximum bending moment, however, occur at the base of the trunk, which experiences the maximum compressive stresses, which is equal to the product of the acceleration of gravity and the total aboveground mass divided by the area of the base of the trunk (i.e. $σ_- = g\, m_{total}/A = P_S/Ai$). (b) A pictorial of a tree subjected to dynamic loading as a result of wind pressure. U = wind speed. The drag force experienced by the tree canopy (with a projected area of S_p) is given by Equation 30.2. The maximum bending moment M_B occurs at the trunk-root junction. M_B equals the drag force times the height of the tree. Mechanical stability requires a rotational (counter) moment M_R such that $M_R \geq M_B$. Flexure of the windward and leeward root system results in compression and tension within the root systems (see (c)). (c) Diagram of root-system flexure experiencing a wind-induced torque. Roots on the windward side bend upward and experience tensile stresses and compression stresses on their upper and lower surfaces ($σ_+$ and $σ_-$, respectively); more distal roots also experience tensile stresses as they are pulled toward the base of the trunk. Roots on the leeward side bend downward and experience compressive stresses and tensile stresses on their lower and upper surfaces ($σ_+$ and $σ_-$, respectively); more distal roots also experience tensile stresses as they are pulled toward the base of the trunk. Root flexure can be reduced if compacted soil resists root displacements and provides compression on the upper surfaces of upwind roots and compression on the lower surfaces of downwind roots (denoted by upward and downward pointing arrows).

distant parts of the root system to be pulled toward the base of the trunk. Gusts of wind coming in different directions can cause the root system to oscillate such that the direction of bending and twisting will change, e.g. root surfaces experiencing compression will experience tension and the direction of torsion will be reversed. Significant and sudden changes in the direction and magnitude of the wind force (drag) can result in root-wood fatigue/failure, some of which may not be visible until uproot occurs.

The extent to which roots bend and twist depends on many factors other than the application of the force and the strength of root-wood. For example, soil depth, compaction, and hydration can contribute to the ability of roots to resist bending, twisting and pulling by virtue of providing counteracting compressive forces near the trunk-root crown and by providing counteracting shearing or suction forces along the surfaces of more distal parts of the root system. Other important factors are:

- the general morphology of the root system, which influences how stresses are distributed along the length and breadth of individual roots (e.g. Crook and Ennos, 1996; Stokes and Mattheck, 1996; Crook et al., 1997);
- the morphology and size of the tree canopy, which influences how drag forces are transmitted to the root system (Ennos, 1993; Edelin and Atger, 1994; Vogel, 1996); and
- soil-type and weather conditions, which influence the extent to which roots can remain anchored and the duration and magnitudes of wind-induced dynamic loadings (e.g. Casada et al., 1980; Marshall and Holmes, 1988).

An important aspect of the transfer of energy from the wind to a tree, or to any large plant, is the damping of oscillations. Damping causes a decrease in the amplitudes of free oscillations and thus reduces the danger of a resonance catastrophe in dynamic winds. Oscillation and oscillation damping in trees have been widely studied (see Mayhead, 1973; Milne, 1991; Peltola et al., 1993; Moore and Maguire, 2004; Jonsson et al., 2007). If friction among different trees, or among different branches, and dissipative mechanisms in the root-soil system are set aside, there are two principal sources of damping: fluid damping and viscous damping within the material (i.e. wood in the case of trees). Fluid damping (dissipation of energy to the surrounding medium) depends on the square of the velocity of the object's movement relative to the surrounding medium. With estimates of the effective projected area and the drag coefficient (see Equation 30.2), fluid damping can be calculated by iteration of the loss of energy during each cycle of the oscillation. Viscous damping in the material (conversion of mechanical energy into heat) is linearly related to the velocity of relative movements between adjoining branches (or systems of branches that move in consort with one another). It can be determined in loading-unloading experiments by measuring the loss of energy in a hysteresis loop. In order to relate these measurements to a real tree, measurements should be performed using green wood, since the water in wood plays a pivotal role in viscous damping.

In complex structures such as trees, certain processes can enhance damping. In gusty winds, branches do not necessarily sway in line or in phase with their subtending stems. Rather, they can perform independent movements relative to one another. In this way, energy is distributed among branches and twigs and is dissipated more effectively than in a structure too stiff to allow relative movements between its elements. This phenomenon is referred to as structural damping (Niklas, 1992) and is not a different mode of damping. It merely emphasizes the enhancement of overall damping by the relative movements of structural elements (branches and twigs), which affect both fluid and viscous damping – processes that are most effective in the periphery of the tree canopy. Structural damping can be caused by the loose coupling mass (termed mass damping;

James, 2003) or by the distribution of mechanical energy through resonance phenomena within the tree (see James et al., 2006; Spatz et al., 2007), a phenomenon well known in the engineering sciences (Holmes, 2001). As shown for a Douglas fir, a tree can react to dynamic wind loads like a system of coupled damped oscillators. This concept has been confirmed theoretically by Rodriguez et al. (2008). The interaction between the different elements leads to a higher damping ratio, and additionally to less strain on the stem as compared with a structure with much stiffer side branches. Multiple resonance damping is therefore essential for the survival of trees and other large plants growing in windy environments. An interesting but a poorly understood topic is the effects of epiphytes and lianas on the damping of the canopies of tropical trees. In theory, the eccentric loads created by epiphytes and the potential for lianas to entangle branches can increase damping ratios and produce coupled damped oscillators.

Buttressed root systems

As noted in the introduction, some trees manifest features that can set them apart from trees growing in other ecosystems. One of these features is the formation of buttressed root systems (Richards, 1952) (Figure 30.9). Various theories have been proposed to account for the formation of these triangular flanges joining the roots to the lower portions of trunks. Black and Harper (1979) suggested that buttresses prevent lianas from climbing trees, a hypothesis that was disproved by Boom and Mori (1982). Senn (1923) and Richards (1952) suggested that buttresses provide mechanical support, a hypothesis that has been supported by indirect observations that correlated tree development with environmental conditions (e.g. Richter, 1984; Lewis, 1988). For example, buttresses rarely if at all develop on the trunks of trees with well-developed tap roots (e.g. Francis, 1924; Corner, 1988), whereas buttressing is correlated with emergent canopy trees (Richards, 1952; Smith, 1972) and with species growing in shallow waterlogged or weak silty soils (Richards, 1952). An additional correlation that tends to support the biomechanical hypothesis is that buttresses on the upwind sides of trunks tend to be more well developed and extensive than those growing on the leeward sides of trunks (Senn, 1923; Baker, 1973; Lewis, 1988; Warren et al., 1988). Perhaps for these and other reasons, Henwood (1973) proposed that buttresses act as tensile elements. Along similar lines, Mattheck (1991, 1993) noted that buttresses are often associated with bayonet-like sinker roots (Jenik, 1978; Baillie and Mamit, 1983) and suggested that sinker roots in tandem with buttresses provide a robust mechanism to resist wind throw (Figure 30.10a).

The biomechanics of sinker roots attached to buttresses was examined in a seminal study provided by Crook et al. (1997) who investigated the anchorage mechanics of the buttressed root systems of *Aglaia* and *Nephelium* and compared it with the anchorage mechanics of the non-buttressed trunks of *Mallotus*. Using winches to simulate wind throw and strain gauges, Crook et al. (1997) exerted bending forces sufficient to rotate the root systems of trees and noted the following:

1 all trees failed in their root systems (and not by trunk failure);
2 despite differences among the species used in the study, windward buttressed laterals with or without sinker roots either pulled out of the ground or delaminated, while leeward buttresses pushed into the ground and broke near or at their ends (Figure 30.10b–c);
3 buttresses with sinker roots resisted simulated wind throw better than those without sinker roots; and

Figure 30.9 An example of a buttressed root system developing at the base of the trunks of *Ficus macrophylla* (National Tropical Botanical Garden, Kauai, Hawaii)

Source: courtesy of Edward D. Cobb (Cornell University)

4 in the non-buttressed root system, upwind roots extended in tension, uprooted, but rarely broke, leeward roots buckled appreciably, and taproots bent and compressed the leeward soil profile that often produced an upwind crevice (Figure 30.10d).

Some of the results reported by Crook et al. (1997) are in disagreement with the Mattheck model (Figure 30.10a). For example, the majority of buttresses examined in the study lacked sinker roots. Likewise, the Mattheck model posits that buttresses strengthen anchorage by preventing delamination at the junction of windward roots and the trunk, whereas windward buttresses with sinker roots delaminated rather than uprooted in the Crook et al. (1997) study. Nevertheless, trees with buttressed root systems had almost twice the anchorage strength as similar sized trees lacking buttressed systems, and leeward and windward buttresses appear to function in compression and tension, respectively.

The question as to why some species regularly produce buttressed root systems while others do not remains unanswered. It is possible that wood density may play a part in this, because tree species with higher wood densities may have thinner trunks and thinner roots that may be insufficient to provide anchorage when individual trees reach threshold critical

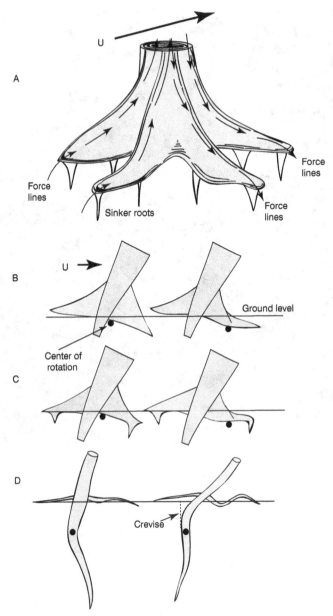

Figure 30.10 Biomechanical behavior of buttressed root systems. (a) Mattheck's model for the distribution of mechanical forces running from sinker roots through buttresses

As the trunk is bent by wind (with speed U), the bending force is transmitted to upwind sinker roots that resist upward forces, whereas leeward roots and buttresses resist downward forces. Compare with Figure 30.8. (b–d) Trunk, buttress, and sinker root displacements resulting from mechanical simulations of wind throw. (b) Displacements of buttressed trunks without sinker roots at two different rotation centers (indicated by black circles). (c) Displacements of buttressed trunks with sinker roots at two different rotation centers (indicated by black circles). (d) Displacements of an unbuttressed trunk with superficial roots subjected to modest and large bending forces (left and right, respectively). For details, see text

Sources: (a) Mattheck (1991, 1993); (b–d) adapted from Crook et al. (1997)

heights. The scaling of stem and root diameters with respect to growth in height is likely to be another factor, because thicker stems and roots can compensate for lower wood densities or greater heights by virtue of magnifying second moments of areas. Clearly, many factors conspire to cope with anchorage requirements and each species 'solves' these requirements using a multi-factorial approach based on its unique combination of functional traits.

A brief comment about palms and bamboos

The preceding treatment of tree biomechanics may give the erroneous impression that all forms of elastic instability are biologically disadvantageous or mechanical failures. The growth and mechanics of arborescent sized palms and bamboos do much to show that this impression is incorrect. For example, provided that stems do not undergo tissue failure, large elastic deflections are permissible as illustrated by the noticeable deflections from the vertical of many palm and bamboo stems (Figure 30.11). The biomechanics of palms and bamboos are also informative because the former illustrates how many plants have accommodated the spatial distribution of bending stresses, whereas the latter illustrates some of the benefits of possessing a hollow, tubular stem.

Anatomical inspection of the stem cross sections through palms shows that the number of primary vascular strands running through stems increases from the center toward the perimeter of cross sections where bending stresses reach their maximum intensities (see Figure 30.2). These strands are extremely strong when tested in tension, and their location

Figure 30.11 An example of a horizontal, structurally stable palm trunk (National Tropical Botanical Garden, Kauai, Hawaii)

Source: courtesy of Edward D. Cobb (Cornell University)

in the stem is optimal for dealing with bending stresses. It is also worth noting that, contrary to popular opinion, many species of palms have secondary growth. Although they lack a vascular cambium, the vascular strands in many species of palms have the capacity to produce secondary phloem fibers, which increase the tensile strength of each strand. Another noteworthy feature of palms is that their fibrous root systems exude organic acids that consolidate sand particles into a cement-like material, which adds ballast to the base of the plant and thus helps it to resist displacements due to wind-induced drag.

Turning to bamboos, recall that hollow tubes can be remarkably economical (because they require less material for their fabrication relative to solid tubes with equivalent diameters) and mechanically advantageous (because they reduce self-loading). However, as noted, hollow tubes can be susceptible to Brazier buckling (see Figure 30.7). One solution to this tradeoff is to shorten tube length without reducing the length of the tube. This sounds contradictory until one realizes that the insertion of transverse plugs or diaphragms into a hollow tube shortens tube length *regionally*. These diaphragms reduce the length-to-diameter ratio of each region of the whole tube and act as restraints against local buckling. This stratagem is seen in the vertical stems of many grasses, but particularly in arborescent size bamboos.

Safety factors

The concepts of safety and reliability are linked. A safe structure functions reliably under the conditions it normally experiences. A safer structure functions reliably under conditions that exceed normal working conditions. The extent to which a structure accommodates unusual working conditions is quantified by the factor of safety. In engineering, this factor is traditionally computed by dividing the working load of a structure or mechanism by its load-bearing capacity (Johnson, 1961; Leitch, 1975). The working (actual) load is the force that a structure normally bears. The load-bearing capacity is the maximum force that the same structure can support without permanently deforming or breaking.

Other measures of reliability can be used to specify the factor of safety. For example, the quotient of a structure's bending (or torsional) working stress and its corresponding bending (or torsional) stress capacity can be used to compute the safety factor. Likewise, the quotient of a working moment and the moment capacity, or the quotient of actual height and critical buckling height, can be used. But, regardless of the features used to compute mechanical reliability, the safety factor invariably reflects some specified criterion for failure – that is, some expectation of how a structure is most likely to fail mechanically. Clearly, the minimum safety factor is one criterion. A structure with a factor of safety less than unity is unreliable even for the working conditions it would normally experience. This lower limit may define the most economical cost of construction, but it invariably defines the workably least safe structure. With few exceptions, structures must have safety factors much greater than one, because every environment can be unpredictable and because structures mechanically fatigue as they age.

In contrast to a well-defined lower limit, the upper limit or 'ceiling' for a particular structure's factor of safety is problematic. It depends on a variety of factors that must be evaluated both in terms of the probability and the consequences of failure (Wainwright et al., 1976). What is the risk that a structure will fail? What are the most likely consequences of failure? These and other questions are typically used by engineers to specify a priori the factor of safety for a particular structure. Unlike engineers, plants are incapable of making these evaluations. Nonetheless, natural selection provides the venue for 'risk and cost management'. Organisms with marginal factors of safety will have a higher probability of dying or reproducing less well than those with higher safety factors. Likewise, organisms with very high factors of safety may be excessively 'overbuilt' and thus squander valuable resources that could be otherwise invested in growth or reproductive effort. Importantly, natural selection cannot anticipate future environmental changes but, over successive generations, it can provide a retrospective 'flywheel' that calibrates the safety factor in terms of the risks and costs a species typically encountered. Because individuals of sexually reproducing species genetically differ, the effects of natural selection on factors of safety must be evaluated at the level of populations (Niklas, 1990a, 1990b, 1998a; Niklas et al., 1999). In this respect, the mechanical capacities of individual trees are known to differ even among individuals of the same size and general appearance. Some trees may produce wood that is intrinsically weaker or stronger than other conspecifics just as some engineering production lines produce materials of varying quality during the same run of production (Weibull, 1939; Niklas, 2000). A population of plants will typically exhibit a range of safety factors, with a frequency distribution manifesting a Weibull frequency distribution. The significance of this distribution is that there are always a few individuals that have high factors of safety such that they will likely survive and perpetuate the species locally after an environmental crisis.

Yet another important consideration is that most plants have a modular construction. That is, they are composed of repeated interconnecting functional units (e.g. shoots and roots) that can have very different load-bearing capacities as a result of growth vagaries or prior damage. The working conditions even of seemingly identical body parts can also differ as a result of micro-environmental conditions (Niklas, 1999b). Stems growing in sheltered locations within the same tree canopy can have lower load-bearing capacities than those growing in exposed locations, because plants respond thigmomorphogenetically to local conditions. That is, individual organ morphology and tissue material properties vary in accordance with the degree of mechanical perturbation (Niklas, 1992). It cannot escape attention that these and other features of plant biomechanics have an obvious bearing on everyday matters of pruning. When stems are exposed by the removal of neighboring portions of a tree, previously sheltered and mechanically reliable body parts may deform or break even under wind conditions that are 'normal.' Pruning also shifts the self-loading conditions of branches or roots. This can stimulate compensatory changes in the growth of other body parts to reacquire overall mechanical stability and reliability. But, during the interim of growth, devastating effects can occur. Yet, in practical terms, it is comparatively easy to measure the load-bearing capacity of organs or entire trees. Individual specimens can be sacrificed and tested in the laboratory to construct a statistical 'picture' of organ load-bearing capacities. This is how engineers typically determine the quality of a production line of steel, aluminum, or nylon thread (Leitch, 1975).

However, engineers have an advantage over the biologist because they know the working conditions of fabricated structures and typically specify the working conditions. In contrast, the biologist must determine the working loads or stresses that organisms sustain, since the conditions of the 'workplace' can vary dramatically, especially for long-lived organisms such as trees. Nonetheless, we can either determine empirically or estimate the general conditions of the workplace, even for large plants (e.g. Niklas and Spatz, 2000). Wind, ice and snow loadings can be measured on a day-by-day or season-by-season basis for a particular site. Alternatively, we can specify a priori the maximum loading conditions that are likely to occur in order to calculate acceptable factors of safety for individual trees growing in particular locations. However, for the reasons stated above, and contrary to one study (Virot et al., 2016), a single critical wind speed at which trees fail cannot be determined (Albrecht et al., 2016).

Concluding remarks

This chapter began with a treatment of the static and dynamic loadings experienced by a vertical cylinder with a radius of 0.03 m and a length (height) of 1.5 m subjected to a uniform wind speed of 11 m/s. It was noted that these loads are far less important than the stresses they produce. The treatment of biomechanical principles that followed allows us to calculate these stresses. As noted, the drag force exerted by a uniform wind speed of 11 m/s was 6.56 kg m σ^{-2} or 6.56 N. The bending moment M_B resulting from this force equals the force times the length of the column, or 6.56 N $\times$ 1.5 m = 9.84 Nm. The maximum stress at the perimeter at the base of this column is given by the formula σ_{max} = 4 M_B /π r^3 = [4 (9.84 Nm)] / [π (0.03 m)^3] = 4.64 $\times$ 10^5 N/m^2. The compressive stress σ_{comp} across the base of the column resulting from static loading is the mass of the column times the acceleration of gravity divided by the cross sectional area of the column. With a density of 844 kg/m^3, we see that σ_{comp} = (ρ pr^2 L g) / (π r^2) = g ρ L = (9.806 m/σ^2) (844 kg/m^3) (1.5 m) = 1.24 $\times$ 10^5 kg m^{-1} σ^{-2} or 1.24 $\times$ 10^5 N/m^2. Thus, the stresses produced by even a modest wind speed are more than 3.5 times the stresses produced by the static loading of the column.

This comparison shows that tree biomechanics is an extremely complex subject because much of what we think we know depends on the questions we ask and how we use equations to answer them. In addition, there are many aspects that are poorly understood, e.g. viscoelastic nonlinear behavior and the effects of lianas on damping. Thus, predicting the mechanical behavior of any particular tree poses many challenges because each tree has its own individual characteristics that reflect its particular genetic composition and growth responses to its particular habitat. The tropics are a huge and diverse ecosystem consisting of many different types of habitats. Likewise, the behavior of a tree observed at any particular time cannot be used to predict the behavior of the same tree a few years hence because trees grow in size and produce new branches and roots and lose branches and roots over the course of their lifetime. Nevertheless, the concepts and equations reviewed here provide some guidance in evaluating the potential mechanical behavior of trees. They provide boundary conditions as to what is possible and what is highly unlikely. However, in the final analysis, predicting the mechanical behavior of any tree requires experience and prudent judgment, attributes that thus far have eluded mathematical description.

References

Albrecht, A., Badel, E., Bonnesoeur, V., Brunet, Y., Constant, T., Défossez, P., de Langre, E., Dupont, S., Fournier, M., Gardiner, B., Mitchell, S. J., Moore, J. R., Moulia, B., Nicoll, B. C., Niklas, K. J., Schelhaas, M.-J., Spatz, H.-C., Telewski, F. W. 2016. Comment on Critical Wind Speed at which Trees Break by Virot et al. 2016. *Physical Review E*, 94: 067001

Baillie, I. C., Mamit, J. D. 1983. Observations on rooting in mixed dipterocarp forest, central Sarawak. *The Malaysian Forester* 46: 369–374.

Baker, H. G. 1973. A structural model of the forces in buttressed tropical rain forest trees (Appendix to Henwood 1973). *Biotropica*, 5: 89–93.

Bastien, R., Douady, S., Moulia, B. 2015. A unified model of shoot tropism in plants: photo-, gravi- and propio-ception. *PLOS Computational Biology*, 11: e1004037. doi:10.1371/journal.pcbi.1004037.

Black, H. L., Harper, K. T. 1979. The adaptive value of buttresses to tropical trees: additional hypothesis. *Biotropica*, 11: 240.

Boom, B. M., Mori, S. A. 1982. Falsification of two hypotheses on liana exclusion from tropical trees possessing buttresses and smooth bark. *Bulletin Torrey Botanical Club*, 4: 447–450.

Braam, J., Davies, R. W. 1990. Rain-, wind-, and touch-induced expression of calmodulin and calmodulin-related genes in *Arabidopsis*. *Cell*, 60: 357–364.

Casada, J. H., Walton, L. R., Swetnam, L. D., 1980. Wind resistance of Burely tobacco as influenced by depth of plants in soil. *Transactions of the American Society of Agricultural Engineering*, 23: 1009–1011.

Chapotin, S. M., Razanameharizaka, J. H., Holbrook, N. M. 2006. A mechanical perspective on the role of large stem volume and high water content in baobab trees (*Adansonia* spp.; Bombacaceae). *American Journal of Botany*, 93: 1251–1264.

Chehab, E. W., Eich, E., Braam, J. 2009. Thigmomorphogenesis: a complex plant response to mechano-stimulation. *Journal of Experimental Botany*, 60: 43-56.

Corner, E. J. 1988. *Wayside Trees of Malaya*. Malaysian Nature Society, Malaysia.

Coutand, C. 2010. Mechanosensing and thigmomorphogenesis, a physiological and biomechanical point of view. *Plant Science*, 179: 168–182.

Coutts, M. P. 1983. Root architecture and tree stability. *Plant and Soil*, 71: 171–188.

Coutts, M. P. 1986. Components of tree stability in Sitka spruce on peaty gley soil. *Forestry*, 59: 171–197.

Crook, M. J., Ennos, A. R. 1996. The anchorage mechanics of mature larch *Larix europea* × *L. japonica*. *Journal of Experimental Botany*, 47: 1507–1517.

Crook, M. J., Ennos, A., Banks, J. R. 1997. The function of buttress roots: a comparative study of the anchorage systems of buttressed (*Aglai* and *Nephileum ramboutan* species) and non-buttressed (*Mallotus wrayi*) tropical trees. *Journal of Experimental Botany*, 48: 1703–1716.

Donaldson, L. A., Singh, A. P. 2016. Reaction wood. In Kim, Y. S., Funada, R., Singh, A. P. (eds), *Secondary Xylem Biology: Origins, Functions and Applications*. Elsevier, Oxford.

Du, S., Yamamoto, F. 2007. An overview of the biology of reaction wood formation. *Journal of International Plant Biology*, 49: 131–143.

Edelin, C., Atger, C. 1994. Stem and root tree architecture: questions for plant biomechanics. *Biomimetics*, 2: 253–266.

Ennos, A. R. 1993. The scaling of root anchorage. *Journal of Theoretical Biology*, 161: 61–75.

Ennos, A. R. 1994. The biomechanics of root anchorage. *Biomimetics*, 2: 129–137.

Francis, W. D. 1924. The development of buttresses in Queensland trees. *Proceedings of the Royal Society of Queensland*, 36: 21–37.

Gardiner, B., Berry, P., Moulia, B. 2016. Review: Wind impacts on plant growth, mechanics and damage. *Plant Science*, 245: 94–118.

Gardiner, B., Barnett, J., Saranpaa, P., Gril, J. 2014. *Biology of Reaction Wood*. Springer-Verlag, Berlin.

Guyette, R. P., Stambaugh, M. 2003. The age and density of ancient and modern oak wood in streams and sediments. *IAWA Journal*, 24: 345–353.

Henwood, K., 1973. A structural model of forces in buttressed tropical rain forest trees. *Biotropica*, 5: 83–89.

Hogan, C. J., Jr., Niklas, K. J. 2004. Temperature and water content effects on the viscoelastic behavior of *Tilia americana* (Tiliaceae) sapwood. *Trees Structure and Function*, 18: 339–345.

Holmes, J. D. 2001. *Wind Loading of Structures*. Spon Press, London.

Jacobs, M. R. 1954. The effect of wind sway on the form and development of *Pinus radiate* D. Don. *Australian Journal of Botany*, 2: 35–51.

Jaffe, M. J. 1973. Thigmomorphogenesis: the response of plant growth and development to mechanical stimulation. *Planta*, 114: 588–594.

Jaffe, M. J., Leopold, A. C., Staples, R. A. 2002. Thigmo responses in plants and fungi. *American Journal of Botany*, 89: 375–382.

James, K. R. 2003. Dynamic loading of trees. *Journal of Arboriculture*, 29: 165–171.

James, K. R., Haritos, N., Ades, P. 2006. Mechanical stability of trees under dynamic loads. *American Journal of Botany*, 93: 1361–1369.

Jenik, J. 1978. Roots and root systems in tropical trees: morphological and ecological aspects. In Tomlinson, P. B., Simmerman, M. H. (eds), *Tropical Trees as Living Systems*. Cambridge University Press, MA, pp. 323–348.

Johnson, R. C. 1961. *Optimum Design of Mechanical Elements*. John Wiley and Sons, New York.

Jonsson, M. J., Froetzi, A., Kalberer, M., Lundström, T., Ammann, W., Stöckli, V. 2007. Natural frequencies and damping ratios of Norway spruce (*Picea abies* [L.] Karst) growing on subalpine forested slopes. *Trees*, 21: 541–548.

Jungikl, K., Goebbels, J., Burgert, I., Fratzl, P. 2009. The role of material properties for the mechanical adaptation at branch junctions. *Trees-Structure and Function*, 23: 79–84.

Knight, T. A. 1803. Account of some experiments on the descent of sap in trees. *Philosophical Transactions of the Royal Society of London*, 96: 277–289.

Knight, T. A. 1811. On the causes which influence the direction of the growth of roots. *Philosophical Transactions of the Royal Society of London*, 1811: 209–219.

Koehler, A. 1933. *Causes of Brashness in Wood*. Technical Bulletin No. 342, United States Department of Agriculture, Washington, DC.

Leitch, R. D. 1975. *Reliability Analysis for Engineers: An Introduction*. Oxford University Press, Oxford.

Lewis, A. R. 1988. Buttress arrangement in *Pterocarpus officinalis* (Fabaceae): effects of crown asymmetry and wind. *Biotropica*, 20: 280–285.

Marshall, T. J., Holmes, J. W. 1988. *Soil Physics*, 2nd edn. Cambridge University Press, Cambridge.

Mattheck, C. 1991. *Trees: The Mechanical Design*. Springer Verlag, Berlin.

Mattheck, C. 1993. *Design in der Natur: Der Baum als Lehrmeister*. Rombach Verlag, Freiburg.

Mayhead, G. J. 1973. Sway periods of forest trees. *Scottish Forestry*, 27: 19–23.

Milne, R. 1991. Dynamics of swaying of *Picea sitchensis*. *Tree Physiology*, 9: 383–399.

Mitchell, C. A., Myers, P. N. 1995. Mechanical stress regulation of plant growth and development. *Horticultural Reviews*, 17: 1–42.

Moore, J. R., Maguire, D. A. 2004. Natural sway frequencies and damping ratios of trees: concepts, review and synthesis of previous studies. *Trees*, 18: 195–203.

Moulia, B., Coutand, C., Lenne, C., 2006. Posture control and skeletal mechanical acclimation in terrestrial plants: implications for mechanical modeling of plant architecture. *American Journal of Botany*, 9: 1477–1489.

Niklas, K. J. 1990a. Safety factors in vertical stems: evidence from *Equisetum hyemale*. *Evolution*, 43: 1625–1636.

Niklas, K. J. 1990b. Determinate growth of *Allium sativum* peduncles: evidence of determinate growth as a design factor for biomechanical safety. *American Journal of Botany*, 77: 762–771.

Niklas, K. J. 1992. *Plant Biomechanics*. University of Chicago Press, Chicago, IL.

Niklas, K. J. 1994. The allometry of safety-factors for plant height. *American Journal of Botany*, 81: 345–351.

Niklas, K. J. 1996. Differences between *Acer saccharum* leaves from open and wind-protected sites. *Annals of Botany*, 78: 61–66.

Niklas, K. J. 1998a. The influence of gravity and wind on land plant evolution. *Review of Palaeobotany and Palynology*, 102: 1–14.

Niklas, K. J. 1998b. A statistical approach to biological factors of safety: bending and shearing in *Psilotum* axes. *Annals of Botany*, 82: 177–187.

Niklas, K. J. 1999a. Variation of the mechanical properties of *Acer saccharum* roots. *Journal of Experimental Botany*, 50: 193–200.

Niklas, K. J. 1999b. Changes in the factor of safety within the superstructure of a dicot tree. *American Journal of Botany*, 86: 688–696.

Niklas, K. J. 2000. Computing factors of safety against wind induced tree stem damage. *Journal of Experimental Botany*, 51: 797–806.

Niklas, K. J., Spatz, H.-C. 1999. Methods for calculating factors of safety for plant stems. *Journal of Experimental Biology*, 202: 3273–3280.

Niklas, K. J., Spatz, H.-C. 2000. Wind-induced stresses in cherry trees: evidence against the hypothesis of constant stress levels. *Trees*, 14: 230–237.

Niklas, K. J., Spatz, H.-C. 2010. Worldwide correlations of mechanical properties and green wood density. *American Journal of Botany*, 97: 1587–1594.

Niklas, K. J., Spatz, H.-C. 2012. *Plant Physics*. University of Chicago Press, Chicago, IL.

Panshin, A. J., de Zeeuw, C. 1980. *Textbook of Wood Technology*, 4th edn. McGraw-Hill, New York.

Peltola, H., Kellomöki, S., Hassinen, A., Lemittinnen, M., Aho, J. 1993. Swaying of trees as caused by wind: analysis of field measurements. *Silva Fennica*, 27: 113–126.

Richards, P. W. 1952. *The Tropical Rain Forest*. Cambridge University Press, Cambridge.

Richter, W. 1984. A structural approach to the function of buttresses of *Quaranbea asterolepis*. *Ecology*, 65: 1429–1435.

Rodriguez, M., De Langre, E., Moulia, B., 2008. A scaling law for the effects of architecture and allometry on tree vibration modes suggests a biological tuning to modal compartmentalization. *American Journal of Botany*, 95: 1523–1537.

Ruellé, J. 2014. Morphology, anatomy and ultrastructure of reaction wood. In Gardiner, B., Barnett, J., Saranpaa, P., Gril, J. (eds), *Biology of Reaction Wood*. Springer-Verlag, Berlin, pp. 13–35.

Senn, G. 1923. Uber die Ursachen der Brettwurzelbildung bei der Pyramiden-Pappel. *Verhandlungen des Naturforschenden Gesellschaft in Basel*, 35: 405–435.

Smiley, E. T. 1989. *Hugo Broken Trees*. Bartlett Tree Research Laboratories Technical Report, F. A. Bartlett Tree Expert Company, Charlotte, NC.

Smiley, E. T., Fraedrich, B. R., Fengler, P. H. 2000. Hazard tree inspection, evaluation, and management. In Kuser, J. E. (ed.), *Handbook of Urban and Community Forestry in the Northeast*. Springer, New York, pp. 243–260.

Smith, A. P. 1972. Buttressing of tropical trees: a descriptive model and new hypothesis. *American Naturalist*, 106: 32–46.

Spatz, H.-C., Niklas, K. J. 2013. Modes of failure in tubular plant organs. *American Journal of Botany*, 100: 332–336.

Spatz, H.-C., Brüchert, F., Pfisterer, J. 2007. Multiple resonance damping or how do trees escape dangerously large oscillations? *American Journal of Botany*, 94: 1603–1611.

Stokes, A., Mattheck, C. 1996. Variation of wood strength in tree roots. *Journal of Experimental Botany*, 47: 693–699.

Stokes, A., Fitter A. H., Coutts M. P. 1995. Responses of young trees to wind and shading: effects on root architecture. *Journal of Experimental Botany*, 46: 1139–1146.

Stokes, A., Ball, J., Fitter, A. H., Brian, P., Coutts, M. P. 1996. An experimental investigation of the resistance of model root systems to uprooting. *Annals of Botany*, 78: 415–421.

Telewski, F. W. 1995. Wind induced physiological and developmental responses in trees. In Coutts, M. P., Grace, J. (eds), *Wind and Tree*. Cambridge University Press, Cambridge, pp. 237–263.

Telewski, F. W. 2006. A unified hypothesis of mechanoperception in plants. *American Journal of Botany*, 93: 1466–1476.

Telewski, F. W. 2012. Is windswept tree growth negative thigmotropism? *Plant Science*, 184: 20–28.

Telewski, F. W. 2016. Flexure wood: mechanical stress induced secondary xylem formation. In Kim, Y. S., Funada, R., Singh, A. P. (eds), *Secondary Xylem Biology: Origins, Functions and Applications*. Elsevier, Oxford, pp. 73–91.

Telewski, F. W., Jaffe M. J. 1986a. Thigmomorphogenesis: field and laboratory studies of *Abies fraseri* (Pursh) Poir, in response to wind and mechanical perturbation. *Physiologia Plantarum* 66: 211–218.

Telewski, F. W., Jaffe, M. J. 1986b. Thigmomorphogenesis: anatomical, morphological and mechanical analysis of genetically different sibs of *Pinus taeda* in response to mechanical perturbation. *Physiologia Plantarum*, 66: 219–226.

Telewski, F. W., Gardner B. A., White, G., Plovanich-Jones, A. 1997. Wind flow around multi-storey buildings and its influence on tree growth. In *Plant Biomechanics, Conference Proceedings, The University of Reading, UK*, vol. I, September 7–12, 1997, pp. 185–192.

Theophrastus. 1976. *De causis plantarum: Volume I: Books 1–2* (trans. Einarson, B., Link, G. K. K.). Harvard University Press, Cambridge, MA.

Timell, T. E. 1986a. *Compression Wood in Gymnosperms Vol.1*. Springer-Verlag, Berlin.

Timell, T. E. 1986b. *Compression Wood in Gymnosperms Vol.2*. Springer-Verlag, Berlin.

Timell, T. E. 1986c. *Compression Wood in Gymnosperms Vol.3*. Springer-Verlag, Berlin.

Valinger, E., Lundqvist, L., Sundberg, B. 1995. Mechanical bending stress applied during dormancy and/or growth stimulates stem diameter growth of Scots pine seedlings. *Canadian Journal of Forest Research*, 25: 886–890.

Virot, E., Ponomarenko, A., Dehandschoewercker, E., Quéré, D., Clanet, C. 2016. Critical wind speed at which trees break. *Physical Review E*, 93: 023001.

Vogel, S. 1981. *Life in Moving Fluids: The Physical Biology of Flow*. Willard Grant, Boston, MA.

Vogel, S. 1996. Blowing in the wind: storm-resisting features of the design of trees. *Journal of Arboriculture*, 22: 92–98.

Wagener, W. W. 1963. *Judging Hazards from Native Trees in California Recreational Amas: A Guide For Professional Foresters*. Research Paper PSW-P, US Forest Service, Washington, DC.

Wainwright, S. A., Biggs, W. D., Currey, J. D., Gosline, J. M. 1976. *Mechanical Design in Organisms*. John Wiley & Sons, New York.

Warren, S. D., Black, H. L., Eastmond, D. A., Wtaaley, W. H. 1988. Structural function of buttresses of *Tachigalia versicolor*. *Ecology*, 62: 532–536.

Weibull, W. 1939. *A Statistical Theory of the Strength of Materials*. Royal Swedish Institute, Stockholm, Sweden.

Wilson, B. F., Archer, R. R. 1977. Reaction wood-induction and mechanical action. *Annual Review of Plant Physiology*, 28: 23–43.

31

TREE RISK ASSESSMENT

E. Thomas Smiley, Nelda P. Matheny and Sharon J. Lilly

Introduction

The benefits of trees in the urban landscape are undisputed. Trees can aid in the management of the urban heat island and global climate change; they can provide human health benefit; and they can provide benefits to commerce. These and many more benefits greatly outweigh the overall low level of risk usually associated with a well-managed urban forest. However, all trees will eventually fail. Therefore, property owners and government agencies need to manage their trees with the knowledge that trees can pose risk and they need to act to manage risk.

Tree risk management is the application of policies, procedures, and practices used to identify, evaluate, mitigate, monitor, and communicate tree risk. Arborists and urban foresters should have policies in place that outline how trees should be assessed for risk and when trees with an unacceptable level of risk should be mitigated. These policies may include a schedule for tree inspection, using a defined methodology during inspections and guidelines for prescribing mitigation, when necessary.

The intent of this chapter is to define terms associated with tree risk in the urban environment, explore options for systematic evaluation of trees, outline one commonly used system and briefly discuss risk mitigation. Much of this chapter has been adapted from the International Society of Arboriculture's *Best Management Practice for Tree Risk Assessment* (Smiley et al., 2011).

Box 31.1 Key principles for tree risk management

It is impossible to maintain trees free of risk; some level of risk must be accepted to experience the benefits that trees provide. The National Tree Safety Group (NTSG), which is a partnership of organizations in the United Kingdom, has drafted a guidance document that identifies five key principles for tree risk management. This document provides a foundation for balancing tree risk and the benefits that trees provide:

- Trees provide a wide variety of benefits to society.
- Trees are living organisms and naturally lose branches or fall.
- The risk to human safety is extremely low.
- Tree owners have a legal duty of care.
- Tree owners should take a balanced and proportionate approach to tree safety management.

What is risk?

Risk has been defined as the combination of the likelihood of an event and severity of the potential consequences (adapted from ISO 31000). In the context of trees, risk is the likelihood of a conflict or tree failure occurring and affecting a target, and the severity of the associated consequences – personal injury, property damage, or disruption of activities.

Key terms used in tree risk assessment include the following:

- Event is the occurrence of a particular set of circumstances. In tree risk assessment, an event is a tree affecting a target.
- Tree risk assessment is the systematic process to identify, analyze, and evaluate tree risk.
- Tree risk evaluation is the process of comparing the assessed risk against given risk criteria to determine the significance of the risk.
- Targets (risk targets) are people, property, or activities that could be injured, damaged, or disrupted by a tree failure.
- Target zone is the area where a tree or branch is likely to land if it were to fail.
- Occupancy rate is the amount of time that a target is within the target zone.
- Tree failure is the breakage of stem, branches, or roots, or the loss of mechanical support in the root system.
- Likelihood is the chance of an event occurring. In the context of tree failures, likelihood refers to (1) the chance of a tree failure occurring, (2) the chance of impacting a specific target, and (3) the combination of the likelihood of a tree failing and the likelihood of impacting a specific target.
- Consequences are the effects or outcome of an event. In tree risk assessment, consequences include personal injury, property damage, or disruption of activities due to the event.
- Hazard is a likely source of harm. In relation to trees, a hazard is the tree part(s) identified as a likely source of harm.
- Hazardous trees are trees that have been assessed and found to be likely to fail and cause an unacceptable degree harm.

Types of risk associated with trees

Trees can pose a variety of risks, which are categorized into two basic groups: conflicts and structural failures. Both types of risk need to be considered by urban forest managers. However, most arboricultural publications, including this chapter, focus primarily on structural failures.

Conflicts

Risk can arise when there are conflicts between trees and human activities. As trees grow, they may produce potentially problematic flowers, fruit, roots, branches and leaves, and these may conflict with pavement, the structures around them, and people. In some situations these conflicts can cause harm. One common conflict in cities is sidewalk lifting by tree roots. This can lead to a tripping hazard, especially for the elderly. Another conflict is between power lines and trees. This contact can interrupt power supplies, start fires, and can injure or kill people.

While the majority of this chapter focuses on tree failure, conflicts should not be ignored. They should be identified and managed in much the same way as are tree failures.

Failures

Tree failure is breakage of stems, branches, or roots, or loss of mechanical support from the root system. Structural failures occur when the forces acting on a tree exceed the strength of the tree structure or soil supporting the tree. Even a structurally strong tree will fail when a load or force is applied that exceeds the strength of one or more of its parts.

Most tree structural failures involve a combination of structural defects and/or conditions, such as the presence of decay or poor structure, and a loading event, such as a strong wind.

Approaches to risk assessment

Before a tree risk assessment takes place, it is important to establish the context of the assignment. Context defines the parameters of the risk assessment including objectives, how risk will be evaluated, communication flow, applicable policies or legal requirements, and limitations of the risk assessment. Tree risk assessment is the systematic process to identify, analyze, and evaluate tree risk. The manner in which this process is applied depends on the context and the methods used to carry out the risk assessment.

The two primary approaches to risk assessment are quantitative and qualitative. Each has advantages and limitations, and each may be appropriate with different objectives, requirements, resources, and uncertainties. Both the quantitative and qualitative approaches are valid when applied properly with reliable data and valid assumptions.

Quantitative risk assessment

Quantitative risk assessment estimates numeric values for the probability and consequences of events, and then produces a numeric value for the level of risk, typically using the formula:

Risk = probability × consequences

An advantage of quantitative assessment is that tree risk can be compared not only to other trees but also to other types of risk, as might be necessary for municipal decisions in which resources must be allocated among departments. However, our ability to quantify probability is often limited with trees because we have little systematically-collected data on which to base probabilities. Since numeric data are not always available and both systematic and statistical uncertainties can be high, full quantitative analysis is often not warranted or practical for tree risk assessment.

Qualitative risk assessment

Qualitative risk assessment is the process of using categorized ratings of the likelihood and consequences of an event to determine a risk level and evaluate the level of risk against qualitative criteria. Categorical ratings can be words or numbers. The tendency with qualitative numeric systems is to add or multiply the numbers associated with different categories. However, risk professionals caution that addition or multiplication of these 'ordinal numbers' is mathematically incorrect.

Categorical systems using named categories typically combine the ratings in a matrix to categorize risk. This is the basis of the tree risk assessment procedure described in the International Society of Arboriculture's (ISA) *Best Management Practice for Tree Risk Assessment* (Smiley et al., 2011).

There is typically a considerable level of uncertainty associated with tree risk assessment due to our limited ability to predict natural processes (rate of progression of decay, response growth, etc.), weather events, traffic and occupancy rates, and potential consequences of tree failure. Sources of uncertainty should be understood and communicated to the risk manager/tree owner. Subjectivity and ambiguity are inherent limitations of the qualitative approach. In order to increase reliability and consistency of application, it is important to provide clear explanations of the terminology and significance of the ratings defined for likelihood, consequences, and risk. Moreover, training and practice with the use of any risk management system is critical to its application. To provide a uniform system and trained assessors, the ISA has developed a *Tree Risk Assessment Qualification* (TRAQ). There are also other systems such as the *Quantitative Tree Risk Assessment* (QTRA) program that provide a system, training and support for assessors using it.

Levels of risk assessment

Tree risk assessments can be conducted at different levels and may employ various methods and tools. The level selected should be specified and/or discussed with the client prior to conducting an assessment. The level(s) should be appropriate for the assignment. Three levels of tree risk assessment have been defined:

- *Level 1:* limited visual;
- *Level 2:* basic; and
- *Level 3:* advanced.

If conditions cannot be adequately assessed at the specified level, the assessor may recommend a higher level or different assessment. In addition to specifying the level of inspection, tree risk assessors also should describe pertinent details regarding the method. For example, a Level 1 assessment can be done by walking by, driving past, or flying above the trees. The method used will greatly influence the cost and reliability of the results.

At any level of assessment, if a situation is encountered where tree failure is imminent and a high-value target is present and likely to be struck by the failure, the situation should be reported to the client as soon as possible. In addition, immediate action may be required to restrict access to the target zone.

Level 1: Limited visual risk assessment

The Level 1 risk assessment is a visual assessment from a specified perspective of an individual tree or a population of trees near specified targets. They are conducted to identify obvious defects or specified conditions. Typically, one or two of the three factors used when performing a tree risk assessment (likelihood of failure, likelihood of impact, consequences) is/are considered as a constant. A Level 1 assessment usually focuses on identifying trees with imminent and/or probable likelihood of failure that are adjacent to specific pre-defined targets and could impact the target(s).

The assessor performs a limited visual assessment, from one side or by an aerial flyover, typically looking for obvious defects such as dead trees, large cavity openings, large dead or broken branches, fungal fruiting structures, large cracks, and severe or uncorrected leans. In addition, the client may specify certain conditions of concern, such as lethal pests or symptoms associated with root decay. When a limited visual assessment is conducted by a trained professional, it can usually provide the manager with an adequate level of information to make decisions.

A constraint of limited visual inspections is that some conditions may not be visible from a one-sided inspection of a tree, nor are all conditions visible on a year-round basis. Also, a Level 1 risk assessment may not be adequate to make a risk mitigation recommendation. The assessor may use the Level 1 inspection to determine which trees require further inspection at the basic or advanced levels after which an appropriate mitigation can be recommended. Limited visual assessments are often done on a specified schedule and/or immediately after storms to rapidly assess a tree population. They are the fastest, but least thorough, means of assessment.

Level 2: Basic assessment

A Level 2 or basic assessment is a detailed visual inspection of a tree and its surrounding site, and a synthesis of the information collected. It requires that a tree risk assessor walk completely around the tree – looking at the site, buttress roots, trunk, and branches. A basic assessment may include the use of simple tools to gain additional information about the tree or defects. This is the level of assessment that is commonly performed by arborists in response to clients' requests for individual tree risk assessments.

The primary limitation of a basic assessment is that it includes only conditions that are detected from a ground-based inspection on the day of the assessment. Internal, belowground, upper-crown conditions, and certain types of decay may be impossible to see or difficult to assess and may remain undetected.

Level 3: Advanced assessment

Advanced assessments are performed to provide detailed information about specific tree parts, defects, targets, or site conditions. They usually are conducted in conjunction with or after a basic assessment if the tree risk assessor needs additional information and the client approves the additional service. Specialized equipment, data collection and analysis, and/or expertise are usually required for advanced assessments. These assessments generally take more time and are more expensive.

Procedures and methodologies should be selected and applied as appropriate, with consideration for what is reasonable and proportionate to the specific conditions and situations. The risk manager/property owner should consider the value of the tree to the owner and community, the possible consequences of failure, and the time and expense to provide the advanced assessment. Advanced assessments can provide additional information that may make the difference between recommending tree retention or removal. The tree risk assessor should identify what additional information is needed and recommend the technique to be used.

While there are many types of advanced assessments that can be conducted including methods to detect internal decay, tree stability, site occupancy, weather conditions and other factors, tree risk assessors are cautioned that all technologies involve some uncertainty. Each technology has limitations; any evaluation of an individual tree or target will not be an accurate measure, but a qualified estimation.

Standard safe work practices/procedures should be applied for all levels of assessment.

Risk categorization

The result of most tree risk assessments is a rating of tree risk. This rating may be used to develop mitigation plans and/or for work prioritization. Tree risk assessment reports should include this risk rating for each tree that was assessed. The risk category is then compared to

Table 31.1 Likelihood matrix (likelihood of the event occurring)

Likelihood of failure	Likelihood of impacting target			
	Very low	*Low*	*Medium*	*High*
Imminent	Unlikely	Somewhat unlikely	Likely	Very likely
Probable	Unlikely	Unlikely	Somewhat unlikely	Likely
Possible	Unlikely	Unlikely	Unlikely	Somewhat unlikely
Improbable	Unlikely	Unlikely	Unlikely	Unlikely

the level of risk that is acceptable to the client, controlling authority, or societal norms. If the risk category defined for the tree risk exceeds the level of acceptable risk, mitigation options should be presented.

The ISA BMP assessment approach

A primary goal of tree risk assessment is to provide information about the level of risk posed by a tree over a specific time period. This is accomplished in qualitative tree risk assessment by first determining the categories for likelihood and consequences of tree failure. The factors to be considered are the likelihood of a tree failure impacting a target and the consequences of the failure. The likelihood of a tree failure impacting a target is determined by considering two factors. First is the likelihood of a tree failure occurring within a specified time frame. Second is the likelihood of the failed tree or branch impacting the specified target. These two factors are evaluated and categorized using a matrix to estimate the likelihood of the combined event: a tree failure occurring and the tree impacting the specified target. The likelihood of that combined event is then compared with the expected consequences of a failure impacting the target to determine a level of risk.

Likelihood of failure

Before assessing the likelihood of failure, a time frame must be specified to put the likelihood rating in context. Without a stated time frame, the rating for likelihood of failure is meaningless. The longer the time frame, the less reliable the rating because conditions that affect failure are prone to change over time. A one to three year time frame for rating tree risk is common. Time frames greater than five years are often not appropriate because the uncertainty within that time frame is excessive. Long time frames may unnecessarily increase the likelihood of failure, which could lead to unnecessary, mitigation.

Assessments may also be conducted using several time frames. For example, an assessment can be done for the next 12 months and for the next five years. In some cases, the results will be different. Obviously, the longer the time frame, the greater the uncertainty. Sometimes, having assessments consider more than one time frame can be helpful for the tree owner/manager in making mitigation choices.

With the ISA BMP methodology, the likelihood of failure is categorized as follows:

- *Improbable:* The tree or branch is not likely to fail during normal weather and may not fail in extreme weather within the specified time frame.
- *Possible:* Failure may be expected in extreme weather, but it is unlikely during normal weather conditions within the specified time frame.
- *Probable:* Failure may be expected under normal weather conditions within the specified time period.
- *Imminent:* Failure has started or is most likely to occur in the near future, even if there is no significant wind or increased load. This is a rare occurrence for a risk assessor to encounter, and may require immediate action to protect people from harm.

Likelihood of impact

The second factor to be considered is the likelihood of the failed tree impacting the target. To estimate this likelihood, the arborist should consider the occupancy rate of any targets within the target zone, the direction of fall, and any protective factors that could affect the impact of the failed tree or part as it falls toward the target.

The likelihood of impacting a target can be categorized using the following guidelines:

- *Very low:* The chance of the failed tree or branch impacting the specified target is remote.
- *Low:* It is not likely that the failed tree or branch will impact the target.
- *Medium:* The failed tree or branch could impact the target.
- *High:* The failed tree or branch is likely to impact the target.

Usually targets are assessed on an individual basis, but they can also be combined to provide the client with a better perspective of what would happen in case of a tree failure.

Consequences

Consequences of a tree failing and impacting a target are a function of the value of the target and the amount of injury, damage, or disruption (harm) that could be caused by the impact of the failure. The amount of damage depends on the part size, fall characteristics, fall distance, and any factors that may protect the target from harm.

Consequences of failures can be categorized as follows:

- *Negligible* consequences are those that involve low-value property damage or disruption that can be replaced or repaired, and do not involve substantial personal injury.
- *Minor* consequences are those that involve low-to-moderate property damage, small disruptions to traffic or a communication utility, or minor personal injury.
- *Significant* consequences are those that involve property damage of moderate-to-high value, considerable disruption, or personal injury.
- *Severe* consequences are those that could involve serious personal injury or death, high-value property damage, or disruption of important activities.

Table 31.2 Risk matrix (categorizing tree risk)

Likelihood of failure and impact	*Consequences of failure*			
	Negligible	*Minor*	*Significant*	*Severe*
Very likely	Low	Moderate	High	Extreme
Likely	Low	Moderate	High	High
Somewhat likely	Low	Low	Moderate	Moderate
Unlikely	Low	Low	Low	Low

Tree risk rating

The ISA BMP methodology uses two matrices to combine likelihood of failure, likelihood of impact, and consequences to determine a risk rating. The matrix approach was selected for use in the ISA BMP because of its broad acceptance in the field of risk assessment, ease of use, and effectiveness for rating tree risk. These matrices were designed specifically for the evaluation of risk posed by tree failures. The limitations associated with using a matrix include the inherent subjectivity and uncertainty associated with the selection of both the likelihood and consequence factors, and the lack of comparability to other types of risk assessed using other means.

In the tree risk assessment matrix, four terms are used to define levels of risk: low, moderate, high, and extreme. These risk ratings are used to communicate the level of risk and to assist in making recommendations to the owner or risk manager for mitigation and inspection frequency. The priority for action depends upon the risk rating and risk tolerance of the owner or manager.

- *Extreme:* The extreme-risk category applies in situations in which failure is *imminent* and there is a *high* likelihood of impacting the target, and the consequences of the failure are *severe.* The tree risk assessor should recommend that mitigation measures be taken as soon as possible. In some cases this may mean immediate restriction of access to the target zone area to avoid injury to people.
- *High:* High-risk situations are those for which consequences are *significant* and likelihood is *very likely* or *likely,* or consequences are *severe* and likelihood is *likely.* This combination of likelihood and consequences indicates that the tree risk assessor should recommend mitigation measures be taken. The decision for mitigation and timing of treatment depends upon the risk tolerance of the tree owner or risk manager. In populations of trees, the priority of high-risk trees is second only to extreme-risk trees.
- *Moderate:* Moderate-risk situations are those for which consequences are *minor* and likelihood is *very likely* or *likely;* or likelihood is *somewhat likely* and consequences are *significant* or *severe.* The tree risk assessor should recommend mitigation. The decision for mitigation and timing of treatment depends upon the risk tolerance of the tree owner or manager. In populations of trees, moderate-risk trees represent a lower priority than high- or extreme-risk trees.
- *Low:* The low-risk category applies when consequences are *negligible* and likelihood is *unlikely;* or consequences are *minor* and likelihood is *somewhat likely.* Some trees with this level of risk may benefit from mitigation or maintenance measures, but immediate action is not usually required. Tree risk assessors may recommend retaining and monitoring these trees or mitigating the risk.

Risk perception and acceptable risk

How people perceive risk and their need for personal safety is inherently subjective; therefore, risk tolerance and action thresholds vary among tree owners/managers. What is within the tolerance of one person may be unacceptable to another. It is impossible to maintain trees completely free of risk – some level of risk must be accepted to experience the benefits that trees provide.

Acceptable risk is the degree of risk that is within the owner/manager's or controlling authority's tolerance, or that which is below a defined threshold. Some countries have a framework directive, which defines thresholds of risk tolerance and acceptability. Municipalities, utilities, and property managers may have a risk management plan that defines the level of acceptable risk. Safety may not be the only basis used by the risk manager to establish acceptable levels of risk; budget, a tree's historical or environmental significance, aesthetics, and other factors also may come into the decision-making process. Tree risk assessors may also assess risk within a population of trees and use that information to prioritize remedial action.

Risk mitigation

Mitigation is the process of reducing risk. Measures to mitigate risk can be arboricultural, to reduce the likelihood of failure or the likelihood of impact; or they can be target-based, to reduce the likelihood of impact or consequences of failure.

In some low-use locations, dead and decaying trees may be retained for wildlife habitat or other uses. Selection of suitable wildlife habitat trees must consider the risk, as well as its value for wildlife. One management strategy is to ensure that wildlife habitat trees are maintained at a height shorter than the distance to the nearest target.

Some species of over-mature trees in natural settings may reconfigure as they age and deteriorate, a process sometimes called 'natural retrenchment.' They may continue to grow trunk diameter while branches die and fail – reducing overall height of the tree and increasing stability. Where tree risk is a concern, tree risk assessors can imitate this process by recommending crown reduction. Eventually, however, even tall stumps without branches will fail when the roots decay. For this reason, the integrity of roots and the trunks of trees with targets that are retained for wildlife habitat should be monitored and the trees removed if the risk exceeds allowable thresholds.

Arboricultural techniques

There are three common arboricultural treatments that can be used to provide an immediate reduction in risk:

- **Pruning.** Dead, dying, and weakly attached branches can be pruned in accordance with the applicable national pruning standards or the ISA BMP for tree pruning (Gilman and Lilly, 2008). Wind resistance can be reduced with reduction pruning and, to some extent, thinning. To maintain the impact of pruning for a longer time period, application of tree growth regulators can be considered.

  Topping is not recommended due to the long-term problems with weak sprouts and the entry of wood decay. Crown raising can eliminate lower branches that could be interfering with structures, pedestrian or vehicle traffic, signs, or safe views. Excessive raising, however, can reduce taper development, change sway patterns, and limit the tree's ability to damp the effect of dynamic wind loading.

- **Installing a structural support system.** Structural support systems can be installed to limit movement of codominant stems and some branches. Various types of hardware are used, depending upon the goals. Examples include:
  - cables (flexible braces) are installed in the upper crown to limit the movement of weak junctions or codominant stems.
  - brace rods (rigid braces) are installed close to or through weak junctions, or through split sections.
  - guys are installed to improve anchorage and stabilize lean.
  - props are installed to support some leaning trees and low branches from below.
  - Information on support systems can be found in the ISA BMP for Support Systems (Smiley and Lilly, 2014).
- **Tree removal, felling, or take down.** The process of removing the entire tree. This is the least desirable option, but it is often necessary especially in the case of dead trees, severe cracks in the trunk, root rot or other incurable defects or conditions.

In addition to these immediate remedial activities, if the tree will be retained, there are treatments that may be able to improve tree health for a longer-term reduction in risk. These include installation of lightning protection systems, fertilization, and other soil treatments that can improve root and tree growth.

Target-based mitigation

In addition to, or as an alternative to arboricultural treatments, targets or target zones may be manipulated in order to reduce risk. Target management can include temporarily or permanently relocating movable targets within the target zone. Mobile targets such as pedestrian or vehicular traffic may be rerouted or restricted from using the space within the target zone. These often are the solutions that will have the lowest impact on the tree and are therefore preferred if tree preservation is a primary management goal.

Work prioritization

With a single, privately owned tree, there is little need for the prioritization of work. The owner will make a decision on the remedial treatment and schedule the work. The exception is with extreme-risk trees, where the tree risk inspector should recommend that the work take place as soon as possible and the target zone may need to be restricted immediately.

With populations of trees, such as in municipal or utility applications, work may need to be scheduled well in advance. The risk assessment can be used to establish the work priority to mitigate the highest risk situations first. If a tree with an extreme risk is discovered, the owner/manager should be notified promptly, and action should be taken as needed. In the rare event that failure is in progress, immediate action may be needed to restrict access to the target zone and notify the client. Trees with lower levels of risk can be scheduled on an appropriate maintenance cycle.

Residual risk

Residual risk is the risk remaining after mitigation. Following any mitigation treatment, there is a residual risk posed by that tree. With tree removal, that residual risk is brought to near zero; however, even stumps can pose a tripping risk. The level of residual risk needs to be acceptable to the risk manager/owner. Some countries follow the principle of 'as low as reasonably practicable' (ALARP; Robens, 1972). To meet this principle, one must demonstrate that the cost involved in further reduction of risk would be grossly disproportionate to the benefit gained. ALARP is sometimes applied in situations where large populations of trees are being managed.

In general, if the residual risk following mitigation would exceed the tolerance of the client, the specified treatment might not be the best course of action. As previously noted, however, tree removal should not be recommended without due consideration of the benefits that would be lost.

Assessment frequency

Inspection interval is the time between assessments. Because site and tree conditions change over time, risk assessments should occur on a regular, recurring basis when justified by the level of risk or target value.

Timing of the initial risk assessment and frequency of future assessments is often not at the discretion of the tree risk assessor. However, after a tree has undergone the initial assessment, an inspection frequency should be recommended based upon the level of risk/ residual risk and the goals of the client. The inspection interval typically ranges between one and five years, but it may be more or less often, depending upon the age of the tree, level of risk, specific conditions, client goals and resources, or regulations. Generally, it is a good idea to inspect trees with known structural weaknesses and/or high-value targets after major storms or other exceptional events on the site (like forest clearing, trenching, or other construction work) to identify damage or changes in condition that may have occurred.

Conclusions

Tree risk assessment is an essential part of arboriculture and urban forestry. In general, trees pose a low risk and provide many benefits to the community. However, all trees will eventually fail, so there needs to a systematic process of assessment and failure can be mitigated by following defined policies and procedures.

References

Gilman, E. F. and Lilly, S. J. 2008. *Best Management Practices: Tree Pruning*. Champaign, IL: International Society of Arboriculture.

International Organization for Standardization (ISO) 31000. 2009. Risk Management – Principles and Guidelines.

International Organization for Standardization ISO Guide 73. 2009. Risk Management – Vocabulary.

National Tree Safety Group. 2011. *Common Sense Risk Management of Trees: Guidance on Trees and Public Safety in the UK for Owners, Manager and Advisers*. Edinburgh: Forestry Commission.

Robens, L. 1972. *Safety and Health at Work: Report of the Committee 1970–1972*. London: HMSO.

Smiley, E. T. and Lilly, S. 2014. *Best Management Practices: Tree Support Systems – Cabling, Bracing Guying, and Propping*. Champaign, IL: International Society of Arboriculture.

Smiley, E. T., Matheny, N. and Lilly, S. 2011. *Best Management Practices: Tree Risk Assessment*. Champaign, IL: International Society of Arboriculture.

32

TOOLS FOR TREE
RISK ASSESSMENT

Steffen Rust and Philip van Wassenaer

Introduction

A wide range of methods and tools are available to aid in the process of tree risk assessment. The vast majority of trees can be assessed visually, but when the likelihood of failure remains uncertain after a visual inspection, a number of tools can be used to measure the extent of decay, loss of strength, or anchorage.

The intensity of assessment is defined at different levels: ISA's *Best Management Practices: Tree Risk Assessment* defines three levels (limited visual, basic and advanced) while guidance in Germany defines only two levels, which correspond to the ISA's 'basic' and 'advanced'.

A limited visual assessment is intended to quickly collect information on large populations of trees, looking for obvious defects while walking, driving, or even flying past the trees. While frequently this level of assessment does not fulfil the requirements for a full risk assessment, it can be very useful after storms, along power lines, or in forests, where legal requirements are often lower than in urban areas.

A basic assessment is a detailed visual inspection of the entire tree, both from a distance and close up, staying on the ground, optionally using simple tools like a mallet. If the basic assessment is not sufficient to evaluate the risk of failure of a tree, advanced assessments can provide specific information about the tree, targets or the site. Only a small fraction of all inspected trees, often less than one in a thousand trees, should require advanced assessment.

Advanced assessment

Advanced assessment does not necessarily involve high-tech instruments. Often, aerial inspection, either from an aerial lift, a ladder, or by climbing the tree, will help to assess features invisible from the ground, such as cracks, included bark, codominant stems, or decay on the upper side of branches.

But most advanced methods are used either to map internal decay, to assess root loss and its effects on anchorage, or to investigate pathogens. They measure physical properties of the wood, the geometry of the stem and the decay, or the reaction of the tree when a load is applied, to assess the likelihood of failure (Figure 32.1).

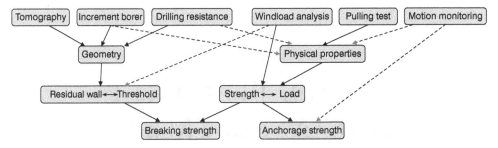

Figure 32.1 Tools for advanced assessment measure the geometry of sound wood within the stem or mechanical properties of the stem or root system. Results are evaluated using thresholds or by comparing limits of strength to estimated wind loads

Decay detection and imaging

The most common way to assess the strength of a tree part is on the basis of the distribution of sound wood (Wagener, 1963; Smiley and Fraedrich, 1992; Mattheck et al., 1994; Kane and Ryan, 2004; Bond, 2006). The size and position of decay, cracks, or included bark in the trunk or branches can be measured with a variety of tools. They vary in terms of damage to the tree, spatial resolution, and parameters measured.

Commonly used tools yield results along one line or across a thin cross-section of the tree. The measuring level on the stem and, in the case of drilling devices, the point or points to assess, must be carefully identified. A visual inspection and a mallet are often used to detect areas of weakness for further assessment.

When a rubber mallet is struck against the bark, the resulting sound can provide some information about the presence and extent of decay. The larger the tree is, the heavier the mallet has to be to give a useful result. For large trees, and those with very thick bark or extensive buttress roots, this method may not be applicable.

Once a preliminary inspection has identified areas of potential weakness, the extent of decay versus solid wood can be mapped using drilling devices or tomographs.

Drilling resistance measurement

Resistance drilling is a technique that uses a flat spade type drill bit with a 3 mm tip diameter (1.5 mm shaft diameter) to drill into targeted areas of wood and measuring the wood resistance (drilling torque) that is encountered as the drill passes through (Eckstein and Saß, 1994; Rinn, 1996). Resistances are recorded electronically or mechanically. Sound wood has higher resistance than decayed wood. Changes in wood resistance as the drill bit passes through the assessment area are documented as a graph with changes in amplitude. The amplitude of resistance readings drops when drilling across cracks or into decayed wood and cavities. The method has been used to map decay in urban trees for more than 20 years.

When the drill bit cuts its way through the wood, wood chips remain in the drilling canal and can cause friction on the rotating drill shaft (Rinn, 2012). Internal stresses in the stem may also push wood against the drill shaft. This shaft friction can cause a continuous rise in drilling resistance as drilling depth increases. Hence drilling resistance is a combination of drilling torque resistance and shaft friction (Weber and Mattheck, 2001). Shaft friction is less pronounced in soft wood species where it usually is of minor importance for the interpretation of the drilling profile. In tree species with higher wood densities, shaft friction can obscure a decrease in resistance, especially when decay is at an early stage of development. Only when the drill bit exits the wood on the other side of the tree or enters a cavity, the amount of shaft friction in the drilling curve can be clearly determined by comparing the final amplitude of the curve with the initial amplitude. In addition to drilling torque, some devices also measure the feed force that the drill bit requires to penetrate into the wood. This measurement is only marginally affected by shaft friction and may better facilitate the identification of decay especially in the presence of shaft friction (Nutto and Biechele, 2015).

In dry wood, drilling resistance correlates closely with wood density (Rinn, 1996), but in standing live trees the correlation can be much weaker, due to effects of wood moisture content and shaft friction. Because strength loss can set in earlier and faster than mass loss, and because density may correlate only weakly with strength, drill resistance readings can only be very approximate indicators of wood strength, especially when considering incipient decay.

In some tree species, there is a large contrast between early wood and late wood density. This density difference allows for the identification of separate tree rings within the resistance profiles when the drill bit passes through the tree rings perpendicularly. This information can sometimes help to date and document a negative incident that previously affected a tree.

Users of resistance drilling devices must look at the whole drilling profile to identify areas of decay and to approximate remaining wall thicknesses. Therefore, it is important to know the general radial trends of wood density in the species being assessed. In many ring-porous species and some conifers, wood density declines towards the cambium, while in many conifers, the opposite pattern can be observed. These patterns must not be mistaken as signs of decay.

The drilling profile obtained when drilling into decayed wood may also yield information about how well or poorly the tree is compartmentalizing fungal infections. An abrupt drop in resistance can be an indication of good compartmentalization.

Drilling resistance profiles are usually evaluated in terms of the ratio between the thickness of the residual wall and the radius of the stem at the measuring point (t/r ratio). Various sources of error, like the stem diameter, bark thickness, start of the decay, and deviations from the radial-horizontal direction result in a rather high degree of uncertainty of this ratio, often in the range of 10–20 per cent. The uncertainty of the t/r ratio steeply increases with decreasing wall thickness.

Effects of the use of drills on the tree

Since the structural form of urban tree stems is so variable, and because the location of a cavity is as important as its size (Figure 32.2), most drilling device users will take more than one measurement per cross-section. Although drilling devices have become less invasive over time, the damage caused by drills, and especially of multiple measurements, is still under debate. All drilling tools can cause discolouration in the sapwood around (and most notably above) the drilling path. Discoloured wood is more likely to be infected by fungi. Drilling bits can also breach defensive zones and can increase the likelihood of existing decay to spread further into the drilled holes. But trees can also compartmentalize the affected wood and slow the spread of decay (Weber and Mattheck, 2005). While most resistance-recording drills produce small canals filled with wood chips, larger drill bits and increment borers can produce empty and often much wider holes. Depending on the combination of fungus and tree species, the spread of decay can be favoured by one or all of the resulting micro-climates that result from different drilling techniques (Kersten and Schwarze, 2005; Weber and Mattheck, 2005).

Wounds caused by various drilling methods may also allow decay organisms to enter the structurally important regions of the tree that are being assessed. However, Weber and Mattheck (2005) did not find any fungal infections that entered the tree through holes made by increment borers and resistance-recording drills. Kersten and Schwarze (2005) found that it seems unlikely that fungi are spread by drilling devices from tree to tree so disinfecting these devices may not be necessary.

An increment borer should not be used on a regular basis and considered an exception because of the large holes and potential damage that the tool creates. Similarly, even when using less invasive drills, the number of measurements should be kept to a minimum. These considerations may become especially important if assessors return to a tree for repeated assessments.

Tomography

Currently, there are two types of tomographs widely used in tree assessment, either based on the time of flight of stress waves or on the electrical resistivity of wood. Their development started in the 1980s (sonic tomography: Tomikawa et al., 1986) and 1990s (electrical resistivity tomography: White, 1996) with studies on utility poles, while in recent years they have been widely used to assess standing trees (Just and Jacobs, 1998; Weihs et al., 1999; Comino et al., 2000; Divos, 2000; Koppán et al., 2000; Rust, 2000; Göcke et al., 2008; Wassenaer and Richardson, 2009; Bieker et al., 2010; Bieker and Rust, 2010a; Brazee et al., 2010; Li et al., 2012; Arciniegas et al., 2014).

When a stem cross-section contains decay, cavities, cracks or included bark, the time of flight of stress waves across the stem increases, because the stress waves must travel around the obstacle. A variety of instruments have been used to measure time of flight and the distance between start and end points of the signal. Today, most devices use several sensors simultaneously.

These sonic tomographs (SOT) measure the time of flight of signals initiated by hammer taps, with varying numbers of sensors that are attached directly to the wood at various points around the stem. The sensors must be in contact with the outermost growth ring and are often installed with nails.

The tomographs are operated from standard computers or as standalone devices, which can collect and analyse stress wave data. They can be used from aerial lifts or by climbers, but this will usually significantly increase the required time to collect images due to the logistics of getting the equipment up high in the tree. Setting up the device involves installing the sensors, recording their positions around the stem, and initiating the stress waves by sequentially tapping on each sensor or nail. Every sensor serves as a sending point and as a receiver, so that a dense array of time of flight data is collected across the cross-section of interest. Depending on the accessibility of the cross-section to be measured and the size and complexity of the tree, acquiring a good tomographic image requires between 30 minutes and several hours.

From the time and distance data measured at the tree, the apparent stress wave velocity can be calculated by assuming that the stress waves travel in a straight line, because their true path through the wood structure is not known. It is very important to measure the distance between sensors exactly, since any error in distance measurements will inevitably cause an equivalent error in the calculated velocity (Arciniegas et al., 2015). Based on research comparing different tomograph models, sufficient accuracy can only be achieved by the careful measurement of the distances between all sensor positions using mechanical or electronic callipers.

Because wood is an anisotropic material, the sound velocity in wood is quite variable, and depends on the angle to the main stem axis. Along the grain, sound velocity is several times higher than across the grain and radial velocity is higher than tangential velocity (Schubert et al., 2009). Therefore, all sensors must be installed in a plane perpendicular to the stem axis or else it is impossible to separate the effects of wood anisotropy from defects in the tomogram.

A further consequence of the anisotropic nature of wood is that every point in a stem cross-section has not one, but a range of sound velocities. Together with the fact that the true path of the stress wave is unknown, this greatly reduces the reliability of information derived from absolute sound velocities obtained in the tomogram. In trees with defects, the correlation between apparent sound velocities and density or hardness of the wood can be very weak (Liang and Fu, 2012; Rust, 2012; Pereira-Rollo et al., 2014; Alves et al., 2015).

Considering the limitations mentioned above and because of the large variation of stress wave velocities within and between individual trees, most tomographs use relative stress wave velocity rather than the absolute stress wave velocity. The result of a sonic tomography assessment therefore is a map of apparent relative stress wave velocities, which will reveal areas of sound wood as long as the wood can be traversed by stress waves on a more or less straight path from one sensor to another.

The accuracy of the tomogram depends very much on the placement of the sensors. Sensor positions should reflect the geometry of the cross-section of the tree, because data are evaluated within a polygon defined by the sensor positions at its vertices. Areas of wood that are not within this polygon will not appear in the tomogram. So, when sensors are placed only adjacent to a buttress, but not on its outside edge, this buttress will not be shown in the tomogram. If measuring points are selected such that there are areas with no wood between a neighbouring pair of sensors (i.e. sensors placed only on the outside edge of two adjacent buttresses), the result will be low velocity readings between those points.

Sensors should also be placed as close as possible to visible defects such as cracks or included bark. When sensors are placed further away from such features, the resulting tomogram will show much wider areas of low velocity adjacent to the features. These could be misinterpreted as large areas of decay and lead to poor management decisions.

Any quantitative mechanical evaluation of the tomogram in terms of the section modulus or the resistance to bending at the evaluation level will suffer from errors if the sensors are not properly located on the stem. This will be the case if areas are omitted because they are outside the polygon of sensors, or areas outside of the stem but within the polygon of sensors are included. Moreover, even apparently small inaccuracies in the measurement of the sensor positions and hence the shape and size of the stem cross-section may result in significant errors in these calculations. This is because the contribution of any part of the cross-section to the section modulus increases with the square of its distance to the centre of the stem.

Despite the cautions above, for practical purposes the accuracy of the tomograms, which is in the range of several centimetres, has proven sufficient for most normal tomograph assessments. This minimally invasive method can be applied to trees ranging from 20 cm to several metres in diameter.

Electrical resistivity tomograms (ERT) display the distribution of resistivity across the stem cross-section (Rust et al., 2008; Bieker et al., 2010; Bieker and Rust, 2010a, 2010b). ERT has been applied to find discoloured wood, ring shake and decay in trees, and even to detect roots in the soil.

While stress wave tomograms of defect-free stems are more or less homogeneous, ERT can show strong species-specific patterns, even in sound trees (Bieker et al., 2010; Bieker and Rust, 2010a, 2010b). These differing patterns correlate with features such as moisture content of the heartwood and wood pH. Three common patterns have been identified so far. Species with dry heartwood, such as pines (*Pinus* sp.), generally will have ERT tomograms with a central area of high resistivity. Species without regular heartwood, such as beech (*Fagus* sp.), often have a homogeneous distribution of resistivity. The most complex patterns are seen in the oaks (*Quercus* sp.) and similar species, which have a homogeneous moisture distribution in their cross-section, but a strong pattern of pH and ion content, resulting in three concentric areas of differing resistivity. Users of the ERT method must have a good working knowledge of these differing patterns seen in tomograms of intact cross-sections before they can properly interpret any tomograms of potentially defective trees.

Similar to sonic measurements the ERT requires point-like contacts to the outermost layer of wood, and can be applied to measure stems of any size with minimal damage to the tested tree. In most assessments 24 measurement points are used. If sonic tomography and ERT are used for the same measuring plane, the nails used for sonic tomography can also be used for a subsequent ERT. The spatial resolution depends on the distance between neighbouring sensors and the tree diameter. In some cases, some measuring points may need to be added to increase the resolution of the ERT tomogram.

Because sonic tomography and electrical resistivity tomography measure very different physical and chemical properties of wood, using a combination the two technologies can greatly improve the assessment of a tree (Rust et al., 2008). Incipient decay is characterized by high stress wave velocities and a reduction in electrical resistivity. Cavities, decay, and cracks are often difficult to distinguish from each other in sonic tomograms. When combined with ERT, this task becomes much easier, because cavities usually have a high resistivity and decay is seen in an ERT tomogram as low resistivity, while the wood around cracks often is unaltered. Where SOT often shows a large area of low stress wave velocity around a crack, which could be misinterpreted as a cavity, ERT can show low resistivity, indicating the presence of sound wood around a crack.

Since ERT and SOT share many parts of their setup and hardware, completing an additional ERT assessment following an SOT can often be done in a relatively short period of additional time.

Evaluation of mapped decay

A common way to evaluate the consequences of mapped decay (detected with drilling or tomography) on the likelihood of failure (stem breakage) is the use of thresholds of residual wall thickness and cavity openings (Figure 32.2). It has been proposed that a tree is considered potentially unsafe if the ratio of the remaining wall thickness compared to the radius of the tree at the measuring level is less than a third (Wagener, 1963; Mattheck et al., 1994), or if the estimated strength loss exceeds one third (Smiley and Fraedrich, 1992). Very often, however, the conditions for the application of these thresholds and formulae are not met. They have been proposed for solitary, unpruned trees with one more or less central cavity, which must not extend to the roots (Mattheck et al., 2014). Since there is an almost infinite number of combinations of size and locations of decay within a tree stem, these rather narrow conditions are rarely found in trees that are assessed with either drilling or tomographic methods.

Trees and decay are rarely round, so the detection of the minimal residual wall thickness with drilling devices can be highly uncertain. Results from such linear measurements, especially when used in simple ratio equations (remaining wall thickness divided by the radius of the tree, t/r), have a wide margin of uncertainty. Experimental and modelling

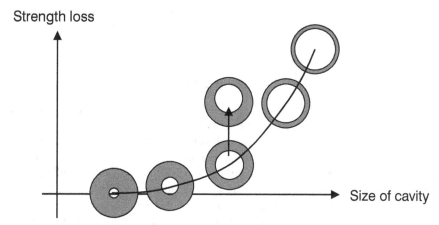

Figure 32.2 The strength of a cross-section decreases rapidly with increasing cavity size. The position of decay is important, because non-central cavities can reduce strength in one direction as much as larger, central cavities. In this figure the stem with an off-centred decay column has a higher strength loss than a stem of the same diameter with a centralized decay column

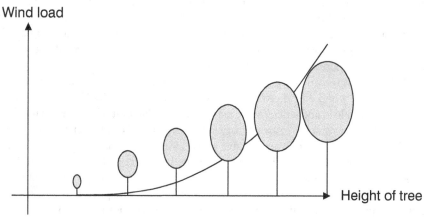

Figure 32.3 Wind load increases more rapidly than tree height. It depends on many factors, including regional climate and sheltering effects by neighbouring structures

studies caution against the use of these formulae (Kane and Ryan, 2004; Bond, 2006; Ruel et al., 2010; Ciftci et al., 2013; Spatz and Niklas, 2013). Consequently, the use of simple equations when attempting to answer complex tree risk questions should be avoided.

The mechanical properties of the wood in a cross-section are rarely taken into account quantitatively in tree risk assessments. Even if mechanical properties can be determined for specific points on a stem, there is no way to extrapolate those properties to the entire cross-section. In practice, decay, whatever its condition, is assumed to have lost its entire strength.

The effects of strength loss cannot be evaluated without considering what kind of wind load is acting on the tree (Figure 32.3). There are some simplified methods to evaluate the effect of mapped decay on tree risk based on prevailing winds, exposure, tree species, tree form and stem diameter (Wessolly and Erb, 1998). They can be used in the form of a set of graphs on paper, or as free or paid web services on the internet. The results of these approaches are expressed as a factor of safety and, in cases where reduced safety margins are discovered, these approaches also provide guidance on how much the tree should be reduced in height in order to achieve the desired factor of safety. Often small reductions in tree height and crown spread can significantly raise the factor of safety.

Pulling tests

Static pulling tests (Ylinen, 1953; Wessolly, 1991) are commonly used to compare the load-bearing capacity of a stem and/or the anchorage strength of the root system with the estimated wind load. The load-bearing capacity is estimated based on the reaction of the tree to a non-destructive static load applied to the tree using a winch. In most cases, tests of the trunk and the root system are carried out at the same time. To date, a static pulling test is the only method that can be used to measure the anchorage strength of trees.

The design wind load is estimated from the projected area of the tree crown (usually derived from a digital image of the tree), precise height measurements, and a range of site factors, including design wind speed, often Beaufort Force 12 (>117 km/h). Adjustments in the analysis can be made to account for sheltering effects of nearby structures and the dynamic properties of the tree, such as resonance and damping.

The assessment of the load-bearing capacity of the stem is based on the assumption that primary failure in bending starts in most cases ($t/r > 0.1$) as compression failure (Wessolly and Erb, 1998; Spatz and Niklas, 2013). During the test, the strain (compression and/or tension) of the outermost fibres is measured at points of the stem where it is suspected to be highest, based on a visual inspection. In cross-sections of similar size and form, the strain will increase as the severity of the defect increases. In practice, a static load test is typically stopped when a strain of at most 1000 μm/m is recorded at any of the measuring points. This amount of strain is within the elastic range of wood and therefore non-destructive. Using Hook's law, the measured relationship between load and strain can be extrapolated to the limit of proportionality in order to estimate the load necessary to cause primary failure. Strength data of green wood are available for many European and North American tree species (Spatz and Pfisterer, 2013). Where these data are lacking, they can be compiled using a new method described by Detter et al. (2013).

To measure the anchorage strength of a tree, tilt sensors monitor the inclination of the root plate as the load is increased. The load test is always stopped below 0.25° of basal inclination, a threshold that ensures that the pulling test will not damage the root system. Recent research has shown that there is a close linear correlation between the load required to incline a tree to 0.25° and the load required to pull the tree to failure (Detter and Rust, 2012). This load is estimated by extrapolation.

The critical load derived from the pulling test is compared with the calculated design wind load for the tree. The ratio of these two parameters is reported as a factor of safety. To account for uncertainties involved in the measurements and estimates, the assessment is based on a worst-case scenario. Therefore, the tree will only be deemed 'safe to retain' if the load-bearing capacity exceeds the design wind load by at least 50 per cent (i.e. the factor of safety is higher than 1.5).

Pulling tests require special equipment and substantial training for those who apply the method. To install the pulling rope in the tree, an aerial lift or two climbers (for safety reasons) are usually needed. The evaluation of the data can also be quite involved and, for these reasons, this more comprehensive method often is more time-consuming and therefore more costly than other risk assessment approaches.

To conduct a pulling test, the direction of pull needs to be established. In cases where the cause and location of specific damages are clear, such as root loss caused by trenching during construction, one load test in one direction is often sufficient. Without a clear indication of damages, or if the root system is weakened from several sides, for example in a *Fagus* with fruiting bodies of *Meripilus giganteus* all around its base, two perpendicular load tests are commonly applied. A rope first needs to be attached to a suitable anchor point on the stem that is strong enough to carry the load from the winch (often in the range of 10 kN to 20 kN). A second anchor point for the winch is required on the ground. The slope of the pulling line should not be too steep, as this will reduce the horizontal force applied to the tree. A forcemeter is installed in the pulling line between the winch and the tree. Strain gauges are attached to the stem parallel to the pulling line in sound wood where the highest strength loss is suspected. The strain gauges can be attached on either the tension or compression side of the tree. They usually have a span of 20 cm to avoid erroneous readings caused by stem bending. A tilt sensor is attached to the base of the tree, perpendicular to the direction of pull. In modern systems, all data from the various instruments are transferred to a computer wirelessly at high sample rates. Many samples (at least 50–100), spanning a wide range of strain (e.g. 20% of strain at primary failure), are required to reliably extrapolate to the load at primary failure. All readings become more stable at higher root plate tilts, but the 0.25° threshold should never be exceeded.

Dedicated commercial software programs are available for the evaluation of pulling tests to derive the current factor of safety of the tree and to model the effect of pruning to increase those safety factors.

Dynamic measurements

One of the most recent methods introduced for assessing the risk of trees monitors the movement of the root plate and the bending of the stem of potentially unsafe trees in natural wind by attaching tilt sensors to the base of the stem. According to one system, trees that tilt more than 0.6° in a wind event with wind speeds exceeding 60 km/h are categorized as requiring additional monitoring, while trees reaching 1° in a wind event with wind speeds of 70 km/h or more would require regular monitoring (James et al., 2013). To date, scientific evidence on the validity of these thresholds is scarce, but recent research indicates that this is the range where plastic deformations of the root system and a decrease in root plate stiffness (primary failure) occur. The maximum load capacity of the root system is not affected by repeated load cycles.

Variations in local wind speeds at the tree location do not necessarily result in simultaneous tree responses, so that long averaging intervals are required to get close correlations between wind speed and tree response. But there is a strong correlation between root plate tilt and peak gust wind speed in a given period (Göcke and Rust, 2015). Since reliable measurements of tree response depend on very high wind speeds, which may not occur while the sensors are attached to the tree, this correlation can be used to extrapolate the reaction of a tree to a desired wind speed from data measured at lower wind speeds. These measurements can also be used to validate the estimated design wind load used for the evaluation of static pulling tests.

Lastly, because there is a strong correlation between stem bending in high wind and stem stiffness derived from static pulling tests, the stem curvature measured by two tilt sensors attached at the base of a tree and at a height of 2 m or more might serve as an indicator of trunk stability against breakage for medium diameter trees.

Conclusions

Several technological innovations and their scientific evaluation help to assess trees in much more detail than just 20 years ago. Although quantitative estimates of the risk of failure are, with the possible exception of pulling 'tests', still not available today, the use of these tools will help to balance the need for safety and the benefits of maintaining mature trees in our cities.

References

Alves, R. C., J. N. Mantilla, C. F. Bremer and E. V. M. Carrasco. 2015. 'Application of acoustic tomography and ultrasonic waves to estimate stiffness constants of muiracatiara Brazilian wood.' *BioResources* 10 (1): 1845–1856.

Arciniegas, A., L. Brancheriau and P. Lasaygues. 2015. 'Tomography in standing trees: revisiting the determination of acoustic wave velocity.' *Annals of Forest Science* 72 (6): 685–691.

Arciniegas, A., F. Prieto, L. Brancheriau and P. Lasaygues. 2014. 'Literature review of acoustic and ultrasonic tomography in standing trees.' *Trees* 28 (6): 1559–1567.

Bieker, D. and S. Rust. 2010a. 'Non-destructive estimation of sapwood and heartwood width in Scots pine (*Pinus sylvestris* L.).' *Silva Fennica* 44 (2): 267–273.

Bieker, D. and S. Rust. 2010b. 'Electric resistivity tomography shows radial variation of electrolytes in *Quercus robur*.' *Canadian Journal of Forest Research* 40 (6): 1189–1193.

Bieker, D., R. Kehr, G. Weber and S. Rust. 2010. 'Non-destructive monitoring of early stages of white rot by *Trametes versicolor* in *Fraxinus excelsior*.' *Annals of Forest Science* 67 (2): 210.

Bond, J. 2006. 'Foundations of tree risk analysis: Use of the t/R ratio to evaluate trunk failure potential.' *Arborist News* December: 31–34.

Brazee, N. J., R. E. Marra, L. Göcke and P. van Wassenaer. 2010. 'Non-destructive assessment of internal decay in three hardwood species of northeastern North America using sonic and electrical impedance tomography.' *Forestry* 84 (1): 33–39.

Ciftci, C., B. Kane, S. F. Brena and S. R. Arwade. 2013. 'Loss in moment capacity of tree stems induced by decay.' *Trees* 28 (2): 517–529.

Comino, E., R. Martinis, G. Nicolotti, L. Sambuelli and L. V. Socco. 2000. 'Low frequency currents tomography for tree stability assessment.' In *Plant Health in Urban Horticulture*, edited by G. F. Backhaus, H. Balder and E. Idczak, 370–378. Berlin: Parey Buchverlag.

Detter, A. and S. Rust. 2012. 'Aktuelle Untersuchungsergebnisse zu Zugversuchen.' *Jahrbuch der Baumpflege* 2012: 9–22.

Detter, A., S. Rust, C. Rust and G. Maybaum. 2013. 'Determining strength limits for standing tree stems from bending tests.' In *18th International Nondestructive Testing and Evaluation of Wood Symposium*, vol. 226, edited by R. J. Ross and X. Wang. General Technical Report. Madison, WI: US Department of Agriculture, Forest Service, Forest Products Laboratory.

Divos, F. 2000. 'Stress wave based tomography for tree evaluation.' In *Proceedings of the 12th International Symposium on Nondestructive Testing of Wood*, 469. Sopron: University of Western Hungary.

Eckstein, D. and U. Saß. 1994. 'Bohrwiderstandsmessungen an Laubbäumen und ihre holzanatomische Interpretation.' *Holz als Roh- und Werkstoff* 52: 279–286.

Göcke, L. and S. Rust. 2015. 'Correlation of wind speed and root plate tilt of trees in urban environment.' Paper presented at ISA Annual Conference, 20–23 August, Orlando, FL.

Göcke, L., S. Rust, U. Weihs, T. Günther and C. Rücker. 2008. 'Combining sonic and electrical impedance tomography for the nondestructive testing of trees.' *Western Arborist* Spring: 1–11.

James, K. R., C. Hallam and C. Spencer. 2013. 'Measuring tilt of tree structural root zones under static and wind loading.' *Agricultural and Forest Meteorology* 168 (January): 160–167.

Just, A. and F. Jacobs. 1998. 'Elektrische Widerstandstomographie zur Untersuchung des Gesundheitszustandes von Bäumen.' In Proc. VII. Arbeitsseminar *Hochauflösende Geoelektrik*, Bucha, Germany. Institut für Geophysik und Geologie der Universität Leipzig.

Kane, B. C. P. and H. D. P. Ryan. 2004. 'The accuracy of formulas used to assess strength loss due to decay in trees.' *Journal of Arboriculture* 30 (6): 347–356.

Kersten, W. and F. W. M. R. Schwarze. 2005. 'Development of decay in the sapwood of trees wounded by the use of decay detecting devices.' *Arboricultural Journal* 28 (3): 165–181.

Koppán, A, M. Kis, S. Szalai, L. Szarka and V. Wesztergom. 2000. 'In vivo electric soundings of standing trees.' In Divos, F. (ed.) *12th International Symposium on Nondestructive Testing of Wood*, 13.-15.9.2000, Sopron, Hungary, 215–218.

Li, L., X. Wang, L. Wang and R. B. Allison. 2012. 'Acoustic tomography in relation to 2D ultrasonic velocity and hardness mappings.' *Wood Science and Technology* 46 (1–3): 551–561.

Liang, S. and F. Fu. 2012. 'Relationship analysis between tomograms and hardness maps in determining internal defects in Euphrates poplar.' *Wood Research* 57 (2): 221–230.

Mattheck, C., K. Bethge and P. W. West. 1994. 'Breakage of hollow tree stems.' *Trees – Structure and Function* 9: 47–50.

Mattheck, C., K. Bethge and K. H. Weber. 2014. *Die Körpersprache der Bäume: Enzyklopädie des Visual Tree Assessment*. Karlsruhe: Karlsruher Institut für Technologie.

Nutto, L. and T. Biechele. 2015. 'Drilling resistance measurement and the effect of shaft friction – using feed force information for improving decay identification on hard tropical wood.' In *19th International Nondestructive Testing and Evaluation of Wood Symposium*, 154–161. General Technical Report Fpl-Gtr-239. Rio de Janeiro: USDA.

Pereira-Rollo, L. C., da Silva Filho, D. F., Filho, M. T., de Oliveira Moraes, S. and do Couto, H. T. Z. 2014. 'Can the impulse propagation speed from cross-section tomography explain the conditioned density of wood?' *Wood Science and Technology* 48 (4): 689–701.

Rinn, F. 1996. 'Resistographic visualization of tree-ring density variations.' *Radiocarbon* 33: 871–878.

Rinn, F. 2012. 'Basics of typical resistance-drilling profiles.' *Western Arborist*, Winter: 30–36.

Ruel, J. C., A. Achim, R. E. Herrera and A. Cloutier. 2010. 'Relating mechanical strength at the stem level to values obtained from defect-free wood samples.' *Trees – Structure and Function* 24: 1127–1135.

Rust, S. 2000. 'A new tomographic device for the non-destructive testing of standing trees.' In *Proceedings of the 12th International Symposium on Nondestructive Testing of Wood*, 233–238. Sopron: University of Western Hungary.

Rust, S. 2012. 'Validation of tomography in standing quercus robur as a tool to study within-tree variability of wood properties.' In *Proc. 2012 IUFRO Conference Division 5 Forest Products*, 8.-13.7.2012, Lisbon: 195.

Rust, S., U. Weihs, T. Günther, C. Rücker and L. Göcke. 2008. 'Combining sonic and electrical impedance tomography for the nondestructive testing of trees.' *Western Arborist*, 1–11.

Schubert, S., D. Gsell, J. Dual, M. Motavalli and P. Niemz. 2009. 'Acoustic wood tomography on trees and the challenge of wood heterogeneity.' *Holzforschung* 63 (1): 107–112.

Smiley, E. T. and B. R. Fraedrich. 1992. 'Determining strength loss from decay.' *Journal of Arboriculture* 18 (4): 201–204.

Spatz, H.-C. and Niklas, K. J. 2013. 'Modes of failure in tubular plant organs.' *American Journal of Botany* 100 (2): 332–336.

Spatz, H. C. and J. Pfisterer. 2013. 'Mechanical properties of green wood and their relevance for tree risk assessment.' *Arboriculture and Urban Forestry* 39 (5): 218–225.

Tomikawa, Y., Y. Iwase, K. Arita and H. Yamada. 1986. 'Nondestructive Inspection of a wooden pole using ultrasonic computed tomography.' *IEEE Transactions on Ultrasonics, Ferroelectrics and Frequency Control* 33 (4): 354–358.

Wagener, W. W. 1963. *Judging Hazard from Native Trees in California Recreational Areas: a Guide for Professional Foresters*. US Forest Service Research Paper. Berkeley, CA: Pacific Southwest Forest; Range Experiment Station, USDA.

Wassenaer, P. van and M. Richardson. 2009. 'A review of tree risk assessment using minimally invasive technologies and two case studies.' *Arboricultural Journal* 32 (4): 275–292.

Weber, K. and C. Mattheck. 2001. *Taschenbuch der Holzfäulen im Baum*. Karlsruhe: Forschungszentrum Karlsruhe.

Weber, K. and C. Mattheck. 2005. 'The effects of excessive drilling diagnosis on decay propagation in trees.' *Trees – Structure and Function* 20 (2): 224–228.

Weihs, U., V. Dubbel, F. Krummheuer and A. Just. 1999. 'Die Elektrische Widerstandstomographie – Ein vielversprechendes Verfahren zur Farbkerndiagnose am stehenden Rotbuchenstamm.' *Forst und Holz* 54: 166–169.

Wessolly, L. 1991. 'Verfahren zur Bestimmung der Stand- und Bruchsicherheit von Bäumen.' *Holz als Roh- und Werkstoff* 49 (3): 99–104.

Wessolly, L. and M. Erb. 1998. *Handbuch der Baumstatik und Baumkontrolle*. Patzer. Verlag, Berlin.

White, N. 1996. 'EIT for the condition monitoring of wood poles carrying overhead power lines.' *IEE Colloquium on Advances in Electrical Tomography*, 1996: 13.

Ylinen, A. 1953. 'Über die mechanische Schaftformtheorie der Bäume.' *Holz als Roh- und Werkstoff* 11 (6): 16–17.

33

MANAGEMENT AND CONSERVATION OF ANCIENT AND OTHER VETERAN TREES

Neville Fay and Jill Butler

Introduction

There are special challenges for owners, arborists and advisors associated with conserving and managing *ancient and other veteran trees* in urban areas. This chapter outlines why such trees should be identified and managed to avoid harm, damage or injury that might induce stress or accelerate the tree aging process and therefore compromise tree longevity and biodiversity.

Ancient trees are in the post-mature life stage. They are old by comparison with other trees of the same species. The term *veteran* is used differently to denote the range and type of wood decay habitats that typically develop in late maturity as the tree approaches the ancient phase, and which are especially valuable for associated colonising species. In urban areas, while ancient trees are generally rare, younger, veteran trees are comparatively more frequently encountered. However, understanding how trees age and the unique values and physiological needs associated with this process is fundamental for effective management of both ancient and veteran trees.

Large, aging and decaying trees represent living history and heritage. Their appreciation encompasses the notions of 'memory' and 'witness' trees that have community and social significance. Their importance is amplified by their age-related ecological history and conservation status. Yet such trees today, worldwide, remain under considerable threat (Lindenmayer and Laurance, 2016).

Young trees are unitary organisms with a relatively simple structure compared to that of older trees. As trees age they become extensively compartmented and have a branching structure which is more an expression of a colony than that of an individual. The range, quantity and quality of habitats develop as a function of the tree's age, size, wood and decay volume. With aging the complexity and scale of foliar and root biomass, soil processes, bark depth and surface area increases as does the continuity and range of *saprotrophic* and *saproxylic* habitats that derive from centuries of growth *in situ*. Advanced trunk and branch hollowing, originating in maturity and developing through the ancient phase, is a highly significant, relatively rare habitat due to the length of time that heartwood or ripewood takes to reach sufficient size or condition for decay to propagate and become extensive. This aging expression is accompanied by other morpho-physiological changes that include natural crown dieback (retrenchment).

Understanding the aging process is important for determining the type and level of management intervention for trees in all age classes. However this is particularly relevant for aging trees due to their increasing vulnerability to rapid changes affecting the crown or root system, leading to critical decline. While such changes may derive from abiotic impacts, they can also result from well-intended, poorly informed interventions.

There is little knowledge of tree soil–root systems, including their biodiversity and below-ground ecology as trees age. However, management and protection of veteran tree root systems require understanding the likely extent of the tree's soil ecosystem (the mycorrhizosphere), land use changes and the function and health of the rooting environment. Crown management needs to take account of current constraints upon healthy growth and that proposed intervention should be gradual and phased. Species' shade tolerance and their light-demanding characteristics need to be taken into account when considering the effects of releasing trees from shade or from climbing shrub competition.

Reducing the risks *to* veteran trees from traumatic branch, stem or root failure is a fundamental aspect of longevity management. Interventions should be the minimum necessary to achieve optimal vitality. However, managing *for tree longevity* together with *reasonable safety* requires special consideration and training to be able to achieve acceptable public safety along with low levels of tree intervention. This needs a sound understanding, not only of the aging process and ancient tree biology, but also the requirements of objective, non-risk-averse, tree risk management.

Where are ancient and other veteran trees in urban areas?

Among the oldest trees in a European context is the yew tree (*Taxus baccata*), which due to its exceptional capacity for longevity and rejuvenation may live for two or more millennia. In some cases yews, nurtured since pre-Christian times (since Celtic–Druidic, early bronze age) in cemeteries and churchyards as druidic symbols of immortality, were adopted into Christianity and planted in churchyards as symbols of everlasting life; many remain today as focal points within villages and towns, especially in the UK, that have been subject to urbanization.

In Europe, treed landscapes, which include former country estates and castles with parklands rich in scattered mature and veteran trees, once incorporated into urban expansion, continue to provide settings for buildings within sites that have acquired new uses. Additionally, common land around town and city margins has over the past century become included within expanding urban settlements (Figure 33.1). These landscape relics, some of medieval origin, today still harbour veteran trees that were once managed within working wood-pasture systems that combined grazing with traditional tree products, such as fodder, fruit and branch material. Traditionally managed pollards once in field margins or hedgerows and orchards have similarly become engulfed, and, as a result, fragmented into green space islands.

The value of street trees has been understood for many decades and those planted in earlier times in Britain and mainland Europe have often been managed as pollards or pruned regularly (Read, 2000) with street veterans being managed for their conservation and succession.

By understanding the history of large old trees, managers and arborists can contribute to their preservation by informing planners, designers and developers about ways to modify development footprints and adapt road systems on sites with mature or veteran trees that may become ancient trees of the future.

Figure 33.1 Ancient oak (*Quercus robur*) in Mildenhall, Suffolk, UK, within new suburban development

Ancient oak pollard, once part of a wood-pasture system, has been incorporated within an urban housing estate. Future management needs to ensure minimal impact upon the root system and avoid risk-averse intervention

Source: © David Humphries

The importance of ancient and other veteran trees to society

The valued characteristics of older generations of trees are due to their age, size, biodiversity and historical associations with people and place.

These qualities are due to:

- *Aesthetic values:* for their stature and appearance, as illustrated in old paintings, poetry and literature.
- *Rarity:* representing a tiny fraction of the tree population as a whole and as representatives of complete life process.
- *Historic features:* such as 'sentinel trees' that have occupied land for centuries, in some cases millennia, and give a sense of time and deepen the sense of place.
- *Cultural presence:* 'worked' trees such as pollards and coppice that have been part of the farming economy and associated with rural living for millennia.
- *Connection to landscape:* the experience of older trees invokes personal connections and a sense of belonging in a landscape.
- *Human cultural features:* evidence of human connections such as historic burial ground yew trees.
- *Commemorative connections:* as 'witness trees' associated with historical events and people; as in the UK memorialising royalty, events marking parliamentary democracy and popular uprisings.

- *Biological diversity:* associated with the range and quality of decaying wood habitat, absent in younger trees. Old and large trees are considered keystone trees, 'arks' that carry colonising species through space and time; and when lost, their species richness and habitat continuity collapses.
- *Ecosystem qualities:* veteran trees are biologically complex comprising *inter alia* ancient soil refugia that contain a myriad of colonizing organisms. They also help generate income for owners such as in tourist venues (e.g. Blenheim Palace).
- *Scientific potential:* ancient trees are some of the oldest organisms on the planet. They provide specialist habitat for assemblages of species within their structure and their surrounding soils, much of which is yet to be researched.
- *Models of arboricultural understanding:* morpho-physiological appreciation of the aging process offers paradigms that apply both for older age classes and to pre-ancient developmental and mature phases.
- *Sustainability indicators:* without understanding population dynamics, arborists are unable to determine the relevance of maintaining the oldest age classes and the number and proportions needed to ensure continuity of habitat.

Studies confirm the importance of veteran trees for species assemblages that exploit their 'host-space'. Such terrestrial habitats are rare. Seven per cent (more than 2,000 species) of British invertebrates are dependent to a significant extent on decaying wood habitats (Alexander, 2012). This species-richness is due to the highly differentiated wood substrate available for colonizing organisms with specialized life styles.

These habitats, from root to the branch tips, are a function of time and the diversity and succession of fungal species that mediate the alteration of the host tissues. Different trees express this diversity according to their morpho-physiological histories and the array of fungal decomposers competing for substrate host-space; processes that affect the local (micro-topographical) environments within the crown, trunk, roots and soil. With age, the bark pH changes producing different habitat types for lichens and other epiphytes. And so too with the root system, as it decays and is replenished and colonised by successions of fungal species, all of which contribute to the soil ecology.

Types and volume of habitat in the late-mature and ancient phases vary greatly; including decaying heartwood, sapwood and fallen branches, dead bark and aerial branches and decaying roots, sap runs, tears, scars and water pools (Fay and de Berker, 1997). Substrate types also vary according to the humidity and light exposure, whether near the trunk outer surface or the inner core, and whether lower or higher on the trunk or the canopy. Through the aging process new habitats are created, and old ones are altered and differently exploited along with colonizing species. In the final stages of growth while stem girth increases by smaller increments, the host-space also alters as inner hollowing expands within the trunk and larger lower branches; also as the bark thickens with its age, depth, rugosity, cracks and flakes, host-space is created for nesting and roosting mammals, birds and reptiles.

Aging differentiates ancients from other trees

The term 'ancient' is an arboricultural classification of the aging process. A tree enters this phase when it grows beyond its peak maturity. There are three age states within this phase, 'early-', 'mid-' and 'late-ancient' (Figure 33.2), characterised by morpho-physiological changes that reflect growth dynamics and habitat qualities; the commencement and duration are ontogenetically influenced and vary according to species and growth conditions (Fay, 2002;

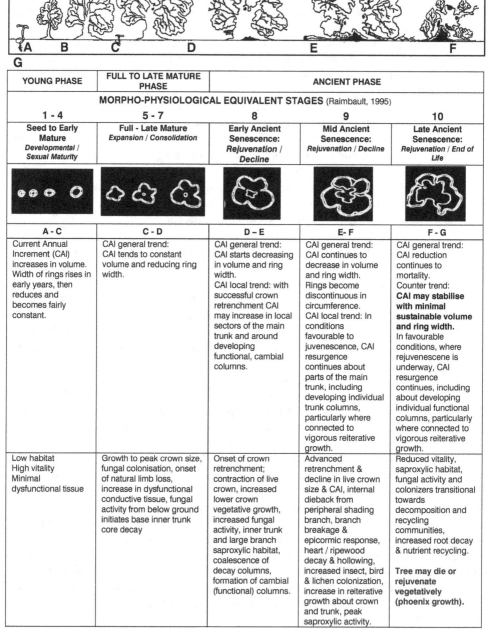

YOUNG PHASE	FULL TO LATE MATURE PHASE	ANCIENT PHASE		
		MORPHO-PHYSIOLOGICAL EQUIVALENT STAGES (Raimbault, 1995)		
1 - 4	**5 - 7**	**8**	**9**	**10**
Seed to Early Mature *Developmental / Sexual Maturity*	Full - Late Mature *Expansion / Consolidation*	Early Ancient Senescence: *Rejuvenation / Decline*	Mid Ancient Senescence: *Rejuvenation / Decline*	Late Ancient Senescence: *Rejuvenation / End of Life*
A - C	**C - D**	**D – E**	**E- F**	**F - G**
Current Annual Increment (CAI) increases in volume. Width of rings rises in early years, then reduces and becomes fairly constant.	CAI general trend: CAI tends to constant volume and reducing ring width.	CAI general trend: CAI starts decreasing in volume and ring width. CAI local trend: with successful crown retrenchment CAI may increase in local sectors of the main trunk and around developing functional, cambial columns.	CAI general trend: CAI continues to decrease in volume and ring width. Rings become discontinuous in circumference. CAI local trend: In conditions favourable to juvenescence, CAI resurgence continues about parts of the main trunk, including developing individual trunk columns, particularly where connected to vigorous reiterative growth.	CAI general trend: CAI reduction continues to mortality. Counter trend: **CAI may stabilise with minimal sustainable volume and ring width.** In favourable conditions, where rejuvenescene is underway, CAI resurgence continues, including about developing individual functional columns, particularly where connected to vigorous reiterative growth.
Low habitat High vitality Minimal dysfunctional tissue	Growth to peak crown size, fungal colonisation, onset of natural limb loss, increase in dysfunctional conductive tissue, fungal activity from below ground initiates base inner trunk core decay	Onset of crown retrenchment; contraction of live crown, increased lower crown vegetative growth, increased fungal activity, inner trunk and large branch saproxylic habitat, coalescence of decay columns, formation of cambial (functional) columns.	Advanced retrenchment & decline in live crown size & CAI, internal dieback from peripheral shading branch, branch breakage & epicormic response, heart / ripewood decay & hollowing, increased insect, bird & lichen colonization, increase in reiterative growth about crown and trunk, peak saproxylic activity.	Reduced vitality, saproxylic habitat, fungal activity and colonizers transitional towards decomposition and recycling communities, increased root decay & nutrient recycling. **Tree may die or rejuvenate vegetatively (phoenix growth).**

Figure 33.2 Schematic model of the tree aging process referenced to morpho-physiological stages. Modelled on a *Quercus* species from seedling to senescence with reference to current annual increment (CAI), development of decay and habitat (veteran) features (Fay & de Berker 1997; updated Fay, 2016)

Source: © Neville Fay

Read, 2000; Lonsdale, 2013b). In Europe, trees with heartwood (e.g. yew, *Taxus* sp.; oak, *Quercus* sp.; sweet chestnut, *Castanea sativa*; Scots pine, *Pinus sylvestris* sp.) tend to have a significantly longer ancient phase than ripewood species (e.g. beech, *Fagus* sp.; hornbeam, *Carpinus betulus*; lime, *Tilia* sp.; birch, *Betula* sp.) whose sapwood ages without forming distinct heartwood (Lonsdale, 2013b).[1]

Veteran trees have characteristics of an ancient tree irrespective of chronological age (Fay, 2002; Read, 2000; Lonsdale, 2013b). While mature (pre-ancient) trees contain decaying crown branches and trunk hollowing, they may be described as veteran trees depending on the extent and quality of their decaying (saproxylic) habitat. Not having passed their peak maturity, such trees are pre-ancient veterans, and as such are valuable for their habitat interests and their potential to become future ancient trees. Conservation management therefore needs to be directed to securing this biological asset.

Veteran trees may also be created by natural or human intervention, rather than through the aging process alone. Storms and snow damage, flooding, erosion, drought and extensive shading may create ('veteranize') habitats, as well as impacts due to tree surgery, animal, root and soil damage, and the effects of pollution. Natural veteran features should be distinguished from 'intentional veteranization' for the creation of certain habitats. Intentional veteranization does not necessarily result in the same quality of habitat as that arising from natural processes and may promote terminal decline. Pollarding is a type of intentional veteranization, which, through pruning, influences veteranizing processes. When regularly practised, pollarding promotes habitat and may contribute to longevity (see further section below on pollarding and coppicing).

The aging process in trees

The life history of a tree is conventionally represented as a linear progression proceeding from *seed to juvenile* through *maturity* to *senescence*, on to *death* (see Figure 33.2). The ontogenetic developmental process describes life stages that universally apply to all trees and that are bound up with the embryonic *meristematic* network, found on shoot and root tips, in the dormant buds and cambial tissue.

This model is somewhat limited as senescence is not inevitably one-way; the 'age-clock' may be 'reversed' to the extent that *rejuvenescence* (negative senescence) may run counter to senescence (Del Tredici, 2000; Fortanier and Jonkers, 1976). The modular structure, regenerative capacity and reiteration processes are mediating factors that contribute to longevity.

The stages of tree development are reflected in the morphology of both above- and below-ground architecture with root decay influencing trunk decay (Raimbault, 1995) (Figure 33.2). Corresponding to the exploratory developmental stages of the trunk and crown, the root system undergoes probing, searching and modification. Apical dominance, light availability and water relations are key influences upon the developmental stages and crown morphology, which in turn *inter alia* influence growth rates and root dynamics.

In the maturing stages (Figure 33.2, stages 1–4) the root system explores below ground space and influences the crown's development, which is directed to achieving height rather than spread. Lignification is relatively low compared to later life stages.

In the mature phase (Figure 33.2, stages 5–7) crown and foliar volume expands as small branches proliferate and branch hierarchy and independent function becomes established between architectural units (branch autonomy). In the *fully mature* stage (stage 7) lower branches begin to 'escape' apical dominance and levels of independence between crown units are amplified. A discernible 'permanent crown' emerges with increasing rounded form.

When *fully mature* (Figure 33.2, stage 7) physiological limitations are imposed on the crown by the root system's capacity to support increased transportational distances to the crown extremity, under the critical influence of twig architecture and the ratio of photosynthesizing to respiring tissues (Thomas, 2013). These limitations are reciprocal and hydraulic flow to the crown extremities are constrained by impeded vascular conduction from larger- to smaller-diameter twig architecture, such that reduced water availability constrains the length of annual extension and results in reduced crown height. This *'crown retrenchment'* is a gradual process occurring over decades and marks the *transition process* from the mature to ancient phase (Figure 33.2, stages 7–8).

In the *early ancient* stage (Figure 33.2, stage 7) the number of declining and dead branches increases accompanied by peripheral dieback and reduced foliar density. Reduced photosynthetic capacity of the retrenched crown results in reduced annual volume of new wood (the Current Annual Increment, CAI). While new wood increases the diameter of the trunk, the CAI ring width becomes smaller compared to earlier growth phases. As long as the CAI is sufficient to maintain growth and vital processes so a new balance is restored between the reciprocal capacities of the root and crown systems (Figure 33.2, stages 8–10). Depending on the conditions, future re-expansion may then occur.

Life stages 8–10 represent the *early-, mid- and late-ancient* stages. In stage 8 the range and quantity of wood decay proliferates according to intensification of trunk hollowing, increased branch diameter, and branch die back and breakage. The crown and roots (Figure 33.3)

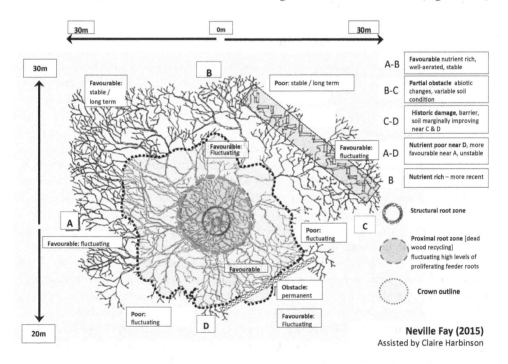

Figure 33.3 Schematic representation of the below-ground environment

Characterisation of the ancient phase root development takes place, in some cases centuries of fluctuating circumstances, where parts of the root system are impacted and impeded leading to local retrenchment while other sectors benefit, explore, develop and expand

undergo extended pulses of decline and regeneration as roots explore, spread, ramify, die off and are regenerated, until dynamic equilibrium is restored between soil and crown systems that conserve energy and water relations.

Reiteration is a key concept in tree and plant morphology. It describes how branch and root systems are built through self-replication. Conversely architectural components can be 'read' as a diminishing scale of self-similar units, whereby the crown is organically constructed as an organised hierarchy of branches with a self-repeating pattern. Branches in this sense *represent young trees growing on older parents that together contribute to a colonial structure.*

The principles of reiteration and modular self-replication are consistent for all trees. The adult tree comprises thousands of higher orders of *partial reiterations* (i.e. only part of the original model is repeated) together with a number of *total* (complete) *reiterations* (total because they mimic the original young tree; conceptually an organic 'fractal' of stages 1–4). Total reiterations in stages 8–10 originate from dormant buds that are 'plumbed' into the vascular system, which form new conductive pathways with internal channels within the sapwood, potentially connected to independent (fascicled) root systems. Paradoxically, until these reiterative units are entirely independent with their own root system, they remain *parasitic upon the parent tree* (F. Hallé, pers. comm., 2016; Lonsdale, 2013a).

Observation of reiterative growth and reiteration units help explain developments in the crown in terms of resource allocation and potential viable architectural units. Such understanding can assist the practitioner to judge how to intervene to support architectural components at various stages of the aging process.

Total reiterations develop in the mid-ancient phase on crown branches that replicate the model of the young tree. In stage 10, reiterations also develop from the main trunk which, being closer to the roots, may more easily develop vascular (cambial) columns connected to sectors of the root system[2] (Raimbault, 1995). When complete, these independent vegetative units confer a survival advantage as, having developed into functional columns, they can break away from and replicate the parent tree, in some cases as a 'phoenix' generation (Lonsdale, 2013a; Fay and de Berker, 2000).

The transitional process of natural *crown retrenchment* reflected in reciprocal changes in the root system, while transitory nonetheless takes place in tree-time, over decades and centuries. While this is only observed in retrospect in the lifetime of the arborist, such transitional states may curiously appear more or less static.

Apart from the very early stages of growth, in the mature and post mature stages the morphological phases that characterise aging do not just apply to the whole tree. Earlier life stages can be present in older age classes and apply to sections of the tree or individual architectural units.

Root aging dynamics

It is particularly difficult to determine the ancient phase, below-ground, root dynamics and morphology, given that the roots are largely entirely hidden. They are however living and dynamic. Root adjustment to change is slow and complex and not all parts are equally old or function comparably.

We here outline our understanding of root system aging and morphological complexity (Figure 33.3). Changes in root dynamics occur over human lifespans. While some parts retrench or become redundant, others opportunistically expand so that new replacement tiers may develop. As with crown branch autonomy, rejuvenated root parts growing on aged components interact and compete. As crown units ebb and flow, so the dynamics of subterranean interactions alter and rebalance. Root growth and distribution is influenced by local conditions that either impede or favour root exploration and the proliferation of new units.

The maturing-phase root system maintains a relatively rudimentary structure influenced by the tap root and the exploratory crown. Through maturity, roots develop a spreading woody, radiating tiered-framework with sinker roots exploiting layers of soil space that ramify into small-dimension fine roots, mycorrhizally connected to outlying soil. In temperate climates in favourable growth circumstances at maturity, tree roots may extend to 2.5 times the radius of an open-grown crown (Perry, 1982), with mycorrhizal associates further extending the influence of the root territory.

There is a shift from reliance on primary (low-order) roots to newer root structures under reciprocal influences between root and crown morphology and organisation. Mature tap root dieback is accompanied by below-ground decay with the pathways into the trunk and root-architecture influenced by the allocation of crown resources and water relations. Natural root and crown retrenchment is counteracted by vegetative influences of reiteration and adventitious rooting while local root extremity decline is balanced with inner-zone renewal. In the ancient phase adventitious roots proliferate within the spread of the crown.

Ancient phase rooting will be variably influenced by events over previous life stages, especially below-ground changes and the tree's physiological capacity for adaptation to adverse impacts. Roots are subject to local influences – in some areas degeneration and mortality occur while in favourable circumstances, rejuvenation and regeneration occur. These ebb and flow pulses occur over extended timespans (beyond human). When contraction is initiated due to fluctuating conditions, exploratory third- and higher-orders of root growth may generate on units closer to the trunk, while in other areas particularly conducive to root extension and exploration, roots ramify and extend the scope of outlying rooting.

As old trees become increasingly vulnerable to rapid change, a precautionary principle is advocated for root protection. Furthermore, as root distribution may be extremely variable (see Figure 33.3) and in some directions may even exceed 2.5 times the radius of the crown spread, ideally veteran trees should have a minimum root protection area that has a radius of 15 times the diameter at breast height (dbh) or 5 m beyond the crown, whichever is the greater. In the case of ancient trees, as their root systems are extremely variable and sensitive to disturbance, non-invasive investigation is required to determine any necessary additional extent of root protection.

Pollards and coppiced trees

Pollarding[3] (including shredding) derives from traditional tree husbandry techniques dating back 4,400 years, practised for tree cropping. A form of 'vernacular arboriculture' practiced in European wood-pasture landscapes and subsistence agricultural systems (Rackham, 1986; Read, 2000). Pollards were created from maiden[4] juvenile trees that were cut back (indeed 'topped' once, at a very early age; approx. 10–20 years) at 2–3m height to produce epicormic sprouts above the reach of browsing animals. Thus tree and animal husbandry were compatibly practised, often in communal grazing areas or extensive field boundary systems. Locally common, mainly deciduous (though also occasionally conifer) species were cut regularly within easy reach according to different produce needs (including leaves for tree-hay fodder, flowers and fruit, and wood of all dimensions for basketry, fencing and construction). Leaf production was beneficial for soil fertility as evident in modern agroforestry systems.

The sprouting 'poles' originating from the top of the reduced trunk or branches represent the pollard crown. Historically, re-cutting and cropping was cyclically carried out by axe according to the size and type of material required, often cutting according to the lunar cycle.

Traditional pollards (see Figure 33.4) are found throughout European rural landscapes and are also found in urban landscapes where relics of former rural land use were engulfed by urbanisation, today surviving as avenue and alleyway pollards managed for shade, ecosystem benefits and amenity (see Figure 33.5) (Treework Environmental Practice, 2015). In urban situations, pollarding methods manage the size and shape of trees according to various amenity values and cultural aesthetics (Coder, 1996; Ferrini, 2006).

Pollarding can contribute to tree vitality and longevity, periodically stimulating and perpetuating juvenile vegetative growth while maintaining reduced transportation distances

Figure 33.4 Traditional beech pollard. Nineteenth-century print of pollard at Burnham Beeches, Buckinghamshire, UK, 25 miles from centre of London

Source: Read (1996)

Figure 33.5 Urban pollards: (a) LHS, (b) centre and (c) RHS

Trees are maintained as tiered pollards whose crowns are formed from stages of pollarding ((a, b), Bristol, UK) to retain trees at a manageable scale for the benefit of urban communities. Pollards are managed for shade and for 'climate control' ((c), Varese, Italy) with crown pruned to high dense foliage according to a cyclical regime

Source: Neville Fay

and physiological balance between crown volume and root system capacity. Trees that are sympathetically 'worked' in this way and that function within their ontogenetic and energetic capacities could theoretically be maintained thus in perpetuity (Coder, 1996; Lonsdale, 2013b; Read, 2000; Fay, 2002; Ferrini, 2006).

The presence of great numbers of rural pollards that survive to the present day, some over a thousand years old, testifies to the relationship between pollard practice and the promotion of longevity. Pollard techniques have their parallel in bonsai practices insofar as both interventions recognise the reciprocal dynamics between the crown and root systems and employ (consciously or otherwise) counter-balancing pruning that interacts with and slows down the aging process (Del Tredici, 2000; Fay, 2002; Ferrini, 2006).

Good pollard practice involves repeat cutting of sprout re-growth. For some products this involves re-cutting from branch ends – especially in the case of fodder production. In other circumstances it involves re-cutting from the top of the pollard trunk (the *'pollard head'* or *'bolling'*) without breaching the branch collar. Repeat cutting promotes a complex robust swollen structure which develops interwoven grain patterns. Despite this fact, in the literature and amenity standards the technique is *inaccurately* referred to as 'topping' (Coder, 1996). Apart from lapsed pollard restoration, pollarding does not normally involve internodal cutting and is *not synonymous* with topping.[5]

Coppicing, another traditional cropping system, practised for millennia in European and Asian landscapes to sustainably produce wood material, is in a sense 'pollarding from near ground level'; it involves cutting *below knee height* to produce epicormic sprouts for various wood uses. Cyclical re-cutting generates an ever-widening ring of poles from the permanent coppice base (*'stool'*). Ancient coppice differs substantially from other trees as the central, long-decayed trunk core is absent and there is a comparatively smaller amount of older wood. Compared with veteran pollards or maidens, large old coppice stools have lower levels of decaying wood or aging bark. However, they have heritage value, not least as they stand upon and conserve an ancient soil ecosystem.

While ancient and other veteran pollards may have the same attributes as old maiden trees (large stem girth, increasingly hollow trunks, aging root systems and bark), pollards do not tend to have large, hollowing, lower horizontal branches, an important saproxylic habitat found in ancient maidens.

It is difficult to determine the relative differences between the root morphology of pollard and maiden trees. Our understanding is that pollard root systems are smaller when pollarding starts from a young age and when practised in a regular cycle that maintains pre-mature aged crown poles. However, mature and ancient-phase lapsed pollard root systems are more likely to reflect those of equivalent maiden trees, (though with a time lag) as, after pollard stems have acquired maturity, they develop into trees in their own right; the top of the trunk simulating ground level, with vascular channels developing through the trunk connecting to independent elements of the root system (see Figure 33.3).

Recent good practice developed from traditional pollard management has provided insights into veteran tree management (BSI, 2010; Lonsdale, 2013b) and should also be applied to large non-veteran trees and to mature trees that may become successors to the current generation of veteran trees, many of which are vulnerable to loss (Fay, 2015).

Many arborists understand the value of alleyway pollards as long-term, structurally stable and sustainable street tree features. Understanding traditional pollard practices contributes transferable knowledge that can be applied to the retention of maiden fully mature and post-mature urban veterans, and methods for future pollard creation, to bridge habitat losses and gaps that arise from shrinking veteran tree populations.

Aging is not just a one-way process

Branch and root retrenchment, breakage and mortality, hollowing and nutrient recycling are intrinsic to veteran trees. However these changes may also be accompanied by a reciprocal *rejuvenescent* counter flow through decay-mediated vegetative responses to changes in water relations and light regimes. These, together with hormone changes, stimulate reiteration above and below ground, to produce new roots, adventitious shoots and aerial root growth within the hollowing trunk and branches.

A linear model does not account for the dynamic nature of aging in post-maturity. Aging can be accelerated, slowed down or even reversed. Tree meristems seem to have the capacity to reset the biological age clock through mitosis (Thomas, in press). As veteran tree restoration relies on working with the '*age clock*' and rejuvenescence in different parts of the tree (Del Tredici, 2000), restoration of veteran trees threatened by decline or breakage entails developing specialised rejuvenation strategies and techniques to stimulate vegetative growth.

Recent developments in conservation arboriculture have explored veteran tree assessment and pruning techniques aimed to influence growth responses to improve crown condition, strengthen architectural elements and improve the dynamic relationship between the crown and root systems (Fay, 2002; Lonsdale, 2013b). As restoring veteran trees depends on interpreting the rejuvenation potential in order to assess a tree's vegetative capacity, the arborist's management decisions are likely to be influenced by their theoretical and operational viewpoint. The aspired 'endgame', whether it is modelled on the tree in peak maturity or the veteran tree with its idiosyncrasies, will influence the management strategy and options employed. The paradigm of peak maturity is measured in terms of practical principles of usefulness and visual amenity, whereas the paradigm of the ancient state is measured in terms of decay, habitat and ecosystem potential. With regard to veteran tree management and restoration, the ancient tree paradigm is the lens that will assist the arborist when engaged in restoration work.

Managing urban veteran tree populations

Typical urban tree populations contain considerably larger numbers of trees of young and early-mature age classes (i.e. without veteran characteristics), compared with those in the fully mature and (the relatively rare) post-mature ancient age classes. To strategically address management implications for urban forestry tree populations, account should be taken of how older age classes and the veteran 'ark' trees fit within the population structure.

New generations of veteran trees are unlikely to develop in urban locations once original individuals and populations of surviving trees have disappeared unless there is active consideration given to succession in the tree population. However, in relation to tree population dynamics, generally significant numbers of trees in younger age classes suffer biotic and abiotic impacts never to reach full maturity. Estimates that point to a *population half-life of around 20 years* and mean life expectancy less than 30 years, suggest a potential crisis for urban forest decision makers.

Younger trees are especially vulnerable to direct and indirect stresses in urban environments (such as from root damage, excavation, local development, installation of services etc.). Significant threats to veteran trees arise from other interventions such as over-pruning precipitating a spiral of decline and from inappropriate risk management.

Changing leisure patterns, increased pedestrian traffic due to appreciation of green space health benefits, and enhanced vehicle access to open spaces all have the unintended potential to detrimentally impact old and veteran trees and their rooting environments. Risk aversion (and inappropriate tree safety management) and planning and development compound the threats to veteran trees (Davis et al., 2000; Lonsdale, 2013b).

Furthermore, managing trees based on a utility model (for 'useful' life expectancy) potentially leads to inbuilt premature obsolescence (weeding out trees when past optimum amenity and timber values). Without an urban forestry policy aimed to conserve and enhance veteran trees and their habitats, we run the risk of eroding larger, mature trees within populations, along with their heritage, habitat and biodiversity potential.

Appropriate management of veteran trees supports the sustainability and diversity of tree populations and the multiple nature conservation and other interests they provide. However, veteran tree protection also fundamentally depends on sympathetically informed legislation and government policies so that professional best practice can be targeted to sector guidance on conservation and maximised longevity. Management for succession needs to take account of population and habitat gaps in order to be able to plan for appropriate measures including the creation of bridge habitat (ATF VETree, 2014) and recruitment of younger veterans (Lonsdale, 2013b).

In order to conserve ancient and other veteran trees managers should:

1 Evaluate the tree population age structure and spatial distribution through a tree survey.
2 Use the tree survey to identify individual or group of trees of increasing or different types of value to inform resource allocation so the most valuable trees are retained appropriately.
3 Assess the soil conditions to include water deficit and waterlogging to identify the need for remediation.
4 Assess, check and improve the soil-rooting environment in relation to condition, structure, aeration as well as biology and chemistry.
5 Assess and manage the light requirements of the tree relative to the species.
6 Assess the overall health and structural condition of the tree to determine appropriate management. This should consider the above- and below-ground conditions.
7 Develop individual tree management plans (ITMPs).
8 Manage the overall tree population to ensure veteran tree continuity.

Veteran trees – reasonable risk management

The 'triangle of risk assessment' includes the target (what might be harmed in falling distance of the tree), the size of part that might fail and the likelihood of failure. Large mature and ancient trees with decay, hollowing trunks, damaged roots and fractured branches may in certain circumstances be considered *hazards* requiring management. However terminology is important. *Hazard-habitats* should be distinguished from so-called *defects of form* (variations from a generalised norm) and *structural defects* (Mattheck and Breloer, 1995) and more importantly from *risks*. It is risks that require safety management (Davis, Fay and Mynors, 2000). Objective risk decision-making should assess and distinguish *real risks* from *perceived risks* to determine sensible, balanced and proportionate risk management (National Tree Safety Group, 2011).

Managing veteran trees in public places needs to take account of human safety. While the arborist manages trees to control the risk of harm, this should be based on rational, rather than risk-averse arboriculture. Tree risk management should be compatible with general polices for conserving veteran trees and their habitats, the benefits of which need to be considered when determining the type and level of management for their conservation and the need to achieve an acceptable level of risk.

Risks should be reasonably controlled rather than necessarily entirely removed. Non-tree risk management should first be explored, which might include moving the target beyond the impact range, before considering measures that may reduce the habitat value. When objectively carried out, veteran tree risk management should be compatible with conservation and other policies that underpin the reasons trees are valued and retained, even when old with structural faults.

Conclusion

Relatively few professionals and practitioners will have had the opportunity to gain direct experience of managing veteran trees – though this is changing. Without this experience there is a knowledge deficit. The loss of older trees in the urban landscape serves to impoverish practitioner knowledge. Understanding survival strategies, morpho-physiological change and rejuvenation systems contributes to the practice of conservation arboriculture.

It should be a major objective of urban forestry to manage trees through younger age classes up to and beyond their peak maturity, when they reach maximum size and in their post-mature decaying state, so that veteran tree populations are maintained and there are older trees providing evidence of complete life processes. However, ensuring mature and veteran tree continuity and sustainable populations with species-rich habitats requires special skills and techniques understandably not yet in mainstream arboriculture. A system that values trees for their veteran characteristics requires a commitment to manage trees not only to maturity but well beyond into ancientness, which in some cases may be the longest life stage. This requires understanding the 'hidden-from-view' colonial properties of trees, the role and function of reiteration, cambial columns and other survival strategies. The ancient tree offers a paradigm in which the tree, its habitat, colonisers, soil and surrounding environment are one ecosystem.

Restoration techniques for crown and soil management are needed that are based on long-term objectives to enhance the longevity of vulnerable trees and recruit new candidates to become successive generations providing landscape scale resilience within the urban context and beyond.

Notes

1 Certain species form distinctive heartwood, impregnated by extractives with decay resistant properties (e.g. *Quercus, Castanea, Robinia* and *Pinus*). In some species heartwood is less durable (e.g. *Fraxinus, Aesculus*), while in yet others inner ripewood forms from the gradual aging of central wood without forming heartwood (e.g. *Fagus, Tilia,* and *Acer*).

2 The embedded vascular cambial column within the trunk may also be thought of as reiterations of the parent trunk.

3 Derivation of the term 'pollard' is derived from 'poll', the crown on the head, which also refers to an animal that has lost its horns or had them removed.

4 A maiden is a tree with its original natural crown; usually with a single central stem from ground level through to the top of the crown (such trees may also be referred to as monopodial).

5 The initial cut to create a pollard, which is made when the tree is young, is typically a small diameter heading cut that avoids breeching the branch collar, though traditionally this was likely to have been internodal. Restoring ancient *lapsed* pollards uses reduction techniques aimed to minimise pruning wound cross-sectional area, and may involve some internodal cuts. However, to describe this as 'topping' is to misunderstand the restorative treatment (which is a phased program, scheduled over decades and based on an Individual Tree Management Plan (ITMP) devised to create a viable tree with a reduced crown and root system).

References

Alexander, K.N.A., 2012. What do saproxylic (wood-decay) beetles really want? Conservation should be based on practical observation rather than unstable theory. In Rotherham, I, et al ed. (2012) *Trees Beyond the Wood*, 33-43. Wildtrack Publishing, Sheffield.

ATF VETree. 2014. Practical management of veteran trees. Retrieved from www.ancienttreeforum. co.uk/resources/videos/veteran-tree-management.

BSI. 2010. *Tree Work: Recommendations*. British Standard 3998:2010. British Standards Institution, London.

Coder, K. 1996. Pollarding: What was old is new again. *Arborist News*, August: 53–59.

Davis, C., Fay, N. and Mynors, C. 2000. *Veteran Trees: A Guide to Risk and Responsibility*. English Nature, Peterborough.

Del Tredici, P. 2000. *Aging and Rejuvenation in Trees*. Arnold Arboretum, Arnoldia.

Fay, N. 2002. Environmental arboriculture, tree ecology and veteran tree management. *Arboricultural Journal*, 26(3): 213–238.

Fay, N. 2015. Der richtige Umgang mit uralten Bäumen: Archebäume und Baumveteranen [Development and management of ancient trees and veteran trees]. *Jahrbuch der Baumpflege*.

Fay, N. and de Berker, N. 1997. *The Specialist Survey Method*. Veteran Trees Initiative, Natural England, Peterborough, pp. 181–197.

Fay, N. and de Berker, N. 2003. *Evaluation of the Specialist Survey Method for Veteran Tree Recording*. Research report no. 529, English Nature, Peterborough.

Ferrini, F. 2006. Pollarding and its effects on tree physiology: a look to mature and senescent tree management in Italy. *Colloque européen sur les trognes*, 1–8. Maison Botanique, Boursay, Vendôme, France.

Fortanier, E. J. and Jonkers, H. 1976. Juvenility and maturity of plants as influenced by their ontogenetical and physiological ageing. *Acta Horticulturae*, 56: 37–44.

Lindenmayer, D. and Laurance, W. F. 2016. The ecology, distribution, conservation and management of large old trees. *Biological Reviews* doi: 10.1111/brv.12290

Lonsdale, D. 2013a. The recognition of functional units as an aid to tree management with particular reference to veteran trees. *Arboricultural Journal*, 35(4): 188–201.

Lonsdale, D. (ed.) 2013b. *Ancient and Other Veteran Trees: Further Guidance on Management*. London, The Tree Council.

Mattheck, C. and Breloer, H. 1995. *The Body Language of Trees. A Handbook for Failure Analysis*. Research for Amenity Trees No.4, Stationery Office, London.

Perry, T. O. 1982. The ecology of tree rots and the practical significance thereof. *Journal of Arboriculture*, 8: 197–211.

Rackham, O. 1986. *The History of the Countryside*. Dent, London.

Raimbault, P. 1995. Physiological diagnosis. *Proceedings of the 2nd European Congress in Arboriculture, Versailles, Société Française d'Arboriculture*. Physiological Diagnosis.

Rayner, A. D. M. 1993. New avenues for understanding processes of tree decay. *Arboricultural Journal*, 17: 171–189.

Read, H. J. (ed.) 1996. *Pollard and Veteran Tree Management II*. City of London, London.

Read, H. J. 2000. *Veteran Trees: A Guide to Good Management*. English Nature, Peterborough.

The National Tree Safety Group, 2011. *Common Sense Risk Management of Trees*. UK Tree Council, London.

Thomas, H. 2013. Senescence, ageing and death of the whole plant. *New Phytologist*, 197: 696–711.

Thomas, H. In press. Plant ageing. Paper for Transformational Nature seminar proceedings. Treework Environmental Practice, Bristol.

Treework Environmental Practice. 2015. *Tree-Lined Routes and the Linear Forest: A New Vision of Connected Landscapes*. Seminar Proceedings, no. 20. Treework Environmental Practice, Bristol.

34

URBAN WOODLANDS AND THEIR MANAGEMENT

Peter N. Duinker, Susanna Lehvävirta, Anders Busse Nielsen and Sydney A. Toni

Introduction

Urban woodlands are distinctive in the context of urban forests, which comprise all the trees in a city (see also Chapter 1), in that they bear some resemblance to rural woodland ecosystems. They are defined here as ecosystems more than about a half a hectare in size (FAO, 2012) with continuous canopy or a wooded savannah-like structure and uncultivated ground vegetation. Therefore, the management considerations presented in this chapter are more about silviculture (i.e. managing stands of trees) than about arboriculture (i.e. managing individual trees). Urban woodlands, situated within or adjacent to cities and towns, may comprise a substantial part of the total urban forest which also includes trees in parks, streets and squares, gardens, cemeteries, and so forth. They can be of natural (e.g. remnants), semi-natural, or human (e.g. plantations) origin. Typically, other ecosystem types such as waterbodies, wetlands, and grasslands are included to form an integrated woodland landscape.

Over time, close and diverse relations between cities and woodlands have developed (Konijnendijk, 2008), and today woodlands constitute the most frequent type of urban green space (in terms of area cover) in many cities worldwide (e.g. Pauleit et al., 2005). While urban woodlands typically account for only a few per cent of a country's total forest area, they play key roles for urban ecosystem functioning and for the health and wellbeing of urban populations (Alvey, 2006; Tyrväinen et al., 2014). For example, urban woodlands account for less than 2 per cent of the Swedish forest cover, but they are estimated to host more than 50 per cent of all recreational visits to woodlands (Rydberg and Falck, 2000).

Compared with manicured parks and other types of urban green space, woodlands provide a wider range of ecosystem services whilst also tending to be more multi-purpose and capable of supporting many uses and users (Bell et al., 2005; Alvey, 2006). At the same time, urban woodlands are under strong, uniquely urban anthropogenic pressures such as fragmentation, high levels of diverse recreational uses, and stressful atmospheric conditions caused by the surrounding urban structures and activities. As a consequence, urban woodlands often lack species and communities evident in their rural counterparts, as well as host species and communities not found in similar roles and compositions outside the urban realm (Williams et al., 2009; Werner and Zahner, 2010).

Despite the long history of urban woodlands, their management is a rather new research theme. When the concept of multiple-use forestry was introduced during the 1970s, it gave rise to studies of how recreational and environmental aspects could be integrated into commercial forestry where the stand is the functional level (e.g. Gundersen and Frivold, 2008). However, wood production is of little or no importance in most urban woodlands, and management usually focuses on stand compositions and structures deemed vital for the provision of desired social and ecological functions (Larsen and Nielsen, 2012). Only since the concept of urban forestry got a foothold in North America from the 1960s onward and Europe during the 1990s have we seen a notable rise in research on urban woodland management.

In this chapter, we first describe the spatial configurations and the social and ecological traits that characterize urban woodlands, and discuss related perspectives and challenges for their management. Although management of urban woodlands is as much about urban design as it is about silviculture, we focus the discussion on the latter. Next, we present a set of good-practice concepts for urban woodland management. We then highlight exemplary management practices in three case reports. Finally, we outline the most urgent challenges as well as the research and development needs associated with managing urban woodlands through the twenty-first century.

The special character of urban woodlands

Many features of urban woodlands call for management approaches that differ from those of rural forests. First, urban woodlands are often relatively small, isolated patches surrounded by other urban green-blue areas and the built environment. Second, the urban context has several profound impacts on woodlands. Third, urban woodlands can differ substantially, in terms of composition, structure, and ecosystem functions and dynamics, from their non-urban counterparts (Lehvävirta, 2007; Pickett et al., 2011).

Landscape context and influences

Many unique characteristics of urban woodlands relate to their size, shape, structure, and position in the landscape. Urban woodlands often have a high edge-to-interior ratio, both from woodland edges and internal dissection by formal and informal paths. The former can be abrupt transitions from woodland into built infrastructure or more-subtle shifts e.g. into backyards with gardens. Further fragmentation can arise intentionally from woodland design that provides, for example, glades, sports fields, or other activity areas. While visitors enjoy semi-open areas, too great an amount of fragmentation and internal open space in woodlands decreases the positive sense of enclosure and escape for forest visitors (Bell et al., 2005; Gundersen and Frivold, 2008).

Most research suggests that fragmentation of urban woodlands also negatively affects their biodiversity. Beninde et al. (2015) confirmed that impacts from patch size and distance to another patch, the key concepts of the theory of island biogeography, apply to patch dynamics in urban settings. The isolation of forest patches can reduce native species richness because it is more difficult for organisms to disperse to some sites (Beninde et al., 2015). For example, species that disperse short distances, such as those with gravity-dispersed seeds, may not be able to reach isolated urban sites due to the intervening built matrix (Williams et al., 2009).

The composition of patches is further affected by edge structure. Open edges allow edge effects to penetrate farther into the stand than would an abundance of trees and shrubs, and create opportunities for urban-adapted and open-habitat species that do well in dry, bright conditions (Hamberg et al., 2009). Understanding the history of fragmentation and previous patch dynamics might help explain the current pattern at an edge, which may reflect an extinction debt of species not yet driven out by edge effects (Ramalho and Hobbs, 2012). The combined influences of small patch size, elongated or amoeba-like patch shapes, and internal fragmentation present challenges but also rich opportunities to managers who may aim for diverse visual and habitat qualities at both the patch and broader woodland levels.

Urban woodlands are frequently subject to an infusion of introduced and sometimes invasive flora and fauna. Humans act as vectors for plant species when seeds adhere to shoes or vehicle tires, and household pets can carry propagules in their fur. Alien species may disperse from nearby gardens or parks but introductions may also be intentional if managers and visitors prefer such species. Sometimes alien species may be well suited to specific conditions, or have cultural significance (Kendle and Rose, 2000). Pets, pests, and other animals that accompany humans may have unwanted effects on the local woodlands but the effects are highly dependent on the human culture and local conditions. For example, high rates of visitation by humans and their pets may hinder the success, and change the spatial distribution, or regeneration, as well as inflict a selective pressure against the most sensitive species (e.g. Lehvävirta et al., 2014).

Impacts of the urban environment on woodlands

Urban woodlands are surrounded by high concentrations of built infrastructure that change the local climate at woodland edges and in small or narrow patches. While precipitation is higher in urban areas, the abundance of impermeable ground cover and grey infrastructure ensures that much rainwater is lost as run-off, rather than absorbed and retained by soil and roots, or in wetlands and surface waters (Pickett et al., 2011). Built infrastructure contributes to the urban heat-island effect that affects plant phenology, such as the timing of flowering, leaf-out, and leaf-drop (White et al., 2002). Furthermore, buildings and their shapes and positions affect the urban wind environment in complex ways (Liu et al., 2016). These changes in climate due to urbanisation can be hypothesized to affect forest dynamics and regeneration.

Urban environments are often dominated by loud and continuous noises with a range of deleterious effects. Noise can adversely affect the fauna as well as the recreational value of a woodland patch. For example, the ability of birds to project and hear songs for mating, territory defense, and communication are affected by urban noise, and they may change the timing, pitch, or volume of their singing (Bermúdez-Cuamatzin et al., 2011). Noise may also cause attentional fatigue and stress in humans (Galea and Vlahov, 2005), and while urbanites appreciate urban woodlands for a feeling of disconnect from the built environment, pervasive noise can undermine this value.

Urban woodland ecosystems are also affected by high light intensities at night. The timing of communication, reproduction, and foraging for animals might change in response to alterations of the photoperiod. Animals such as frogs may become disoriented due to an inability to navigate using night vision, or some birds and bats may increase their foraging time by feeding on insects attracted to artificial lights (Longcore and Rich, 2004). Artificially lit woodlands invite human visitors for potentially much longer periods of the day, and the presence of people in urban woodlands at night may negatively affect some scotobiotic processes such as night-time movement and feeding by animals.

The air quality in urban woodlands can be low due to pollution from abundant vehicles, fossil-fuel-burning buildings, and industrial facilities. This is reflected in the absence of many pollution-intolerant lichen species from urban areas, but there is also some evidence for the return of lichens after the improvement of urban air quality (see Nash, 2008). Urban woodlands may help improve air quality by absorbing pollutants (Escobedo et al., 2011).

Urban woodland structure

In urban settings, unique stand structures may occur due to the influences presented above. Different kinds of deviations from rural woodlands may be found and depend on both the above-presented urban effects as well as the chosen management regimes. Consider High Park in Toronto, Canada, an oak-dominated urban-core woodland of some 160 ha. Despite the original donor's nineteenth-century demand to keep the woodland in a natural state, fire suppression and ingress of invasive non-native plant species threatened to obliterate the natural oak (*Quercus velutina*) savannah ecosystem. A recent program of prescribed burning has had remarkable success in achieving much more natural conditions in the park woodland (High Park Nature, 2016).

Regrowth of seedlings and saplings in the lower canopy layers may not follow the abundance pattern of the dominant trees, raising questions about whether the canopy species dominance pattern will change over time (Lehvävirta and Rita, 2002). Replacement of current dominant species by local invasive or alien species well adapted to urban disturbance regimes has been observed in many places. Urban woodlands may also struggle with regeneration failure as the result of human trampling, as seen along woodland edges near residential areas or in woodlands closer to the city centre where visitation rates are higher (Hedblom and Söderström, 2008; Pickett et al., 2011).

The composition of urban woodlands varies tremendously across the globe. In some places, remnant forests are rich in indigenous species and may even host endangered taxa. In other places, remnant forest communities within cities are dominated by generalists, though some specialist species that are susceptible to fragmentation may occur if they established prior to the emergence of conditions making their persistence unlikely or impossible (Ramalho and Hobbs, 2012). In yet other places, there may be many ornamental tree species planted by humans, in addition to spontaneously occurring native and non-native, sometimes invasive, species.

Some urban woodlands may host a relatively greater richness of flora compared to rural woodlands. This is partly due to the escape of urban-adapted non-native species planted in gardens (Breuste et al., 2008; Pickett et al., 2011), examples including Scotch broom (*Cytisus scoparius*) and Persian hogweeds (*Heracleum persicum* group) in Europe. The heterogeneity of urban woodlands contributes to a high beta diversity (Breuste et al., 2008). Many animal species are absent from urban woodlands, but this is highly variable. Large herbivores may be completely missing as permanent residents (e.g. moose, *Alces alces*) or, quite the contrary, so common in urban areas that they become a nuisance (e.g. deer, *Odocoileus* spp.). The hypo- or hyperabundance of herbivores has an effect on the browsing regime of tree regeneration, which has important implications for forest succession.

Top predators are typically absent from urban areas, partly because their habitat needs for extensive areas may not be met by the small size of urban woodlands, which in turn may lead to high populations of large prey species (Fischer et al., 2012). Predators may also be absent because humans do not tolerate their presence in the city. Avian species are affected by the overall size and structure of habitat more than the spatial arrangement of habitat patches (Werner and Zahner, 2010). Too little suitable habitat, or the absence of a prey base, can make the urban forest unsuitable for many bird species. Many specialist phytophagous birds are absent due to a low abundance of the native invertebrate community (Raupp et al., 2010). For micro-fauna, impoverishment may relate to a relatively high abundance of non-native flora. Native plant species typically carry a greater richness and abundance of insects, as they are a co-evolved food source for native insect species.

Urban woodland management

The management of urban woodlands has both landscape and stand-level perspectives, and even individual trees can be of concern due to conservation, landscape experience, or safety reasons. Usually, though, the practices of silviculture prevail, with patches or stands of trees as the unit of management. While aesthetic and other cultural ecosystem services have traditionally dominated the agenda, the focus is increasing on ecological or environmental qualities of the urban forest. This considers the role of urban woodlands in providing habitat for regional species, challenges whether or not greenways function as corridors, and has spawned a naturalization movement in urban forest management (e.g. Toni and Duinker, 2015).

To begin, how do managers consider what sizes and shapes their urban woodlands should have? Large areas might better satisfy human desires for a feeling of immersion in the natural world and support species that typically avoid urban or edge conditions (Hauru et al., 2012; Beninde et al., 2015). However, abundant small patches may be more feasible in an urban setting than is a single large area. Compact patch shapes allow a maximisation of interior to edge ratio, so planning for compact shapes is a wise option when interior habitat is a target. However, long slim woodland patches may be more appropriate as human movement corridors. Dense and soft edges (i.e. transitioning into another natural or vegetated area rather than bordering impermeable surfaces) help decrease the spatial extent of edge effects, and therefore additional green space around small woodland patches may support the maintenance of interior habitat while sustaining the psychological benefits of forest visitation.

Woodlands may be included in greenways or corridors, greenbelts, or wedges/fingers. Corridors have social functions (e.g. providing space and linkages for recreation), planning functions (e.g. linking neighbourhoods), and ecological functions (e.g. providing linkages between habitat patches) (Ignatieva et al., 2011). Consistent with the notion that urban woodlands are vital for their non-timber forest values (Konijnendijk et al., 2006), they should be managed to provide recreational opportunities, foods (e.g. berries and mushrooms) for foraging, and spaces for children to explore (Rydberg and Falck, 2000). After all, urban woodlands are exceptional spaces for exploration and discovery, and contribute to formative experiences for developing an appreciation and concern for the natural world alongside their much-touted restorative psychological benefits (Savard et al., 2000; Pickett et al., 2011; Tyrväinen et al., 2014).

Close-to-nature management that relies on natural processes as the foundation is increasingly seen as a most promising approach (Bell et al., 2005; Toni and Duinker, 2015). We say 'close to nature' rather than 'wholly natural' because users often prefer woodland where a modest amount of 'tidying up' of undergrowth and deadwood has occurred (Edwards et al., 2012). Emphasizing natural dynamics in urban woodland management can be cost-effective. It can also deliver well on such ecosystem services as education, local identity, contact with nature, and aesthetic and recreational benefits. Promoting a high abundance of large veteran trees allows some urban residents to forge and enjoy connections to the past (Rydberg and Falck, 2000). Furthermore, an abundance of dead and decaying wood offers unique habitats for wildlife and there is some evidence that visitors can accept or value dead wood in urban woodlands (Gundersen and Frivold, 2008; Hedblom and Söderström, 2008). Naturally, safety is paramount, so snags can be cut down and left on the forest floor, or trimmed back to drop the branches and leave the valuable trunk standing for as long as possible.

Management includes more than just trees – other flora as well as fauna and soils are all important components of woodlands. Many of the social ecosystem services of urban woodlands relate to experienced variation. Users of any urban woodland are diverse and so are their preferences for stand compositions and structures. The habitat requirements of flora and fauna are correspondingly diverse. Therefore, to fully develop the recreational and ecological potential of urban woodlands, they should include a wide structural diversity potentially ranging all the way from single-species stands resembling the pillar hall structure, through form- and species-rich stand types associated with close-to-nature forest management, to historical/cultural forest management regimes such as coppice (Larsen and Nielsen, 2012; Figure 34.1). Unmanaged/untouched zones further assist in diversifying woodlands.

Opinions about specific animal species can diverge – some residents may dislike deer due to the risk of vehicular collisions or the nuisance they pose in gardens, whereas others find deer attractive and a cherished indicator of wildness in the city. Urban woodlands can even support populations of predatory species such as coyotes (*Canis latrans*) in North America. The abundance of urban wildlife may affect its reception – too many pigeons (*Columba* spp.) or starlings (*Sturnus vulgaris*) are considered a nuisance, for example, while single or small groups of the same birds may be considered attractive (Savard et al., 2000).

Managing where and how people visit an urban woodland affects the quality of the services provided by the woodland. Easily accessible trails enable elderly and disabled citizens to enter woodlands and thus gain access to ecosystem services and opportunities for nature appreciation (Rydberg and Falck, 2000). Trails can be designed so that visitors experience the sensation of moving through woodland interiors under closed canopies, between scattered

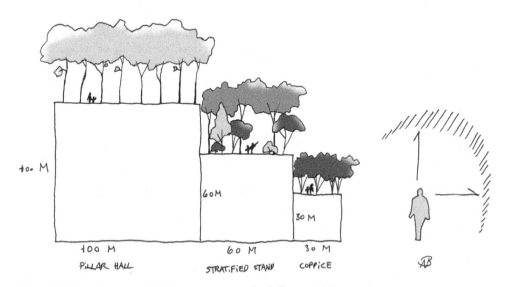

Figure 34.1 Stand structure and size requirements for different visitor experiences

The areal requirement to make a stand interior perceptible depends on the visibility at eye level, such as pillar hall structures with extensive visibility and possibility for free movement; stratified structures where the room unfolds gradually when moving between well-developed understory and shrubs layers; coppice with intimate rooms below the low canopy

Source: Anders Busse Nielsen

trees in semi-open areas, inside small intimate glades, along the edges between wooded and open parts, and out on larger open areas framed by the woodland (Larsen and Nielsen, 2012). A well designed and maintained path system with benches may help direct and limit the spatial extent of human and pet trampling of the forest floor. However, many activities such as dog-walking, berry-picking, orienteering, mountain-biking, geocaching, and children's play may be seen as sub-optimal if restricted to trails, so managers should allow for off-trail activities.

Another layer in the social environment determining management direction is the diverse approaches taken by the various professions and stakeholders (e.g. Özgüner et al., 2007), as well as the leverages held by different professions in the urban-forest administration, planning, and management. Viewpoints of different actors and stakeholders may be far apart, and sometimes a lack of understanding of the ecological consequences may result in internally contradictory wishes (e.g. when a person simultaneously wishes to clear the bushes and preserve the songbirds). The situation is further complicated when there is a lack of, or poorly implemented, participatory planning. Since urban forest management is carried out in full view of the public and under constant scrutiny, it follows that to be successful, management should be transparent and accountable to the public while also being consistent with sound scientific principles (van Gadow, 2002).

Examples of concept and idea development for urban woodlands in North America and Europe

Among the thousands of examples of noteworthy urban woodlands and conceptual developments around the world, we selected a mere three to profile in this chapter. The choices were driven largely by our personal experiences with research and management associated with these most intriguing treed ecosystems. We acknowledge that the three cases all represent north-temperate conditions with considerable annual precipitation, which means that we do not discuss urban woodlands in a great many other climatic conditions around the world. We offer the cases as glimpses into the complex world of managing woodlands in the heart of the city.

Point Pleasant Park, Halifax, Canada

The City of Halifax, Canada (44°38'N; 63°34'W) was founded in 1749 to provide Britain with a defence headquarters for its interests in northeastern North America. The southern point of the peninsula was quickly abandoned as a potential site for the new settlement because of shallow waters offshore and scant fresh water onshore. However, Point Pleasant immediately became a key site in the defence network of the new city. The majestic forest of beech (*Fagus grandifolia*), birch (*Betula* spp.), oak (*Quercus rubra*), maple (*Acer* spp.), white pine (*Pinus strobus*), hemlock (*Tsuga canadensis*), and red spruce (*Picea rubens*) was cleared and fortifications installed (NIPpaysage Landscape Architects et al., 2008).

During the eighteenth and nineteenth centuries, residents spent much of their recreational time along the Point Pleasant shoreline, enjoying the sights of a harbour busy with ships. The site was established as a 70-hectare city park in 1866, but extensive military use of the site continued past World War II. Through the twentieth century, natural regeneration of the woodland was permitted, and frequent tree plantings using alien species – for example, Norway maple (*Acer platanoides*), European beech (*Fagus sylvatica*), Scots pine (*Pinus sylvestris*) – occurred. By the late twentieth century, the Park was covered with a dense, mature canopy dominated by red spruce and white pine. The Park received roughly a million day-visits per year, with users mainly jogging, dog-walking, or simply ambling on the numerous trails throughout the Park.

Figure 34.2 View to the Halifax Outer Harbour during a Remembrance Day celebration shortly after Hurricane Juan devastated the woodlands of Point Pleasant Park

The hurricane occurred in September 2003, the blown-down forest was cleaned up in winter 2004, and this photo dates from November 2005. This view is unavailable today on account of dense forest regeneration

Source: photo by P. Duinker

Three disturbances hit the Park within a decade. In the early 1990s, an ice-storm damaged many of the red-spruce trees. Then the red spruce took another assault from the brown spruce long-horned beetle (*Tetropium fuscum*), a non-native species that arrived a decade or two earlier in unsanitized wood packaging in the marine container terminal adjacent to the park. Finally, the coup de grace came in September 2003 in the form of Hurricane Juan, a category-2 storm that blew down roughly three-quarters of all the mature trees in the Park. That night, the Park's woodland experienced a major successional reset.

Following harvest and clean-up of the downed wood during the winter of 2004, the public re-entered the Park in June 2004. Many tears flowed, and the people expressed clear hopes for the newly denuded landscape – just give us back the woods, they said! A major planning process ensued, and the Park's first ever comprehensive plan was developed with the aim of creating an attractive and resilient recreational forest at Point Pleasant. The plan (NIPpaysage Landscape Architects et al., 2008) presented a vision for the forest with (a) greater age–class diversity, (b) greater native tree-species diversity with reduced populations of alien tree species, (c) retention of abundant snags (except adjacent to roads and trails) and downed woody debris, (d) a windthrow resilience plan with predominantly deciduous tree species on south-facing (i.e. ocean-facing) slopes, and (e) climate-change adaptation based on anticipatory plantings. At this time of writing, the recovering forest has experienced 13 growing seasons, park visitation has increased, and visitors find the new forest much to their liking (Andrade et al., 2015).

Several lessons arise from the natural and management history of Point Pleasant Park. First, unless managers are attentive to wind-throw risk, a woodland can become quite vulnerable to storms – at the time of the hurricane, almost the entire woodland was mature in a silvicultural sense, with little natural regeneration on the forest floor. Second, natural regeneration can be a powerful source of new tree growth, and despite having planted over one hundred thousand seedlings in the Park in 2007–2008, most of today's new trees are naturally regenerated. Unfortunately, the survival rates of planted seedlings are completely unknown, although it is suspected by some as being very low. Third, the Park's visitors desire a naturalized woodland, and while the current woodland is nowhere near the venerated condition of the pre-hurricane forest, people are happy to have the landscape well-treed again. The demand for a large inner-city naturalized woodland is strong, and even a woodland dominated by young trees can satisfy this demand. As cruise-ship visitors to the city get a prolonged view of the park when they approach, they can well agree with the notion that Point Pleasant Park is one of the city's great civic places.

Alnarp Västerskog, southern Sweden

People generally perceive young woodlands, whether resulting from afforestation or reforestation, as being unattractive and requiring some time or a special treatment before they can be appreciated for recreational uses. Despite this, woodland management is often directed towards enhancing the recreational qualities in the mature stage primarily, while qualities in the young stages are less explored. That managers withhold interest during the young stages of a new woodland stand might seem logical; 'The establishment phase is over, and everyone can see everything grow, and everyone can relax', as Gustavsson (2002, p. 290) phrases it. However, management approaches that develop experiential values in newly established or regenerated woodlands can expand the borders for recreational use of urban woodlands and help achieve a sustainable balance between recreational use and biodiversity conservation by redirecting users away from old – and often crowded – woodlands.

In the landscape laboratory at the campus of the Swedish University of Agricultural Sciences, Alnarp, Sweden (55°39'N; 13°04'E), different management approaches are combined to create experiential qualities and variation in the young woodland. A part of the landscape laboratory is Alnarp Västerskog (AVS), planted in 1994 (9 ha) and 1998 (4 ha) on former farmland. AVS has an elongated shape where two wooded belts enclose an open meadow around a stream with three ponds. Here and there the meadow intersects the woodland, offering views to the surrounding landscape. The path system is laid out so visitors walk under the closed canopies of the woodland interiors, along the woodland edge, and out on the open meadow with views to the surrounding landscape.

The woodland is composed of 32 stands. The species mixtures of the stands follow so-called complexity ladders where key species, e.g. silver birch (*Betula pendula*), European beech, wild cherry (*Prunus avium*) and English oak (*Quercus robur*) are planted in monocultures, through mixtures with one or two other species to species-rich mixtures with up to 15 species (Gustavsson, 2002). Different thinning regimes (thinning from above and below, and coppicing) are applied to support the long-term development of: pillar-hall structures with extensive visibility and possibility for free movement; stratified structures where the visible room unfolds gradually when moving between well-developed understory and shrub layers; coppice with intimate rooms below the low canopy; and savannah-like structures where the canopy of scattered trees creates a mosaic of sun and shade on the field layer.

Figure 34.3 A 'pocket park' inside a dark, young beech stand with larch as a nurse tree

The small-scale interventions create immediate place identity and turn a walk into a sequence of existing and revealing views of places and no-places where you utter to yourself the unspoken words 'I am outside it, I am entering it, I am in the middle of it'

Source: photo by Ines Isabel Monteiro de Vasconcelos Luís

What was not included in the initial design and management planning for AVS is small intimate glades, semi-open zones, and individual trees/vegetation elements with special character that create surprises and place identity. These elements were implemented through *creative management*, conceptualized by Gustavsson (2002) as design-oriented interventions aimed at turning space into place. The creative management was initiated in parallel with the first thinning of nurse trees 7–10 years after establishment. In the field, students of landscape architecture and art identified places along the path system where thinning, pruning, and trimming of the vegetation could articulate (or contrast) site-specific characters that are not apparent from conventional woodland maps and management plans. The interventions are positioned along the path system, where their character and management level contrast against the surrounding uniform and extensively managed vegetation. To support the sense of place, each intervention is restricted to an area that can be grasped by the human eye, typically a few hundred square metres. Altogether, the creative management sites cover less than 5 per cent of the woodland area.

Conceptually, the creative management has many similarities with the design concept of 'stops' (Stationenkonzept) for perennial plantings developed by the German landscape architect Hans Luz and the theory of urban acupuncture of derelict areas in the cityscape. While the origin, form, and contexts vary, the concepts of creative management, stations, and urban acupuncture can all be regarded as small-scale intervention aimed at turning space into place by experimental and creative use of what is already there. As such, the creative management erases the traditional divide between woodland design and woodland management. Further, the focus on the immediate effect of punctual management interventions on place identity supplements conventional management where the focus is on long-term development at the stand level.

Urban woodlands in Helsinki capital region

In Finland, urban woodlands are mostly remnants of the pre-urbanization forest landscapes. While some green wedges have been preserved successfully through several revisions of city planning, even in the capital region of Helsinki (60°10'N; 24°56'E), the current densification trend is turning many remnant urban woodlands into residential and commercial areas. City planners and foresters face serious questions such as whether the regeneration in urban forests is sufficient, what is the minimum size of a viable woodland patch, and how to conserve biodiversity. Furthermore, managers want to avoid complaints from forest users. Here we outline how recent research has addressed these issues in the Helsinki capital region. The research-based insights below emanate from papers by Hamberg et al. (2009), Hauru et al. (2012), Lehvävirta (2007), Lehvävirta and Rita (2002), Lehvävirta et al. (2014), and others by these authors.

Regeneration is the number-one issue, as without tree seedlings and saplings there will be no future woodlands. At the end of the 1990s, a common worry among Helsinki's urban foresters was that urban pollution, mischief, and human trampling were decreasing the amount of regeneration, especially in stands dominated by Norway spruce (*Picea abies*). In response, managers implemented artificial regeneration of these stands. At the same time, the local nature conservation non-government organization advocated benign neglect as a way to support recreational and biodiversity values. Inspired by these contradicting views, the potential for natural, spontaneous, self-sustained regeneration of these woodlands was studied.

A surprisingly diverse regrowth was revealed, where the tree populations exhibited stem-diameter distributions skewed to the left, meaning sufficient regeneration. As compared to rural woodlands, a high proportion of the regrowth consisted of rowan (*Sorbus aucuparia*). In currently spruce-dominated woodlands, several non-conifer species have abundant regeneration towards the edge while rowan saplings peak at 25–30 m from the edge. Tall-growing species (e.g. Norway maple; grey alder, *Alnus incana*; English oak) were found as saplings even though they were not discovered among the canopy species.

Because woodlands are more than just trees, the microbial community, the forest-floor vegetation, invertebrates, and human perceptions were also explored. Changes in them were evident, both due to trampling and the edge effect. The edge effect extends, as a rule of thumb, 50 m into the woodland from the edge. It decreases the abundance of interior forest species at this edge zone as well as the prospects for psychological restoration. A minimum diameter of 150 m, by educated guess, is needed to preserve some interior habitat. Luckily, edge effects can be further decreased by preserving dense edges, or by lush greening outside the woodland. Dense edges also support psychological restoration as they close the view from the woodland into the urban matrix, leaving the visitor visually surrounded by the woodland. As the research findings concerning humans and other biota parallel each other, restorative and ecological values can be managed together.

Even a moderate amount of human trampling is lethal to most plants. In experimental research, only 35 repeat tramplings were needed to create a new desire line that caused a 10–30 per cent decrease in vegetation cover. Furthermore, the effects of trampling are devious: even in the seemingly untrampled vegetation, subtle changes in microbial and plant communities occur. How to safeguard the forest floor, plants, and tree regeneration against trampling seemed to be a tricky thing as accessibility and safety have been traditionally interpreted to require thinning throughout the woodland patches. A subtle way to guide walkers and offer safe sites for regeneration is the use of natural barriers such as fallen logs and sheltering groups. The idea of sheltering groups is that groups of saplings and small trees survive best as small thickets, and should not be thinned for optimal growth as done in commercial forests. In Helsinki, fallen logs are left on site, when possible, to act as natural barriers and to support biodiversity because they do not provoke negative feelings or decrease the recreational value of urban woodlands and were well accepted by forest users.

The core idea in research on Helsinki's urban woodlands has been to combine visitor perspectives with ecological realities and search for synergies and conflicts. It appears possible to find approaches that simultaneously benefit humans, biodiversity, and conservation values. To sum up, the list of innovative approaches includes that of:

- sheltering groups (i.e. letting tree seedlings and saplings grow as small thickets that fend off trampling);
- logs on the forest floor that foster native biodiversity and offer safe sites for tree regeneration;
- thick edges with abundant trees and bushes; and
- lush greening outside the woodland edges to increase and sustain interior conditions.

Challenges and opportunities for innovative management of urban woodlands

Management of rural woodlands, where timber production is often a key value, is one thing. Management of urban woodlands, where the values are decidedly not timber-related and challenges abound in relation to high densities of people, pets, and built infrastructure, is another. For success in delivering well on the diverse expectations city people have for their urban woodlands, managers need to be clever indeed in their choice and implementation of approaches.

One might ask first what the long-term future can be for urban woodlands. Unless they are protected by regulation, they may be under intense pressure to be developed with yet more built infrastructure as cities worldwide continue to grow and densification is called for. Moreover, with more people in the world's cities, the existing woodlands may see even greater human use. The advent of climate change represents a potentially massive challenge to urban forest managers (Ordóñez, 2015). Not only will urban woodlands be called upon to help mitigate climate change through carbon sequestration, but they will also be managed to help cool urban ecosystems as part of a city's adaptation to climate change. Of course, an urban woodland's ability to deliver on such ecosystem services will depend on it being managed so that it too can adapt to the changing climate.

In our view, three avenues of work are needed. The first is active consideration of what we know works. The case reports above reveal that much is known about the dynamics of urban woodlands as well as how people relate to, and how they value and appreciate, or generate undue pressures on urban woodlands. The management concepts that we know work in some places should be widely used in operational-scale experiments. The second avenue of work is to engage with the people, and to do that with energy and insight. We know that there are both good and bad ways to try to get citizens' views on management of public assets, whether they are streets, sports facilities, or trees. We also know that people have strong views about trees and woodlands, sometimes with valid underpinnings and sometimes based on serious fallacies about how the world really works in biophysical terms. Mutual learning mechanisms are sorely needed where citizens and woodland managers, along with other professionals, engage productively so that everybody learns and everybody has opportunity to voice views in respectful productive dialogue. The third avenue of work is continued research to discover how urban woodlands can be managed to deliver on a plethora of potentially conflicting values under the duress of the harsh urban environment. We know that trees present dozens of important ecosystem services and benefits to urban dwellers, whether in solitary situations or in woodlands. Well-targeted research by keen and competent investigators generates findings certainly worth the investment.

Trees – the dominant species in woodlands – are rather resilient life forms that give us a chance to retain and enjoy wooded ecosystems even in the hearts of our cities. To maintain the multiple values that urban woodlands offer, continuous lobbying for preservation and expansion of woodland territory in cities is needed. Success in such work will go a long way toward enhancing the sustainability of cities worldwide.

References

Alvey, A. A. (2006) 'Promoting and preserving biodiversity in the urban forest', *Urban Forestry and Urban Greening* vol. 5, pp. 195–201.

Andrade, M., Bethune, E., Cardella, C., Klimek, J. J., Mitukiewicz, K., Smith, J. (2015) *User Viewpoints on Point Pleasant Park*, Faculty of Management, Dalhousie University, Halifax, Canada.

Bell, S., Blom, D., Rautamäki, M., Castel-Branco, C., Simson, A., Olsen, I. A. (2005) 'Design of urban forests', in C. C. Konijnendijk, K. Nilsson, T. B. Randrup, J. Schipperijn (eds), *Urban Forests and Trees*, Springer, Berlin, pp. 149–186.

Beninde, J., Veith, M., Hochkirch, A. (2015) 'Biodiversity in cities needs space: a meta-analysis of factors determining intra-urban biodiversity variation', *Ecology Letters* vol. 18, pp. 581–592.

Bermúdez -Cuamatzin, E., Ríos-Chelén, A. A., Gill, D., Garcia, C. M. (2011) 'Experimental evidence for real-time song frequency shift in response to urban noise in a passerine bird', *Biology Letters* vol. 7, pp. 36–38.

Breuste, J., Niemelä, J., Snep, R. P. (2008) 'Applying landscape ecological principles in urban environments', *Landscape Ecology* vol. 23, pp. 1139–1142.

Edwards, D. M., Jay, M, Jensen, F. S., Lucas, B., Marzano, M., Montagné, C., Peace, A., Weiss, G. (2012) 'Public preferences across Europe for different forest stand types as sites for recreation', *Ecology and Society* vol 17, p. 27.

Escobedo, F. J., Kroeger, T., Wagner, J. (2011) 'Urban forests and pollution mitigation: analyzing ecosystem services and disservices', *Environmental Pollution* vol. 159, pp. 2078–2987.

FAO (2012) *FRA 2015 Terms and Definitions*, Forest Resources Assessment Working Paper 180, Food and Agriculture Organization of the United Nations, Rome.

Fischer, J. D., Cleeton, S. H., Lyons, T. R., Miller, J. R. (2012) 'Urbanization and the predation paradox: the role of trophic dynamics in structuring vertebrate communities', *BioScience* vol. 62, pp. 809–818.

Galea, S., Vlahov, D. (2005) 'Urban health: evidence, challenges, and directions', *Annual Review of Public Health* vol. 26, pp. 341–365.

Gundersen, V. S., Frivold, L. H. (2008) 'Public preferences for forest structures: a review of qualitative surveys from Finland, Norway and Sweden', *Urban Forestry and Urban Greening* vol. 7, pp. 241–258.

Gustavsson, R. (2002) 'Afforestation in and near urban areas: dynamic design principles and long-term management aspects. Landscape laboratories as reference and demonstration areas for urban and urban–rural afforestation', in T. B. Randrup, C. C. Konijnendijk, T. Christophersen, K. Nilsson (eds), *Urban Forests and Trees, Proceedings No. 1. COST Action E12*, Office for Official Publications of the European Communities, Luxembourg, pp. 286–315.

Hamberg, L., Lehvä,virta, S., Kotze, J. (2009) 'Forest edge structure as a shaping factor of understorey vegetation in urban forests in Finland', *Forest Ecology and Management* vol. 257, pp. 712–722.

Hauru, K., Lehväirta, S., Korpela, K., Kotze, D. J. (2012) 'Closure of view to the urban matrix has positive effects on perceived restorativeness in urban forests in Helsinki, Finland', *Landscape and Urban Planning* vol. 107, pp. 361–369.

Hedblom, M., Söderström, B. (2008) 'Woodlands across Swedish urban gradients: status, structure and management implications', *Landscape and Urban Planning* vol. 84, pp. 62–73.

High Park Nature. (2016) 'Prescribed burns', retrieved from www.highparknature.org/wiki/wiki. php?n=Restore.PrescribedBurns (accessed 24 April 2016).

Ignatieva, M., Stewart, G. H., Meurk, C. (2011) 'Planning and design of ecological networks in urban areas', *Landscape and Ecological Engineering* vol. 7, pp. 17–25.

Kendle, A. D., Rose, J. E. (2000) 'The aliens have landed! What are the justifications for "native only" policies in landscape plantings?', *Landscape and Urban Planning* vol. 47, pp. 19–31.

Konijnendijk, C. C. (2008) *The Forest and the City: The Cultural Landscape of Urban Woodland*, Springer, Berlin.

Konijnendijk, C. C., Ricard, R. M., Kenney, A., Randrup, T. B. (2006) 'Defining urban forestry – a comparative perspective of North America and Europe', *Urban Forestry and Urban Greening* vol. 4, pp. 93–103.

Larsen, J. B., Nielsen, A. B. (2012) 'Urban forest landscape restoration – applying forest development types in design and planning', in J. Stanturf, D. Lamb, P. Madsen (eds), *Forest Landscape Restoration: Integrating Natural and Social Sciences*, Springer, Netherlands, pp. 177–199.

Lehvävirta, S. (2007) 'Non-anthropogenic dynamic factors and regeneration of (hemi)boreal urban woodlands – synthesising urban and rural ecological knowledge', *Urban Forestry and Urban Greening* vol. 6, pp. 119–134.

Lehvävirta, S., Rita, H. (2002) 'Natural regeneration of trees in urban woodlands', *Journal of Vegetation Science* vol. 13, pp. 57–66.

Lehvävirta, S., Vilisics, F., Hamberg, L., Malmivaara-Lämsä, M., Kotze, D. J. (2014) 'Fragmentation and recreational use affect tree regeneration in urban forests', *Urban Forestry and Urban Greening* vol. 13, pp. 869–877.

Liu, H., Ma, W., Qian, J., Cai, J., Ye, X., Li, J., Wang, X. et al. (2016) 'Effect of urbanization on the urban meteorology and air pollution in Hangzhou', *Journal of Meteorological Research* vol. 29, pp. 950–965.

Longcore, T., Rich, C. (2004) 'Ecological light pollution', *Frontiers in Ecology and the Environment* vol. 2, pp. 191–198.

Nash, T. H., III. (2008) 'Lichen sensitivity to air pollution', in T. H. Nash III (ed), *Lichen Biology*, Cambridge University Press, New York, pp. 301–316.

NIPpaysage Landscape Architects, Ekistics Planning and Design, Duinker, P. N., Black Spruce Heritage Services, Form:Media, LandDesign Engineering Services. (2008) *Point Pleasant Park Comprehensive Plan*, Halifax Regional Municipality, Halifax, Canada.

Ordóñez, C. (2015) 'Adopting public values and climate change adaptation strategies in urban forest management: a review and analysis of the relevant literature', *Journal of Environmental Management* vol. 164, pp. 215–221.

Özgüner, H., Kendle, A., Bisgrove, R. J. (2007) 'Attitudes of landscape professionals towards naturalistic versus formal urban landscapes in the UK', *Landscape and Urban Planning* vol. 81, pp. 34–45.

Pauleit, S., Jones, N., Nyhuus, S., Pirnat, J., Salbitano, F. (2005) 'Urban forest resources in European cities', in C. C. Konijnendijk, K. Nilsson, T. B. Randrup, J. Schipperijn (eds), *Urban Woodlands and Trees*, Springer, Berlin, pp. 49–80.

Pickett, S. T. A., Cadenasso, M. L., Grove, J. M., Boone, C. G., Groffman, P. M., Irwin, E., Kaushal, S. S., Marshall, V., McGrath, B. P., Nilon, C. H., Pouya, R. V., Szlavecz, K., Troy, A., Warren, P. (2011) 'Urban ecological systems: scientific foundations and a decade of progress', *Journal of Environmental Management* vol. 92, pp. 331–362.

Ramalho, C. E., Hobbs, R. J. (2012) 'Time for a change: dynamic urban ecology', *Trends in Ecology and Evolution* vol. 27, pp. 179–188.

Raupp, M. J., Shrewsbury, P. M., Herms, D. A. (2010) 'Ecology of herbivorous arthropods in urban landscapes', *Annual Review of Entomology* vol. 55, pp. 19–38.

Rydberg, D., Falck, J. (2000) 'Urban forestry in Sweden from a silvicultural perspective: a review', *Landscape and Urban Planning* vol. 47, pp. 1–18.

Savard, J., Clergeau, P., Mennechez, G. (2000) 'Biodiversity concepts and urban ecosystems', *Landscape and Urban Planning* vol. 48, pp. 131–142.

Toni, S. A., Duinker, P. N. (2015) 'A framework for urban-woodland naturalization in Canada', *Environmental Reviews* vol. 23, pp. 321–336.

Tyrväinen, L., Ojala, A., Korpela, K., Lanki, T., Tsunetsugu, Y., Kagawa, T. (2014) 'The influence of urban green environments on stress relief measures: a field experiment', *Journal of Environmental Psychology* vol 38, pp. 1–9.

van Gadow, K. (2002) 'Adapting silvicultural management systems to urban forests', *Urban Forestry and Urban Greening* vol. 1, pp. 107–113.

Werner, P., Zahner, R. (2010) 'Urban patterns and biological diversity: a review', in N. Muller, P. Werner, J. G. Kelcey (eds), *Urban Biodiversity and Design*, John Wiley & Sons, Chichester, pp. 145–173.

White, M. A., Nemani, R. R., Thornton, P. E., Running, S. W. (2002) 'Satellite evidence of phenological differences between urbanized and rural areas of the eastern United States deciduous broadleaf forest', *Ecosystems* vol. 5, pp. 260–273.

Williams, N. S. G., Schwartz, M. W., Vesk, P. A., McCarthy, M. A., Hahs, A. K., Clemants, S. E., Corlett, R. T., Duncan, P., Norton, B. A., Thompson, K., McDonnell, M. J. (2009) 'A conceptual framework for predicting the effects of urban environments on floras', *Journal of Ecology* vol. 97, pp. 4–9.

INDEX

Italic page numbers indicate tables; **bold** indicate figures.